Thermal
Analysis
and Control
of Electronic
Equipment

Thermal Analysis and Control of Electronic Equipment

Allan D. Kraus
University of South Florida, Tampa

Avram Bar-Cohen
Ben-Gurion University of the Negev

● HEMISPHERE PUBLISHING CORPORATION   *Washington   New York   London*

McGRAW–HILL BOOK COMPANY   *New York   St. Louis   San Francisco*
Auckland   Bogotá   Hamburg   Johannesburg   London   Madrid   Mexico
Montreal   New Delhi   Panama   Paris   São Paulo   Singapore   Sydney
Tokyo   Toronto

THERMAL ANALYSIS AND CONTROL
OF ELECTRONIC EQUIPMENT

1 2 3 4 5 6 7 8 9 0   D O D O   8 9 8 7 6 5 4 3

This book was set in Press Roman by Hemisphere Publishing
Corporation. The editors were Anne Shipman and Lynne
Lackenbach; the designer was Sharon Martin DePass; the
production supervisor was Miriam Gonzalez; and the typesetter
was Wayne Hutchins.
R. R. Donnelley & Sons Company was printer and binder.

Library of Congress Cataloging in Publication Data

Kraus, Allan D.
  Thermal analysis and control of electronic equipment.

  Bibliography: p.
  Includes index.
  1. Electronic apparatus and appliances—Cooling.
2. Bar-Cohen, Avram, date. II. Title.
TK7870.25.K73 1983   621.3815'1   82-15546
ISBN 0-07-035416-2

For our departed parents
    Raymond Kraus
    and
    Dorothy Markowitz

"לכל זמן ועת לכל חפץ תחת השמים"

(קהלת ג:ו)

To every thing there is a season, and a time to every purpose under the heaven.

Ecclesiastes 3:1

■ contents

■ preface

The cooling of modern electronic components is one of the prime areas for the application of advanced thermal control techniques. Much of the improvement in the reliability and packaging density of electronic logic and microwave devices can be traced to advances in thermal design. Consequently, the choice of thermal control technology and the particular decisions made in the course of evolving the thermal packaging design often have far-reaching effects on both the reliability and cost of the electronic system or assembly.

Rational thermal design, commencing at the earliest stages of product or system development, is thus essential if a successful laboratory brassboard or prototype unit is to become a competitive commercial product. Insufficient attention to thermal control may lead to low or even erratic mean time before failure, failure to meet customer specifications or the relevant military specifications, and/or excessive replacement and maintenance costs. On the other hand, thermal "overkill" may well result in prohibitively high product costs.

These considerations have spurred widespread recognition in the industrial and academic electronic thermal control community that staying competitive in the 1980s and 1990s will necessitate rapid dissemination and application of advanced thermal control concepts and predictive techniques. The National Science Foundation sponsored research workshop on "Directions of Heat Transfer in Electronic Equipment" in Atlanta, Georgia, in 1977, and the proliferation of university and professional society short courses in electronic cooling attest to the validity of this view.

This book was prepared in response to the electronic thermal control challenge of the 1980s. The content and form grew out of discussions at the 1977 NSF workshop and Summer Session courses at the Massachusetts Institute of Technology during 1978 and 1979. As such, this book seeks to refine the physical insight and mathematical modeling capability of the thermal analyst and electronic packaging specialist, working in both conventional and advanced thermal control configurations. It is divided into four primary sections: "Thermal Considerations in the Design of Electronic Equipment," "Fundamentals of Heat Transfer and Fluid Mechanics," "Thermal Control Techniques," and "Electronic System Applications."

In its entirety, this book provides a self-contained treatment of thermal analysis and design of electronic equipment and can be used by packaging engineers and as a basis for a two-semester advanced undergraduate or graduate course for electrical and electronic engineering students. Mechanical engineers and/or students with previous exposure to heat transfer and fluid mechanics may find it preferable to focus on Parts I, III, and IV, for example, in a one-semester advanced undergraduate course.

During the past decade, it has become apparent that the efficiency of the design and synthesis of engineering systems can be substantially improved by studying the stages of the design process and exercising the skills needed to assure its success. In response to this development and the growing demand for the application of formal design and/or optimization procedures in product development, it appears essential that a thermal design work, such as this, include a detailed treatment of the subject, as has been done in Chapter 3.

In formulating the philosophy and style of this book, the authors were guided by the African proverb that states, "Give a starving man a fish and you have satisfied him for a single day; teach him how to fish and he will never starve again." Echoing this proverb, it is the authors' humble hope that this book will provide the practicing engineer and engineering student with the tools needed to determine the range and limits of available thermal control technology and to establish performance goals for more novel techniques. It is our belief that to achieve these goals it is necessary first to define the electronic thermal control problem and physical as well as conceptual constraints to its solution (Part I) and then to explore the thermal mechanisms active in the dissipation of heat from electronic components (Part II). With this foundation it is possible to develop the rationale for mathematical models and/or empirical correlations appropriate to the various thermal configurations and to establish the opportunities for, as well as the mathematical form of, the relevant thermal optimizations (Part III). The application of thermal control technology to various electronic systems, such as microwave equipment, microelectronics on printed circuit boards, and inertial systems, is presented in Part IV.

Preparation of this book would have been impossible without the overt and covert, explicit and implicit contributions of our teachers, our many colleagues, and all those who have preceded us in this field; wherever possible, we have tried to reference their efforts. We wish to acknowledge, in particular, the significant influence of our many discussions with Professors Warren M. Rohsenow and Arthur E. Bergles—teachers, colleagues, and friends—on the style, form, and content of this book. Valuable suggestions were made by Richard Chu, Erwin Fried, Robert Lott, and William A. Smith. Special thanks are also due to Professors Bergles, B. B. Mikic, Michael Kovac, and Michael Yovanovich as well as Mr. Steven Davidson for specific aid in the preparation of Chapters 19, 4, 23, 9 and 1, respectively; to Mr. Donald L. Cochran for launching ABC (then Avram Markowitz) on a career in electronic cooling and to Mr. Shimshon Markowitz for teaching ABC to strive for self-realization. We were fortunate that Hemisphere Publishing, by the luck of the draw, assigned Anne Shipman, a truly superb editor, to this project and we also note most appreciatively the careful and dedicated efforts of Linda Federspiel, Lili Lang, and Orna Nirenberg in converting a manuscript into legible type and ink drawings. The publication of a work of this magnitude could not have been accomplished without the patience, understanding, support, and encouragement of our wives, Ruth Kraus and Annette Bar-Cohen. Finally, we acknowledge the dedicated efforts of Ruth Kraus, fondly known as "the buckeye," in assisting with a monumental proofreading endeavor. She has already ordered her straitjacket (size 36) and will not be seen or heard from until the next manuscript is ready.

Allan D. Kraus
Avram Bar-Cohen

I

■ thermal considerations in the design of electronic equipment

1

■ component characteristics and the thermal environment

1.1 INTRODUCTION

Thermal control of electronic components has been, for the past 50 years, one of the primary areas of application of advanced heat transfer techniques. Many of the benefits associated with improved reliability, increased power capability, and physical miniaturization can be traced directly to improved thermal analysis and design, which has allowed component temperatures to be stabilized at a desired level despite variations in ambient conditions or the presence of a hostile environment.

The publication of a method for determining the temperature distribution in the laminated core of a transformer by Cockroft in 1925 [1] served to introduce electronic component cooling. Cockroft's work was followed by several key papers by Mouromtseff [2, 3] dealing with air and liquid cooling of high-power vacuum tubes and a classic 1942 study by Elenbaas [4] of the design and optimization of natural convection fin arrays.

The growing use of electronics, in both the military and civilian sectors following World War II, led to widespread recognition of the need for thermal packaging and design of electronic components and to the flowering of thermal control technology. To evolutionary improvements in vacuum tube cooling, including enhancement of external convective heat transfer and the development of integral coolant (liquid) passages for high-power traveling wave tubes, were added totally new areas of application. The design and development of so-called cold plates for the mounting and thermal servicing of miniaturized components began to occupy many workers in the field. The stabilization of transistor junction temperature, necessitated by the inverse temperature dependence of efficiency and reliability, imposed new thermal constraints on the design of electronic equipment despite the dramatic reduction in overall power dissipation obtained by the use of transistors. Similarly, the extreme demands for isothermality in gyroscopes used for inertial navigation of satellites, missiles, aircraft, and submarines offered a significant technical challenge to the thermal packaging community during the 1950s [5].

Although the scope of electronic cooling applications continued to expand during the 1960s and 1970s, much of the technical effort was devoted to consolidating and documenting previous achievements [6] and standardizing conventional air and liquid cooling techniques. In parallel with this activity, substantial funding was devoted to the research and development of more novel techniques, including immersion cooling, augmented flow boiling, heat pipes, and thermoelectric devices, but most of these remained at the prototype stage and only rarely found commercial use. This situation has continued into the early part of the 1980s, but the rapid improvements in microelectronics and large-scale integration (LSI) technologies, coupled with the growing demand for reduced maintenance of electronic equipment, are once again pushing thermal control technology to its limits.

In the three chapters that follow, we shall explore the thermal characteristics and cooling requirements of standard electronic components, provide the tools needed for determining the desired component temperatures or the reliability to be expected for a specified temperature matrix, and finally examine the structure inherent in the synthesis of new products in general, and thermal packaging concepts in particular.

1.2 AIMS OF THERMAL CONTROL

In the thermal control of modern electronic devices, examples of which are shown in Figs. 1.1 and 1.2, it is necessary to provide an acceptable "microclimate" for a diversity of components that, while frequently in close proximity to each other, often display substantially different sensitivities to environmental factors and widely varying heat dissipation rates. Some knowledge of the possible range of component and

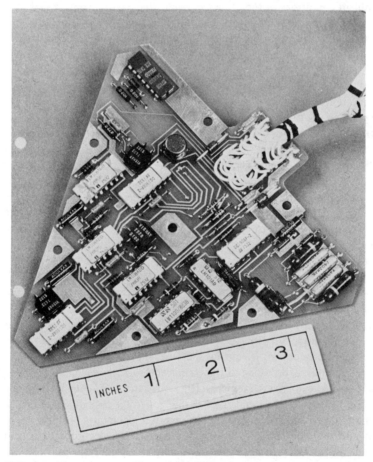

FIG. 1.1 Multicomponent printed circuit board. (Courtesy of Honeywell, Inc.)

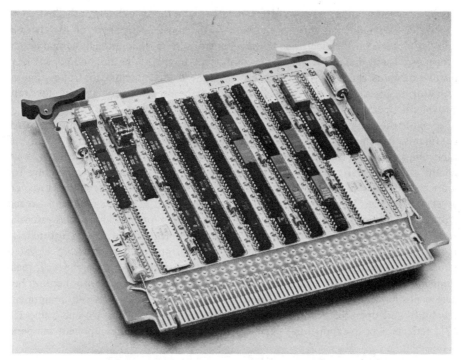

FIG. 1.2 Printed circuit board with dual-in-line packages. (Courtesy of Honeywell, Inc.)

environmental variations is thus a necessary prerequisite for successful design and development of thermal control systems and the rational interpretation of bench and prototype tests. A review of the aims of thermal control, commonly encountered thermal environments, and the thermal characteristics of several classes of components is presented in this chapter.

The prevention of catastrophic thermal failure, defined as an immediate, thermally induced, total loss of electronic function in a specified component, must be viewed as the primary and foremost aim of electronic thermal control. Such failure can result from melting or even vaporization of some part of the component, but more often is related to thermal fracture of the mechanical support element (e.g., case, substrate) or separation between the leads and the external electrical network. In microelectronic devices in particular, catastrophic failure may also result from plastic flow of printed circuits and the deleterious migration of dopants in semiconductor materials.

Catastrophic failure is by its nature generally dependent to some extent on the details of the local temperature field, as well as on the operating history and modes of the component. Consequently, it is difficult to determine precisely the temperature at which such failure can be expected to occur. Nevertheless, test and operating experience combined with even rudimentary failure analysis can be used to generate a catastrophe-free, upper temperature bound for most common components. This maximum allowable component temperature can generally be found in manufacturers' catalogs as well as other trade bulletins; together with the specified heat dissipation of the various components, this temperature is the most determinate of the thermal criteria imposed on the electronic thermal control system. In a rational thermal design procedure, these maximum allowable component temperatures and component heat dissipation rates can be expected to establish the overall thermal control configuration—for example, the mode of heat transfer to be used, the coolant, its flow rate and inlet temperature (if appropriate).

With the primary aim achieved by selection of an appropriate thermal control configuration, attention can be turned to meeting the desired level of reliability of each component in the electronic device. Individual electronic components are inherently reliable and can typically be expected to operate at room temperature for more than 10,000 yr—that is, with a base failure rate less than $0.01/10^6$ h. However, as shown in Fig. 1.3, the number of components used in each circuit can exceed 250,000 in LSI technology; consequently, to avoid equipment failure, it is essential to minimize the failure rate of each and every component and device. Unfortunately, most electronic components are prone to failure from prolonged use at elevated temperatures, resulting from creep in the bonding materials, parasitic chemical reactions in switches and connectors, and diffusion in solid state devices, to mention but a few possibilities. These failure modes establish a direct tie between component reliability and operating temperature, which, as illustrated in Fig. 1.4 and Table 1.1 [7], reflects a near exponential dependence of failure rate on component temperature.

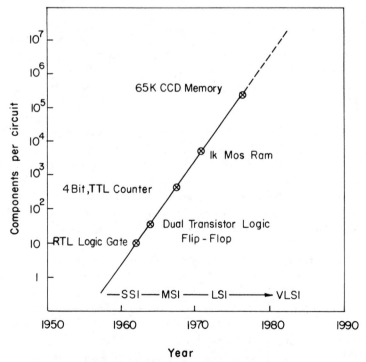

FIG. 1.3 State of the art in circuit complexity [8].

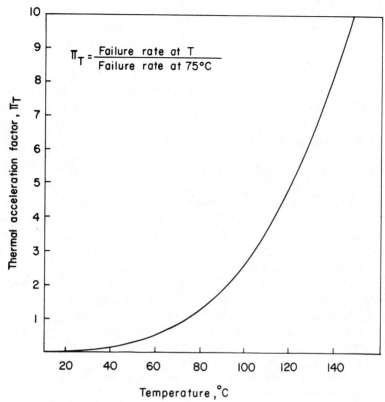

$$\pi_T = \frac{\text{Failure rate at } T}{\text{Failure rate at } 75°\text{C}}$$

FIG. 1.4   Thermal acceleration factor for bipolar digital devices [7].

TABLE 1.1   Failure Rates [7]

Part description	λ_b, failures per million hours; base failure rate		ΔT (°C)	Ratio of high to low failure rate
	High temperature	Low temperature		
PNP silicon transistors	0.063 at 130°C and 0.3 stress	0.0096 at 25°C and 0.3 stress	105	7:1
NPN silicon transistors	0.033 at 130°C and 0.3 stress	0.0064 at 25°C and 0.3 stress	105	5:1
Glass capacitors	0.047 at 120°C and 0.5 stress	0.001 at 25°C and 0.5 stress	95	47:1
Transformers and coils, MIL-T-27, class Q	0.0267 at 85°C	0.0008 at 25°C	60	33:1
Resistors carbon composition	0.0065 at 100°C and 0.5 stress	0.0003 at 25°C and 0.5 stress	75	22:1

Temperature cycling in excess of $\pm 15°C$ around a specified average temperature has also been found to reduce reliability, almost independent of the influence of the temperature level [9]. In this U.S. Navy-sponsored study, an eightfold increase in failure rate was encountered in equipment subjected to a deliberate temperature cycling of more than $20°C$. Thermal stress-induced fatigue in metal connectors and delamination of layered composites due to differential thermal expansion are at least two possible causes for accelerated failure, as are rupture of partially ductile bonding materials and reduced operating efficiency due to electrical property mismatch brought on by time delays in temperature-sensitive properties.

The temperature sensitivity of key electrical properties is also of prime concern in the thermal control of frequency-modulated microwave systems, where frequency drift of low-power oscillators is often detrimental to the operation of both transmitters and receivers. Automatic frequency control employing, for example, discriminators and a stable reference device, can be used to limit the frequency error, but direct thermal "compensation" or passive thermal regulation can substantially reduce system complexity and cost [10]. Similarly, in many memory units the maintenance of a relatively narrow temperature range at the core, despite variations in properties and power dissipation among neighboring components, facilitates precise determination of the required drive current and the associated power supplies. This not only results in reduced power input to the system and decreased heat dissipation within the core memory [11] but can also lead to the elimination of complex (parasitic) circuitry otherwise needed to compensate for thermal drift [12].

The need for long-term reliability in electronic systems subject to time-varying environmental conditions and experiencing various operating modes or duty cycles introduces yet another degree of complexity in the thermal analysis and design of electronic thermal control equipment. In lieu of more detailed consideration, continuous operation at the extremes of the parametric range—for example, maximum ambient temperature and maximum heat dissipation—is often used to establish design constraints, despite the low probability of occurrence or minimal forecasted operating periods at these conditions. As a result, the specified system may well be substantially overdesigned for more typical operating conditions, and the inevitable cascading of such excess requirements through the hierarchy of "architectural" levels (component, device, system, or vehicle) can at times overwhelm the bounding values initially prescribed.

Much of this difficulty can be alleviated by the proper assessment of component failure rate. Alternately, the use of thermal control techniques that include a capability for "graceful," as opposed to abrupt, degradation in heat removal rate can not only aid in streamlining the reliability analysis but can, in addition, result in dramatic reductions in the required redundancy of both cooling equipment and electronic circuitry. No single thermal control measure can guarantee such performance, but ways can often be found to introduce a potential for graceful degradation into the overall cooling configuration. Thus, in forced air cooling, use of relatively widely spaced fins can provide considerable heat dissipation capability in natural convection and/or radiation transfer following a blower failure [13], and the sensible heat rise or even partial vaporization of the circulating coolant in a liquid-cooled unit can provide a considerable operating margin despite pump or heat exchanger malfunction.

It should be noted that many of the secondary aims of component thermal servicing, which require spatial and temporal temperature stability in the components, can be achieved by the calculated addition of external heat to a specified component, a subassembly, or an entire device during part of its operating cycle [13, 14]. It is largely for this reason that the terms "thermal control" and "thermal servicing" are preferred to the more commonly used "cooling" of electronic equipment.

1.3 THERMAL ENVIRONMENT

One of the major constraints in the design of electronic thermal control systems is the variability of the thermal environment both within and among various application categories. The average external thermal conditions under which large-frame computers operate have very little in common with those experienced by missile-borne electronic and microwave components, which must themselves contend with substantially different ambient temperatures and pressures within a specified flight envelope. Similar arguments can be advanced relative to mechanical vibration, electromagnetic, and other aspects of the environment, but detailed exploration of these variables is beyond the present scope.

In discussing the thermal environment, it is convenient to examine the range encountered in the ultimate heat sink, conditioned fluid, or local heat sink, and the duration of required operation for airborne, seagoing, and ground vehicles, including their respective subcategories. In both the first and third of these categories, atmospheric air constitutes the ultimate heat sink. Structures, equipment shelters, and ground vehicles are influenced primarily by the atmospheric boundary layer, and temperatures of $-50°C$ to $+50°C$, representing the polar regions at one extreme and the subtropical deserts at the other, coupled with pressure variations from 11 psi (75.8 kN/m^2) to 15.5 psi (106.9 kN/m^2), corresponding to high plateau lands and deep rift valleys, can be said to span the likely parametric range. Incident solar fluxes up to 1 kW/m^2, outgoing long-wave radiation of the order of 0.01 to 0.1 kW/m^2 [15], and convective heat transfer coefficients from 6 W/m^2 K for still air to 75 W/m^2 K at wind velocities approaching 100 km/h (27.8 m/s) serve to complete the environmental definition for this category.

The environmental boundaries for airborne vehicles—missiles as well as low- and high-level aircraft—are determined largely by the aerodynamic flow around the vehicle. At low velocities near the earth's surface, the conditions approximate those outlined earlier, with the possible exception of elevated pressures in deep rift valleys. However, as transition is made toward and beyond supersonic speeds, thermal external recovery in the boundary layers can lead to relatively high skin temperatures, approaching $130°C$ at low Mach numbers near sea level and similar temperatures at high supersonic velocities at 10–20 km above sea level. Under the latter conditions, the large dynamic pressure coupled with a relatively low static pressure can often result in a maximum pressure slightly above 15.5 psi (106.9 kN/m^2) and a minimum pressure significantly below the previously mentioned value, leading to a dramatic increase in the pressure range likely to be encountered.

Seagoing vessels, whether surface or submarine, are of course confronted with a far more benign environment. In this category, the external ambient temperatures are unlikely to exceed 35°C in equatorial waters or fall below 0°C on the surface of open bodies of water or below pack ice in the polar regions. Incident solar radiation and convective heat transfer coefficients for surface vehicles can be expected to approximate those cited earlier for ground structures/vehicles, whereas submarines approaching top cruising speeds may experience heat transfer coefficients for exchange with sea water approaching 10^5 W/m^2 K. At this transfer rate, any wetted surface can be expected to attain a temperature nearly identical to that of the ambient sea water.

In establishing the thermal control needs of various categories of electronic equipment, considerable variation must also be noted in the requirements for continuous operation and mean time to first failure (MTFF), depending on the vehicle and its mission. Although the MTFF requirement must necessarily reflect a system-level reliability decision, the limits on excursion time appropriate to each vehicle are derived from such considerations as the volume of fuel carried by the vehicle, the maximum distance of communication and control, and the size of the arena. These limits apply as well to the electronic equipment carried aboard the vehicle and, consequently, whereas components in ground-based radar systems and passenger ships may need to operate diurnally, missile components will generally be in the range of 30 to 200 s (exclusive of captive flight conditions), those aboard aircraft from 3 to perhaps 24 h, and equipment in armored ground vehicles typically from 6 to 24 h.

Recognizing the wide diversity represented in the environmental ranges described and the detrimental effects on the reliability and survivability of electronic components exposed to such ambiences, the electronic thermal control industry has generally resorted to the use of a conditioned fluid to act as an intermediate or local heat sink for most components of interest. Standard U.S. practice in the late 1970s generally involved the use of air supplied at 55°C inlet temperature to airborne assemblies and approximately 25°C in ground structures and surface vehicles. Submarine and seagoing vessels more often employed a liquid coolant, typically supplied at 40°C to specified terminal points. The precise flow rates and allowable temperature rise of the coolant in its passage through the controlled unit must once again involve a system-level decision and can be expected to vary from application to application.

The many advances in electronic technology and the concomittant rapid growth of electronic applications over the past few decades makes it most difficult to define in absolute terms the thermal environment for all electronic components. The preceding has aimed at acquainting the reader with the salient environmental features of the major application categories.

In the selection of electronic components for military equipment, the environmental requirements are often reduced to a set of standard military specifications, as outlined for example in Table 1.2. Compliance of each component with the thermal limits established in the classification imposed by the customer is taken as tantamount to safe and reliable operation in the anticipated environment. Furthermore, standard test procedures must be followed in establishing component compliance with the environmental specifications and component configurations known to meet certain standards are in turn codified in detailed component specifications. The sample specifications listed in Table 1.3 serve to identify some of the more common military component specifications.

TABLE 1.2 Environmental Specifications for Airborne Electronic Components (MIL-E-5400 R)

Environment	Classification[a]
50,000 ft altitude and continuous sea-level operation over the temperature range −54 to +55°C (+71°C intermittent operation)	Class 1
30,000 ft altitude and continuous sea-level operation over the temperature range −54 to +55°C (+71°C intermittent operation)	Class 1A
15,000 ft altitude and continuous sea-level operation over the temperature range −40 to +55°C (+71°C intermittent operation)	Class 1B
70,000 ft altitude and continuous sea-level operation over the temperature range −54 to +71°C (+95°C intermittent operation)	Class 2
100,000 ft altitude and continuous sea-level operation over the temperature range −54 to +95°C (+125°C intermittent operation)	Class 3
100,000 ft altitude and continuous sea-level operation over the temperature range −54 to +125°C (+150°C intermittent operation)	Class 4

[a]The addition of the letter x (e.g., Class 2x) identifies the equipment as operating in the ambient environment of the class but requiring cooling from a source external to the equipment.

1.4 THERMAL CHARACTERISTICS OF TYPICAL COMPONENTS

The opening sections of this chapter reviewed the aims of electronic component thermal control and the range of environmental conditions under which standard components may need to operate. To complete the definition of the tasks confronting the thermal specialist, it is now appropriate to examine in some detail the inherent temperature limits and heat dissipation characteristics of various electronic components.

As the reader is undoubtedly aware, the rate of obsolescence of specific component technologies is extremely high, and radically new concepts for providing, for example, a capacitance function, diode behavior, or signal storage capability are proposed each day. Nevertheless, the use of some components, notably paper dielectric capacitors and metal cap transistors, has persisted for an extended period. Therefore, although it is hoped that the information provided will serve to quantify the thermal control needs of electronic components, the values presented are probably best used only to initiate system calculations and are understood not to absolve the thermal designer of the obligation to obtain more precise values for a particular set of components.

1.4.1 Semiconductor Devices

The energy dissipation associated with the passage of electrons through the transistor or diode emitter-base junction in a semiconductor device gives rise to the characteristic

TABLE 1.3   Governing Specifications for Some Electronic Components

Component	Specification
Capacitors	
Paper/plastic	
Fixed, plastic	MIL-C-19978
Fixed, paper-plastic	MIL-C-14157
Fixed, metalized	MIL-C-39022
Mica	
Fixed	MIL-C-5
Fixed, button style	MIL-C-10950
Fixed, established reliability	MIL-C-39001
Glass	MIL-C-23269
Ceramic	
Fixed, temperature-compensating	MIL-C-20
Fixed, general-purpose	MIL-C-11015
Fixed, established reliability	MIL-C-39014
Electrolytic	
Fixed (DC, aluminum, dry, polarized)	MIL-C-62
Fixed, nonsolid, tantalum	MIL-C-3965
Connectors	
Circular	MIL-C-5015; -26482; -38999; -81511; -83723
Coaxial, Rf	MIL-C-3607; -3643; -3650; -3655; -25516; -39012
Printed wiring board	MIL-C-21097; -55302
Power	MIL-C-3767
Rack and panel	MIL-C-24308; -28748; -83733
Hook-up wire	
Vinly/polyamide (nylon) jacket	MIL-W-16878
	Type B/N-600 V
	Type C/N-1000 V
	Type D/N-300 V
Polyethylene/polyamide (nylon) jacket	MIL-W-16878
	Type J/N-600 V
Fluorocarbon-TFE (Teflon)	MIL-W-22759
	MIL-W-16878
	Type E-600 V
	Type EE-1000 V
	Type ET-250 V
Fluorocarbon-FEP	MIL-W-16878
	Type K-600 V
	Type KK-1000 V
	Type KT-250 V
Polyalkene/polyvinylidene fluoride (Kynar) jacket	MIL-W-81044
Silicone	MIL-W-16878
	Type F-600 V
	Type FF-1000 V
Rf cable	MIL-C-17
Flat cable	NAS-729
Flexible, printed wiring	IPC-FC-240
Microcircuits, digital	MIL-M-38510
TTL, NAND gates, monolithic silicon	MIL-M-38510/1B
TTL, flip-flops, monolithic silicon	MIL-M-38510/2E

TABLE 1.3   Governing Specifications for Some Electronic Components
(*Continued*)

Component	Specification
Microcircuits, digital	
TTL, NAND buffers, monolithic silicon	MIL-M-38510/3B
TTL, multiple NOR gates, monolithic silicon	MIL-M-38510/4C
TTL, AND-OR-INVERT, monolithic silicon	MIL-M-38510/5A
TTL, binary full address, monolithic silicon	MIL-M-38510/6C
TTL, exclusive-OR gates, monolithic silicon	MIL-M-38510/7B
TTL, buffers/drivers, monolithic silicon	MIL-M-38510/8B
TTL, shift registers, monolithic silicon	MIL-M-38510/9C
TTL, decoders, monolithic silicon	MIL-M-38510/10B
TTL, arithmetic logic units, monolithic silicone	MIL-M-38510/11C
TTL, monostable multivibrators, monolithic silicone	MIL-M-38510/12E
TTL, counters, monolithic silicone	MIL-M-38510/13C
TTL, data selectors/multiplexers, monolithic silicone	MIL-M-38510/14B
TTL, bistable latches, monolithic silicone	MIL-M-38510/15
DTL, NAND gates, monolithic silicone	MIL-M-38510/30A
DTL, flip-flops, monolithic silicone	MIL-M-38510/33
CMOS, NAND gates, monolithic silicone	MIL-M-38510/50C
CMOS, flip-flops, monolithic silicone	MIL-M-38510/51C
CMOS, NOR gates, monolithic silicone	MIL-M-38510/52B
ECL, flip-flops, monolithic silicone	MIL-M-38510/61A
ECL, AND/NAND gates, monolithic silicone	MIL-M-38510/62
ECL, QUAD translator, monolithic silicone	MIL-M-38510/63
Schottky TTL, NAND gates, monolithic silicone	MIL-M-38510/70
Schottky TTL, flip-flops, monolithic silicone	MIL-M-38510/71A
Schottky TTL, shift registers, monolithic silicone	MIL-M-38510/76
Linear, voltage regulator, monolithic silicone	MIL-M-38510/102A
Linear, transistor arrays, monolithic silicone	MIL-M-38510/108
Linear, precision timers, monolithic silicone	MIL-M-38510/109
Prom, 512-bit bipolar programmable, read-only memory (P-ROM), monolithic silicone	MIL-M-38510/201A
Resistors	
Composition, fixed	
Insulated	MIL-R-11
Established reliability	MIL-R-39008
Film, fixed	
High stability	MIL-R-10509
Insulated	MIL-R-22684
Established reliability	MIL-R-39017
Power type	MIL-R-11804
Wire-wound, fixed	
Power type	MIL-R-26
Accurate	MIL-R-93
Power type, chasis mounted	MIL-R-18546
Thermistor	
Insulated	MIL-T-23648
Composition, variable	
Standard	MIL-R-94
Lead screw actuated	MIL-R-22097
Wire-wound, variable	
Low operating temperature	MIL-R-19

TABLE 1.3   Governing Specifications for Some Electronic Components
(*Continued*)

Component	Specification
Resistors	
Wire-wound, variable	
Power type	MIL-R-22
Precision	MIL-R-12934
Lead screw actuated	MIL-R-27208
Established reliability	MIL-R-39015
Relays, electrical, hermetically sealed	MIL-R-5757
6 PDT contacts, low level and 2 A	MIL-R-5757/1
3 stud mounting, DPSTNO & DPDT contacts	MIL-R-5757/3
DPDT, 2 A, 40 mW	MIL-R-5757/13
SPDT, 15 A RMS, 12 kV	MIL-R-5757/35
Current responsive, 1.5 A, inductive	MIL-R-5757/46
AC operated, DPDT contacts, 5 A	MIL-R-5757/49
Subminiature, DPDT contacts, 5 A	MIL-R-5757/77
Time-delay, fixed (300–500 ms)	MIL-R-5757/81
Transformers and inductors	MIL-T-27
Coils, radiofrequency	MIL-C-15305
Transformers, pulse, low power	MIL-T-21038

release of heat in such devices and constitutes the primary motivation for the thermal control of semiconductor elements. Most ceramic semiconductors are processed at temperatures exceeding 1000°C and can thus operate much above the catastrophic failure temperatures of most other components. However, the failure rate for long-term operation of, for example, silicon dioxide components at 450°C approaches 90% [16] and, consequently, if commonly accepted failure rates of 0.5 to 2% per 1000 h of operation are to be achieved, a far more benign thermal environment must be provided. Common practice in the early 1980s appears to suggest that junction temperatures of 110 to 125°C are generally compatible with this requirement. Nevertheless, in systems with especially long life or low maintenance and replacement-cost constraints, as well as in systems subjected to frequent power surges, average junction temperatures as low as 60°C may be desirable.

In the commonly employed monolithic semiconductor devices, appropriate combinations of active and passive elements are formed into discrete "chips," which are bonded to a "chip carrier" or substrate. The substrate is then packaged in one of several ways, with the TO-5 and the hybrid package shown in Figs. 1.5 and 1.6 common configurations. Individual chips, containing many emitter-base junctions in logic circuits but perhaps only a single junction in power transistors [17], are typically 1.5 mm (60 mils) wide by 1.5 mm (60 mils) long and 0.1 to 0.2 mm (4 to 8 mils) thick. The energy dissipation in such a chip generally occurs within 0.04 mm (1 mil) of the top surface, and its time-average value can vary from 0.1 W to approximately 1.5 W, depending on the function and technology of the device [18–21]. Instantaneous dissipation levels in power transistors can approach 30 W under some circumstances, and even at modest duty cycles attain average values of 3 to 5 W [17]. The resulting heat flux at the surface of the chip, averaged in time and space, can thus vary from approximately 5 to 65 W/cm² in logic circuits and approach or even exceed 200 W/cm² in high-power transistors.

The monolithic semiconductor chips are generally bonded to a ceramic or metallic substrate, which may either constitute the external package or be bonded to it. Each such package provides the necessary electrical connections to other devices and may carry one or several to perhaps as many as a dozen chips. The heat dissipated in the chip is conducted into the substrate and transferred by some combination of thermal conduction, convection, and radiation to the cover of the flat pack or can. The relatively high and often nonuniform chip heat flux, the thermal resistance of the bonding material between the chip and the substrate, and the frequently torturous conduction path between the chip and the base of the substrate or flange of the can combine to yield a significant overall thermal resistance between an individual junction in the chip and the exterior of the integrated circuit package.

This so-called junction-to-case resistance is strongly influenced by the size of the chip, the bonding material(s), the substrate/case material(s), and package geometry.

FIG. 1.5   TO-5 configuration. (Courtesy of Honeywell, Inc.)

FIG. 1.6 Hybrid package. (Courtesy of Honeywell, Inc.)

In monolithic devices the junction-to-case resistance can vary from $11°$C/W in a power transistor package [17] to approximately 40 to $80°$C/W in a gold-soldered flat pack or TO-5 can [22] and to as high as $150°$C/W in a glass-bonded flat pack [22], with a range of 40 to $100°$C/W being typical of commercially available flat packs [22].

Similar junction-to-case resistances are encountered in hybrid microelectronic packages, which, as shown in Fig. 1.2, combine integrated circuit devices, semiconductor components, and conventional elements interconnected by appropriate electrical conduction paths (leads, metalized strips, or thick and thin film resistors) on a single substrate. Whereas epoxy-bonded chips in hybrid devices may display resistances as high as $120°$C/W [23], and resistance in a dual-in-line (DIP) plastic package in a hybrid driver may approach $135°$C/W [24], most hybrid units provide more effective heat transfer between the chips and the exterior surfaces of the package, and values of 25 to $40°$C/W are more typical of this application [24, 25].

As may be readily ascertained from the design junction temperatures cited above and the earlier characterization of the thermal environment, much of the available temperature difference between the environment and the device junction is needed to overcome junction-to-case resistance. Consequently, significant thermal improvements in semiconductor devices, that is, lower junction temperatures and higher dissipation rates, must involve a reduction in this resistance toward a target of 3 to $5°$C/W [22, 26]. At this value the internal thermal resistance in a high-power device represents approximately one-fourth to one-third of the overall thermal resistance between the junction and inlet temperature of the coolant.

To complete the thermal characterization of semiconductor devices, it is of interest to note that a single integrated circuit package, containing one to as many as a dozen chips, is approximately 3 cm by 3 cm and 0.5 cm thick in the case of flat packs, 2 cm by 5 cm by 0.5 cm thick in the case of dual-in-line packages, and approximately 1 to 3 cm in diameter and 1 cm high for transistor cans. Under special circumstances, individual large-scale integration and hybrid packages contain as many as 100 chips and may approach 8 cm on each side [27]. Because of the combined thermal contribution of all the chips carried in a single package, heat dissipation at this level generally varies from 1 to 25 W, yielding a surface heat flux of 0.1 to 3 W/cm^2. Significantly, the distribution of the chip heat dissipation over the envelope area of the microelectronic package is thus seen to result in more than an order-of-magnitude reduction in surface heat flux.

1.4.2 Multilayer/Printed Circuit Boards

Although many distinct methods are available for mounting and interconnecting semiconductor and conventional devices into electronic subassemblies, multilayer boards (MLBs)/printed circuit boards (PCBs), such as those shown in Fig. 1.1, are in common use. These boards, or cards as they are frequently termed, are generally 10 cm by 15 cm in plan and 1 to several millimeters in thickness and are typically composed of thermoplastic and glass epoxy materials with one or several metallic layers (or claddings) added to provide common electrical reference planes and thermal conduction paths. Heat dissipation on a ML/PC board consists primarily of the heat released by semiconductor packages mounted on the board, augmented to a minor extent by heat loss in discrete and printed resistors, and can vary from approximately 5 to 10 W in standard cards to as much as 20 to 30 W in hybrid and advanced avionic modules [21, 23]. The heat flux on one of the board surfaces is thus not greater than 0.2 W/cm^2, representing a further order-of-magnitude reduction relative to the heat flux on a single microelectronic package.

However, in many standard applications, system constraints dictate heat flow in the plane of the ML/PC board to metallic connectors along the perimeter of the board. Because of the thinness and relatively low thermal conductivity of the basic ML/PC board materials, thermal resistance in this direction, in the absence of a metallic layer, may be quite substantial, for example, 20 to 60°C/W. The addition of copper cladding, even 0.01 cm thick, can dramatically increase heat transfer in the plane of the board and reduce the device-to-card perimeter resistance to a more typical value of 5 to 10°C/W [22].

Although, as noted above, much of the heat dissipation associated with a printed circuit card originates in semiconductor devices, the allowable board temperatures may, under some circumstances, be constrained by the thermal requirements of capacitors and resistors in the electronic subassembly. Many high-performance capacitors consist of metalized plastic and paper within metal cases, and even at 40°C have a continuous operating life expectancy of only 1 yr. Teflon and plastic film capacitors can operate at up to 200°C, but for reliable operation most capacitors are rated below 125°C [16]. Resistors are generally more tolerant of high-temperature operation, but it is of interest to recall that whereas metal film and ceramic resistors can be operated at 150 to 200°C, carbon resistors are typically rated at 130 to 150°C.

1.4.3 High-Power Microwave Equipment

Because of the inefficiencies inherent in electrical-to-microwave energy conversion and the management of radiofrequency (RF) energy, the operation of radar equipment results in significant heat dissipation within the microwave components. In high-power civilian and military radar systems, this heat dissipation can lead to unacceptable component temperatures and thus necessitate active cooling to ensure satisfactory system operation.

In conventional, mechanically steered radar systems, much of the total power dissipation occurs in the klystron tubes, with secondary dissipation in the rotary joints and wave guides. Heat release along the collector surfaces in the klystron tubes is proportional to the RF power of the system and can average from 1 kW to approximately 20 kW at a peak local heat flux of 3 to 2000 W/cm² [11, 28, 29]. Prevention of melting and mechanical deformation of the collector structure and attainment of temperature stability in the tube [10], to within several degrees Celsius, are the primary aims of klystron tube thermal control. Their achievement often requires the use of pool or flow boiling of appropriate fluids around the collector (see Fig. 1.7) or through especially designed channels, as in the liquid-cooled vanes of cross-field amplifiers [11] or traveling-wave tube collectors as shown in Fig. 1.8.

Similar, though less severe, thermal problems are encountered in the Rf rotary joints and wave guides. Heat fluxes of 5 to 10 W/cm² and maximum allowable temperatures of 150°C are typical of these oxygen-free high-conductivity (OFHC) copper components in high-power systems, and again provision must generally be made for liquid cooling [30].

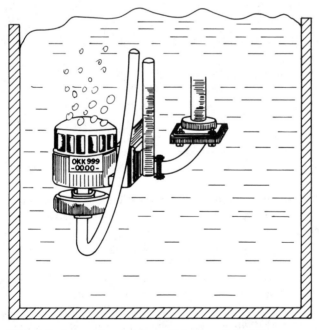

FIG. 1.7 Immersion cooling of klystron tube [5].

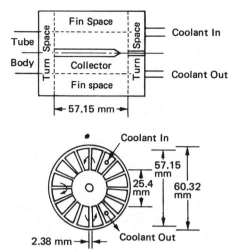

FIG. 1.8 Cooling arrangement for traveling-wave tube collector. The material is copper and there are four fins in each of the four coolant passes, each 2.38 mm wide by 15.9 mm high.

The advent of active antenna array elements, utilizing amplifier diodes, and the growing use of ferrite phase shifters, to provide electronic steering of the radiated RF beam, have introduced new dimensions in the thermal control of microwave equipment. In radar systems employing amplifier diodes, high-power klystron tubes and high-power waveguides can be eliminated and only low-power RF needs to be delivered to each antenna element, on which are mounted one, two, or several amplifier diodes. Regretably, the inherent low conversion efficiency of the amplifier diodes results in diode heat dissipation that is typically between 1 and 10 W [31] and local heat fluxes comparable to the power transistor chip previously discussed. Maintenance of the diode junction temperature below 200°C is, however, generally sufficient to ensure proper microwave operation [31].

When mounted on a radiator element, as in Fig. 1.9, and activated by an impressed electrical voltage, the ferrite phase shifters serve to delay the RF wave traversing the radiator element. Appropriate combination of many similarly activated

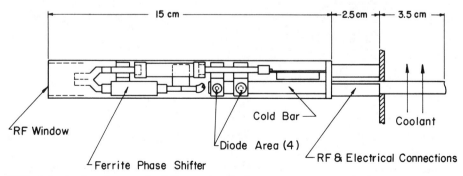

FIG. 1.9 Active element for RF antenna [31].

elements can thus facilitate precise shaping of the RF beam wave front and provide the capability for electronic focusing and steering of the radiated microwave beam. In contrast to the amplifier diodes, heat dissipation in the ferrite phase shifters is generally small if not negligible, but precise control of the radiated wave front can be achieved only in the absence of significant temperature excursions and/or variations. Consequently, ferrite temperatures in phased-array antennas must generally be maintained below 90°C with a maximum variation across the antenna array of 10°C [31, 32].

In many electronically steered and hybrid (both mechanical and electronic steering) radar systems, final distribution of the RF power to the radiator elements is accomplished by the use of stripline board attached to the rear surface of the antenna array. In such systems, array thermal dissipation occurs primarily in the stripline power divider, approximately 0.3 db of incident RF power in one typical example, but the stripline material is relatively insensitive to temperature and can generally perform successfully at temperatures approaching 230°C [31].

1.5   THERMAL CONTROL TECHNOLOGY

The foregoing discussion has established the basic thermal control requirements of electronic components and set the stage for a detailed exploration of thermophysics, presented in Part II, and its application to the thermal stabilization of electronic components, the subject of Parts III and IV. Although it is thus somewhat premature to discuss specific thermal control techniques at this point, a brief review of the available technology will serve to define the options available to the thermal packaging engineer.

1.5.1   Heat Transfer Relations

Heat transfer through solids occurs primarily by conduction and is governed by Fourier's law, which, in integral form, is expressed as

$$q = \frac{kA(T_1 - T_2)}{L} \tag{1.1}$$

The temperature difference resulting from the flow of a specified heat flux (or equivalently, necessary to transfer a specified heat flux) is thus related to the thermal conductivity of the material and the path length, or

$$\Delta T_{\text{cond}} = \frac{L}{k} q'' \tag{1.2}$$

The convective heat transfer rate is given by

$$q = hA(T_1 - T_f) \tag{1.3}$$

The differences between natural and forced convection, as well as heat exchange variations among various fluids, are reflected in the range of values of h, the heat transfer coefficient. The temperature difference associated with the convective transport of a specified heat flux is equal to

$$\Delta T_{\text{conv}} = \frac{1}{h} q'' \tag{1.4}$$

Unlike conduction and forced convection, radiative heat transfer between a black surface and an effectively black environment is not linearly dependent on the temperature difference and is expressed instead as

$$q = \sigma A (T_1^4 - T_2^4) \tag{1.5}$$

Thermal transport associated with the boiling process displays a similarly complex dependence on the temperature difference between the boiling surface and the saturation temperature of the liquid and can be approximated by a relation of the form

$$q = C'_{sf} A (T_1 - T_{\text{sat}})^3 \tag{1.6}$$

When the temperature differences are small, both Eqs. (1.5) and (1.6) can be linearized and reexpressed in the form of the convection relation, Eq. (1.3). Following the appropriate algebraic manipulation, the radiative heat transfer coefficient h_r is found to equal $4\sigma T_m^3$ and the boiling heat transfer coefficient h_b equal to $C'_{sf}(T_1 - T_{\text{sat}})^2$. In the range of applicability of h_r and h_b, the respective radiative and boiling temperature differences are given by Eq. (1.4). It is of interest to note that for modest temperature levels and black surfaces, h_r is found approximately to equal the heat transfer coefficient in natural convective heat exchange with air.

1.5.2 Thermal Control Options

When the heat flux dissipated by the electronic component, device, or assembly is known and the allowable temperature rise above the local ambient condition is specified, the equations in the previous subsection together with the value of the junction-to-case temperature difference, ΔT_{j-c}, can be used to determine which heat transfer process or combination of processes (if any) can be employed to meet the desired performance goals. To aid in this process, Eqs. (1.2), (1.4), and (1.6) are plotted in Fig. 1.10 for a variety of conditions and working fluids.

Examination of Fig. 1.10 reveals that for a typical allowable temperature difference of 60°C between the component surface and the ambient, "natural" cooling in air—relying on both free convection and radiation—is effective only for heat fluxes below approximately 0.05 W/cm². Although forced convection cooling in air offers approximately an order-of-magnitude improvement in the heat transfer coefficient, this thermal configuration is unlikely to provide a heat removal capability in excess of 1 W/cm² even at an allowable temperature difference of 100°C. To facilitate the transfer of moderate and high heat fluxes from component surfaces, the thermal designer must choose between the use of finned air-cooled heat sinks and direct or indirect liquid cooling. Finned arrays and sophisticated techniques for improving convective heat transfer coefficients can extend the effectiveness of air cooling to progressively higher component heat fluxes but often at ever-increasing weight, cost, and volume penalties. Alternately, reliance on heat transfer to liquids flowing at high velocity through so-called cold plates can offer a dramatic improvement in the trans-

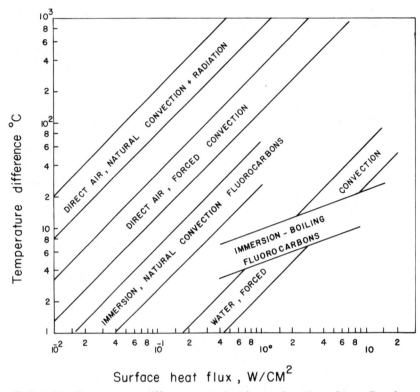

FIG. 1.10   Temperature differences attainable as a function of heat flux for various heat transfer modes and various coolant fluids.

ferrable heat flux even at temperature differences as low as 10°C, when the conduction resistance in the cold plate wall is negligible.

A similarly high heat flux capability is offered by ebullient heat transfer to the fluorochemical liquids. The high dielectric properties of these liquids make it possible to immerse most electronic components directly in the fluorochemical liquids and thus remove component heat fluxes in excess of 10 W/cm² with saturated pool boiling at temperature differences typically less than 20°C. Immersion cooling, employing natural convection heat transfer, can also offer significant advantages and, as seen in Fig. 1.10, serves to bridge the gap between direct air cooling and cold-plate technology.

1.6   NOMENCLATURE

Roman Letter Symbols

A area normal to heat flow path (cross-sectional area), m²
C'_{sf} constant in Eq. (1.6), W/m² K³
h heat transfer coefficient, W/m² K
k thermal conductivity, W/m K
L length of heat flow path, m

q	heat flow, W
q''	heat flux, W/m^2
T	temperature, K

Greek Letter Symbols

Δ	indicates change in variable

Subscripts

b	indicates boiling mode or designates boiling coefficient
cond	indicates conduction mode
conv	indicates convection mode
f	indicates fluid
$j\text{-}c$	indicates junction to case property or value
r	designates radiation coefficient
sat	indicates saturation condition

1.7 REFERENCES

1   Cockroft, J. D., The Temperature Distribution in a Transformer or Other Laminated Core of Rectangular Cross-Section in Which Heat Is Generated at a Uniform Rate, *Proc. Cambridge Philos. Soc.*, vol. 22, pp. 759–772, 1925.

2   Mouromtseff, I. E., and Koyanowski, H. N., Analysis of Operation of Vacuum Tubes as Class C Amplifiers, *Proc. IRE*, vol. 23, pp. 731–769, 1935.

3   Mouromtseff, I. E., Water and Forced Air Cooling of Vacuum Tubes, *Proc. IRE*, vol. 30, pp. 190–205, 1942.

4   Elenbaas, W., Heat Distribution of Parallel Plates by Free Convection, *Physica*, vol. 9, no. 1, pp. 665–671, 1942.

5   Kaye, J., Review of Industrial Applications of Heat Transfer to Electronics, *Proc. IRE*, vol. 44, pp. 977–991, 1956.

6   Bergles, A. E., Chu, R. C., and Seely, J. H., Survey of Heat Transfer Techniques Applied to Electronic Equipment, ASME paper 72-WA/HT-39, 1972.

7   *Reliability Prediction of Electronic Equipment*, U.S. Dept. Defense, MIL-HDBK-217B, NTIS, Springfield, Va., 1974.

8   Hanneman, R., Digital Equipment Corporation, Maynard, Mass., personal communication, Aug. 1979.

9   Hilbert, W. F., and Kube, F. H., Effects on Electronic Equipment Reliability of Temperature Cycling in Equipment, Final Report, Grumman Aircraft Engr. Corp., Report No. EC-69-400, Bethpage, N.Y., Feb. 1969.

10   Gucker, G. B., Long Term Frequency Stability for a Reflex Klystron Without the Use of External Cavities, *Bell System Tech. J.*, vol. 41, pp. 945–958, May 1962.

11   Cochran, D. L., Boiling Heat Transfer in Electronics, paper presented at NEPCON '68 East, New York, June 1968, Raytheon Paper No. P 246-1, MSD, Bedford, Mass., 1968.

12   Hughes Aircraft Company (R. W. Freeman), Environmental Stabilization of Data Processors Program—Final Report, FR-64-10-269, Fullerton, Calif., Sept. 1964.

13   Chertoff, A. B., and Forti, J. J., Problems of Heat Removal Chill Progress in IC's, *Electronics*, vol. 40, pp. 129–136, Sept. 18, 1967.

14   Greenhouse, H. M., and McGill, R. L., Design of Temperature Controlled Substrates for Hybrid Microcircuits, *IEEE Trans. Parts, Hybrids and Packaging*, vol. PHP-10, no. 2, pp. 137–145, 1974.

15   Rambach, C., and Bar-Cohen, A., Nocturnal Heat Rejection by Skyward Radiation, in *Future Energy Production Systems*, ed. J. C. Denton and N. Afgan, vol. II, pp. 713–726, Hemisphere, Washington, D.C., 1976.

16   Black, E. P., and Daley, E. M., Thermal Design Considerations for Electronic Components, ASME publication 70-DE-17, 1970.

17   Baxter, G. K., Thermal Response of Microwave Transistors Under Pulsed Power Operation, *IEEE Trans. Parts, Hybrids and Packaging*, vol. PHP-9, no. 3, pp. 185–193, 1973.

18   Baxter, G. K., and Anslow, J. W., High Temperature Thermal Characteristics of Microelectronic Packages, *IEEE Trans. Parts, Hybrids and Packaging*, vol. PHP-13, no. 4, pp. 385–389, 1977.

19   Wilson, E. A., Thermal Analysis of Integrated Circuit Packaging Techniques, *Proc. 23d Annual Electronic Components Conference, San Francisco, May 14–16*, pp. 328–333, 1973.

20   Watson, D., Thermal Study of Circuit Card Assembly, *Microelectron. Reliab.*, vol. 12, pp. 531–534, 1973.

21   Token, K. H., A New Avionics Thermal Control Concept, ASME publication 77-ENAS-14, 1977.

22   Plizak, B. T., Thermal Parameters and Analytical-Experimental Procedures for Integrated Circuit Devices and Systems, Report No. NADC-AE-6849, Naval Air Development Center, Warminster, Pa., 1969.

23   Pogson, J. T., and Franklin, J. L., Analysis and Temperature Control of Hybrid Microcircuits, *J. Eng. Ind.*, vol. 95, pp. 1048–1052, Nov. 1973.

24   Mandel, A. P., and Vahey, P. K., Thermal Design Criteria for Hybrid Microelectronic Modules, Technical Report, Microtek Electronics, Inc., Cambridge, Mass., 1969.

25   Boucher, S., Computer-Aided Thermal Design of LSI Packages, Technical Report, Raytheon Company, Missile Systems Division, Bedford, Mass., 1972.

26   Donegan, M., Aircraft and Spacecraft Electronics, in *Proceedings, NSF Workshop on Directions of Heat Transfer in Electronics, Atlanta*, Oct. 1977.

27   Wilson, E. A., Applying Plastics in a Highly Reliable, Low Cost Cooling System for Microelectronics, Honeywell Information Systems, Phoenix, 1977.

28   Paradis, L. R., Simplified Transmitter Cooling System, paper presented at 8th Int. Circuit Packaging Symp. (WESCON), San Francisco, Aug. 1967.

29   Thompson, J. C., Raytheon Company, MSD, Bedford, Mass., personal communication, 1972.

30   Hartop, R., and Bathker, D. A., A 400-KW Long Pulse X-Band Planetary Radar, *Proc. IEEE-MTT Symp.*, vol. 24, pp. 136–138, 1976.

31   Markowitz, A., Thermal Control of SHF Antennas—Conceptual Design, Raytheon Company, MSD, Internal Memo BSH-53, Bedford, Mass., 1971.

32   Joachim, R. J., and Pitasi, M. J., IR and Thermal Evaluation of a Phased Array Antenna, *Materials Evaluation*, vol. 29, no. 9, pp. 193–204, 1971.

2

■ reliability

2.1 INTRODUCTION

Reliability is the probability of a device or system performing without failure for the period intended under the operating conditions encountered. The ultimate goal of thermal analysis and the prediction of component operating temperatures is the assessment of reliability.

Reliability is intimately linked with failure, failure rates, and component and system mortality. Figure 2.1 displays the so-called bathtub mortality curve with the failure rate of a particular component plotted against component age. Such a curve may be constructed from data taken from a large sample of homogeneous components all of which are placed in operation at $t = 0$. At the outset, the population of components under test will exhibit a high failure rate or infant mortality during the so-called break-in or burn-in period. This period, which ends at t_b, sees failures due to poor quality control during the manufacturing process.

The period between t_b and t_w is the useful life of the component, during which the failure rate is relatively constant. Failures during this period occur at random and are completely unpredictable. It is this period that is of the greatest concern because, in the electronic industry, components should never be used unless they have been properly "burned in" and have reached age t_b.

At time t_w, only a small fraction of the component will have failed if the components are to be called reliable. After time t_w, the failure rate increases as the components begin to age or wear out. The period of increasing failure rate after time t_w is called the "wear-out" period. Time t_w can be considered to be the useful life, since, with proper quality control, testing, and burn-in procedures, $t_b = 0$.

2.2 PROBABILITY, RELIABILITY, AND FAILURE RATE

Consider a fixed number of components n_0, subjected to test at $t = 0$. At time t, a certain number will have survived, $n_s(t)$, and a certain number will have failed, $n_f(t)$. The functions $n_s(t)$ and $n_f(t)$ are mutually exclusive; a component can either survive or fail but not both. With

$$n_s(t) + n_f(t) = n_0 \tag{2.1}$$

the probability of survival, which is the reliability $R(t)$, is

$$R(t) = \frac{n_s(t)}{n_0} = \frac{n_s(t)}{n_s(t) + n_f(t)} \tag{2.2}$$

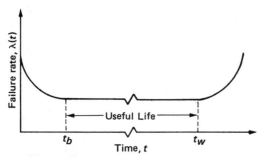

FIG. 2.1 "Bathtub" mortality curve showing variation of failure rate with component age.

and the probability of failure $Q(t)$ is

$$Q(t) = \frac{n_f(t)}{n_0} = \frac{n_f(t)}{n_s(t) + n_f(t)} \tag{2.3}$$

Observe that

$$R(t) + Q(t) = \frac{n_s(t) + n_f(t)}{n_s(t) + n_f(t)} = 1 \tag{2.4}$$

and that rearrangement of Eq. (2.2) gives

$$R(t) = \frac{n_s(t)}{n_s(t) + n_f(t)} = \frac{n_0 - n_f(t)}{n_0} = 1 - \frac{n_f(t)}{n_0} \tag{2.5}$$

The rate at which components fail is obtained from readjustment of Eq. (2.5) after differentiation with respect to t:

$$\frac{dR(t)}{dt} = -\frac{1}{n_0} \frac{dn_f(t)}{dt}$$

or $$\frac{dn_f(t)}{dt} = -n_0 \frac{dR(t)}{dt} \tag{2.6}$$

The survival rate is

$$\frac{dn_s(t)}{dt} = \frac{d}{dt} [n_0 - n_f(t)] = -\frac{dn_f(t)}{dt} \tag{2.7}$$

which shows that rate at which components survive is equal to the negative of the rate at which components fail. It is also apparent that at $t = 0$,

$$R(0) = 1 \quad Q(0) = 0$$

and that after a long period of time ("forever"),

$$R(\infty) = 0 \quad Q(\infty) = 1$$

If both sides of Eq. (2.6) are divided by $n_s(t)$, the instantaneous failure rate $\lambda(t)$ is obtained:

$$\frac{1}{n_s(t)} \frac{dn_f(t)}{dt} = -\frac{n_0}{n_s(t)} \frac{dR(t)}{dt} = \lambda(t) \tag{2.8}$$

Then with Eq. (2.2) substituted into Eq. (2.8), a simple first-order differential equation comes forth,

$$\lambda(t) = -\frac{1}{R(t)} \frac{dR(t)}{dt} \tag{2.9}$$

which after separation of the variables and integration from $t = 0$ to $t = t$ yields

$$R(t) = e^{-\int_0^t \lambda(t)\,dt} \tag{2.10}$$

If $\lambda(t)$ is constant as inferred during the useful life period shown in Fig. 2.1, then

$$R(t) = e^{-\lambda t} \tag{2.11}$$

because in this case $\lambda(t)$, the failure rate, is merely λ.

2.3 REPLACEMENT AND NONREPLACEMENT TESTING

Consider a test of n_0 components in which k components fail after t hours, but as each failure occurs, the failed component is replaced by a component that functions. In this case, which is called a *replacement test*, the failure rate is the number of failures divided by the number of component hours that have accumulated,

$$\lambda = \frac{k}{n_0 t} = \frac{k}{T} \tag{2.12}$$

where $T = n_0 t$ is called the accumulated test time.

Now consider the *nonreplacement test*, where components are not replaced after they fail. For failures at $t_1, t_2, t_3, \ldots$, the failure rate is

$$\lambda = \frac{k}{t_1 + t_2 + t_3 + \cdots + (n_0 - k)t} \tag{2.13}$$

where t_k is the time at which the kth failure occurs. In this case, the accumulated test time is

$$T = \sum_{i=1}^{k-1} t_i + (n_0 - k)t$$

■ EXAMPLE

In a test of 500 components, single failures occur at 200, 400, 600, and 800 h. Determine the failure rates for the replacement and nonreplacement tests in a 1000-h test.

SOLUTION

Replacement test, Eq. (2.12):

$$n_0 = 500 \quad k = 4 \quad t = 1000$$

$$\lambda = \frac{4}{(500)(1000)} = 8 \times 10^{-6}$$

or 8 failures per million hours.

Nonreplacement test, Eq. (2.13):

$$(n_o - K)$$
$$T = 200 + 400 + 600 + 800 + 496(1,000) = 498,000$$

$$\lambda = \frac{4}{498,000} = 8.032 \times 10^{-6}$$

2.4 MEAN TIME BETWEEN FAILURES

The mean time between failures (MTBF) may be defined as the accumulated test hours divided by the number of failures,

$$\text{MTBF} = \frac{T}{k} \tag{2.14}$$

and for a constant failure rate, using Eq. (2.12),

$$\text{MTBF} = \frac{1}{\lambda} \tag{2.15}$$

This makes Eq. (2.11) read

$$R(t) = e^{-t/\text{MTBF}} \tag{2.16}$$

where it is assumed that all of the components under test have failed.

Looking at this in another way, observe that for the last component failing after a very long time, the accumulated test time is

$$T = \int_0^\infty n_s(t)\,dt$$

Here, the upper limit of integration could easily have been taken as t_n, the time at which the last component failed. In this case, because no components have survived after t_n,

$$T = \int_0^{t_n} n_s(t)\, dt + \int_{t_n}^{\infty} (0)\, dt = \int_0^{\infty} n_s(t)\, dt$$

which is the same thing.

Now, in Eq. (2.14), k is the number of failures and at $t = \infty$, all components have failed and $k = n_0$. Then

$$\text{MTBF} = \int_0^{\infty} \frac{n_s(t)}{n_0}\, dt$$

and by Eq. (2.2),

$$\text{MTBF} = \int_0^{\infty} R(t)\, dt$$

Use of Eq. (2.11) and performance of a simple integration yields

$$\text{MTBF} = \int_0^{\infty} e^{-\lambda t}\, dt = -\frac{1}{\lambda} e^{-\lambda t} \Big|_0^{\infty} = \frac{1}{\lambda}$$

which confirms Eq. (2.15).

2.5 SYSTEM RELIABILITY

While the failure rates and MTBF of components can be measured and used to provide a yardstick of sorts, it is the reliability of a system composed of many components that is the ultimate concern. Fortunately, the reliability of a system may be determined from its individual components.

2.5.1 Series Reliability

Consider the system shown in Fig. 2.2. In this *series system*, the system can function only if *both* subsystem A and subsystem B function. The probability of operation of the system, which is the reliability of the system, is the total probability

$$P_T = R_T = R_A R_B$$

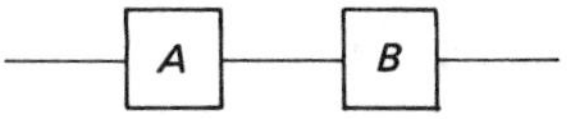

FIG. 2.2 System composed of two subsystems arranged in series.

If the reliability of system A is 0.9962 and that of system B is 0.9973, then the system reliability of the total system is

$$R_T = R_A R_B = (0.9962)(0.9973) = 0.9935$$

This is easily extended to n systems in series, where the total reliability is

$$R_T = \prod_{i=1}^{n} R_i \qquad (2.17)$$

■   **EXAMPLE**

A simple series circuit is composed of a resistor with a MTBF of 80,000 h, an inductor with a MTBF of 100,000 h, and a capacitor with a MTBF of 125,000 h. What is the circuit MTBF and its reliability for the passage of current for a 20-h operating time?

SOLUTION

Failure rates:

Resistor: $\lambda_1 = \dfrac{1}{80,000} = 1.25 \times 10^{-5}$

Inductor: $\lambda_2 = \dfrac{1}{100,000} = 1.00 \times 10^{-5}$

Capacitor: $\lambda_3 = \dfrac{1}{125,000} = 8.00 \times 10^{-6}$

In accordance with

$$R_T(t) = R_1(t)R_2(t)R_3(t) = e^{-\lambda_1 t}e^{-\lambda_2 t}e^{-\lambda_3 t} = e^{-(\lambda_1 + \lambda_2 + \lambda_3)t}$$

the equivalent failure rate of the system is the sum of the failure rates of the components:

$$\lambda_T = (1.25 + 1.00 + 0.80) \times 10^{-5} = 3.05 \times 10^{-5}$$

This makes the system MTBF

$$\text{MTBF}_T = \frac{1}{\lambda_T} = 32,786.9 \text{ h}$$

and the system reliability for 20 h of operation is

$$R(20) = e^{-\lambda_T t} = e^{-3.05 \times 10^{-5}(20)} = 0.999390$$

Observe that the individual reliabilities for 20 h of operation are

Resistor: $R_1(20) = e^{-1.25 \times 10^{-5}(20)} = 0.999750$

Inductor: $R_2(20) = e^{-1.00 \times 10^{-5}(20)} = 0.999800$

Capacitor: $R_3(20) = e^{-8.00 \times 10^{-6}(20)} = 0.999840$

and that the system reliability is, of course,

$$R_T(20) = R_1(20)R_2(20)R_3(20)$$
$$= (0.999750)(0.999800)(0.999840) = 0.999390$$

2.5.2 Parallel Reliability

For the parallel situation containing system A and system B as shown in Fig. 2.3, the entire system can operate if either A or B or both are operating. Thus, the only way that the system can fail to operate is in the eventuality that both system A and system B are not operating. Thus for total failure,

$$Q_T = Q_A Q_B$$

and from Eq. (2.4),

$$Q = 1 - R$$
$$Q_T = (1 - R_A)(1 - R_B) = 1 - R_A - R_B + R_A R_B$$

or $1 - R_T = 1 - R_A - R_B + R_A R_B$

and finally, for the parallel system composed of system A and system B,

$$R_T = R_A + R_B - R_A R_B$$

A similar development does not extend to n systems in parallel. For example, take Fig. 2.4 with systems A, B, and C in parallel. Here

$$Q_T = Q_A Q_B Q_C$$

or $1 - R_T = (1 - R_A)(1 - R_B)(1 - R_C)$

$$= 1 - R_A - R_B - R_C + R_A R_B + R_B R_C + R_A R_C - R_A R_B R_C$$

Finally,

$$R_T = R_A + R_B + R_C - R_A R_B - R_B R_C - R_A R_C + R_A R_B R_C$$

FIG. 2.3 System composed of two subsystems arranged in parallel.

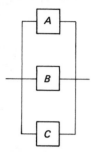

FIG. 2.4 System composed of three subsystems arranged in parallel.

With the implied assumption that all units operate simultaneously,

$$1 - R_T = \sum_{i=1}^{n} (1 - R_i) \qquad\qquad\qquad (2.18)$$

SHOULD BE PRODUCT NOT SUMMATION

$$1 - R_T = \prod_{i=1}^{n} (1 - R_i)$$

is the relationship to be used to determine the total reliability.

■   *EXAMPLE*

If three subsystems, all having identical reliabilities, are put in parallel and the satisfactory functioning of the entire system depends on the operation of any or all of the subsystems, determine the allowable time of operation of the system to achieve a reliability of 0.9962 if the MTBF of each subsystem is 2500 h.

SOLUTION

To get $R_T = 0.9962$, Q_T is restricted to $1 - 0.9962 = 0.003800$. Hence, by a form of Eq. (2.18),

$$Q_T = Q_1 Q_2 Q_3 = Q^3$$

where $Q = Q_1 = Q_2 = Q_3$. This makes

$$Q = (0.003800)^{1/3} = 0.156049$$

and the reliability of each subsystem

$$R = 1 - Q = 1 - 0.156049 = 0.843951$$

To find the time of operation, use Eq. (2.16):

$$R = 0.843951 = e^{-t/2500}$$

or $$\frac{t}{2500} = 0.169661$$

Thus $t = 424.15$ h, which shows that the luxury of redundant components and subsystems is glorious: All it takes is cost, weight, volume, and complexity to pay the price.

2.6   SERIES–PARALLEL SYSTEMS

Some redundancy is necessary in large systems where, if all components or subsystems were arranged in series, if any one were to fail, the system would fail. Thus, the less reliable components in complex systems must be redundant. It is not uncommon to see parallel-series, mixed-parallel, and series-parallel arrangements of components as shown in Fig. 2.5.

It is not difficult to derive, from Eqs. (2.17) and (2.18) for components of reliability R, the total reliability R_T of a parallel-series system containing n series elements in m parallel paths.

$$R_T = 1 - (1 - R^n)^m \tag{2.19}$$

EXAMPLE

Consider the six components in Fig. 2.5 and assume that they are identical, with reliability $R = 0.903621$. Determine the system reliability R_T of each system.

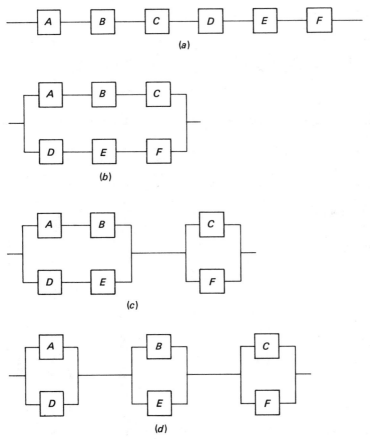

FIG. 2.5   Six subsystems arranged in (a) series, (b) parallel-series, (c) mixed parallel, and (d) series-parallel.

SOLUTION

(a) For the series arrangement of Fig. 2.5a, use Eq. (2.17).

$$R_T = R^6 = (0.903621)^6 = 0.544400$$

(b) For the parallel-series arrangement of Fig. 2.5b, use Eq. (2.19).

$$m = 2 \quad n = 3 \quad 1 - R = 0.096379$$
$$R_T = 1 - (1 - R^n)^m$$
$$= 1 - [1 - (0.903621)^3]^2$$
$$= 1 - 0.068731 = 0.931269$$

(c) For the mixed-parallel arrangement of Fig. 2.5c, use Eq. (2.19) twice for the cascade.

$$m_1 = 2 \quad n_1 = 2 \quad m_2 = 2 \quad n_2 = 1$$
$$R_T = [1 - (1 - R^2)^2][1 - (1 - R^1)^2]$$
$$= \{1 - [1 - (0.903621)^2]^2\}\{1 - [1 - (0.903621)^1]^2\}$$
$$= (0.966339)(0.990711) = 0.957363$$

TABLE 2.1   Hybrid Circuit Element Base Failure Rates in Percent per Thousand Hours [2, 3]

Element	Temperature (all values $\times 10^6$)				
	25°C	50°C	75°C	100°C	125°C
Thick film resistor	500	1,000	1,500	2,000	2,500
Chip capacitor	1,000	1,500	2,500	6,000	25,000
Wire bonds					
Au-Al ball	5	20	100	100	6,000
Al-Au	10	10	10	10	50
Al-Al	10	10	10	10	10
Au-Au	4	4	4	4	4
Crossovers	5	5	6	8	10
Transistor chips					
Low power	100	300	900	2,700	7,000
Power	5,000	10,000	30,000	90,000	270,000
Diode chips	100	300	900	2,700	7,000
Microcircuits					
Quad gate	2,000	3,600	18,000	82,000	240,000
Dual flip-flop	4,000	7,200	36,000	164,000	480,000
SSI (25 gates)	12,500	22,500	112,500	512,000	
MSI (50 gates)	25,000	45,900	225,000		
LSI (100 gates)	50,000	90,000	450,000		

(d) For the series-parallel arrangement of Fig. 2.5*d*, use Eq. (2.19).

$$m = 2 \quad n = 1 \quad \text{three times}$$

$$R_T = [1 - (1 - R^n)^m]^3$$
$$= [1 - (1 - 0.903621)^2]^3$$
$$= (0.990711)^3 = 0.972391$$

From the foregoing, one observes that alternative (d) is the most attractive because it has the greatest degree of what reliability engineers call *cross strapping* [1].

2.7 FAILURE RATES AND COMPONENT TEMPERATURE

A vast literature displays curves and tables of failure rate as a function of component operating temperature. The conclusion is inescapable: Failure rate varies directly with component operating temperature.

Table 2.1 lists the failure rates for hybrid circuit elements in thick film microcircuit modules as a function of temperature. The numbers demonstrate the necessity for precise thermal control to a maximum operating temperature dictated by the reliability of the components and system. Indeed, as discussed in Chap. 1, the goal of much of the thermal analyses of electronic equipment is to provide data for the assessment of system reliability.

2.8 NOMENCLATURE

Roman Letter Symbols
k failure index or number of failures
MTBF   mean time between failures
n number of components
p probability
Q probability of failure
R probability of survival or reliability
t time
T accumulated test time

Greek Letter Symbols
λ failure rate

Subscripts
b indicates end of break-in or "burn-in" period
f indicates failure
k indicates time at which kth failure occurs
n indicates time at which last component fails

s	indicates survival
T	indicates total
w	indicates beginning of "wear-out" period
0	indicates a fixed number

2.9 REFERENCES

1 Weaver, L., personal communication, July 1979.
2 Thornell, J. W., Fahley, W. A., and Alexander, W. L., Hybrid Microcircuit Design and Procurement Guide, Boeing Company, document available from National Technical Information Service, Springfield, Va., *DOC* AD 705974, 1972.
3 Harper, C. A., *Handbook of Thick Film Hybrid Microelectronics*, McGraw-Hill, New York, 1974.

3

■ concept formulation

3.1 INTRODUCTION

The spectacular success of the synergism of science and engineering in providing systems of nearly unimagined capabilities and immense destructive power has obscured the contribution of engineering design to the technological growth of modern society. The transfer of information from science to engineering has been accompanied to a large extent by a transfer of approaches and tools that has elevated scientific rigor and mathematical analysis to a predominant position in many areas of engineering practice. Yet examination of the past and recent history of technology fails to justify the ascendency of science in this domain. Although many common objects have been influenced by science, the form, function, dimensions, and appearances of objects and systems are largely the products of nonscientific modes of thought [1].

Much of the early history of Western technology can be related to nonverbal or visual thinking. Illustrated technical works, including Book Seven of Francesco di Giorgio Martini's *Trattato di Architectura* (composed around 1475) and Georg Agricola's classic 1566 book on mining and metallurgical processes, *De re Metallica*, served both to suggest novel mechanisms and to diffuse established techniques [1]. In these texts, and others published throughout the sixteenth and seventeenth centuries, primary emphasis was placed on detailed drawings and explanatory notes were often terse or totally absent. During this era the practitioners of the "mechanic arts" (as engineering was called) were generally trained as artists and in their work relied heavily on a sense of form and pictorial catalogs of machine elements [1]. Concepts and approaches were thus generally developed in pictorial form and tested in scale models and prototype devices. The development of isometric views, exploded views, and the ordinary graph did much to ease pictorial communication and facilitated the considerable achievements of Renaissance engineering.

Despite the common mythology to the contrary, the importance of the nonmathematical aspects of engineering development has persisted, and perhaps even grown, in the modern era. Many prominent nineteenth-century inventors, including Robert Fulton and Samuel Morse, were artists turned technologists, and others, such as James Watt, credited significant breakthroughs to mental pictures or visual inspirations [1]. More recently, studies of major technological innovations, for example, in military hardware, have strongly suggested that conceptual inspiration, rather than the application of a new or more sophisticated analytical technique, is most responsible for sudden improvements in conventional equipment or the development of new approaches [2].

These findings are reinforced by a growing body of literature on the mental processes involved in problem solving. This literature suggests that successful problem solving requires highly effective cooperation between the analytical and visual or integrative faculties of the human mind and that it is usually the failure to use available and stored information (rather than inaccurate analysis) that leads to "disaster" [3–5]. Furthermore, when engineering design and development of products and systems is viewed objectively, analysis is seen to represent but a single step of a complex process that, when successful, commences with task, problem, and constraint definition, progresses through the development and evaluation of alternative solutions, and terminates, after many iterations, with a product or system specification [6]. In many cases this process proceeds on an intuitive plane, but, like most human activities, the efficacy of the design process can be substantially improved by studying its various stages and exercising the skills needed to assure its success [5].

In recognition of this fact and the realization that in a world of shrinking economic horizons some evidence of the application of a formal design and/or optimization procedure must accompany every proposed technical solution, it appears desirable to include a detailed discussion of the design process in the present treatment of thermal control techniques for electronic components. In the course of this discussion, and following a general presentation of the design process, attention will focus on concept formulation—its importance, obstacles to its success, and techniques for overcoming these obstacles.

3.2 THE DESIGN PROCESS

3.2.1 Synthesis

Although engineering development is commonly associated with mathematical analysis, the mental processes involved in technological problem solving are largely different from, if not totally opposite to, analytical procedures. The "synthesis" of engineered products and systems requires the specification of "black boxes," each capable of producing a desired result when acted upon by a known stimulus. Alternatively, conventional analysis involves determination of the behavior of a known system or component when acted upon by a specified external stimulus. Thus, the design of a toaster—that is, the synthesis of a device that, when loaded with slices of bread and activated, will produce slices of toast—is an inherently different activity than determining the maximum permissible loading on a column, that is, the failure analysis of a specified component. Nonetheless, it must be realized that synthesis is often aided by iterative analysis of possible concepts and although in principle this analysis can be based on testing of physical models, the availability of sophisticated mathematical and computer techniques substantially increases the number and accuracy of such analyses. In this perspective, synthesis or concept formulation is seen to lie at the heart of the design process and serves to establish the structure—that is, boundaries, variables, and realm—within which mathematical analysis is performed. The stages of this design process and their interrelations are described in the subsections that follow.

3.2.2 Need Identification
and Problem Definition

The design process, as suggested in Fig. 3.1, generally begins with the perception of a "societal" need. The scope of this need is, of course, established by the control volume in use and may be a national concern, such as improving the balance of payments, a

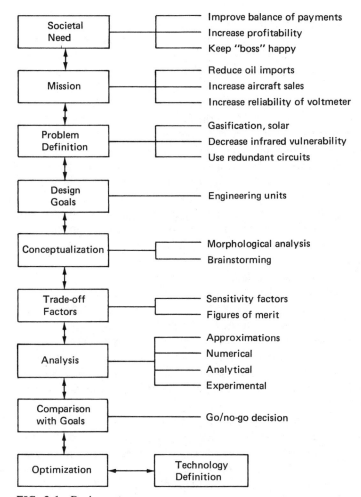

FIG. 3.1 Design process.

company concern, such as increasing profitability, or, on a more immediate level, keeping the "boss" happy. The need may be identified in a variety of ways, including observation, surveys, and referrals from colleagues, but must in all cases be such that "society" will recognize the need and be willing to pay for its satisfaction.

Once the need is established, it is possible to specify various missions whose successful completion might answer the need. Thus, in response to national need, it may be appropriate to reduce oil imports, acquire overseas territories, or revalue the currency. In an aerospace company, increased aircraft sales, decreased overhead, and diversification in the investment portfolio may all result in increased profits. Finally, on the personal level, the boss's pleasure might be solicited by wearing (or not wearing) green polkadot bow ties, announcing that you've decided to quit, or increasing the reliability of the prototype digital voltmeter under development in your group. There is clearly no single approach to the alleviation of the perceived need and, in fact, a single approach may answer several needs; for example, an improved voltmeter might also improve the company's profitability and perhaps even affect the national balance of payments. More important, however, the missions specified are not in themselves amenable to engineering solution, and additional mental effort or logical processing is required to define the engineering tasks or problems involved in assuring the success of the mission.

Problem definition is one of the critical elements in the design process and involves translation of the societal need, via the defined mission(s), into technical terms. If the national need is to be met by reducing oil imports, the engineering community might be called upon to develop an efficient process for coal gasification, an economically viable solar power station, or a price-competitive electric car. The engineering supervisor who has responded to the prospect of an improved digital voltmeter might direct the engineering staff to consider changing critical components, adding redundant circuits, or lowering the operating temperature of key components. The aerospace company's aircraft sales could perhaps be improved by reducing the plane's vulnerability to heat-seeking missiles, increasing its payload, or simplifying the engine's maintenance procedure.

In each of these categories, it is clearly possible to define an entire range of technical problems requiring definition (by no means exhausted with the modest examples cited), and this must be done before the traditional engineering activity of synthesis can begin. Furthermore, in order to provide the design team with necessary guidelines, it is essential that specific design goals accompany the problem definition. These goals, expressed in engineering quantities (such as failures/time or pounds mass of gas/pounds mass of coal) are often an integral part of the problem definition and can be derived from the more qualitative goals implied by the mission statement or the societal need.

Although the activity involved in problem definition is rather different than that required in the next few stages, separation of problem definition from concept formulation and analysis—as is common in many large development laboratories—may result in excellent paper solutions to the wrong problem and failure to complete the specified mission. Product and system development, by their synthetical nature, require constant iteration between the various stages depicted in Fig. 3.1, and improved understanding of the limits of a given technology, gained during conceptualization and analysis, can often exert a pivotal influence on the problem definition and design goals. Thus, for example, the realization that the improvement in system reliability sought through thermal management may require the case of the critical component to be below ambient temperature or the possible finding that the energy required to fabricate and construct a solar power station could temporarily increase oil imports, could serve to modify the problem definition in two of the previously cited examples.

3.2.3 Conceptualization

The preceding stages in the design process have delimited the multidimensional space within which engineering solutions are to be sought. It is now necessary to formulate concepts that appear capable of solving the defined problem(s). This conceptualization stage is at the heart of the design process and the one that will determine the quality of the engineer's response to the perceived need. If this need can be thought to answer the question "why" and problem definition the question "what," conceptualization involves the response to the question "how."

The analysis that follows provides a measure of the viability of a given concept and ultimately establishes the values of the relevant engineering parameters, but analysis alone cannot generate innovative solutions. Under the best circumstances, proper interpretation of analytical (and experimental) results can aid the evolutionary growth of existing products and systems and lead, for example, to increased conversion efficiency, reduced weight, or increased speed. However, it is a conceptual breakthrough, a "brainstorm," combining previously separate elements or applying a familiar idea in a new setting, that accounts for most abrupt or revolutionary changes in technology. Thus, it is unlikely that a massive investment in the analysis of windsail interactions would have led to the design of the steamboat; the analyses that could have been performed to determine the optimum shape or grease properties for the runners of prehistoric sleds would not have resulted in the discovery of the wheel; nor would a better understanding of arrow-bow dynamics have led to the invention of the gun. In each of these examples, and countless others, innovation began with the formulation of a new concept, a mental image of an as yet nonexistent object.

Because of the critical importance of this stage in the design process, increasing attention has been directed in the literature to commonly encountered obstacles to successful conceptualization and ways of overcoming these "blocks." These aspects of conceptualization are examined in separate sections of this chapter.

3.2.4 Analysis

The analysis and evaluation of prospective solutions constitute one of the major "bottlenecks" in the design process, and a creative design team can often generate many more "reasonable" schemes than it can possibly examine in detail. As a consequence, accelerated engineering development of products and systems requires that considerable effort be devoted to analysis, and it is this stage that has benefited most from the availability of high-speed computers and sophisticated mathematical techniques. Whereas the Renaissance engineer's analysis consisted largely of reviewing his own experience and examining the results of scale-model tests, today's engineer has at his or her disposal a wide array of analytical, numerical, and experimental tools and can thus evaluate in depth many more concepts than was previously possible.

Nevertheless, the scope, complexity, and budgetary constraints of most modern engineering projects necessitate a preliminary selection of the most viable alternatives prior to detailed analysis. This preliminary selection can be performed on the basis of first-order mathematical or physical models of the proposed concepts, but can, in addition, be aided considerably by use of figures of merit (FOMs) representing the relationship among key properties or characteristics of the candidate concepts. Overall thermal conductances between the two fluids in heat exchangers, the Mouromtseff number in coolant selection, and the so-called stagnation temperature (attained in the absence of coolant flow) in liquid-heating solar collectors, typify figures of merit in common use by thermal analysts and designers.

When, as may sometimes occur, the design goals are specified in terms of figures of merit, the viability of the concepts examined can be determined by comparison of the calculated and initial (design) FOMs. More commonly, however, FOMs and other modes of preliminary analysis serve to identify the most promising candidates, which are then subjected to a more rigorous analysis. The results of these analyses are then compared to the design goals to determine the adequacy of the proposed solutions. At this stage in the design process, it is not uncommon to find that none of the initial concepts meets all the requirements specified and that modification of both the design goals and proposed concepts is desirable.

Several such iterations generally serve to focus the design team's efforts on the crucial aspects of the engineering mission and either to define the preferred solution or to pinpoint the breakthrough required to bridge the gap between the available technology and the desired capability. In the latter event, recommendations for applied research can be channeled to the appropriate company, government, or independent research laboratory, whereas in the former and more likely event, the concept can be subjected to more refined analysis, aimed at optimizing the dimensions and performance of the proposed solution.

In an engineering development project, this analysis stage terminates in the engineering definition of the solution, such as drawings and parts lists. The project is then transferred to production engineering for construction and fabrication.

3.3 OBSTACLES
TO CREATIVE CONCEPTUALIZATION

Creative thought, though commonly associated with genius alone, is one of the basic processes of the human mind and one that engages some fraction of every individual's day. Many biological functions, such as maintaining constant body temperature or proper oxygen concentration in the blood, are generally performed without conscious intervention. Others, including walking erect or head scratching, require a relatively brief learning period, after which occasional mental monitoring suffices to ensure that the function is being properly performed. However, most higher-order functions, from aardvark hunting to zeroing a potentiometer, require extensive training and constant practice to ensure proficiency. Creative thought, whether in synthesizing a combination of musical sounds, unpuzzling a maze, or integrating a complex mathematical expression, falls in this last category and can therefore not only benefit from learning and practice but can also suffer from acquired prejudices and patterned responses [5]. Unfortunately, social pressures toward conformity and limited opportunities for creative thought in daily life result in a steady erosion of most people's creative ability [7], and consequently it is necessary to overcome mental petrification and compartmentalization before the inherent human capacity for innovative thought can reemerge. Following J. L. Adams and his colleagues at the Stanford University Design Division [5, 7], it is convenient to group these obstacles to successful conceptualization into several major categories, including cultural, intellectual, environmental, emotional, and expressive blocks. Several of these are explored briefly in the following subsections.

3.3.1 Cultural Blocks

Cultural imprinting is a basic component of the socialization process, and in Western society often involves the labeling of fantasy and playfulness or humor as childish occupations, unbecoming of professional adults. Since at least one approach to the synthesis of as yet nonexistent objects involves imagining a "better" world and asking "why not?", a bias against fantasizing may be burdensome. Curiously, it may well be the weakness of this block in children, the "uneducated," and science fiction writers that explains the ability of these individuals occasionally to spark an idea in the mind of a professional engineer charged with a problem-solving task.

The essence of humor has been defined by many writers as the juxtaposition of mutually exclusive contexts [5]. The verbal synthesis of a joke or humorous story is thus not far different from the innovative thinking desired in the conceptualization stage of the design process and, indeed, many novel solutions may at first appear absurdly funny. Thus, for example, the possible use of a hand-made ice lens to provide a life-saving fire for an isolated individual in the polar region by focusing the sun's rays on an available fuel might easily be dismissed by a design team seeking only "serious" solutions.

Similarly, many, if not most, scientifically trained professionals have been made to believe that intuition or feelings cannot be trusted and that only rational thought can bring results. Even a casual examination of the design process suffices to reveal the need for total mental involvement in the search for alternative solutions, and the failure to use the integrative pattern-recognition capability of the human brain may unnecessarily prolong or totally destroy a potentially promising line of thought.

The political conservatism of most practicing engineers, and even engineering students, is a well-recognized societal factor (at least) in the United States and contrasts sharply with the mental restlessness and openness often characteristic of highly creative individuals. The projection of such attitudes onto the design process, in the form of a preference for "tried-and-true" approaches or the all-powerful "right way," while necessary for the prevention of the premature introduction of novel technology, is undoubtedly anathematic to innovative conceptualization. A design team in which both an appreciation for the state of the art and a desire to seek new horizons are represented might well obtain the best results. However, since—by virtue of biological and/or psychological factors—the more experienced engineer can be expected to adopt a more conservative posture, it falls to the younger engineer to supply the innovative impulse. The frequently observed successful design team model of several individuals personifying the "corporate memory," or the company's "bag-of-tricks," joined together with one or more young "hot shots," who have not yet stopped asking "why not?", may be seen as supporting the foregoing contention.

Whereas the preceding blocks are among the more visible obstacles to creative thought, those associated with taboos imposed early in the socialization process are far better hidden and affect conceptualization in a more fundamental way. These "Johnny: Don't you dare" prohibitions against socially unacceptable actions involving biological functions, religious objects, game rules, and, in Western society, the integrity of money, can often remove entire families of solution from the ready grasp of the problem solver [5].

The existence of these blocks can be easily demonstrated by exercises that require breaking, even if only in thought, one of the common taboos. Thus, for example, confronted by a situation in which it is necessary to cool a small box of electronics for several minutes while using it to transmit a coded message from a remote desert location, only a few individuals would readily suggest urinating on the device. Similarly, if it were desired to stop the fall of a 3-cm-diameter steel ball from a height of 30 to 50 cm and the collection of available materials included a half-a-dozen $5 bills, most participants in the exercise would only reluctantly suggest cushioning the ball by allowing it to rip through several of the $5 bills. These "anti-taboo" concepts may not, of course, be the best possible solutions to the problems described, but they are sufficiently viable to deserve more initial attention than they would normally get from a design team working at the home office.

Significantly, under the pressure of field conditions, physical danger, or even an impending proposal deadline—when other social barriers begin to dissolve—many such measures will and have been adopted. Consequently, the identification and overcoming of these cultural blocks can be seen as part of an attempt to impart a "real-life" flavor to the design activity in the office setting.

3.3.2 Environmental Blocks

As a result of their interdisciplinary nature and wide scope, most engineering design activities are today performed in groups organized both vertically—that is, novice to expert—and laterally—that is, experts from different fields—within a single industrial concern. Not surprisingly, interactions among individuals in the design team and the physical environment within which the design activity occurs have a profound influence on the modes of thought and creativity of the individual designers. In recognition of these facts, significant attention is and has been devoted to group creativity [8], and some of the environmental obstacles to innovative conceptualization identified by J. L. Adams and his colleagues [5, 7] can be briefly reviewed.

The need to meld the individual contribution of many designers, analysts, draftsmen, and production personnel forces most product development organizations to establish prescribed hours, areas, and procedures for the performance of assigned activities. As a consequence, the eccentric behavior commonly accepted in the creative artist is not generally tolerated in engineers, who are often expected to create on demand while sitting erect and wearing a three-piece suit in a windowless, fluorescent-lighted cubicle. Rather amazingly, it seems, some engineers are able to work creatively in such an environment, but many others become near-sterile conceptualizers or leave engineering to seek more creative outlets. There is probably no single environment that is universally conducive to creativity. Some individuals work best under circumstances with phones ringing, people yelling and coffee spilling on blueprints that would totally frustrate others. It is important, however, to recognize the influence of the physical setting on creativity and the dysfunction induced by an inappropriate setting.

The importance of a successful conclusion to the activities of a mission-oriented design team leads frequently to behavior and communication patterns that inhibit creative thought. The free expression by team members of hunches or "half-baked" ideas is crucial to the success of group conceptualization and yet is frought with emotional and intellectual risk for the individual. The presence of a group leader who is quick to find fault with every idea but his own, team members who insult the intelligence of the individual whose idea disintegrates under scrutiny, or colleagues who quickly claim a promising approach as their own proprietary concept, all act to undermine the constructive interchange required at this stage of the design process. Furthermore, when free discussion and technical exchange are absent, the mantle of a well-conceived, thoroughly explored solution may be mistakenly attached (by those outside the group) to the approach that emerges as a result of the "bullying" of team members by the group leader. In such situations the success of the design activity is almost entirely dependent on the ability of a single person, and the time as well as effort invested by other team members bears no fruit.

Somewhat similar results are obtained when all team members share a common perspective and conceptual framework. Discussions in such groups tend to avoid taboos and sacred cows and are generally devoted to confirming, by example and implication, the prevailing view. Although the design process can be quite brief and is especially effective in evolutionary product improvements, the near absence of criticism in such an "old-boy" environment can occasionally lead to gross errors [9]. Political and military history abounds with examples of such "group think." The design and manufacture of the Edsel, on the one hand, and solar cookers for villagers who traditionally cook their food after sunset, on the other, may suffice to establish the occurrence of this phenomenon in engineering practice.

Though the availability of continuing financial support does not directly influence concept formation, the existence of obstacles to the development of apparently viable concepts—whether imposed by the group leader, the company, or the society—works not only to deprive the marketplace of new technology, but may also inhibit the future conceptualizing ability of the design engineer. The manipulation of financial incentives is surely a management prerogative and can be legitimately practiced on all levels. However, differences in philosophy among various industrial concerns and even among nations, as well as the consequent levels of support offered to inventive individuals and/or novel ideas, can thus largely dictate the pace and quality of technological innovation. Although some measure of such control is essential, if a trauma-free transition from one technology to the next is to be assured, the intervention of vested interests seeking to maintain control of profitable economic sectors (or technical disciplines in the company setting) may well deny the society (or the company) the potential benefits of new approaches and products [5, 10].

3.3.3 Additional Blocks

In addition to the cultural and environmental blocks discussed earlier, the engineering design educators at Stanford University have identified several other categories of obstacles to creative conceptualization [5, 7]. Among these are perceptual blocks—including difficulty in isolating the problem and delimiting the task too narrowly; emotional blocks—including fear of making mistakes, lack of interest or excessive zeal; and intellectual blocks—including failure to search for or incorrect use of information. These and others are discussed in great detail by J. L. Adams in *Conceptual Blockbusting* [5].

3.4 OVERCOMING OBSTACLES
TO CONCEPTUALIZATION

As in many other realms of human activity, the recognition that obstacles to creative conceptualization exist can, in itself, contribute to overcoming these blocks. In addition, a deeper understanding of the mental barriers erected by the individual, in response to "normal" social interactions in the crowded, competitive conditions typical of Western society, can serve to identify specific, creativity-enhancing strategies. Most of these involve techniques that prevent the individual from settling too soon on the "obvious" (but perhaps inadequate) solution and aiding him or her in viewing the problem from many possible perspectives. Some of these strategies can be applied in a conscious and straightforward manner; others require a change in the representational system; and still others rely on soliciting the cooperation of the subconscious [3, 5, 7]. Several such "blockbusting" strategies are reviewed in the subsections that follow.

3.4.1 Conscious Techniques

Regardless of the particular techniques used to stimulate creativity, there is no ready substitute for the predispositional influence of curiosity and discontent. Since, at least among the vested groups in Western society, maturity is often associated with an acceptance of the status quo and a reluctance to display ignorance, conscious effort may be required in adults to sustain the questioning attitude readily apparent in children and adolescents.

List making, though a near-cliche in creativity stimulation, remains a powerful and yet simple technique. In the absence of a more rational commitment to creative thought, the conceptualizing activity of many professional engineers can be directed into innovative channels by establishing the number of concepts listed as a quantitative measure of creativity and thus harnessing the competitive instinct to the generation of many diverse solutions. The suspension of self-criticism, which often results from the desire to obliterate the empty spaces on the sheet of paper, the visual stimulus offered by the listed ideas, and the possibilities for mental rearrangement of the generated concepts perhaps explain the surprising degree of success in identifying both novel and useful concepts often attained by use of this method.

As a result of its success, simple list making has spawned several generations of more sophisticated listing techniques, which attempt to converge more rapidly on the particular characteristic of the problem and to facilitate the "discovery" of unexpected combinations. These techniques, often referred to as *morphological analysis*, are highly successful in translating known technology from one field to another field and of uncovering long neglected state-of-the-art concepts. This success is attained by a rigorous assessment of the characteristics desired in the solution, an exhaustive review of the known ways of achieving or providing these characteristics, and the use of one of various algorithms for synthesizing systems as combinations of previously identified technologies.

Returning to an earlier example, it is possible to gain some appreciation of the nature and efficacy of morphological analysis by tackling the problem of reducing the vulnerability of a specified aircraft to heat-seeking missiles. In an appropriate setting, exploration of this subject would commence with specified design goals, stating the desired reduction in lost aircraft per flight hour or strike, the duration of or number of sortees during which this protection must be available, and some limitation on the added weight, volume, or cost. Assuming that these constraints have indeed been stated and, furthermore, than an infrared missile detection system is already deployed on the aircraft, consideration of jet aircraft and infrared missile characteristics suggests that reduced aircraft vulnerability can be based on a reduction in the detectability of the aircraft, an improvement in the aircraft's ability to evade the missile, and its ability to survive a "hit" by the missile. Clearly, any one of these could provide the desired protection, but in recognition of the limitations inherent in the various technologies, it is more likely that an appropriate combination of these functions will be required.

Following the "recipe" of morphological analysis, it is now necessary to identify as many "reasonable" ways as possible of providing each of the desired functions. The design team can begin this intensive list-making activity either jointly or on an individual basis, but will generally convene as a group to review a composite listing, bearing some resemblance to that shown in Fig. 3.2. The reduction in detectability can

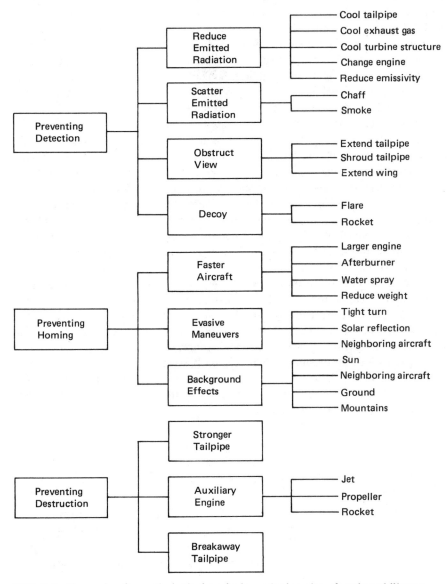

FIG. 3.2 Example of morphological analysis—reducing aircraft vulnerability to heat-seeking missiles.

perhaps be achieved by reducing emitted radiation, scattering the radiation on a large angle area, or offering a second, more attractive target (using decoys). Among other possibilities, an increase in aircraft velocity (under combat conditions), clever utilization of background radiation, (the sun), and evasive maneuvers can all contribute to preventing a missile homing on the aircraft. Finally, structurally strengthening the engine tail-pipe, constructing a breakaway tailpipe or adding an auxiliary engine can make it possible for the aircraft to survive a missile strike.

Interestingly enough, morphological analysis can now be applied again to some of the proposed approaches to try and identify various ways of, for example, reducing emitted radiation or increasing the aircraft's velocity. Figure 3.3, representing the results of this secondary analysis in a subgroup of the design team, reveals that the engine tailpipe can be cooled by the flow of fuel or water, the temperature of the exhaust gases can be reduced by mixing with ambient air or water spray, and the tailpipe can be shrouded to reduce its visibility. Similarly, use of a larger engine, reducing aircraft weight, use of water spray, firing of an afterburner or an auxiliary rocket engine could each provide the sought-after improvement in aircraft velocity.

Although some of the concepts presented in Fig. 3.2 may later prove to be unworkable or of limited use, the generation of as thorough a listing as possible helps to ensure not only that all the conventional solutions are represented but that unusual and previously neglected solutions can also emerge. Thus, whereas the combined use of decoys and evasive maneuvers to prevent detection and homing can be considered standard technology, and the use of chaff and intelligent engine placement coupled with larger aircraft thrust-to-weight ratios may be viewed as the present state of the art, an extended "breakaway" tailpipe, water spray for cooling and extra thrust, as well as return to propeller or turbopropeller engines, are perhaps nonobvious possibilities worthy of further study.

To conclude this by no means exhaustive review of conscious "blockbusting" techniques, it is of interest to mention several other approaches that force the designer to answer thought-provoking questions appearing on a checklist. Some, for example, those due to Alex Osborn [11], focus on possible perturbation of the recommended solution, such as magnify (stronger, longer), multiply (two, many), and reverse (right to left, inside out). Others rely on "manipulative" words such as integrate, eliminate, repel, distort, and soften to suggest new conceptualizing directions, and still others are content to remind the designer of the various stages in the design process via a series of provocative statements [5].

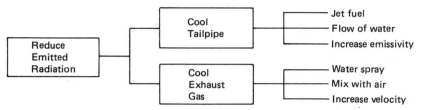

FIG. 3.3 Example of secondary morphological analysis—reducing aircraft vulnerability to heat-seeking missiles.

3.4.2 Changed Representational System

In contrast to the rather mechanical and "plodding" quality of the various list-making techniques described in the previous subsection, approaches that rely on changes in the "language" or representational system are often both elegant and enjoyable. The importance of selecting the appropriate coordinate system and/or redefined variable as well as the utility of well-known composite functions, such as error and Bessel functions, can serve to illustrate this approach in mathematical analysis. The frequent reference in technical discussions to graphical representations of important phenomena, including the pump curve, the boiling curve, and the hysteresis loop, as well as reliance, when convenient, on graphical solutions and geometric approximations, appears to confirm the common acceptance of hybrid representational systems in engineering analysis. Not surprisingly, this technique has also found application in engineering design and can be especially effective in stimulating innovative conceptualization.

The historical role of visual thinking and communication in the growth of Western technology places mathematical analysis in the context of the (historical) alternative language. There is no doubt that the availability of sophisticated analytical and numerical techniques has had a wide impact on engineering design, so much so, in fact, that today it is visual thinking that must be viewed as the alternative to the common "language" of analysis.

The anthropomorphic inversion, associated with ascribing human (your own) qualities to the machine to be designed or function to be performed, is a powerful though less common example of this "blockbusting" approach. Many interesting concepts for a mountain climbing machine, for example, can be generated by considering the methods of human climbers and, similarly, efficient control algorithms can often be established by performing a thought experiment in which the human brain is allowed to control a complex process.

3.4.3 Subconscious Techniques

No discussion of creativity would be complete without mention of "brainstorming" and other subconscious association techniques in which design team members are encouraged to "free associate" concepts, ideas, and approaches. Though brainstorming can be done individually, it is far more effective when done in groups, and together with list making accounts for much of the creativity stimulation effort in industrial settings. Brainstorming is especially effective in overcoming obstacles to creativity caused by fear of criticism or fear of appearing foolish or "dumb" [7]. In addition to "natural" solutions, participants in brainstorming sessions are expected to generate ridiculous, fantastic, humorous, absurd, and apparently bad ideas, which can then serve as a stimulus for the synthesis of more viable, better matched concepts from other participants. To achieve its aim, brainstorming must be done in an environment that is free of judgment (for the duration of the session) and in which each idea has equal weight and import. Sample brainstorming sessions are presented in several texts, including [11–13].

3.5 REFERENCES

1   Ferguson, E. S., The Mind's Eye: Nonverbal Thought in Technology, *Science*, vol. 197, no. 4306, pp. 827–836, 1977.

2   Sherwin, C. W., and Isenson, R. S., Project Hindsight, *Science*, vol. 156, pp. 1571–1577, 1967.
3   Hyman, R., and Anderson, B., Solving Problems, in *The R&D Game: Technical Men, Technical Managers and Research Productivity*, ed. David Allison, M.I.T. Press, Cambridge, Mass., pp. 90–105, 1969.
4   Kueller, P., *The Art and Science of Creativity*, Holt, New York, 1965.
5   Adams, J. L., *Conceptual Blockbusting*, Freeman, San Francisco, 1977.
6   Dixon, J. R., *Design Engineering*, McGraw-Hill, New York, 1966.
7   Roth, B., Design Process and Creativity, class notes, Department of Mechanical Engineering, Ben-Gurion University of the Negev, Beer Sheva, Israel, 1973.
8   Adams, J. L., Individual and Small Group Creativity, *Eng. Educ.*, vol. 62, pp. 100–105, Nov. 1972.
9   Janis, I. L., *Victims of Group Think*, Houghton Mifflin, Boston, 1968.
10 Beakley, G. C., and Leach, H. W., *Engineering: An Introduction to a Creative Profession*, Macmillan, London, 1967.
11 Osborn, A., *Applied Imagination*, Scribners, New York, 1953.
12 Alger, J. R. M., and Hays, C. V., *Creative Synthesis in Design*, Prentice-Hall, Englewood Cliffs, N.J., 1964.
13 Gordon, W. J. J., *Synectics*, Harper & Row, New York, 1961.

II

■fundamentals of heat transfer and fluid mechanics

4

■ conduction—steady state

4.1 INTRODUCTION

The mechanism of heat transfer by conduction has been postulated as occurring in two distinct ways: by kinetic energy interchange between the molecules of a substance and by electron drift. Conduction takes place when there is a difference of temperature between points in the same body or when two bodies at different temperatures are in contact. The flow of heat by conduction is commonly thought of as an interchange of kinetic energy between rapidly vibrating molecules of a substance and less rapidly vibrating neighboring molecules. It is common knowledge that good conductors of electricity are good conductors of heat as well. Because electricity is actually electron flow, some schools of thought point out that the flow of heat can also be attributed to electron flow. According to this theory, heat conduction is quite naturally a function of the mobility of the free or valence electrons in the molecules of the substance in question.

Note that in the two mechanisms described to explain conduction heat transfer, energy transfer occurs by way of microscopic phenomena. This contrasts with convective heat transfer (see Chap. 6), in which macroscopic effects play an important role.

In the study of heat conduction a distinction is made among uniform, nonuniform, steady, and unsteady or transient heat flow. Usually the words "uniform" and "nonuniform" refer to spatial variation, whereas "steady" and "transient" refer to the time domain. Steady-state heat flow exists when the temperature at each point in the heat flow path does not vary with time. Conversely, transient heat flow occurs when the temperature at any point in the heat flow does vary with time. This chapter considers conduction in the steady state. Transient conduction is discussed in Chap. 5.

4.2 GENERAL EQUATION FOR HEAT CONDUCTION

It was J. B. J. Fourier who proposed [1] that the rate of heat flow by conduction through a material is proportional to the area of the material normal to the heat flow path and to the temperature gradient along the heat flow path. These facts may be written mathematically as

$$q \sim -A \frac{dT}{dx}$$

where the minus sign is used to allow for a positive heat flow in the presence of a negative temperature gradient.[1] Insertion of a proportionality constant gives what is sometimes called the *Fourier law*,

[1] The second law of thermodynamics says that heat will not flow uphill, that is, from a region of lower temperature to a region of higher temperature, unless external work is done.

$$q = -kA \frac{dT}{dx} \tag{4.1}$$

which serves to define the proportionality constant k as the thermal conductivity, a property of the material:

$$k \equiv \frac{q}{A(dT/dx)} \quad \text{(W/m °C, Btu/ft h °F)}$$

With Eq. (4.1) in hand, it is now possible to derive the general equation of heat conduction by considering the differential volume shown in Fig. 4.1. An energy balance is written[2] equating the rate of heat flow entering the subvolume plus the rate of heat generation within the subvolume to the rate of heat flow leaving the sub-volume plus the rate of change of internal energy (actually the heat stored) in the subvolume. Hence, one may write

$$q_{in} + q_G = q_{out} + q_s$$

or $$q_{in} - q_{out} + q_G = q_s \tag{4.2}$$

[2] The first law of thermodynamics says that energy cannot be created or destroyed but can be transformed from one form to another.

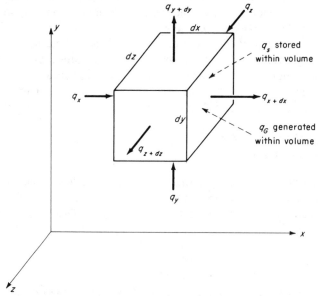

FIG. 4.1 Differential volume for the derivation of the general equation of heat conduction.

Consider the x direction first. Heat will enter the left face in accordance with Eq. (4.1),

$$q_x = -kA \frac{\partial T}{\partial x} = \left(-k \frac{\partial T}{\partial x} \right) dy\, dz$$

where the temperature gradient must be expressed as a partial derivative because the temperature T will also be a function of y, z, and the time t. Heat will leave at the right face at $x + dx$,

$$q_{x+dx} = \left[\left(-k \frac{\partial T}{\partial x} \right) + \frac{\partial}{\partial x} \left(-k \frac{\partial T}{\partial x} \right) dx \right] dy\, dz$$

so that the difference between the heat entering at x and that leaving at $x + dx$ is

$$q_x - q_{x+dx} = \frac{\partial}{\partial x} \left(k \frac{\partial T}{\partial x} \right) dx\, dy\, dz$$

The same procedure may be used for the y and z directions, with the result that

$$q_y - q_{y+dy} = \frac{\partial}{\partial y} \left(k \frac{\partial T}{\partial y} \right) dx\, dy\, dz$$

and $\quad q_z - q_{z+dz} = \dfrac{\partial}{\partial z} \left(k \dfrac{\partial T}{\partial z} \right) dx\, dy\, dz$

Thus, $q_{\text{in}} - q_{\text{out}}$ in Eq. (4.2) is now accounted for and is

$$q_{\text{in}} - q_{\text{out}} = \left[\frac{\partial}{\partial x} \left(k \frac{\partial T}{\partial x} \right) + \frac{\partial}{\partial y} \left(k \frac{\partial T}{\partial y} \right) + \frac{\partial}{\partial z} \left(k \frac{\partial T}{\partial z} \right) \right] dx\, dy\, dz \qquad (4.3)$$

The heat generated within the subvolume can come from an electric or electronic heat dissipation such as an $I^2 R$ heat input or from a chemical reaction. With q_i as the volumetric rate of heat generation, q_G in Eq. (4.2) is

$$q_G = q_i(dx\, dy\, dz) \qquad (4.4)$$

Finally, the heat stored within the subvolume is related to the rate of change in internal energy,

$$\frac{du}{dt} = c_v\, dm\, \frac{dT}{dt}$$

where u is internal energy in joules (Btu) and c_v is the specific heat at constant volume, J/kg$^\circ$C(Btu/lb$^\circ$F). With c_v replaced by[3] c and with $dm = \rho\, dx\, dy\, dz$, where ρ is the density of the material,

[3] There is no need to distinguish between specific heats at constant volume and constant pressure when dealing with a solid.

$$q_s = \frac{du}{dt} = \rho c \frac{\partial T}{\partial t} \, dx \, dy \, dz \qquad (4.5)$$

where a partial derivative is now employed.

When Eqs. (4.3), (4.4), and (4.5) are inserted into Eq. (4.2), the result is

$$\frac{\partial}{\partial x} k \left(\frac{\partial T}{\partial x} \right) + \frac{\partial}{\partial y} \left(k \frac{\partial T}{\partial y} \right) + \frac{\partial}{\partial z} \left(k \frac{\partial T}{\partial z} \right) + q_i = \rho c \frac{\partial T}{\partial t}$$

where the common $dx \, dy \, dz$ terms have been canceled. Then if k, c, and ρ are assumed to be independent of temperature, direction, and time, one obtains the general equation of heat conduction,[4]

$$\frac{\partial^2 T}{\partial x^2} + \frac{\partial^2 T}{\partial y^2} + \frac{\partial^2 T}{\partial z^2} + \frac{q_i}{k} = \frac{1}{\alpha} \frac{\partial T}{\partial t} \qquad (4.6)$$

where α is known as the thermal diffusivity of the material:

$$\alpha \equiv \frac{k}{c\rho} \qquad (4.7)$$

If the system contains no heat sources, Eq. (4.7) becomes the *Fourier equation*,[5]

$$\frac{\partial^2 T}{\partial x^2} + \frac{\partial^2 T}{\partial y^2} + \frac{\partial^2 T}{\partial z^2} = \frac{1}{\alpha} \frac{\partial T}{\partial t} \qquad (4.8)$$

If steady state is considered, that is, if the temperature distribution does not vary with time, the *Poisson equation* results:

$$\frac{\partial^2 T}{\partial x^2} + \frac{\partial^2 T}{\partial y^2} + \frac{\partial^2 T}{\partial z^2} + \frac{q_i}{k} = 0 \qquad (4.9)$$

Finally, in the absence of heat sources and in the steady state, Eq. (4.6) reduces to

$$\frac{\partial^2 T}{\partial x^2} + \frac{\partial^2 T}{\partial y^2} + \frac{\partial^2 T}{\partial z^2} = 0 \qquad (4.10)$$

which is called the *Laplace equation* and is frequently written as

$$\nabla^2 T = 0$$

[4] The general equation of heat conduction in cylindrical and spherical coordinates is listed in Sec. 4.13.

[5] Not the Fourier law.

4.3 SIMPLE SHAPES
WITHOUT HEAT GENERATION

For the plane slab with faces maintained at T_1 and T_2 as shown in Fig. 4.2, Eq. (4.10) using only the x coordinate is employed:

$$\frac{d^2T}{dx^2} = 0$$

A pair of integrations yields

$$\frac{dT}{dx} = C_1$$

and $T = C_1 x + C_2$

Then the use of the boundary conditions,

$$T(x = x_1) = T_1 \qquad T(x = x_2) = T_2$$

permits the evaluation of the constants of integration so that

$$T = T_1 - \frac{x}{L}(T_1 - T_2) \tag{4.11}$$

where $L = x_2 - x_1$.

Equation (4.11) gives the temperature distribution across the slab. The heat flow can be determined from Eq. (4.1):

$$q = -kA \frac{dT}{dx} \tag{4.1}$$

or $q = -kA \left[\frac{-(T_1 - T_2)}{L} \right] = \frac{kA}{L}(T_1 - T_2) \tag{4.12}$

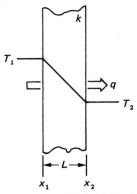

FIG. 4.2 Heat transfer by conduction through a plane slab.

One may not follow the same procedure for the hollow cylinder shown in Fig. 4.3 because Eq. (4.10) is *not* applicable, without modification, to a cylindrical coordinate system.[6] Instead, one may begin with the Fourier law,

$$q = -kA \frac{dT}{dr} \tag{4.1}$$

and note that A is a function of r, $A = 2\pi r L$, so that

$$q = -2\pi k L r \frac{dT}{dr} \tag{4.13}$$

This is a very simple, first-order, linear differential equation, which is solved by separation of variables.

$$\frac{-q}{2\pi k L} \frac{dr}{r} = dT$$

whereupon integration from r_1 to r_2, where the temperatures are T_1 and T_2,

$$\frac{-q}{2\pi k L} \int_{r_1}^{r_2} \frac{dr}{r} = \int_{T_1}^{T_2} dT$$

yields[7]

$$\frac{-q}{2\pi k L} \ln r \Big|_{r_1}^{r_2} = T \Big|_{T_1}^{T_2}$$

or $\quad q = \dfrac{2\pi k L}{\ln (r_2/r_1)} (T_1 - T_2) \tag{4.14}$

[6] See Sec. 4.13.
[7] In this book ln will be used to designate the natural logarithm.

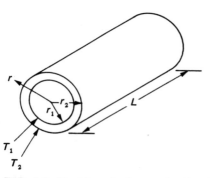

FIG. 4.3 Heat transfer by conduction through a hollow cylinder.

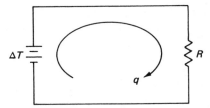

FIG. 4.4 Electrothermal analog circuit. The thermal resistance R may take various forms.

4.4 THE ELECTROTHERMAL ANALOG

The form of Eqs. (4.12) and (4.14) indicates a direct analogy between heat flow and electrical current governed by *Ohm's law*,

$$I = \frac{V}{R} \tag{4.15}$$

In this analogy, the analogous quantities are

Current $I \longleftrightarrow$ heat flow q

Potential $V \longleftrightarrow$ temperature difference ΔT

Resistance $R \longleftrightarrow$ thermal resistance R

It is easily seen that for the plain slab, the thermal resistance is

$$R = \frac{\Delta T}{q} = \frac{L}{kA} \tag{4.16a}$$

and for the hollow cylinder,

$$R = \frac{\Delta T}{q} = \frac{\ln (r_2/r_1)}{2\pi k L} \tag{4.16b}$$

Electrothermal analog circuits for the plain slab and hollow cylinder are displayed in Fig. 4.4. Configurations vastly more complex are possible, and these may be treated using all of the techniques of steady-state electrical network theory [2].

4.5 SIMPLE SHAPES
WITH HEAT GENERATION

Suppose that the plain slab shown in Fig. 4.2 is subjected to a uniformly distributed heat generation rate of q_i throughout its interior. In this case, the temperature profile will *not* be as shown in Fig. 4.2. However, the temperature profile is easily obtained from a form of Eq. (4.6) that considers only the x coordinate:

$$\frac{d^2 T}{dx^2} + \frac{q_i}{k} = 0$$

Double integration quickly yields the general solution

$$\frac{dT}{dx} = -\frac{q_i}{k}x + C_1$$

and $$T = -\frac{q_i}{2k}x^2 + C_1 x + C_2$$

Then application of the two boundary conditions,

$$T(x = x_1) = T_1 \quad \text{and} \quad T(x = x_2) = T_2$$

shows that

$$C_1 = \frac{T_2 - T_1}{x_2 - x_1} + \frac{q_i}{2k}(x_2 + x_1)$$

$$C_2 = \frac{x_2 T_1 - x_1 T_2}{x_2 - x_1} - \frac{q_i}{2k}x_1 x_2$$

Substitution of these into the general solution gives

$$T = \left[T_1 + \frac{T_2 - T_1}{x_2 - x_1}(x - x_1) \right] + \left[\frac{q_i(x_2 - x_1)^2}{2k} \right]\left[\frac{x - x_1}{x_2 - x_1} - \left(\frac{x - x_1}{x_2 - x_1} \right)^2 \right]$$

(4.17)

which indicates a parabolic temperature distribution.

   Indeed, if the temperatures at the faces are equal, $T_1 = T_2 = T_0$, and if $x_2 - x_1 = L$, then with $x_1 = 0$,

$$T - T_0 = \frac{q_i L^2}{2k}\left[\left(\frac{x}{L} \right) - \left(\frac{x}{L} \right)^2 \right]$$

(4.18)

which is the parabola with vertex at the center of the slab.

   In order to study the same phenomenon in a hollow cylinder, one begins with a form of Eq. (4.6) in cylindrical coordinates or formulates the energy balance directly. The configuration is shown in Fig. 4.3; of interest is a differential element of width dr.

   The heat that enters by conduction in the r direction is

$$q_r = -kA_r \frac{dT}{dr}\bigg|_r = -2\pi kLr \frac{dT}{dr}\bigg|_r$$

and the heat leaving by conduction at $r + dr$ is

$$q_{r+dr} = -kA_{r+dr}\frac{dT}{dr}\bigg|_{r+dr} = -2\pi kL(r + dr)\frac{dT}{dr}\bigg|_{r+dr}$$

The heat generated in dr is

$$q_G = 2\pi q_i L r\, dr$$

and an energy balance gives, after simplification,

$$k\left[(r+dr)\frac{dT}{dr}\bigg|_{r+dr} - r\frac{dT}{dr}\bigg|_r\right] + q_i r = 0$$

Then an application of the mean value theorem provides the differential equation

$$r\frac{d^2T}{dr^2} + \frac{dT}{dr} = -\frac{q_i r}{k}$$

which is solved by noting that

$$\frac{d}{dr}\left(r\frac{dT}{dr}\right) = r\frac{d^2T}{dr^2} + \frac{dT}{dr}$$

Thus integration gives

$$r\frac{dT}{dr} = -\frac{q_i r^2}{2k} + C_1$$

or $$\frac{dT}{dr} = -\frac{q_i r}{2k} + \frac{C_1}{r}$$

in which C_1 must be equal to zero in order to give a finite dT/dr at $r = 0$. Thus

$$\frac{dT}{dr} = -\frac{q_i r}{2k}$$

and another integration provides the general solution

$$T = -\frac{q_i r^2}{4k} + C_2$$

The constant of integration C_2 is evaluated from the condition

$$T(r = r_1) = T_1$$

so that

$$C_2 = T_1 + \frac{q_i r_1^2}{4k}$$

and hence

$$T - T_1 = \frac{q_i r_1^2}{4k}\left[1 - \left(\frac{r}{r_1}\right)^2\right]$$

(4.19)

4.6 COMPOSITE SHAPES WITHOUT HEAT GENERATION

Figure 4.5a shows a plane wall composed of three different slabs each with a different thermal conductivity. The temperatures at the outermost faces are designated as T_1 and T_2 and the interface temperatures are designated as T_a and T_b.

In the steady state, the heat flowing from left to right must pass through each slab in turn. Hence, Eq. (4.12) shows that

$$q = \frac{k_1 A_1}{L_1}(T_1 - T_a) = \frac{k_2 A_2}{L_2}(T_a - T_b) = \frac{k_3 A_3}{L_3}(T_b - T_2)$$

The individual temperature drops are easily determined.

$$T_1 - T_a = \frac{qL_1}{k_1 A_1}$$

$$T_a - T_b = \frac{qL_2}{k_2 A_2}$$

and $$T_b - T_2 = \frac{qL_3}{k_3 A_3}$$

whereupon a simple addition shows that

$$T_1 - T_2 = \frac{qL_1}{k_1 A_1} + \frac{qL_2}{k_2 A_2} + \frac{qL_3}{k_3 A_3}$$

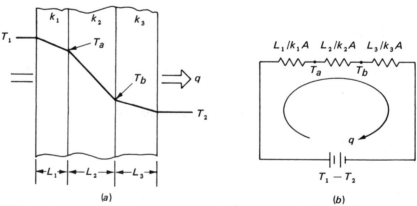

(a)   (b)

FIG. 4.5 Composite plane wall showing (a) temperature distribution and (b) electrothermal analog.

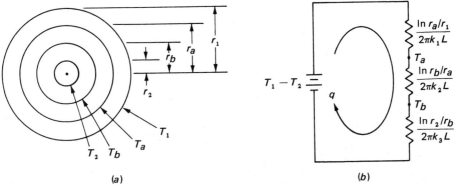

FIG. 4.6 Composite cylinder of three materials showing (*a*) configuration and (*b*) electrothermal analog.

or

$$q = \frac{T_1 - T_2}{R_T} \qquad (4.20)$$

where in this case,

$$R_T = \frac{L_1}{k_1 A_1} + \frac{L_2}{k_2 A_2} + \frac{L_3}{k_3 A_3} \qquad (4.21)$$

and where in general for n resistances in *series*,

$$R_T = \sum_{i=1}^{n} \frac{L_i}{k_i A_i} \qquad (4.22)$$

The electrothermal analog of the three plane slab composite is shown in Fig. 4.5*b*.

For the composite hollow cylinder, a cylindrical coordinate example, the configuration is shown in Fig. 4.6*a*. Note that three distinct materials are depicted and that this may be easily generalized to consider n materials. The heat must flow through each cylinder in turn. Hence, the individual temperature drops can be determined from Eq. (4.14)

$$T_1 - T_a = \frac{q \ln (r_a/r_1)}{2\pi k_1 L}$$

$$T_a - T_b = \frac{q \ln (r_b/r_a)}{2\pi k_2 L}$$

and $\quad T_b - T_2 = \frac{q \ln (r_2/r_b)}{2\pi k_3 L}$

Addition yields

$$T_1 - T_2 = \frac{q \ln (r_a/r_1)}{2\pi k_1 L} + \frac{q \ln (r_b/r_a)}{2\pi k_2 L} + \frac{q \ln (r_2/r_b)}{2\pi k_3 L}$$

or $$T_1 - T_2 = q \left[\frac{\ln (r_a/r_1)}{2\pi L k_1} + \frac{\ln (r_b/r_a)}{2\pi L k_2} + \frac{\ln (r_2/r_b)}{2\pi L k_3} \right]$$

which in terms of the analog $q = \Delta T/R_T$ shows that

$$R_T = \frac{\ln (r_a/r_1)}{2\pi L k_1} + \frac{\ln (r_b/r_a)}{2\pi L k_2} + \frac{\ln (r_2/r_b)}{2\pi L k_3} \tag{4.23}$$

This result is easily generalized to n cylindrical shell resistances; the electro-thermal analog is shown in Fig. 4.6*b*.

4.7 THE OVERALL COEFFICIENT OF HEAT TRANSFER

Conduction is often accompanied by other modes of heat transfer. A most important case is that of conduction and convection occurring simultaneously. Each mode has an effect on the overall transfer of heat, and it is important to develop a single expression that incorporates the effect of both modes.

Convection is treated in detail in Chap. 6. All that is needed here, however, is the statement of fact that heat flow by convection is directly proportional to the surface area normal to the heat flow path and to the temperature difference between the bulk of the convective fluid and the confining surface:

$$q \sim S \, \Delta T$$

The proportionality constant is h, the coefficient of heat transfer,[8]

$$q = hS \, \Delta T$$

which serves to define h:

$$h \equiv \frac{q}{S \, \Delta T} \quad \text{W/m}^2 \, {}^\circ\text{C (Btu/ft}^2 \, \text{h} \, {}^\circ\text{F)}$$

Figure 4.7*a* shows a composite plane wall of two slabs with T_m designated as the interface temperature. The composite is presumed to be exposed to warm fluid at T_h on the left and a cold fluid at T_c on the right and the heat transfer coefficients are designated as h_h and h_c. The heat must flow from hot fluid to cold fluid through the slab in *series*:

$$q = h_h S(T_h - T_1) = \frac{k_1 A}{L_1} (T_1 - T_m) = \frac{k_2 A}{L_2} (T_m - T_2) = h_c S(T_2 - T_c)$$

---

[8] $q = hS \, \Delta T$ is known as *Newton's equation.*

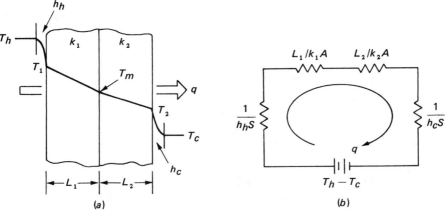

FIG. 4.7   Composite plane wall exposed to hot and cold fluids showing (*a*) temperature distribution and (*b*) electrothermal analog. Note that $S = A$.

If this is put into terms of temperature difference,

$$T_h - T_1 = \frac{q}{h_h S}$$

$$T_1 - T_m = \frac{qL_1}{k_1 A}$$

$$T_m - T_2 = \frac{qL_2}{k_2 A}$$

and $$T_2 - T_c = \frac{q}{h_c S}$$

Simple addition gives, with $A = S$ for the plane wall configuration,

$$T_h - T_c = \frac{q}{A}\left(\frac{1}{h_h} + \frac{L_1}{k_1} + \frac{L_2}{k_2} + \frac{1}{h_c}\right) \tag{4.24}$$

Now propose that the effect of the four entities can be combined into what is called an *overall coefficient of heat transfer* such that

$$q = UA\,\Delta T \tag{4.25}$$

Comparison of Eqs. (4.24) and (4.25) shows that for the plane wall,

$$U = \frac{1}{(1/h_h) + (L_1/k_1) + (L_2/k_2) + (1/h_c)} \tag{4.26}$$

and that the units of U are W/m^2 °C (Btu/ft^2 h °F).

Note that U is *not* a conductance because it does *not* contain the area. Observe that the electrothermal analog shown in Fig. 4.7b has been constructed from Eq. (4.24) written as

$$T_h - T_c = qR_T = q\left(\frac{1}{h_h A} + \frac{L_1}{k_1 A} + \frac{L_2}{k_2 A} + \frac{1}{h_c A}\right)$$

which serves to define R_T as

$$R_T = \sum R = \frac{T_h - T_c}{q} = \frac{1}{h_h A} + \frac{L_1}{k_1 A} + \frac{L_2}{k_2 A} + \frac{1}{h_c A}$$

The reciprocal of the resistance is the conductance,

$$K = \frac{1}{R_T} \quad \text{W/}^\circ\text{C (Btu/h}^\circ\text{F)}$$

which, in this case, is

$$K = \frac{1}{(1/h_h A) + (L_1/k_1 A) + (L_2/k_2 A) + (1/h_c A)}$$

and when this is compared to Eq. (4.26), one sees that K is *not* U. It must be emphasized that U is an overall coefficient and *not* an overall conductance.

The composite hollow cylinder exposed to hot and cold convective fluids at T_h and T_c (through h_h and h_c) is shown in Fig. 4.8a. Here the heat flow is equal in the four entities:

$$q = h_h S_h(T_h - T_1) = \frac{2\pi k_1 L(T_1 - T_m)}{\ln(r_m/r_1)} = \frac{2\pi k_2 L(T_m - T_2)}{\ln(r_2/r_m)} = h_c S_c(T_2 - T_c)$$

In terms of the temperature differences,

$$T_h - T_1 = \frac{q}{h_h S_h} = \frac{q}{2\pi k_h r_1 L}$$

$$T_1 - T_m = \frac{q \ln(r_m/r_1)}{2\pi k_1 L}$$

$$T_m - T_2 = \frac{q \ln(r_2/r_m)}{2\pi k_2 L}$$

and $$T_2 - T_c = \frac{q}{h_c S_c} = \frac{q}{2\pi h_c r_2 L}$$

Now simple addition gives

$$T_h - T_c = q\left[\frac{1}{2\pi h_h r_1 L} + \frac{\ln(r_m/r_1)}{2\pi k_1 L} + \frac{\ln(r_2/r_m)}{2\pi k_2 L} + \frac{1}{2\pi h_c r_2 L}\right]$$

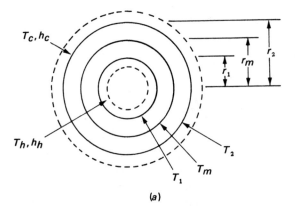

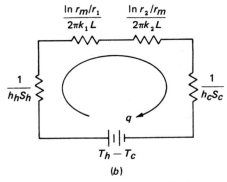

FIG. 4.8 Composite hollow cylinder exposed to hot and cold fluids showing (*a*) configuration and (*b*) electrothermal analog.

which defines R_T, *not* U, as shown in Fig. 4.8*b* as

$$R_T = \frac{1}{2\pi h_h r_1 L} + \frac{\ln (r_m/r_1)}{2\pi k_1 L} + \frac{\ln (r_2/r_m)}{2\pi k_2 L} + \frac{1}{2\pi h_c r_2 L} \tag{4.27}$$

The definition of U in accordance with Eq. (4.25) requires a reference area. For example, if Eq. (4.25) is written with the hot surface as the reference area,

$$q = U_h S_h \Delta T$$

then Eq. (4.27) must be modified by multiplying throughout by $S_h = 2\pi r_1 L$. Thus

$$U_h = \frac{1}{R_T S_h} = \frac{1}{(1/h_h) + [r_1 \ln (r_m/r_1)/k_1] + [r_1 \ln (r_2/r_m)/k_2] + (r_1/h_c r_2)}$$

If Eq. (4.25) is written with the cold surface as the reference area,

$$q = U_c S_c \Delta T$$

then Eq. (4.27) is modified by multiplying throughout by $S_c = 2\pi r_2 L$.

$$U_c = \frac{1}{R_T S_c} = \frac{1}{(r_2/h_h r_1) + [r_2 \ln (r_m/r_1)/k_1] + [r_2 \ln (r_2/r_m)/k_2] + (1/h_c)}$$

4.8 SIMPLE SHAPES–VARIABLE THERMAL CONDUCTIVITY

Whereas thermal conductivity varies with temperature, the effect of this variation is minimized when the temperature extremes under consideration are not large or when the temperature dependence of the thermal conductivity is minimal.

For many materials of interest, the variation of thermal conductivity is linear and is represented by

$$k(T) = k_0(1 + bT) \tag{4.28}$$

where b is the temperature coefficient of thermal conductivity. With this relationship it is possible to investigate the effect of variable thermal conductivity in plane slab and hollow cylinder configurations.

For the plane slab shown in Fig. 4.2, consider that k is given by Eq. (4.28). In this case, the heat flow is given, not by Eq. (4.1), but by

$$q = -Ak(T) \frac{dT}{dx}$$

With variables separated,

$$\frac{q}{A} dx = -k_0(1 + bT) dT$$

and integration between x_1 and x_2, where the temperatures are T_1 and T_2,

$$\frac{q}{A} \int_{x_1}^{x_2} dx = -k_0 \int_{T_1}^{T_2} dT - k_0 b \int_{T_1}^{T_2} T \, dT$$

leads to

$$\frac{q}{A} (x_2 - x_1) = -k_0(T_2 - T_1) - \frac{1}{2} k_0 b(T_2^2 - T_1^2)$$

This may be rewritten with $x_2 - x_1 = L$,

$$q = \frac{A}{L} (T_1 - T_2)k_0 \left[1 + b\left(\frac{T_1 + T_2}{2}\right) \right]$$

which when compared to

$$q = \frac{k_m A}{L} (T_1 - T_2)$$

shows that the problem may be treated by using a mean thermal conductivity defined by

$$k_m = k_0 \left[1 + \frac{b}{2} (T_1 + T_2) \right] \tag{4.29}$$

For the hollow cylinder shown in Fig. 4.3,

$$q = -Ak(T) \frac{dT}{dr} = -2\pi r L k(T) \frac{dT}{dr}$$

is the governing equation with the thermal conductivity a function of temperature. With variables separated and using Eq. (4.28) for $k(T)$,

$$\frac{q}{2\pi L} \frac{dr}{r} = -k_0 (1 + bT) dT$$

Now the integration shows

$$\frac{q}{2\pi L} \int_{r_1}^{r_1} \frac{dr}{r} = -k_0 \int_{T_1}^{T_2} dT - k_0 b \int_{T_1}^{T_2} T \, dT$$

and hence

$$\frac{q}{2\pi L} \ln \frac{r_2}{r_1} = k_0 (T_1 - T_2) + \frac{k_0 b}{2} (T_1^2 - T_2^2)$$

A similar result occurs,

$$q = \frac{2\pi L}{\ln (r_2/r_1)} k_0 \left(1 + b \frac{T_1 + T_2}{2} \right) = \frac{2\pi k_m L}{\ln (r_2/r_1)} (T_1 - T_2)$$

which shows that a mean thermal conductivity defined by Eq. (4.29) may be used here as well.

4.9 MORE COMPLICATED SHAPES— ANALYTICAL METHOD

In many real-world situations, the geometry of the system under consideration does not permit a simple one-dimensional treatment. For example, when the shape of the system is irregular or when the temperature along a boundary is not uniform, the temperature distribution may be a function of two or perhaps even three coordinates. Because of the added complexity, analytical solutions are not always obtainable.

Certain simpler problems can be treated by an analytical method known as *separation of variables*, which is restricted to linear partial differential equations such as the Laplace equation, Eq. (4.10). This method will now be illustrated in a consideration of the thin plate with faces insulated shown in Fig. 4.9. Let its edges at $x = 0$, $x = L$, and $y = \infty$ be held at some steady, yet uniform, arbitrary temperature T_0, and let the edge at $y = 0$ have the temperature distribution $T = T_0(1 + x/L)$. The faces and edges of this semiinfinite plate, so called because one of the dimensions is unlimited, are perfectly insulated, and there is no heat interchange between the plate and the surroundings. The steady-state temperature distribution is governed by Eq. (4.10), the Laplace equation. Since the plate is thin, the temperature gradient in the z direction may be neglected and only the first two terms of Eq. (4.10) need be used. Hence with $\theta = T - T_0$, the equation to be solved is

$$\frac{\partial^2 \theta}{\partial x^2} + \frac{\partial^2 \theta}{\partial y^2} = 0 \tag{4.30}$$

and the solution will be subject to the boundary conditions

$$\theta(x = 0) = T_0 - T_0 = 0 \tag{4.31a}$$

$$\theta(x = L) = T_0 - T_0 = 0 \tag{4.31b}$$

$$\theta(y = \infty) = T_0 - T_0 = 0 \tag{4.31c}$$

$$\text{and} \quad \theta(y = 0) = T_0 \left(1 + \frac{x}{L}\right) - T_0 = \frac{T_0 x}{L} \tag{4.31d}$$

Equation (4.30) is solved by the method of separation of variables. Here a product solution of the form $\theta(x, y) = X(x) \cdot Y(y)$ is assumed. This says that the temperature distribution will be a product of two solutions, one of x alone and one of y

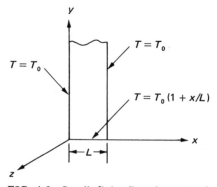

FIG. 4.9 Semiinfinite flat plate with faces insulated and with base at a prescribed temperature distribution.

alone. With double primes used to designate the derivatives, $\partial^2\theta/\partial x^2 = X''Y$ and $\partial^2\theta/\partial y^2 = Y''X$. Thus Eq. (4.30) is reduced to

$$\frac{X''}{X} = -\frac{Y''}{Y} \tag{4.32}$$

For Eq. (4.32) to be satisfied for all values of x and y, both sides must be equal to a constant, say, $-\lambda^2$. Thus two ordinary differential equations are obtained,

$$X'' + \lambda^2 X = 0$$
$$Y'' - \lambda^2 Y = 0$$

and these have general solutions

$$X = C_1 \sin \lambda x + C_2 \cos \lambda x \tag{4.33a}$$
$$\text{and} \quad Y = C_3 e^{\lambda y} + C_4 e^{-\lambda y} \tag{4.33b}$$

The general solution of Eq. (4.30) is therefore

$$\theta(x,y) = (C_1 \sin \lambda x + C_2 \cos \lambda x)(C_3 e^{\lambda y} + C_4 e^{-\lambda y})$$

In order to have a solution $\theta = 0$ at $x = 0$, the first term of this equation given by Eq. (4.33a) must vanish at $x = 0$. Thus $C_2 = 0$. Similar reasoning for $\theta = 0$ at $y = \infty$ requires that the second term given by Eq. (4.33b) must also vanish at $y = \infty$ so that $C_3 = 0$. The general solution then reduces to

$$\theta = Ce^{-\lambda y} \sin \lambda x$$

where $C = C_1 C_4$.

For the boundary condition at $x = L, \theta = 0$ to hold, it is necessary that $\sin \lambda L = 0$ or $\lambda = n\pi/L$ with $n = 1, 2, 3, \ldots$. The general solution then becomes

$$\theta = \sum_{n=1}^{\infty} C_n e^{-n\pi y/L} \sin \frac{n\pi x}{L}$$

where the use of C_n shows that a different value of C must be used for every value of n. To determine the values of C_n, the boundary condition at $y = 0$ as given by Eq. (4.31d) is employed, $\theta(x, 0) = T_0 x/L$ and

$$\theta(x, 0) = \frac{T_0 x}{L} = \sum_{n=1}^{\infty} C_n \sin \frac{n\pi x}{L}$$

This is recognized to be a Fourier sine series expansion of the function $\theta(x) = T_0 x/L$ over the interval 0 to L. Then

$$C_n = \frac{2}{L} \int_0^L T_0 \left(\frac{x}{L} \right) \sin \frac{n\pi x}{L} \, dx$$

or $C_n = \dfrac{2T_0}{(n\pi)^2} (\sin n\pi - n\pi \cos n\pi) \qquad n = 1, 2, 3, \ldots$

This makes the particular solution with $\theta = T - T_0$,

$$T = T_0 + 2T_0 \sum_{n=1}^{\infty} \left[\frac{\sin n\pi - n\pi \cos n\pi}{(n\pi)^2} \right] e^{-n\pi y/L} \sin \frac{n\pi}{L} x \qquad (4.34)$$

4.10 MORE COMPLICATED SHAPES— CONFORMAL MAPPING

The collector of a traveling wave tube is a thick-walled cylinder, capped at the end, with the center bored to accommodate the collection of the electrons contained in the beam. An end view of a collector merely shows two concentric circles of significantly different radii. The conversion of kinetic energy contained in the beam to heat in the collector, often of the order of several kilowatts, causes a significant temperature differential across the metal, usually copper, and if the exterior of the cylindrical collector is cooled, the interior will be at a considerably higher temperature than the exterior. The cooling of these devices is a delicate problem, because if the interior wall temperature exceeds a prescribed maximum, the metal will "outgas" and the vacuum inside the tube will degrade and the device itself will become useless.

If, by chance, the interior is bored "off center," an even greater thickness of metal is encountered, and the interior temperature is even higher in the region where the metal thickness is greater. It is an interesting problem with a very useful solution to account for the interior temperature of the inside of a traveling wave tube collector as a function of manufacturing tolerance.

Conformal mapping is an analytical method that permits the transformation or mapping of a complex configuration to a simpler configuration where the given problem is easier to solve. The mapping is accomplished by what is called a mapping function, and these are tabulated in several works [3].

In order to get at the basis for conformal mapping, consider the two complex domains shown in Fig. 4.10. In both, a real and an imaginary axis can be noted. In Fig. 4.10*a*, which will be called the *z* plane, a point *z* is located by considering the coordinates *x* and *y* so that the complex variable *z* may be written as

$$z = x + iy$$

where $i = \sqrt{-1}$. Figure 4.10*b* shows the so-called *w* plane, where the complex variable *w* is written as

$$w = u + iv$$

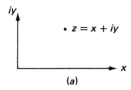

(a)

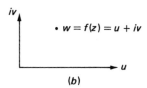

(b)

FIG. 4.10   (a) The z plane where $z = x + iy$ and (b) the w plane where $w = u + iv$. The function w is a function of the complex variable z, where $w = f(z)$.

The variable w is a function of the complex variable z, and u and v are both functions of x and y.

For example, consider the complex variable $w = z^2 + 2z$. With $z = x + iy$, it is easy to see that

$$w = u + iv = (x + iy)^2 + 2(x + iy)$$
$$= x^2 + i2xy - y^2 + 2x + i2y$$

or $\quad w = x^2 + 2x - y^2 + i(2y + 2xy)$

and $\quad u = x^2 + 2x - y^2$

$$v = 2x(1 + y)$$

Indeed, $w = f(z)$ may be thought of as a mapping from the z plane to the w plane. Furthermore, an inverse mapping will exist that will map in reverse, $z = f^{-1}(w)$:

$$z = -1 \pm \sqrt{1 - w}$$

which is seen to be a double-valued function because a single w yields two values of z. Care must be exercised by the analyst in specifying the region that is applicable to a particular inverse mapping.

Conformal mappings can be obtained only from analytic functions of a complex variable. A complex variable is said to be analytic in a region if it has a derivative at every point in the region. To determine whether the complex variable $w = f(z) = u + iv$ is analytic, the Cauchy-Riemann equations,

$$\frac{\partial u}{\partial x} = \frac{\partial v}{\partial y} \qquad \frac{\partial u}{\partial y} = -\frac{\partial v}{\partial x}$$

are employed. If the Cauchy-Riemann equations are satisfied, then the function w is analytic.

For example if $w = x - iy$,

$$u = x \qquad \frac{\partial u}{\partial x} = 1 \qquad \frac{\partial u}{\partial y} = 0$$

$$v = -y \qquad \frac{\partial v}{\partial x} = 0 \qquad \frac{\partial v}{\partial y} = -1$$

it is seen that w is not analytic because

$$\frac{\partial u}{\partial x} \neq \frac{\partial v}{\partial y}$$

and $w = f(z)$ cannot yield a conformal mapping.

However, $w = x + iy$ is analytic because

$$u = x \qquad \frac{\partial u}{\partial x} = 1 \qquad \frac{\partial u}{\partial y} = 0$$

$$v = y \qquad \frac{\partial v}{\partial x} = 0 \qquad \frac{\partial v}{\partial y} = 1$$

and the Cauchy-Riemann equations are satisfied.

To see how conformal mapping is applied, consider the annular region composed of the two cylinders shown in Fig. 4.11a. Here the scaled temperature $\theta = 0$ on the

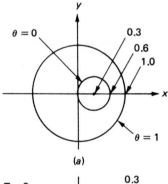

(a)

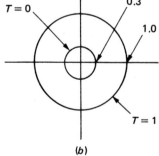

(b)

FIG. 4.11   Example of a conformal mapping from (a) the z plane to (b) the w plane using the mapping function $w = (1 - 3z)/(z - 3)$.

inner cylinder and $\theta = 1$ on the outer cylinder. It is desired to determine the temperature distribution within the annular region. This is the traveling wave tube collector problem, and it is quite difficult. If, however, some way can be found to make the two cylinders concentric as shown in Fig. 4.11b, it becomes a very easy problem in cylindrical coordinates.

The desired transformation or mapping function is given by Saff and Snider [3],

$$w = \frac{1-3z}{z-3} \qquad (4.35)$$

and its employment results in Fig. 4.11b, a configuration that can be handled nicely by Laplace's equation in cylindrical coordinates[9] with no angular and axial dependency:

$$\frac{1}{r}\frac{d}{dr}\left(r\frac{dT}{dr}\right) = 0 \qquad (4.36)$$

This equation is integrated twice and the constants of integration are evaluated from the boundary conditions:

$$T(r = 0.3) = 0 \qquad (4.37a)$$

and $\quad T(r = 1.0) = 1 \qquad (4.37b)$

Double integration gives

$$\frac{dT}{dr} = \frac{C_1}{r}$$

and $\quad T = C_1 \ln r + C_2$

Use of Eqs. (4.37) shows that $C_2 = 1$ and $C_1 = -1/\ln 0.3$, so that

$$T = 1 - \ln\frac{r}{0.3} \qquad (4.38)$$

With $r = w = u + iv$, it is seen that

$$T(w) = T(u, v) = \frac{\ln|3w|}{\ln 3}$$

which satisfies the boundary conditions of Eqs. (4.37) exactly.

The final step is to use Eq. (4.35) to express $|w|$ in terms of x and y

$$|w| = \left[\frac{(1-3x)^2 + 9y^2}{(x-3)^2 + y^2}\right]^{1/2} \qquad (4.39)$$

[9] See Sec. 4.13.

which then makes

$$\theta(z) = \theta(x,y) = \frac{1}{\ln 3} \left\{ \ln 3 + \frac{1}{2} \ln [(3x-1)^2 + 9y^2] - \frac{1}{2} \ln [(x-3)^2 + y^2] \right\}$$

(4.40)

Observe that because solutions to the Laplace equation

$$\nabla^2 T = 0$$

regardless of the coordinate system involved remain solutions to the Laplace equation after a conformal mapping (as long as the function is analytic), vast mathematical simplification can be obtained if a mapping function can be found.

4.11   MORE COMPLICATES SHAPES–DISCRETE HEAT SOURCES

Mikic [4] has proposed an interesting method for the determination of the temperature at discrete heat sources on the face of a conducting medium. Notice in Fig. 4.12 that the heat will diverge from the concentrated area under the heat source. This, for a heat source of constant magnitude, gives rise to what is called a constriction effect, so that the temperature just under the discrete heat source T_j is given by

$$T_j = T_0 + \Delta T_c$$

(4.41)

Here T_0 is the temperature that would be obtained if all of the heat were spread over the entire area and ΔT_c is the constriction effect as shown in Fig. 4.13. Mikic has provided relationships for ΔT_c in a variety of cases.

For a circular contact on an infinite conducting medium of thermal conductivity k as shown in Fig. 4.14,

$$\Delta T_c = \frac{q}{2\sqrt{\pi a k}} \qquad \text{Note} : \quad R_{cs} = \frac{1}{2\sqrt{\pi}\, a\, K}$$

(4.42)

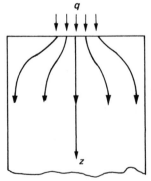

FIG. 4.12   Discrete heat source on the face of a conducting medium.

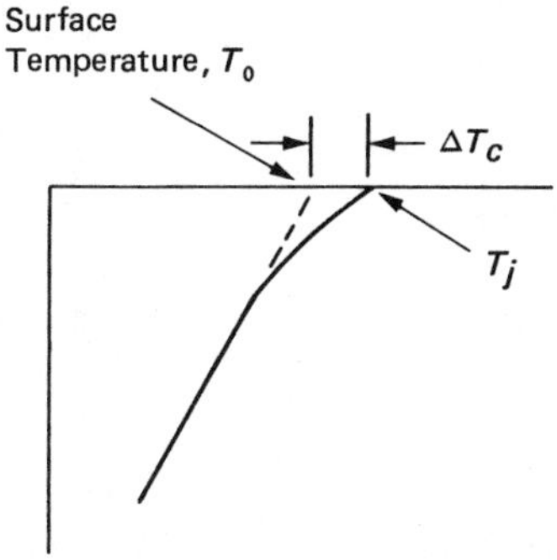

Surface
Temperature, T_0

ΔT_c

T_j

FIG. 4.13 The constriction effect allows the calculation of the temperature directly under the heat source by adding ΔT_c to the average surface temperature obtained by spreading the heat over the entire surface.

For a circular contact on a finite conducting medium, Fig. 4.15,

$$\Delta T_c = \frac{q}{2\sqrt{\pi a}k} \left(1 - \frac{a}{b}\right)^{3/2} \tag{4.43}$$

For a long, strip heat source, Fig. 4.16,

$$\Delta T_c = \frac{q}{\pi Lk} \ln\left[\frac{1}{\sin(\pi a/2b)}\right] \tag{4.44}$$

If the strip is not long as shown in Fig. 4.17,

$$\Delta T_c = \Delta T_{c1} + \Delta T_{c2} + \Delta T_{c3} \tag{4.45}$$

where

$$\Delta T_{c1} = \frac{q}{2\pi^2 k} \cdot \frac{b}{ac} \sum_{m=1}^{\infty} \frac{\sin(m\pi a/b)}{m^2}$$

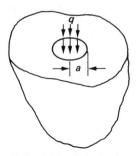

FIG. 4.14 Circular heat source on infinite conducting medium.

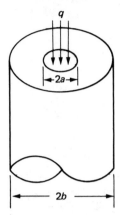

FIG. 4.15   Circular heat source on finite conducting medium.

$$\Delta T_{c2} = \frac{q}{2\pi^2 k} \cdot \frac{c}{db} \sum_{m=1}^{\infty} \frac{\sin (m\pi d/b)}{m^2}$$

and $$\Delta T_{c3} = \frac{q}{2\pi^2 k} \cdot \frac{2}{ad} \sum_{m=1}^{\infty} \sum_{n=1}^{\infty} \frac{\sin (n\pi d/c) \sin (m\pi a/b)}{mn[(m\pi/b)^2 + (n\pi/c)^2]^{1/2}}$$

Finally, for a circular contact on a finite conducting medium with finite thickness as shown in Fig. 4.18,

$$\Delta T_c = \frac{4q}{\pi ak} \cdot \frac{b}{a} \sum_{m=0}^{\infty} \tanh \left(\lambda_m \frac{L}{b} \right) \frac{J_1^2 \left[\lambda_m(a/b) \right]}{\lambda_m^3 J_0^2(\lambda_m)} \tag{4.46}$$

where the λ_m's are those values of the argument of the Bessel function J_1 that make $J_1(\lambda_m) = 0$.

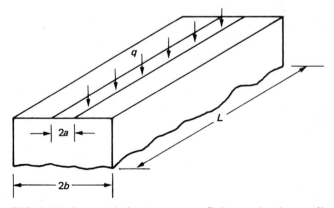

FIG. 4.16   Long strip heat source on finite conducting medium.

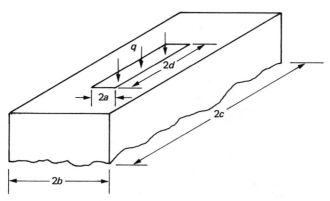

FIG. 4.17 Short strip heat source on finite conducting medium.

Mikic has also proposed an additional correction, ΔT_f, which accounts for fluctuations of the heat source. This will not be considered here because consideration of a steady heat source yields a worst case and a conservative result when Eq. (4.41) is employed.

4.12 ADDITIONAL METHODS

Landis and Zupnick [5] have shown that the temperature profiles (isotherms) in complex geometric shapes can be determined with good accuracy by cutting a "model" of the shape from Teledeltos paper. This paper possesses an electrically conductive coating on one side and, with appropriate end treatment so that electrical potentials (voltages) can be applied uniformly and consistent with the temperature boundary conditions in the problem at hand, a voltmeter may be used to "probe" the paper model to find lines of constant voltage that are analogous to the isotherms.

Another analog approach is that of graphic field mapping. This is a technique that electrical engineers and physicists frequently employ in evaluating electric and electromagnetic fields. The technique is well described by Kraus and Carver [6].

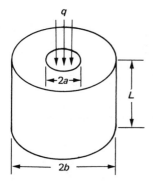

FIG. 4.18 Circular heat source on finite conducting medium with finite thickness.

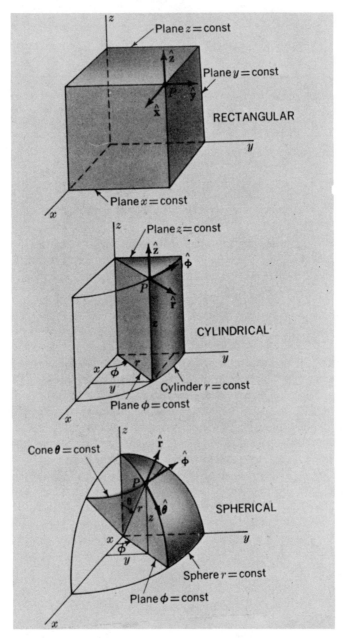

FIG. 4.19 Rectangular, cylindrical, and spherical coordinate systems. (From Kraus and Carver [6], with permission.)

4.13 OTHER COORDINATE SYSTEMS

Figure 4.19 displays elemental volumes and lengths in three coordinate systems—rectangular, cylindrical, and spherical. In each system, a point P and unit vectors emanating from P are shown.

The general equation of heat conduction in the cylindrical coordinate system is

$$\frac{\partial^2 T}{\partial r^2} + \frac{1}{r}\frac{\partial T}{\partial r} + \frac{1}{r^2}\frac{\partial^2 T}{\partial \phi^2} + \frac{\partial^2 T}{\partial z^2} + \frac{q_i}{k} = \frac{1}{\alpha}\frac{\partial T}{\partial t} \tag{4.47}$$

TABLE 4.1 Thermal Conductivities of Various Materials ($80°$F, $26.7°$C)

Material	Btu/ft h °F	W/m °C
Aluminum		
Pure	125	216.34
356 T6	87	150.52
2024 T4	70	121.11
5052	83	143.60
6061 T6	90	155.71
7075 T6	70	121.11
Alumina, 96% pure	17.0	29.41
Alumina, 90%	7.0	12.11
Beryllia, 99.5% pure	140	242.21
Beryllia, 95%	90	155.71
Beryllium	95	164.36
Beryllium copper	47.8	82.70
Brass		
Red	63.7	110.21
Yellow	54.6	94.46
70 Cu-30 Zn	58.0	100.35
Copper	220	380.62
Epoxy fiberglass	0.15	0.260
Glass (soft)	0.569	0.984
Gold	171	295.85
Iron		
Pure	43.2	74.74
Cast	32	55.36
Wrought	34	58.82
Lead	18.9	32.70
Magnesium	91	157.44
Mica	0.341	0.590
Molybdenum	77.4	133.91
Monel	20.5	35.47
Phenolic (paper base)	0.159	0.275
Silicon	88.7	153.46
Steel		
Kovar	9	15.57
SAE 1010	34	58.82
SAE 1020	32	55.36
SAE 1045	26	44.98
SAE 4130	24	41.52
Teflon	0.145	0.251
Titanium	9	15.57

In spherical coordinates, the general equation of heat conduction is

$$\frac{1}{r^2} \frac{\partial}{\partial r} \left(r^2 \frac{\partial T}{\partial r} \right) + \frac{1}{r^2 \sin \theta} \frac{\partial}{\partial \theta} \left(\sin \theta \frac{\partial T}{\partial \theta} \right) + \frac{1}{r^2 \sin^2 \theta} \frac{\partial^2 T}{\partial \phi^2} + \frac{q_i}{k} = \frac{1}{\alpha} \frac{\partial T}{\partial t}$$

(4.48)

In either Eq. (4.47) or (4.48), care must be exercised to assure adherence to the direction of the unit vectors shown in Fig. 4.19.

4.14 THERMAL CONDUCTIVITY

Table 4.1 provides a table of thermal conductivity for materials commonly used in electronic equipment and microelectronics packaging. These values have been extracted from the work of Harper [7, 8].

4.15 NOMENCLATURE

Roman Letter Symbols

A	area normal to heat flow path (cross-sectional area), m^2
b	temperature coefficient of thermal conductivity, $\text{W/m} \, ^\circ\text{C}^2$
c	specific heat, $\text{J/kg} \, ^\circ\text{C}$
C	constant of integration
h	heat transfer coefficient, $\text{W/m}^2 \, ^\circ\text{C}$
I	current, A
k	thermal conductivity, $\text{W/m} \, ^\circ\text{C}$
K	thermal conductance, $\text{W/}^\circ\text{C}$
L	length of heat flow path, m
m	mass, kg
n	a number of physical entities or a number of values
q	heat flow, W
r	radius or radial length coordinate, m
R	electrical resistance, Ω, or thermal resistance, $^\circ\text{C/W}$
S	surface
t	time, h
T	temperature, $^\circ\text{C}$
u	internal energy, J
U	overall heat transfer coefficient, $\text{W/m}^2 \, ^\circ\text{C}$
V	voltage or difference in potential, V
w	function of a complex variable
x	length coordinate, m
y	length coordinate, m
z	length coordinate, m, or complex variable

Greek Letter Symbols

α	thermal diffusivity, m^2/h
Δ	indicates change in variable

θ	temperature difference or excess, °C, coordinate in spherical and cylindrical coordinate systems, radians
λ	a constant (eigenvalue)
ρ	density, kg/m^3
ϕ	coordinate in spherical coordinate system, radians

Subscripts

a	indicates position in a composite configuration
b	indicates position in a composite configuration
c	indicates constriction effect or cold fluid
f	indicates fluctuation effect
G	indicates heat generated
h	indicates hot fluid
i	indicates volumetric rate of heat generation
m	indicates midpoint or mean value
r	indicates radial direction
s	indicates surroundings or heat stored
T	indicates total
v	indicates constant volume
0	indicates nominal value

4.16 REFERENCES

1 Fourier, J. B. J., *The Analytical Theory of Heat*, transl. Alexander Freeman, Cambridge University Press, London, 1878.

2 Kraus, A. D., The Use of Steady State Electrical Network Analysis in Solving Heat Flow Problems, *Second Natl. ASME-AIChE Heat Transfer Conf., Chicago, Aug. 1958.*

3 Saff, E. B., and Snider, A. D., *Fundamentals of Complex Analysis for Mathematics, Science and Engineering*, Prentice-Hall, Englewood Cliffs, N.J., 1976.

4 Mikic, B. B., course notes, M.I.T. Special Summer Session on Thermal Control of Modern Electronic Components, 1978 and 1979.

5 Landis, F., and Zupnick, T., The Effectiveness of Stub-Fins as Determined by the Teledeltos Paper Analog, *First Natl. ASME-AIChE Heat Transfer Conf., University Park, Pa.,* 1957.

6 Kraus, J. D., and Carver, K. R., *Electromagnetics*, McGraw-Hill, New York, 1973.

7 Harper, C. A., *Handbook of Thick Film Hybrid Microelectronics*, McGraw-Hill, New York, 1974.

8 Harper, C. A., *Handbook of Electronic Packaging*, McGraw-Hill, New York, 1969.

5

■ conduction—transient

5.1 INTRODUCTION

Unfortunately, the simple shapes for which transient temperature solutions exist are seldom encountered in electronic hardware. Solutions for such shapes have been cataloged in the works of Carslaw and Jaeger [1], Schneider [2], Arpaci [3], and Myers [4]. These solutions are termed *classical* or *analytical* because they derive from a solution to a partial differential equation.

When the configuration is only moderately complicated, analog, numerical, or graphical techniques must be employed. For example, if a warm-up analysis for an entire package of electronics is required, there is no question that the solution will come forth from a computer; if simple enough, possibly from an analog computer and if complicated, from a digital computer.

It is the intent of this chapter to discuss four techniques of solving a transient heat flow problem. These techniques are:

Classical or analytical
Analog
Numerical
Graphical

A simple configuration, a plane wall or slab subjected to a sudden change of temperature on its faces, has been chosen to illustrate these four techniques for solution, not only because a solution can be obtained from the governing partial differential equation but because a comparison of results for the four methods can be easily accomplished. This exposition begins in Sec. 5.4 after an introductory consideration of the warmup of a small component.

5.2 THE WARMUP OF A SMALL COMPONENT

Consider a small component originally at some *soak* temperature T_0. Let the component be sufficiently small so that a spacial variation of its temperature as a function of time is precluded. This is a realistic situation, which might pertain to the substrate of an integrated circuit.

If the component is energized with a heat generation q_G in watts (Btu/h), then the equation that governs derives from a simple energy balance,

$$q_G = Wc \frac{dT}{dt} \tag{5.1}$$

where W is the weight of the component in kilograms (lb) and c is its specific heat in J/kg $^\circ$C (Btu/lb $^\circ$F).

The solution to Eq. (5.1), which insists that the component at all times is at the same temperature throughout, is quickly obtained by a separation of variables,

$$\frac{q_G}{Wc} dt = dT$$

and an integration from time $t = 0$ when the temperature is at $T = T_0$ to time $t = t$ when the temperature is $T = T$:

$$\frac{q_G}{Wc} \int_0^t dt = \int_{T_0}^T dt$$

The result is

$$T = T_0 + \frac{q_G}{Wc} t \tag{5.2}$$

and it is seen that the temperature of the component will increase without bound as time marches on unless the heat is dissipated to the environment or to a heat sink as the component temperature increases.

The idea of a small component suspended in space by thin nonconducting cords is cute but not too realistic. Of more significance is the inclusion of a heat flow to or from the component to or from an environment held at T_0, which in this particular study is the temperature throughout the component at the start of the warm-up period. In this case, the energy balance is composed of the difference between the heat entering and leaving the component,

$$q_G - K(T - T_0)$$

where K is some conductance *to* the environment,[1] and the increase in component internal energy is

$$Wc \frac{dT}{dt}$$

Hence the differential equation to be solved is the energy balance

$$q_G - K(T - T_0) = Wc \frac{dT}{dt}$$

or $$\frac{dT}{dt} + \frac{K}{Wc} T = \frac{1}{Wc} (q_G + KT_0) \tag{5.3}$$

---

[1] Conductive or convective but not radiative, because radiation is governed by a fourth power on the temperatures.

One way to solve Eq. (5.3) is to note that the solution is composed of a complementary function and a particular integral. The complementary function T_c derives from a solution of

$$\frac{dT_c}{dt} + \frac{K}{Wc} T_c = 0$$

and it is easy to see that

$$T_c = C_1 e^{-(K/Wc)t}$$

where C_1 is a constant of integration.

The particular integral is obtained from

$$\frac{dT_p}{dt} + \frac{K}{Wc} T_p = \frac{1}{Wc} (q_G + KT_0)$$

by an application of what is known as the method of undetermined coefficients, in which one assumes that $T_p = A$ (a constant). After it is noted that $dA/dt = 0$, one obtains, by substituting A for T_p,

$$\frac{K}{Wc} A = \frac{1}{Wc} (q_G + KT_0)$$

or $\qquad A = \frac{q_G}{K} + T_0$

Thus, with $T(t) = T_c + T_p$,

$$T(t) = C_1 e^{-(K/Wc)t} + \frac{q_G}{K} + T_0$$

where C_1 is to be evaluated from the condition at $t = 0$ where $T = T_0$. Hence

$$T_0 = C_1 + \frac{q_G}{K} + T_0$$

and $\quad C_1 = -\frac{q_G}{K}$

The result is

$$T(t) = T_0 + \frac{q_G}{K} (1 - e^{-(K/Wc)t}) \tag{5.4}$$

and it is seen that the component will eventually reach a steady-state temperature $T(\infty)$, given by

$$T(\infty) = T_0 + \frac{q_G}{K} \tag{5.5}$$

Observe that if $q_G = 100$ W, $T_0 = 100°C$, $K = 1$ W/$°C$, and $Wc = \frac{1}{2}$ J/$°C$, then Eqs. (5.4) and (5.5) reduce respectively to

$$T(t) = 100 + 100(1 - e^{-2t})$$

and $T(\infty) = 100 + 100 = 200°C$

5.3 A SMALL COMPONENT SUBJECTED TO VARIOUS FORMS OF HEAT INPUT

The heat source q_G applied to the small component considered in the previous section was a constant, applied at time $t = 0$. The temperature response of this component to other forms of heat input is of interest, and this section examines the temperature response when the component is subjected to the ramp, sinusoidal, and pulsed heat inputs shown in Fig. 5.1. In all cases the differential equation to be solved is Eq. (5.3):

$$\frac{dT}{dt} + \frac{K}{Wc} T = \frac{1}{Wc} [q(t) + KT_0] \tag{5.3}$$

with $q(t)$ taking the form $q_G f(t)$ depending on the case considered.

Solutions utilizing the Laplace transformation[2] are of advantage here, because all will derive from the transformed equation. With

$$\mathcal{L}[T(t)] = T(s)$$

$$\mathcal{L}[f(t)] = F(s)$$

and $\mathcal{L}\left(\frac{dT}{dt}\right) = sT(s) - T(0)$

a transformation, term by term, of Eq. (5.3) yields

$$sT(s) - T(0) + \frac{K}{Wc} T(s) = \frac{1}{Wc} \left[q_G F(s) + \frac{KT_0}{s} \right]$$

and with the values $q_G = 100$ W, $T_0 = 100°C$, $Wc = \frac{1}{2}$ J/$°C$, and $K = 1$ W/$°C$ substituted,

$$sT(s) - T(0) + 2T(s) = 200F(s) + \frac{200}{s}$$

[2] Further remarks on the Laplace transformation and its use in equation solving may be found in Sec. 5.5.

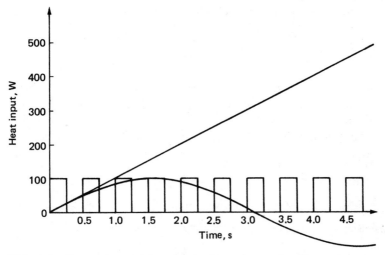

FIG. 5.1 Ramp, sinusoidal, and pulsed heat inputs for illustrative example. The sinusoid has a frequency of $1/2\pi$ hz.

or with a little algebra,

$$T(s) = \frac{sT(0) + 200sF(s) + 200}{s(s + 2)} \tag{5.6}$$

Equation (5.6) may be used for all of the cases to be considered. Observe, however, that T_0, which is the environmental temperature, may differ from $T(0)$, which is the uniform temperature throughout the component at the start of the warm-up period.

Before investigation of the ramp, sinusoidal, and pulsed heat inputs is undertaken, it is of interest to show that the solution to Eq. (5.3) for a step heat input ($q_G = 100$ W) by the Laplace transform method is identical to Eq. (5.4). In this case $q(t) = q_G f(t)$, $f(t) = 1$, and $F(s) = 1/s$. Then with $T(0) = T_0 = 100°C$, Eq. (5.6) can be shown to reduce to

$$T(s) = \frac{100(s + 4)}{s(s + 2)}$$

and $T(t)$ can be obtained from a table of Laplace transforms or from a partial fraction expansion as follows:

$$T(s) = \frac{100(s + 4)}{s(s + 2)} = \frac{C_1}{s} + \frac{C_2}{s + 2}$$

$$C_1 = \frac{100(s + 4)}{s + 2}\bigg|_{s=0} = 200$$

$$C_2 = \frac{100(s + 4)}{s}\bigg|_{s=-2} = -100$$

$$T(s) = \frac{200}{s} - \frac{100}{s + 2}$$

One can consult a table of Laplace transform pairs and obtain the temperature as a function of time:

$$T(t) = 200 - 100e^{-2t} = 100 + 100(1 - e^{-2t}) \qquad (5.7)$$

This may be compared to Eq. (5.4) with the appropriate values of q_G, K, Wc, and T_0 inserted. Indeed, the steady-state temperature will be $200°C$, which has been obtained from Eq. (5.5) using the values of T_0, q_G, and K.

For a ramp heat input, $q(t) = q_G t$, $f(t) = t$, and $F(s) = 1/s^2$. In this case, Eq. (5.6) becomes

$$T(s) = \frac{100s + (200/s) + 200}{s(s + 2)}$$

or, with some algebraic adjustment,

$$T(s) = 100 \frac{s^2 + 2s + 2}{s^2(s + 2)}$$

and this time the partial fraction expansion sets up as

$$T(s) = \frac{C_1}{s} + \frac{C_2}{s^2} + \frac{C_3}{s + 2} = 100 \frac{s^2 + 2s + 2}{s^2(s + 2)}$$

An algebraic method can be employed to determine C_1, C_2, and C_3.

$$\frac{C_1 s(s + 2) + C_2(s + 2) + C_3 s^2}{s^2(s + 2)} = 100 \frac{s^2 + 2s + 2}{s^2(s + 2)}$$

or

$$\frac{(C_1 + C_3)s^2 + (2C_1 + C_2)s + 2C_2}{s^2(s + 2)} = \frac{100s^2 + 200s + 200}{s^2(s + 2)}$$

When the coefficients of like powers in the numerators are equated, three equations in three unknowns result:

$$C_1 + C_3 = 100$$
$$2C_1 + C_2 = 200$$
$$2C_2 = 200$$

from which $C_2 = 100$, $C_1 = 50$, and $C_3 = 50$. Hence

$$T(s) = \frac{50}{s} + \frac{100}{s^2} + \frac{50}{s + 2}$$

and the temperature as a function of time is

$$T(t) = 50 + 100t + 50e^{-2t}$$

or $T(t) = 50(1 + e^{-2t}) + 100t$ (5.8)

Next take a sinusoidal input where the positive peak heat input takes place every 2π seconds. In this case $q(t) = q_G \sin t = 100 \sin t$ watts. Here $f(t) = \sin t$ and

$$F(s) = \frac{1}{s^2 + 1}$$

Now Eq. (5.6) with $T(0) = 100°F$ has the form

$$T(s) = \frac{100s + 200s/(s^2 + 1) + 200}{s(s + 2)}$$

and, with some algebraic manipulation,

$$T(s) = \frac{100s(s^2 + 1) + 200(s^2 + 1) + 200s}{s(s + 2)(s^2 + 1)}$$

or $$T(s) = \frac{100s^3 + 200s^2 + 300s + 200}{s(s + 2)(s^2 + 1)}$$

In this case, the partial fraction expansion is

$$T(s) = \frac{C_1}{s} + \frac{C_2}{s + 2} + \frac{As + B}{s^2 + 1}$$

and the method of algebra that equates like coefficients in the numerators gives

$$C_1(s + 2)(s^2 + 1) + C_2 s(s^2 + 1) + (As + B)s(s + 2)$$
$$= 100s^3 + 200s^2 + 300s + 200$$

from which four equations in four unknowns are obtained:

$$C_1 + C_2 + A = 100$$
$$2C_1 + 2A + B = 200$$
$$C_1 + C_2 + 2B = 300$$
$$2C_1 = 200$$

The solution of this set gives $C_1 = 100, C_2 = 40, A = -40$, and $B = 80$, so that

$$T(s) = \frac{100}{s} + \frac{40}{s + 2} - \frac{40s}{s^2 + 1} + \frac{80}{s^2 + 1}$$

A table of Laplace transforms then yields

$$T(t) = 100 + 40e^{-2t} - 40 \cos t + 80 \sin t$$ (5.9)

A consideration of the pulsed input with duration of $\frac{1}{4}$ s and repetition rate of $\frac{1}{2}$ s can also employ Eq. (5.6), but care must be exercised to account properly for the heat source and the initial temperature $T(0)$. In all cases, T_0 will remain at 100°C. During the heating cycle, $q_G = 100$ W and $f(t) = 1$, so that $F(s) = 1/s$. Thus, when the heat is on, Eq. (5.6) becomes

$$T(s) = \frac{sT(0) + 400}{s(s + 2)}$$

with inverse transform to the time domain,

$$T(t) = 200(1 - e^{-2t}) + T(0)e^{-2t} \qquad (5.10)$$

and with $T(0)$ the temperature of the body at the start of the heating cycle.
    When the heat goes off, $f(t) = 0$ and $F(s) = 0$. In this case, Eq. (5.6) becomes

$$T(s) = \frac{sT(0) + 200}{s(s + 2)}$$

The time domain solution is

$$T(t) = 100(1 - e^{-2t}) + T(0)e^{-2t} \qquad (5.11)$$

where $T(0)$ is the temperature at the start of the "off" cycle.
    Equations (5.7) through (5.11), along with their respective heat input functions, are plotted in Figs. 5.2 through 5.5.

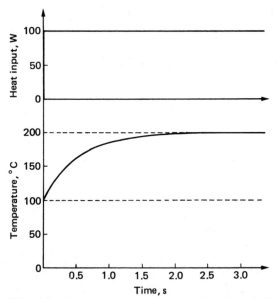

FIG. 5.2   Response of component to step heat input of 100 W.

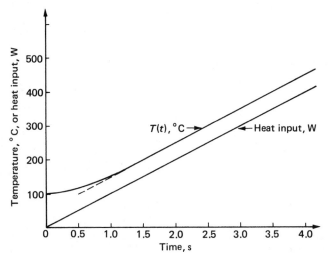

FIG. 5.3 Response of component to ramp heat input of 100 W.

5.4 CLASSICAL (ANALYTICAL) METHOD

If no heat sources or sinks are present and if the thermal properties of thermal conductivity and specific heat and the material specific weight are independent of temperature, the transfer of heat by conduction in the unsteady state is represented by the Fourier equation

$$\frac{\partial^2 T}{\partial x^2} + \frac{\partial^2 T}{\partial y^2} + \frac{\partial^2 T}{\partial z^2} = \frac{1}{\alpha} \frac{\partial T}{\partial t} \tag{4.8}$$

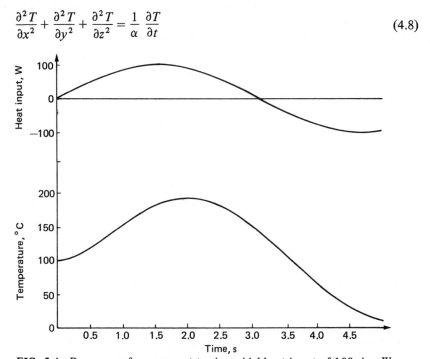

FIG. 5.4 Response of component to sinusoidal heat input of 100 sin *t* W.

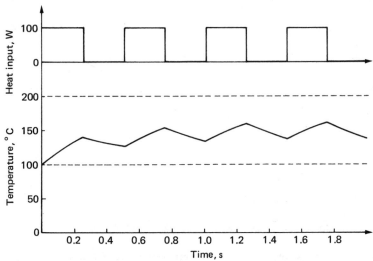

FIG. 5.5 Response of component to pulsed heat input of 100 W. The pulse width is $\frac{1}{4}$ s and the pulse repetition rate is $\frac{1}{2}$ s.

This equation must be solved to obtain a general solution. A particular solution may then be obtained from the boundary conditions and an initial condition.

Consider a slab of material of width L and at a constant temperature throughout, T_i at $t = 0$. Suppose that this slab is of infinite extent in the y and z directions as shown in Fig. 5.6. If the slab is suddenly immersed at $t = 0$ in a cold bath at temperature $T = 0$, Eq. (4.8) will govern but there will be no second partial derivatives of temperature with respect to y or z because *infinite extent in the y and z directions* means that y and z are ignorable coordinates. Hence, for Fig. 5.6, Eq. (4.8) reduces to

$$\frac{\partial^2 T}{\partial x^2} = \frac{1}{\alpha}\frac{dT}{dt} \tag{5.12}$$

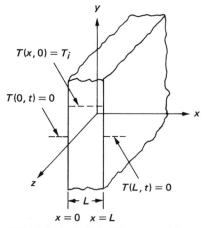

FIG. 5.6 Slab of infinite extent in y and z directions. The slab is initially at $T = T_i$ throughout, and at time $t = 0$, the faces of the slab are instantaneously reduced from $T = T_i$ to $T = 0$.

The boundary conditions are

$$T(x = 0, t = t) = T(0, t) = 0 \qquad (5.13a)$$
$$T(x = L, t = t) = T(L, t) = 0 \qquad (5.13b)$$

and the single initial condition is

$$T(x = x, t = 0) = T(x, 0) = T_i \qquad (5.14)$$

Equation (5.12) may be solved by the *method of separation of variables*, which involves assuming that the solution, $T(x, t)$, is a product of two solutions, one of x alone, $X(x)$, and one of T alone, $\theta(t)$. Thus

$$T(x, t) = X(x) \cdot \theta(t) \qquad (5.15)$$

$$\frac{\partial^2 T(x, t)}{\partial x} = X''(x)\, \theta(t) \qquad (5.16a)$$

and $\qquad \dfrac{\partial T(x, t)}{\partial t} = X(x)\, \theta'(t) \qquad (5.16b)$

where the primes designate derivatives. When Eqs. (5.16) are put into Eq. (5.12), the result is

$$X''\theta = \frac{1}{\alpha} X\theta' \qquad (5.17)$$

where the functional notation $X(x)$ and $\theta(t)$ has been omitted.
Equation (5.17) may be rearranged to yield

$$\frac{X''}{X} = \frac{1}{\alpha}\frac{\theta'}{\theta} \qquad (5.18)$$

and it is observed that Eq. (5.18) cannot be satisfied for all values of x and t unless both sides of the equation are equal to a constant, say, $-\lambda^2$, so that

$$\frac{X''}{X} = \frac{1}{\alpha}\frac{\theta'}{\theta} = -\lambda^2$$

from which two ordinary differential equations are obtained,

$$X'' + \lambda^2 X = 0 \qquad (5.19a)$$

and $\quad \theta' + \alpha\lambda^2\theta = 0 \qquad (5.19b)$

which have solutions

$$X(t) = C_1 \sin \lambda x + C_2 \cos \lambda x \qquad (5.20a)$$

and $\theta(t) = C_3 e^{-\alpha \lambda^2 t}$ (5.20b)

Equations (5.20) may be multiplied in accordance with Eq. (5.15) to give the general solution

$$T(x, t) = e^{-\alpha \lambda^2 t}(A \sin \lambda t + B \cos \lambda t) \tag{5.21}$$

where $A = C_1 C_3$ and $B = C_2 C_3$, both still arbitrary. These arbitrary constants as well as the value of λ must be determined from the boundary and initial conditions.
Use of Eq. (5.13a) in Eq. (5.21) results in

$$T(0, t) = 0 = e^{-\alpha \lambda^2 t}[A \sin \lambda(0) + B \cos \lambda(0)]$$

which shows that $B = 0$. With $B = 0$, Eq. (5.21) reduces to

$$T(x, t) = A e^{-\alpha \lambda^2 t} \sin \lambda t \tag{5.22}$$

Next, if Eq. (5.13b) is used in Eq. (5.22), the result is

$$T(L, t) = 0 = A e^{-\alpha \lambda^2 t} \sin \lambda L$$

which requires *either* $A = 0$ *or* $\sin \lambda L = 0$. If $A = 0$, the entire solution is trivial, so the only alternative is to require that $\sin \lambda L = 0$. This means that $\lambda = n\pi/L$ for $n = 1, 2, 3, \ldots$. Thus Eq. (5.22) becomes

$$T(x, t) = A_n e^{-\alpha(n\pi/L)^2 t} \sin \frac{n\pi}{L} x \quad n = 1, 2, 3, \ldots \tag{5.23}$$

where each value of n yields a different value of A.
To satisfy the initial condition of Eq. (5.14), Eq. (5.23) must reduce to

$$T_i = \sum_{n=1}^{\infty} A_n \sin \frac{n\pi}{L} x$$

and this is the Fourier series expansion of the constant T_i, where the coefficients A_n are given by

$$A_n = \frac{2}{L} \int_0^L T_i \sin \frac{n\pi}{L} x \, dx$$

When the indicated integration is performed,

$$A_n = -\frac{2T_i}{n\pi} (\cos n\pi - 1) \quad n = 1, 2, 3, \ldots \tag{5.24}$$

The particular solution is therefore Eq. (5.23) with Eq. (5.24) incorporated,

$$T(x, t) = \sum_{n=1}^{\infty} \frac{4T_i}{n\pi} e^{-\alpha(n\pi/L)^2 t} \sin \frac{n\pi}{L} x \quad n = 1, 3, 5, \ldots \tag{5.25}$$

because

$$\cos n\pi - 1 = \begin{cases} 0 & n = 0, 2, 4, 6, \ldots \\ -2 & n = 1, 3, 5, \ldots \end{cases}$$

If, on the other hand, the faces are suddenly increased to a temperature of T_0 at $t = 0$, one may employ an identical solution technique to arrive at

$$T(x, t) - T_0 = \sum_{n=1}^{\infty} \frac{4(T_i - T_0)}{n\pi} e^{-\alpha(n\pi/L)^2 t} \sin \frac{n\pi}{L} x \quad n = 1, 3, 5, \ldots \tag{5.26}$$

Olson and Schultz [5] have evaluated the infinite series of Eq. (5.26) and, for the midplane where $x = L/2$, have developed the so-called *plate series* in terms of the *Fourier modulus for the midplane*,[3]

$$\text{Fo} \equiv \frac{4\alpha t}{L^2} \tag{5.27}$$

The plate series is then defined by

$$P(\text{Fo}) \equiv \frac{4}{\pi} \sum_{n=1}^{\infty} e^{(n\pi/2)^2 \text{Fo}} \sin \frac{n\pi}{2} \tag{5.28}$$

so that the temperature at the midplane is

$$T\left(\frac{L}{2}, t\right) = T_0 + (T_i - T_0) P(\text{Fo}) \tag{5.29}$$

This development is based on a sudden change in temperature on the faces of the slab. If heat is transmitted to or from the faces of the slab, the boundary conditions for the solution of

$$\frac{\partial^2 T}{\partial x^2} = \frac{1}{\alpha} \frac{dT}{dt} \tag{5.12}$$

---

[3] In general, the Fourier modulus is a dimensionless grouping of parameters, $\text{Fo} = \alpha t/\Delta x^2$.

which is also applicable to Fig. 5.6, with the origin of the x coordinate system at the slab midplane, change to

$$\frac{\partial T(x, t)}{\partial x}\bigg|_{x=0, t=t} = 0 \tag{5.30a}$$

and $\quad \dfrac{\partial T(x, t)}{\partial x}\bigg|_{x=L/2, t=t} = \dfrac{-h}{k} T\left(\dfrac{L}{2}, t\right) \tag{5.30b}$

The initial condition remains the same and is given by Eq. (5.14).

$$T(x, 0) = T_i \tag{5.14}$$

The method of separation of variables may also be employed in this case where it is desired to determine the temperature at any point in the slab at any time, when the slab is immersed in a hot fluid at temperature T_0 at time $t = 0$ with heat transfer to the slab from the fluid governed by $q = h[T_0 - T(L/2, t)]$ for unit face area of slab. The result is

$$T(x, t) = 2(T_0 - T_i) \sum_{n=1}^{\infty} e^{-(\lambda_n L/2) \text{Fo}} \left[\frac{\sin (\lambda_n L/2) \cos (\lambda_n x)}{(\lambda_n L/2) + \sin (\lambda_n L/2) \cos (\lambda_n L/2)} \right]$$

$$n = 1, 2, 3, \ldots \tag{5.31}$$

where the λ_n's are roots of a transcendental equation.

Complicated expressions such as Eq. (5.31) and others for a cylinder and a sphere have been charted by many, including Heisler [6], Boelter et al. [7], Gröber [8], Schneider [9], and Gurney and Lurie [10]. These are charts of actual solutions and cannot be considered as solutions by a graphical method. The charts of Heisler [6], which are based on the Fourier modulus defined by Eq. (5.27) and the Biot modulus,

$$\text{Bi} \equiv \frac{hL}{2k} \tag{5.32}$$

are displayed in Figs. 5.7 through 5.12. Use of these charts will yield reasonably accurate temperature values and will eliminate the tedium of evaluating infinite series such as the one shown in Eq. (5.31).

5.5 ANALOG METHOD

The analog method is based on the similarity between the two forms of the diffusion equation,

$$\frac{\partial^2 T}{\partial x^2} = \frac{1}{\alpha} \frac{\partial T}{\partial t} \tag{5.12}$$

and $\quad \dfrac{\partial^2 V}{\partial x^2} = R'C' \dfrac{\partial V}{\partial t} \tag{5.33}$

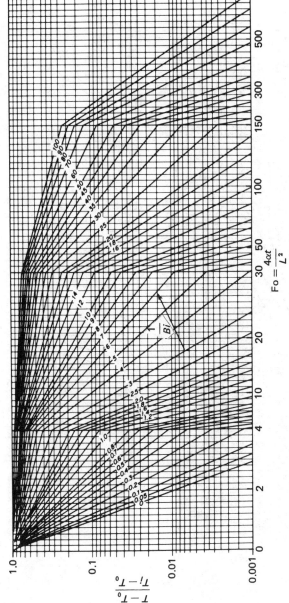

FIG. 5.7 Heisler's chart [6] for the midplane temperature of a slab of infinite extent. (From Gebhart [12], with permission.)

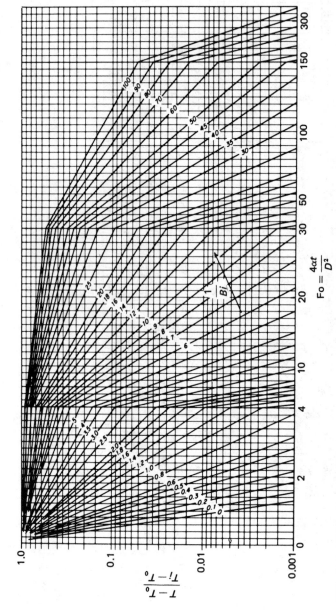

FIG. 5.8 Heisler's chart [6] for the axis temperature of an infinite cylinder. (From Gebhart [12], with permission.)

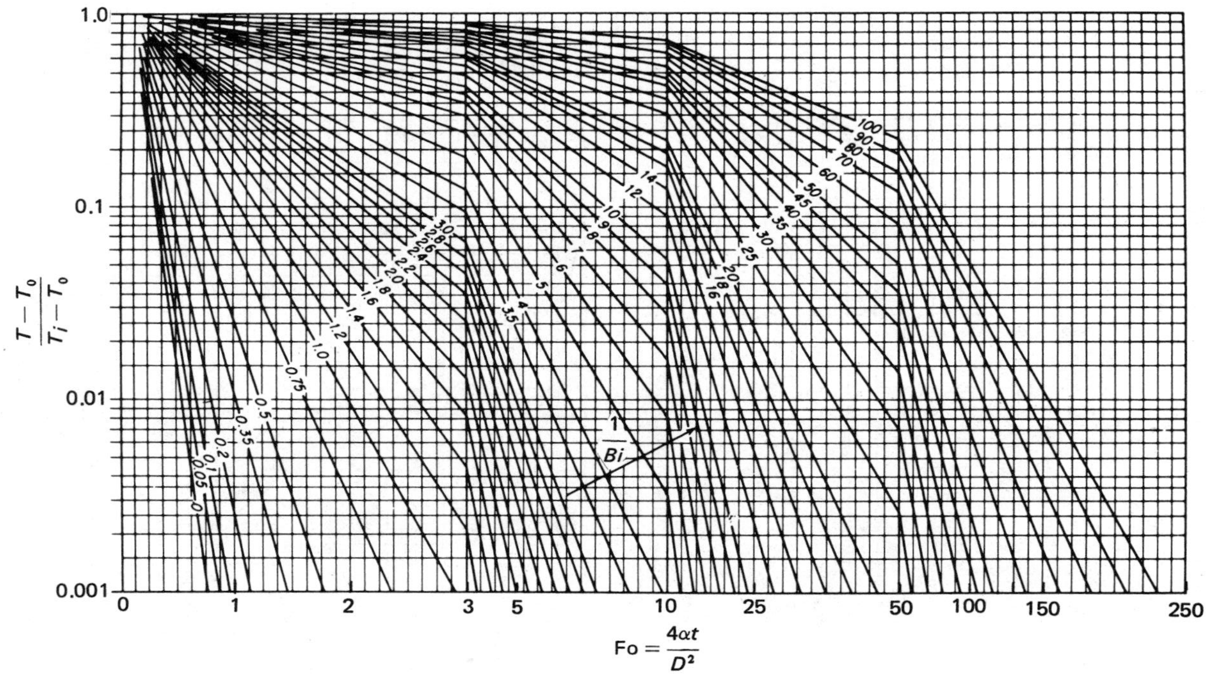

1.0
0.1
0.01
0.001
$\frac{T - T_0}{T_i - T_0}$
0
1
2
3
5
10
25
50
100
150
250
$Fo = \frac{4\alpha t}{D^2}$
$\frac{1}{Bi}$
0
0.05
0.1
0.2
0.35
0.5
0.75
1.0
1.2
1.4
1.6
1.8
2.0
2.2
2.4
2.6
2.8
3.0
3.5
4
5
6
7
8
9
10
12
14
16
18
20
25
30
35
40
45
50
60
70
80
90
100

FIG. 5.9 Heisler's chart [6] for the center temperature of a sphere. (From Gebhart [12], with permission.)

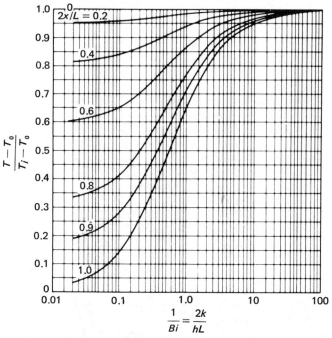

FIG. 5.10 Heisler's position correction chart [6] for the slab of infinite extent. (From Gebhart [12], with permission.)

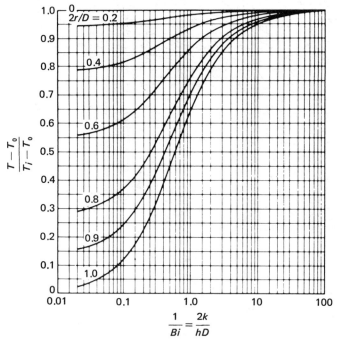

FIG. 5.11 Heisler's position correction chart [6] for the infinite cylinder. (From Gebhart [12], with permission.)

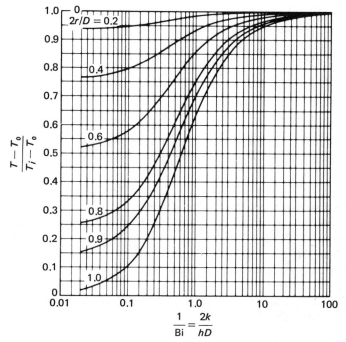

FIG. 5.12 Heisler's position correction chart [6] for the sphere. (From Gebhart [12], with permission.)

where V is voltage and R' and C' are electrical resistance and capacitance per unit length, respectively. From a comparison of Eqs. (5.12) and (5.33), it is observed that the analogous quantities are voltage-temperature, current-heat flow, and electrical time in seconds-thermal time in seconds (hours). These analogous quantities permit the establishment of the thermal resistance[4]

$$R = \frac{L}{kA} \tag{5.34}$$

and thermal capacitance

$$C = \gamma Vc \tag{5.35}$$

With these definitions, it is easy to see that the product of R and C, which are on a per-length basis, is indeed $1/\alpha$:

$$RC = \left(\frac{L/kA}{L}\right)\left(\frac{\gamma Vc}{L}\right) = \left(\frac{1}{kA}\right)\left(\frac{\gamma LAc}{L}\right) = \frac{\gamma c}{k} = \frac{1}{\alpha}$$

The transient analog method is formed by using a number of lumped *RC tee sections* with subsequent employment of the method of node analysis. Figure 5.13

---

[4] See Chap. 4.

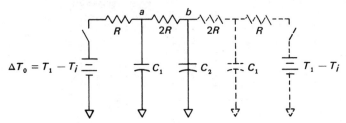

FIG. 5.13 Three-tee representation of the infinite slab. The slab is initially at $T = T_i$ throughout, and at time $t = 0$, the faces of the slab are suddenly elevated to $T = T_1$.

shows a three-tee representation of a slab that is initially at temperature T_i throughout and whose face temperatures are suddenly raised to T_1. The problem is analogous to the sudden addition of identical steps of voltage at opposite ends of a symmetrical RC network with subsequent charging of the capacitors. It can be solved as a problem in linear transient analysis or on an analog computer. The analytical solution of the analog circuit will be discussed first.

A pair of node equations can be written for nodes a and b in Fig. 5.13. These equations are based on Kirchkoff's current law and are

$$\frac{1}{R}(T_a - \Delta T_0) + C_1 \frac{dT_a}{dt} + \frac{1}{2R}(T_a - T_b) = 0$$

$$\frac{1}{2R}(T_b - T_a) + C_2 \frac{dT_b}{dt} = 0$$

where $\Delta T_0 = T_1 - T_i$. With a rearrangement, a more usable form, with

$$\frac{1}{R} + \frac{1}{2R} = \frac{3}{2R}$$

results:

$$\left(C_1 \frac{dT_a}{dt} + \frac{3}{2R} T_a\right) - \frac{1}{2R} T_b = \frac{\Delta T_0}{R} \tag{5.36a}$$

$$-\frac{1}{2R} T_a + \left(C_2 \frac{dT_b}{dt} + \frac{1}{2R} T_b\right) = 0 \tag{5.36b}$$

Equations (5.36) are a pair of first-order simultaneous linear differential equations and may be solved using the Laplace transformation. Define the Laplace transform as

$$\pounds[T(t)] \equiv T(s) \equiv \int_0^\infty T(t)e^{-st}\, dt$$

From this definition it is easy to see that the Laplace transform of the constant ΔT_0 is

$$\Delta T_0 = \int_0^\infty \Delta T_0 e^{-st}\, dt = \frac{\Delta T_0}{-s} e^{-st} \Big|_0^\infty = \frac{\Delta T_0}{s}$$

and it can be shown that

$$\pounds\left(\frac{dT}{dt}\right) = sT(s) - T(0)$$

where $T(0)$ represents the value of the temperature at time $t = 0$.

With these definitions in hand, Eqs. (5.36) can be transformed term by term for the case of $T_a(0) = T_b(0) = T_i = 0$,

$$\left(sC_1 + \frac{3}{2R}\right)T_a(s) - \left(\frac{1}{2R}\right)T_b(s) = \frac{\Delta T_0}{Rs} \tag{5.37a}$$

$$\left(-\frac{1}{2R}\right)T_a(s) + \left(C_2 s + \frac{1}{2R}\right)T_b(s) = 0 \tag{5.37b}$$

and these are a pair of algebraic equations that may be solved for $T_a(s)$ and $T_b(s)$. If attention is to be focused on the midplane temperature, then

$$T_b(s) = \frac{\begin{vmatrix} sC_1 + \dfrac{3}{2R} & \dfrac{\Delta T_0}{Rs} \\[3mm] -\dfrac{1}{2R} & 0 \end{vmatrix}}{\begin{vmatrix} sC_1 + \dfrac{3}{2R} & -\dfrac{1}{2R} \\[3mm] -\dfrac{1}{2R} & sC_2 + \dfrac{1}{2R} \end{vmatrix}}$$

or, with the omission of some algebra,

$$T_b(s) = \frac{\Delta T_0/2(RC_1)(RC_2)}{s(s^2 + \{[(3/2)/RC_1] + [(1/2)/RC_2]\}s + 1/2(RC_1)(RC_2))} \tag{5.38}$$

Equation (5.38) may be put in the form

$$T_b(s) = \frac{\Delta T_0/2(RC_1)(RC_2)}{s(s - \alpha_1)(s - \alpha_2)} \tag{5.39}$$

where α_1 and α_2 are the roots of the quadratic equation

$$s^2 + \left(\frac{3/2}{RC_1} + \frac{1/2}{RC_2}\right)s + \frac{1}{2(RC_1)(RC_2)} = 0 \tag{5.40}$$

Now a partial fraction expansion or consultation of a table of Laplace transforms shows that

$$T_b(t) = \frac{\Delta T_0}{2(RC_1)(RC_2)}\left[\frac{1}{\alpha_1\alpha_2} - \frac{1}{\alpha_1(\alpha_2-\alpha_1)}e^{\alpha_1 t} - \frac{1}{\alpha_2(\alpha_1-\alpha_2)}e^{\alpha_2 t}\right] + T_i$$

(5.41)

An analog computer may also be used to solve Eqs. (5.36). However, the analog computer has two limitations. There are a finite number of amplifiers in a computer installation, so the number of tees employed must be held below a certain maximum that depends on computer capacity. In addition, the analog computer must be scaled to work in the electrical domain. This scaling involves a delicate balance of time units and degrees of temperature to volts, and considerable effort must be expended to assure that amplifiers do not overload because of high voltage. In any event, Fig. 5.14 shows an analog computer patch diagram for the solution of Eqs. (5.36) after they have been rearranged to isolate dT_a/dt and dT_b/dt.

5.6 NUMERICAL METHOD

The numerical method also involves the writing of node equations. Figure 5.15 shows the friendly plane slab divided into eight subplanes; if the faces of the slab are to be

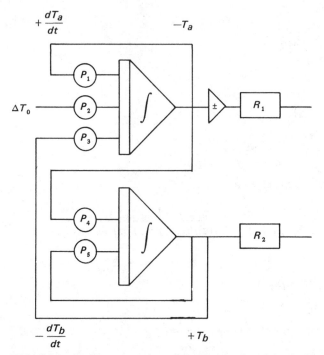

FIG. 5.14 Analog computer patch diagram for the solution of Eqs. (5.36). The P's indicate potentiometers and the R's indicate recording devices.

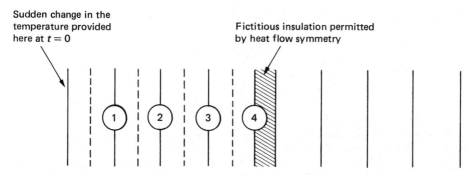

FIG. 5.15 Plane slab divided into eight sublayers for numerical solution.

subjected to a sudden increase in temperature from T_i to T_0, symmetry permits the use of only four of the subplanes.

Consider unit face area $(A = 1 \text{ m}^2)$, so that the resistance to the flow of heat is given by $R = L/kA = L/k$. If all subplanes are of equal thickness, $L = \Delta x$ and $R = \Delta x/k$ °C/W (h °F/Btu). Four node equations may be written based on an energy balance equating the heat entering minus the heat leaving with the heat stored. With T^+ designating the temperature after the time increment has elapsed, the four node equations are, with $K = 1/R = k/\Delta x$ W/°C (Btu/h °F),

$$K(T_0 - T_1) - K(T_1 - T_2) = c\gamma V \frac{T_1^+ - T_1}{\Delta t} \tag{5.42a}$$

$$K(T_1 - T_2) - K(T_2 - T_3) = c\gamma V \frac{T_2^+ - T_2}{\Delta t} \tag{5.42b}$$

$$K(T_2 - T_3) - K(T_3 - T_4) = c\gamma V \frac{T_3^+ - T_3}{\Delta t} \tag{5.42c}$$

$$K(T_3 - T_4) = c\gamma \frac{V}{2} \frac{T_4^+ - T_4}{\Delta t} \tag{5.42d}$$

Equations (5.42) may be solved for the temperatures at the end of the time increment. With $V = A \Delta x = \Delta x$ m³ (ft³) and $A = 1$ m² (1 ft²),

$$\frac{K \Delta t}{c\gamma V} = \frac{k \Delta t}{c\gamma \Delta x^2} = \frac{\alpha \Delta t}{\Delta x^2} = \text{Fo}$$

The result is

$$T_1^+ = \text{Fo} \left[\left(\frac{1}{\text{Fo}} - 2 \right) T_1 + T_2 + T_0 \right] \tag{5.43a}$$

$$T_2^+ = \text{Fo} \left[\left(\frac{1}{\text{Fo}} - 2 \right) T_2 + T_1 + T_3 \right] \tag{5.43b}$$

$$T_3^+ = \mathrm{Fo}\left[\left(\frac{1}{\mathrm{Fo}} - 2\right)T_3 + T_2 + T_4\right] \tag{5.43c}$$

$$T_4^+ = 2\,\mathrm{Fo}\left[\left(\frac{1}{2\,\mathrm{Fo}} - 1\right)T_4 + T_3\right] \tag{5.43d}$$

Equations (5.43) are the equations that are used to solve the problem. Their use, however, is constrained by the fact that, physically, there cannot be a flow of heat of its own accord from regions of lower temperature to regions of higher temperature. The coefficient of the first term to the right of the equal sign in each node equation must be examined. This yields the inequalities

$$\frac{1}{\mathrm{Fo}} - 2 \geqslant 0$$

and $$\frac{1}{2\,\mathrm{Fo}} - 1 \geqslant 0$$

and the constraint on Fo comes from either (or both)

$$\mathrm{Fo} \leqslant \tfrac{1}{2}$$

Because the space increment Δx is usually selected on the basis of an integer number of subplanes, it is Δt, the time increment, that is only partially at the analyst's discretion, $\Delta t \leqslant \Delta x^2/2\alpha$.

Equations (5.43) are amenable to hand calculator solution or may be solved on any computer. Unfortunately, however, one rarely encounters a shape as simple as a plane slab with no end or edge effects.

5.7 GRAPHICAL METHOD

If, in Eqs. (5.43), Fo is set equal to $\frac{1}{2}$, then

$$T_1^+ = \frac{T_0 + T_2}{2}$$

which shows that the temperature at the first node, after each time increment, is equal to the arithmetic average of the temperatures at the start of the time increment. This is the basis for the *Schmidt plot* [11], which is a graphical method that can be employed for simple shapes to yield an approximate temperature-time profile.

Alternate layer boundaries may be connected by straight lines to represent a linear temperature gradient through the layer. The series of straight lines approximates the temperature gradient in the entire slab, and the temperature gradient comes closer to the actual temperature gradient as the layer space increments, Δx, are made smaller. The temperature at the center of each layer at the time the temperature gradient exists can be read directly from the graph.

5.8 COMPARISON OF TRANSIENT CONDUCTION COMPUTER METHODS

The classical (analytical), analog, numerical, and graphical methods of analysis discussed in this chapter will now be compared in a simple example.

EXAMPLE

A slab with plane faces is 1.2 m thick and has the following properties:

Thermal conductivity $k = 250$ W/m °C.
Specific heat $c = 1000$ J/kg °C
Specific weight $\gamma = 2500$ kg/m^3

The slab is initially at a uniform temperature of 100°C, and at time zero the faces of the slab are instantaneously raised to 500°C. It is desired to obtain the midplane temperature after 900 s ($\frac{1}{4}$ h) by each of the four methods.

SOLUTION

(a) Classical (analytical) method:

$$T_0 = 500°C$$

$$T_i = 100°C$$

$$\alpha = \frac{k}{c\gamma} = \frac{250}{(1000)(2500)} = 10^{-4} \text{ m}^2/\text{h}$$

$$L = 1.20 \text{ m}$$

$$\text{Fo} = \frac{4\alpha t}{L^2} = \frac{4(10^{-4})(900)}{(1.2)^2} = 0.25$$

Equation (5.29) will be used, and this equation requires that the plate series, $P(\text{Fo} = 0.25)$ be evaluated. Using the tabulation in Schneider [2], $P(\text{Fo} = 0.25) = 0.68546$. With this in Eq. (5.29),

$$T(0.6 \text{ m}, 900 \text{ s}) = T_0 + (T_i - T_0) P(\text{Fo} = 0.25)$$

$$= 500 + (100 - 500)(0.68546) = 225.82°C$$

(b) Analog method: In Fig. 5.13, which is a three-tee representation of the slab, each subplane is 40 cm thick. Hence, with $A = 1$ m^2,

$$R = \frac{\Delta x}{k} = \frac{0.2}{250} = 8 \times 10^{-4} \text{ °C/W}$$

$$C_1 = \gamma \Delta x c = 2500(0.4)(1000) = 10^6 \text{ J/°C}$$

and C_2 is just half of C_1:

$$C_2 = 0.5 \times 10^6 \text{ J/°C}$$

Then

$$RC_1 = 800 \text{ s}$$
$$RC_2 = 400 \text{ s}$$

and Eq. (5.40) reads

$$s^2 + (3.125 \times 10^{-3})s + (1.5625 \times 10^{-6}) = 0$$

which has roots

$$\alpha_1 = -6.25 \times 10^{-4}$$

and $\alpha_2 = -2.50 \times 10^{-3}$

These values of α_1 and α_2 may be placed into Eq. (5.41) with $\Delta T_0 = 400°C$ and $T_i = 100°C$.

$$T_b(t) = 400(1 - \tfrac{4}{3}e^{-6.25 \times 10^{-4}t} + \tfrac{1}{3}e^{-2.50 \times 10^{-3}t}) + 100$$

where it is seen that $T_b(0) = 100°C$, as it should. Then, at $t = 900$ s,

$$T_b(t = 900 \text{ s}) = 210.17°C$$

(c) Numerical method: With a four subplane model so that node 4 is at the geometric center of the slab, $\Delta x = 0.15$ m. The time increment must be constrained by Fo $< \tfrac{1}{2}$ to

$$\Delta t = \frac{\Delta x^2}{2\alpha} = \frac{(0.15)^2}{2(0.0001)} = 112.5 \text{ s}$$

Suppose that Δt is set at 56.25 s. Then after $16(56.25) = 900$ s or 16 iterations, the result can be obtained. A summary of the iterative procedure using Eqs. (5.43) with

$$\text{Fo} = \frac{\alpha \, \Delta t}{\Delta x^2} = \frac{(0.0001)(56.25)}{(0.15)^2} = 0.25$$

is shown in Table 5.1. The result for node 4 after 16 iterations is $T_4 = 230.20°C$.

(d) Graphical method: With Fo $= \tfrac{1}{2}$, the temperature at the first node, after each time increment, is equal to the arithmetic mean of the temperatures at the start of the time increment. With Fo $= \tfrac{1}{2}$, $\Delta t = 112.5$ s and the graphical plot must proceed for eight time increments to reach 900 s. Such a plot is shown in Fig. 5.16, where it is seen that after eight iterations, the temperature at the midplane is $241°C$.

(e) Summary: A summary of the results of the four methods of analysis is displayed in Table 5.2. The agreement appears quite good except for the three-tee

TABLE 5.1 Summary of Iterations for Plane Slab Example

Fo $= 0.25$, $T_0 = 500°C$
$T_1^+ = 0.25(2T_1 + T_2 + 500)$
$T_2^+ = 0.25(2T_2 + T_1 + T_3)$
$T_3^+ = 0.25(2T_3 + T_2 + T_4)$
$T_4^+ = 0.5(T_4 + T_3)$

Iteration no.	T_1	T_2	T_3	T_4
0	100.00	100.00	100.00	100.00
1	200.00	100.00	100.00	100.00
2	250.00	125.00	100.00	100.00
3	281.25	150.00	106.25	100.00
4	303.13	171.88	115.63	103.13
5	319.53	190.63	126.56	109.38
6	332.42	206.84	138.28	117.97
7	342.92	221.09	150.34	128.13
8	351.73	233.86	162.48	139.23
9	359.33	245.48	174.51	150.85
10	366.04	256.20	186.34	162.68
11	372.07	266.20	197.89	174.51
12	377.58	275.59	209.12	186.20
13	382.69	284.47	220.01	197.66
14	387.46	292.91	230.54	208.84
15	391.96	300.95	240.71	219.69
16	396.22	308.64	250.51	230.20

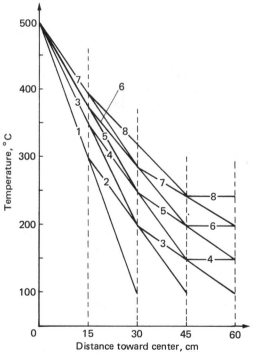

FIG. 5.16 Schmidt plot for plane slab example.

TABLE 5.2 Summary of Results—Four Solution Methods: Plane Slab Example, 900 s Duration

Solution method	Midplane temperature (°C)
Analytical	225.8
Three-tee analog	210.2
Numerical	230.2
Graphical	241.0

analog, which has the coarsest spacial increment. This, of course, could be improved by using a five- or seven-tee analog.

5.9 NOMENCLATURE

Roman Letter Symbols

A	area normal to heat flow path (cross-sectional area), m^2; a constant; a constant of integration
B	a constant of integration
Bi	Biot modulus, dimensionless
c	specific heat, cal/kg °C
C	thermal capacitance, J/°C; a constant of integration
C'	electrical capacitance per unit length, farads/m
Fo	Fourier modulus, dimensionless
h	heat transfer coefficient, W/m^2 °C
k	thermal conductivity, W/m °C
K	thermal conductance, W/°C
L	length of heat flow path or width of slab, m
n	number of physical entities or a number of values
$P(\text{Fo})$	indicates plate series as a function of Fourier modulus
q	heat flow, W
R	thermal resistance, °C/W
R'	electrical resistance per unit length, Ω/m
t	time, h
T	temperature, °C
V	voltage, V; volume, m^3
W	weight, kg
x	length coordinate, m
y	length coordinate, m
z	length coordinate, m

Greek Letter Symbols

α	thermal diffusivity, m^2/s; roots of a quadratic equation
γ	specific weight, kg/m^3
Δ	indicates change in variable
λ	a constant (eigenvalue)

Subscripts

a	indicates position in a composite configuration or a node in a nodal analysis
b	indicates position in a composite configuration or a node in a nodal analysis
c	indicates complementary function
G	indicates heat generated
i	indicates initial value
p	indicates particular integral
0	indicates initial value

5.10 REFERENCES

1 Carslaw, H. S., and Jaeger, J. C., *Conduction of Heat in Solids*, Oxford University Press, London, 1947.

2 Schneider, P. J., *Conduction Heat Transfer*, Addison-Wesley, Reading, Mass., 1955.

3 Arpaci, V. S., *Conduction Heat Transfer*, Addison-Wesley, Reading, Mass., 1966.

4 Myers, G. E., *Analog Methods in Conduction Heat Transfer*, McGraw-Hill, New York, 1971.

5 Olson, F. C. W., and Schultz, O. T., Temperatures in Solids during Heating or Cooling, *Ind. Eng. Chem.*, vol. 34, pp. 874–877, 1942.

6 Heisler, M. P., Temperature Charts for Induction and Constant Temperature Heating, *Trans. ASME*, vol. 69, p. 227, 1947.

7 Boelter, L. M. K., Cherry, H., Johnson, H. A., and Martinelli, R. C., *Heat Transfer Notes*, McGraw-Hill, New York, 1965.

8 Gröber, H., The Heating and Cooling of Simple Geometric Bodies, *Z. ver Deut. Ing.*, vol. 69, pp. 705–711, 1925.

9 Schneider, P. J., *Temperature Response Charts*, Wiley, New York, 1965.

10 Gurney, H. P., and Lurie, J., Charts for Estimating Temperature Distributions in Heating and Cooling Solid Shapes, *Ind. Eng. Chem.*, vol. 15, p. 1170, 1923.

11 Schmidt, E., On the Application of the Calculus of Finite Differences to Technical Heating and Cooling Problems (in German), in *Festschrift zum Siebzigsten Geburtstag August Foeppls*, pp. 179–184, Julius Springer, Berlin, 1924.

12 Gebhart, B., *Heat Transfer*, 2d ed., McGraw-Hill, New York, 1971.

6

■ convection

6.1 INTRODUCTION

Heat transfer at the interface between a solid and a fluid at a different temperature is significant in many common applications, ranging from preparing pasta or swimming to cooling vacuum tubes and hot water heating. When the fluid is stationary relative to the solid, thermal transport is primarily by conduction, and the relations, as well as methods, of Chaps. 4 and 5 can be applied. The presence of fluid motion, however, distorts the temperature field and, under these circumstances, heat exchange results from the conduction of heat into the fluid layers adjacent to the surface coupled with the subsequent removal of this fluid by the prevailing fluid circulation pattern. The transfer of heat in this "convective" fashion (see Fig. 6.1) can yield significantly higher heat transfer rates than pure conduction, but analysis of convective heat transfer requires detailed knowledge of fluid motion in the presence of and adjacent to solid surfaces.

Fluid motion may result from forced circulation by pump, blower, or stirrer, as well as from density differences associated with temperature or solute concentration variation within the fluid. The details of the velocity field encountered in each category are substantially different, and a distinction is thus made between forced-convection heat transfer and natural or free-convection heat transfer. Furthermore, whereas for relatively low velocities and short flow paths, the fluid generally flows in layers or laminae along the surface, at higher velocities and longer flow paths, a churning motion may develop. A second distinction is therefore generally made between "laminar" and "turbulent" convection.

The fluid mechanics of convection are examined in greater detail in the next section and then form the basis for the theoretical development of convective heat transfer relations. Presentation of empirical correlations for free and forced, laminar and turbulent convective heat transfer rates completes this chapter.

6.2 FLUID MECHANICS OF CONVECTION

Convective heat transfer is, by definition, concerned with heat exchange between a solid surface and a circulating viscous fluid. Determination of the temperature field and heat flow rate encountered in this thermal mode must therefore involve definition of fluid behavior adjacent to the surface. For at least some configurations of interest, this behavior can be analyzed in detail by rigorous solution of the three-dimensional fluid momentum (Navier-Stokes) and continuity relations, as is done in many leading textbooks [1–3]. Such an approach is, however, beyond the scope of the present discussion, which focuses instead on conceptual development and relies on basic physical reasoning and integral rather than differential fluid mechanic formulations.

CONDUCTION   +   FLUID   MOTION

FIG. 6.1   Conceptual representation of convective heat transfer.

6.2.1   Viscous Shear Stress

A flat plate moving through an otherwise quiescent viscous fluid accelerates the fluid molecules adjacent to its surface while leaving more distant molecules unaffected by its passage. Similarly, as shown in Fig. 6.2, the velocity of a fluid flowing past a stationary flat plate approaches zero near the surface, whereas the velocity of fluid far removed from the surface asymptotically approaches the velocity of an unconstrained stream. In both cases frictionlike forces develop at the fluid/solid interface, inducing fluid circulation or retarding the flow, according to the sign of the relative motion between the plate and the fluid.

The adhesion of fluid molecules and the resulting "no-slip" condition at an

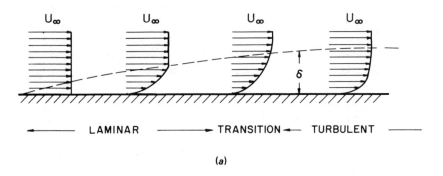

(a)

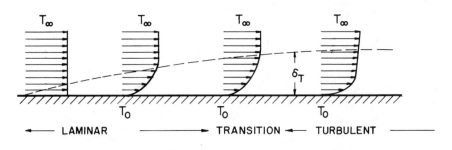

(b)

FIG. 6.2   Boundary-layer flow along a flat plate (*a*) velocity boundary layer and (*b*) temperature boundary layer.

immersed surface is a basic postulate of fluid flow for all but rarefied gases. Although this postulate, expressed in terms of a monomolecular layer, may be justified by the action of intermolecular and surface tension forces, an additional mechanism is needed to explain the smooth transition from the wall to the "free stream" velocity and the generation, as well as transmission, of surface forces into the fluid as a consequence of the fluid/solid interaction.

In liquids, at temperatures considerably above their freezing points, and in gases, this additional mechanism—the resistance to shearing offered by a flowing or contained fluid—is generally related to momentum exchange between fluid molecules. Thermally excited, random molecular motion in the fluid affects the movement of a finite number of molecules from the wall layer to an adjacent layer. In this layer, collisions between fast, oncoming molecules and the nearly stationary wall-layer molecules result in a momentum exchange and a reduction in the average velocity of the oncoming molecules. Repetition of this process in fluid layers further removed from the solid surface leads to the characteristic velocity profile shown in Fig. 6.2.

Following Newton's law of motion, the change in momentum experienced by the fluid can be related to a force acting in a direction opposite to that of fluid flow, and it is then this shear force that embodies the fluid's viscous resistance to shearing deformation. In keeping with the above model, the shear stress, or shear force per unit area, must be governed by the velocity gradient in the fluid and a fluid parameter related to the diffusion rate of molecules in the fluid. This fluid parameter is termed the *dynamic viscosity* μ, and is extensively tabulated for most conventional fluids. When the dynamic viscosity and velocity gradient are known, the *shear stress* τ is expressible as

$$\tau = -\mu \frac{du}{dy} \tag{6.1}$$

where u is the velocity[1] parallel to the surface and y is the coordinate normal to the surface. Equation (6.1) is known as *Newton's equation of viscosity*, and fluids that obey this relation are known as *Newtonian fluids*. Even in Newtonian fluids, however, Eq. (6.1) cannot be expected to apply in turbulent flow, where the presence of swirling, eddy motion overshadows molecular-scale momentum exchange.

Because of the frictionlike character of viscous shear stress, it is often convenient to relate the forces acting at a solid/fluid interface to a friction factor or coefficient. In pipe flow this friction factor can be simply related to the shear stress at the wall and the velocity head (or dynamic pressure) of the flow, according to [1], by

$$f \equiv \frac{\tau_0}{\rho U^2 / 2} \tag{6.2}$$

An identical relation is used to define the skin friction coefficient on a flat plate, C_f, for laminar and turbulent flow, with U_∞ replacing U.

---

[1] A lowercase u will be used for the variable, velocity. Specific values of velocity, such as the free stream velocity, U_∞, will be designated by an uppercase U.

6.2.2 Boundary Layers

Section 6.2.1 suggests that viscous effects become progressively less significant as the velocity gradients decrease. In view of the characteristic velocity profile shown in Fig. 6.2, it might thus be expected that viscous effects dominate fluid behavior near solid surfaces, whereas inviscidlike flow prevails in regions far removed from these surfaces. This perception led Prandtl in 1904 [4] to propose that the flow field, in fluid/solid interactions, be divided into two regions: a relatively viscous thin layer adjacent to the solid boundary, to be called the boundary layer, and an inviscid fluid core.

Prandtl's division implies that the boundary-layer thickness δ be determined on the basis of a stated value of the velocity gradient. For convenience, however, δ is generally taken to equal the distance from the surface at which the fluid velocity (parallel to the wall) is within 1% of the unconstrained or free stream velocity. Other definitions of δ are also possible and are discussed in [1–3].

As shown in Fig. 6.2, the temperature profile in a fluid flowing past a cold surface is generally similar to the fluid velocity profile discussed earlier. This similarity also exists in other configurations, and thus suggests that a thermal boundary layer, analogous to the velocity boundary layer, can be defined. External to the thermal boundary layer, temperature gradients and thermal conduction effects are considered negligible, whereas within the boundary layer thermal conduction is of significance. The thermal boundary-layer thickness δ_T is generally defined as the distance from the surface at which the fluid temperature is within 1% of the free stream temperature. It must be noted that although δ is generally not equal to δ_T, this is often a satisfactory approximation for convective exchange between solids and gases.

The analysis of hydrodynamic and thermal fluid/solid interactions is greatly simplified by use of Prandtl's boundary-layer approach, which limits the effects of viscosity and conductivity to relatively thin layers adjacent to the solid boundaries. It is the velocity and temperature fields within these boundary layers that are of primary importance in convective heat transfer, and subsequent attention focuses almost exclusively on these boundary layers.

6.2.3 Flow Transition

The flow within the boundary layer near the leading edge of a flat surface is generally smooth and well ordered. The fluid appears to move past the surface along clearly defined streamlines, which can be easily visualized by the addition of trace amounts of dye, and is said to be in laminar flow. As the flow path lengthens, however, minute disturbances in the flow field begin to grow, and a churning, eddying motion becomes superimposed on the laminar flow field. Somewhat further downstream, the laminar pattern is completely destroyed and replaced by large-eddy turbulent flow.

The transition to turbulent flow is dependent on fluid properties, the flow distance, and the level of turbulence in the free stream. The downstream distance at which transition occurs can be expressed in terms of the *Reynolds number* Re, a non-dimensional parameter that unifies the effects of flow, distance, and fluid properties:

$$\mathrm{Re} \equiv \frac{\rho U_\infty x}{\mu} \tag{6.3}$$

The transition value of the Reynolds number varies from approximately 2×10^5 for flows with large disturbances to 10^6 for extremely smooth conditions, but is typically taken as 3×10^5 for both commercial and laboratory equipment.

Boundary-layer flow within pipes and channels displays a phenomenologically similar transition from laminar to turbulent flow. Although transition here too can be related to a Reynolds number, the pipe Re is defined in terms of pipe diameter rather than the flow distance, that is,

$$\text{Re} \equiv \frac{\rho U d}{\mu} \qquad (6.4)$$

and 2300 is taken as a typical transition value.

6.2.4 Fully Developed Flow

In the flow of a fluid parallel to a solid surface, the boundary-layer thickness increases smoothly and continuously in the direction of flow. This trend is shown qualitatively in Fig. 6.2, and a rigorous analysis would show δ to vary with the flow distance x to the $\frac{1}{2}$ power in laminar flow and $\frac{4}{5}$ in turbulent flow [1].

In channel or pipe flow, boundary-layer growth must eventually result in the merging of boundary layers from opposite channel walls and the establishment of a velocity profile that is no longer dependent on the flow path x. This x-invariant velocity profile, as well as the analogous temperature profile, represents the *fully developed* condition, and flow at smaller values of x is referred to as *developing flow*. Fully developed flow may be laminar or turbulent in nature, depending on the Reynolds number of the pipe or channel, and can be predicted to occur when δ and/or δ_T is approximately equal to one-half of the pipe diameter.

Because of the relative simplicity of the governing equations for thermally and hydrodynamically developing flow, this regime is at the focus of most analyses and many semiempirical correlations. Caution must thus be used in the application of these expressions to situations involving significant departures from the fully developed condition.

6.2.5 Noncircular Cross Section

Flow and boundary-layer development in ducts and channels, of various aspect ratio and shape, is not necessarily similar to that occurring in circular pipes. It has nevertheless been found possible to relate noncircular geometry channels to circular pipes by use of an equivalent diameter D_e, defined by

$$D_e \equiv \frac{4A}{p} = \frac{4(\text{cross-sectional area})}{\text{wetted perimeter}} \qquad (6.5)$$

For a circular pipe, D_e reduces to the actual internal diameter.

In turbulent channel flow, an adequate first approximation for the friction factor and the heat transfer coefficient can be obtained by equating the equivalent channel diameter with the diameter of the circular pipe in existing formulations and correlations. A similar substitution yields adequate results in laminar flow through ducts of triangular and trapezoidal cross section [5], but can result in errors of as much as 35% in rectangular channels. The correct relationships for this latter configuration are plotted in [1] from data appearing in [6].

6.3 CONVECTIVE HEAT TRANSFER COEFFICIENT – THEORETICAL BACKGROUND

As discussed in Sec. 6.1, at the interface between a solid and a fluid, heat is transferred by conduction and must obey *Fourier's law*,

$$q_0 = -kA \left(\frac{\partial T}{\partial y} \right)_0 \tag{6.6}$$

Unfortunately, however, it is generally difficult to predict or measure accurately the fluid temperature gradient at the wall and, as a consequence, the literature abounds with empirical relations for convective thermal transport. Following a suggestion first made by Newton in 1701 [7], that the surface heat transfer rate be related to the product of surface area and the temperature difference between the surface and the fluid, much effort has been devoted to determining and correlating the proportionality factor h, in Newton's equation:

$$q_0 = hA(T_0 - T_f) \tag{6.7}$$

This proportionality factor has become known as the surface heat transfer coefficient. The fluid temperature T_f in Eq. (6.7) is generally taken as the free stream—that is, the undisturbed fluid—temperature in essentially infinite media and as the mixed mean temperature for confined flows.

Comparison of Newton's equation with Fourier's law reveals that the transfer coefficient in convective heat transfer can be related to the thermal conductivity and wall temperature gradient of the fluid and the surface fluid temperature difference:

$$h \equiv \frac{q_0/A}{(T_0 - T_f)} = \frac{-k(\partial T/\partial y)_0}{(T_0 - T_f)} \tag{6.8}$$

Consequently, any correlation of heat transfer coefficients must reflect the direct dependence of h on the fluid's thermal conductivity and on the ratio of the wall temperature gradient to the temperature difference. Analytic evaluation of this latter quantity can thus serve to develop precise formulations for the heat transfer coefficient in simple configurations and guide the form of the empirical correlations for more complex configurations.

6.3.1 Laminar Flow on Flat Plate

THERMAL BOUNDARY LAYER

The situation depicted in Fig. 6.2 represents one of the simplest and most commonly idealized convective heat transfer configurations. For heat exchange between a cold surface and a hot fluid, the thermal boundary-layer thickness can be taken to equal the distance from the wall at which the fluid temperature nearly equals the free stream temperature; that is,

$$\delta_T \equiv y(T = 0.99T_\infty) \tag{6.9}$$

With the boundary-layer thickness defined, it is possible to prescribe boundary conditions on the temperature profile for $0 < y < \delta_T$. At the wall the fluid temperature must equal the wall temperature, whereas at $y = \delta_T$ the fluid temperature must, by definition, equal the free stream temperature and be essentially invariant with y. Although several different temperature profiles can be found to meet these boundary conditions, the relatively simple function shown in Eq. (6.10) is seen to have the requisite characteristics and will be accepted as a plausible, though clearly nonunique, boundary-layer temperature profile.

$$\frac{T - T_0}{T_\infty - T_0} = \frac{3}{2} \frac{y}{\delta_T} - \frac{1}{2} \left(\frac{y}{\delta_T} \right)^3 \tag{6.10}$$

$$y = 0 \quad \frac{T - T_0}{T_\infty - T_0} = 0, T = T_0$$

$$y = \delta_T \quad \frac{T - T_0}{T_\infty - T_0} = 1, T = T_\infty$$

$$\frac{d}{dy} \left(\frac{T - T_0}{T_\infty - T_0} \right) = 0, \frac{dT}{dy} = 0$$

Differentiation of Eq. (6.10) and evaluation of the fluid temperature gradient at the surface yields

$$\left. \frac{\partial T}{\partial y} \right|_0 = \frac{3}{2} \frac{T_\infty - T_0}{\delta_T} \tag{6.11}$$

Thus, if δ_T is known, $(\partial T/\partial y)_0$ can be determined and an analytic function found for the heat transfer coefficient h:

$$h = \frac{-k(\partial T/\partial y)_0}{T_0 - T_\infty} = \frac{3k}{2\delta_T} \tag{6.12}$$

Various techniques, each at a different level of sophistication, can be employed to find the functional form of δ_T. Among these is an integral procedure based on a heat balance on a specified control volume chosen to conform to the boundary-layer geometry. This technique addresses the essential physical features of the thermal boundary layer but avoids mathematical complexities and is therefore adopted in the present discussion.

For the control volume shown in Fig. 6.3, the requisite heat balance can be stated as

$$dq = I_{x+\Delta x} - I_x - I_\delta \tag{6.13}$$

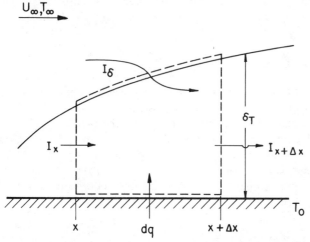

FIG. 6.3   Control volume and key parameters for thermal boundary-layer analysis. The I's are enthalpy fluxes per unit depth.

where the symbols are as shown in the figure. The wall heat flux is, as before, obtainable from Fourier's law and with the assumed temperature profile is found to equal

$$dq = -3k\ \frac{\Delta x(T_\infty - T_0)}{2\delta_T} \tag{6.14}$$

The enthalpy flux at any point x in the boundary layer of a constant-property fluid can be expressed as

$$I_x = \rho c_p \int_0^{\delta_T} uT\,dy \tag{6.15}$$

The enthalpy flux at a nearby point, $x + \Delta x$, can then be obtained by a two-term Taylor series expansion, which yields

$$I_{x+\Delta x} = \rho c_p \int_0^{\delta_T} uT\,dy + \frac{d}{dx}\left(\rho c_p \int_0^{\delta_T} uT\,dy\right)\Delta x \tag{6.16}$$

Finally, the enthalpy flux into the boundary layer from the free stream fluid can be set equal to the product of the free stream enthalpy and the incoming mass flow. This latter quantity must, by flow continuity considerations, equal the difference between the boundary-layer mass flow at points $x + \Delta x$ and x. Consequently,

$$I_\delta = \rho c_p T_\infty \frac{d}{dx}\left(\int_0^{\delta_T} \rho u\,dy\right)\Delta x \tag{6.17}$$

Substituting Eqs. (6.14) through (6.17) into Eq. (6.13), the control-volume heat balance is seen to yield

$$\frac{3\alpha(T_\infty - T_0)}{2\delta_T} = \frac{d}{dx}\left[\int_0^{\delta_T} u(T_\infty - T)\,dy\right]$$ (6.18)

Equation (6.18) is an implicit relation for δ_T which, however, requires specification of the boundary-layer velocity profile—$u(y)$—in order to effect a solution. This aspect of boundary-layer flow is examined in the next section.

VELOCITY BOUNDARY LAYER

As with the previously discussed thermal boundary layer, the definition of the velocity boundary layer imposes inherent boundary conditions on the velocity profile. At the wall the velocity must equal zero, whereas at $y = \delta$ the velocity is essentially equal to the free stream value and the velocity gradient is vanishingly small. These boundary conditions can be met by a variety of functions, but the profile represented by Eq. (6.19) possesses the requisite characteristics and is thus a plausible (though again nonunique) laminar boundary-layer profile.

$$\frac{u}{U_\infty} = \frac{3}{2}\left(\frac{y}{\delta}\right) - \frac{1}{2}\left(\frac{y}{\delta}\right)^3$$ (6.19)

$$y = 0 \qquad u = 0$$

$$y = \delta \qquad u = U_\infty$$

$$y = \delta \qquad \frac{du}{dy} = 0$$

Although the use of Eq. (6.19) is limited by the dependence of u on an as yet unknown boundary-layer thickness, δ can be found by requiring that momentum be conserved in the control volume shown in Fig. 6.4. Using the symbols indicated in the figure, and assuming pressure forces or the streamwise pressure gradient to be negligible, the balance between the shear force at the wall and the net momentum flux through the boundary-layer control volume can be expressed as

$$\tau_0\,\Delta x = M_{x+\Delta x} - M_x - M_\delta$$ (6.20)

The wall shear stress τ_0 can be found by differentiating the assumed velocity profile (Eq. (6.19) at the wall and inserting the derivative into the shear stress expression appropriate to laminar flow, Eq. (6.1), to obtain

$$\tau_0 = \frac{-3\mu}{2}\frac{U_\infty}{\delta}$$ (6.21)

The momentum flux entering the control volume is expressible as

$$M_x = \int_0^\delta \rho u^2\,dy$$ (6.22)

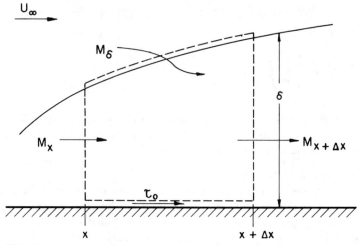

FIG. 6.4   Control volume and key parameters for velocity boundary-layer analysis. The M's are momentum fluxes per unit depth.

Applying a two-term Taylor series expansion, the momentum flux at $x + \Delta x$ is given by

$$M_{x+\Delta x} = \int_0^\delta \rho u^2 \, dy + \frac{d}{dx}\left(\int_0^\delta \rho u^2 \, dy \right) \Delta x \tag{6.23}$$

Finally, the momentum flux crossing into the boundary-layer control volume can be obtained by multiplying the incoming mass flow (i.e., the difference between the boundary-layer mass flow at $x + \Delta x$ and x) by the free stream velocity to obtain

$$M_\delta = U_\infty \frac{d}{dx}\left(\int_0^\delta \rho u \, dy \right) \Delta x \tag{6.24}$$

Combining Eqs. (6.20) through (6.24), momentum conservation in the control volume is seen to require that

$$\frac{3}{2} \mu \frac{U_\infty}{\delta} = \rho \frac{d}{dx}\left[\int_0^\delta (U_\infty - u)u \, dy \right] \tag{6.25}$$

Solution of Eq. (6.25) with the assumed velocity profile of Eq. (6.19) yields the sought-after expression for the thickness of the laminar velocity boundary layer as a function of the flow distance,

$$\delta = \frac{4.64x}{\sqrt{\rho U_\infty x / \mu}} \tag{6.26a}$$

or, in more common form,

$$\frac{\delta}{x} = \frac{4.64}{\sqrt{\mathrm{Re}}} \qquad (6.26b)$$

This expression for δ can now be used in Eq. (6.19) to provide an analytic expression for the boundary-layer velocity profile as a function of the free stream velocity and the flow distance.

It is interesting to note that the exact analytic solution for the boundary-layer thickness is $\delta/x = 5.0/\sqrt{\mathrm{Re}}$ and thus differs by less than 10% from the approximate integral solution.

THEORETICAL EXPRESSION FOR h

The solution of the implicit δ_T relation derived in the section entitled "Thermal Boundary Layer" (pp. 120-123) requires specification not only of the temperature profile but also of the velocity profile in the boundary layer and thus motivates the determination of u and δ. These can now be inserted in Eq. (6.18) to provide an analytic expression for the thermal boundary-layer thickness and, via Eq. (6.12), the heat transfer coefficient.

Returning to Eq. (6.18) and substituting the assumed velocity and temperature profile yields

$$\frac{3\alpha(T_\infty - T_0)}{2\delta_T} = U_\infty(T_\infty - T_0)\frac{d}{dx}\int_0^{\delta_T}\left[\frac{3}{2}\left(\frac{y}{\delta}\right) - \frac{1}{2}\left(\frac{y}{\delta}\right)^3\right]$$

$$\times \left[1 - \frac{3}{2}\left(\frac{y}{\delta_T}\right) + \frac{1}{2}\left(\frac{y}{\delta_T}\right)^3\right] dy \qquad (6.27)$$

The presence of the two boundary-layer thickness, δ and δ_T, somewhat complicates the integration on the right side of Eq. (6.27) but, with $\zeta \equiv \delta_T/\delta$, the integration yields

$$\frac{3\alpha}{2U_\infty\delta_T} = \frac{d}{dx}\left[\delta_T\zeta\left(\frac{3}{20} - \frac{3}{280}\zeta^2\right)\right] \qquad (6.28)$$

For values of ζ less than unity (as is the case for most gases and liquids), the ζ^2 term is negligible and Eq. (6.28) can be written as

$$\frac{10\alpha}{U_\infty\zeta\delta} = \frac{d}{dx}(\delta\zeta^2) \qquad (6.29)$$

Substitution of Eq. (6.26) for δ and some algebraic manipulation now yields

$$\zeta^3 + \frac{4}{3}x\frac{d}{dx}(\zeta^3) = \frac{13}{14}\left(\frac{\alpha}{\nu}\right) \qquad (6.30)$$

and, following integration,

$$\zeta = \frac{1}{1.026} \left(\frac{\alpha}{\nu} \right)^{1/3} \tag{6.31}$$

The ratio ν/α, relating the kinematic viscosity or *diffusivity* of momentum to the thermal diffusivity, is encountered frequently in convection analyses and is commonly referred to as the *Prandtl number* Pr. Inserting this definition into Eq. (6.31), the desired expression for the thermal boundary-layer thickness is found as

$$\delta_T = 0.975 \left(\frac{\delta}{Pr^{1/3}} \right) \tag{6.32}$$

Interestingly, it is now seen that for $Pr \geqslant 1$, the earlier assumption of $\delta_T < \delta$ (or $\zeta < 1$) can be met.

The availability of Eq. (6.32) removes the last obstacle to the analytic determination of the heat transfer coefficient. Combining Eqs. (6.12), (6.26), and (6.32), the value of h for laminar flow along a flat plate at constant wall temperature is given by

$$h_x = 0.323 \left(\frac{k}{x} \right) Pr^{1/3} Re^{1/2} \tag{6.33}$$

A nondimensional heat transfer coefficient can be obtained by dividing h_x by k/x. This operation yields the *Nusselt number* $Nu \equiv hx/k$, which is then expressible as

$$Nu_x = 0.323 \, Pr^{1/3} \, Re^{1/2} \tag{6.34}$$

The average Nu over the length of the plate can be obtained by integration of Eq. (6.34):

$$Nu \equiv \frac{1}{L} \int_0^L Nu_x \, dx = 0.646 \, Pr^{1/3} \, Re^{1/2} \tag{6.35}$$

Although Eq. (6.34) was derived in an approximate fashion, relying on an integral rather than a differential formulation, the resulting Nu values are nearly identical to those obtained by the far more rigorous solution of Pohlhausen [8]:

$$Nu_x = 0.332 \, Pr^{1/3} \, Re^{1/2} \tag{6.36}$$

When the fluid properties are evaluated at the average boundary-layer temperature, both Eqs. (6.34) and (6.36) are in close agreement with experimental data for all but liquid metals, where Pr is generally less than 0.1 and $\delta_T \gg \delta$ [18].

REYNOLDS ANALOGY
The similarity in the form of the boundary-layer velocity and temperature profiles suggests that a close relationship exists between the exchange of momentum and

exchange of heat within the boundary layer. The existence of such a relationship was first recognized by Reynolds [9], and its mathematical statement bears his name. Reynolds' analogy was, however, limited to the case of $Pr = 1$, and more modern treatments by Prandtl and Colburn have attempted to extend the range of this reciprocal formulation. The *Colburn analogy* [10] is the most widely applicable, and its development is outlined below.

The wall shear stress for laminar flow along a flat plate can be found by differentiating the velocity profile expressed in Eq. (6.19) at the wall and multiplying the result by the fluid viscosity to obtain

$$\tau_0 = \frac{\frac{3}{2} \mu U_\infty}{4.64 \sqrt{\nu x / U_\infty}} \tag{6.37}$$

Inserting this expression for τ_0 into Eq. (6.2) and rearranging terms yields an equation for the skin friction coefficient C_f:

$$C_f = 0.646 \sqrt{\frac{\nu}{U_\infty x}} = 0.646 \, Re^{-1/2} \tag{6.38}$$

The heat transfer coefficient for this same configuration is given by Eq. (6.33), which may be rewritten as

$$\frac{Nu}{Re \, Pr} = \frac{h}{\rho c_p U_\infty} = 0.323 \, Pr^{-2/3} \, Re^{-1/2} \tag{6.39}$$

Thus,

$$\frac{h}{\rho c_p U_\infty} Pr^{2/3} = 0.323 \, Re^{-1/2} \tag{6.40}$$

Comparison of Eqs. (6.38) and (6.40) shows that for the stated conditions,

$$\frac{h}{\rho c_p U_\infty} Pr^{2/3} = \frac{C_f}{2} \tag{6.41}$$

This statement is known as the *Colburn analogy*, and although it is strictly valid only for laminar flow over flat plates, it has been found to apply as well to turbulent flow along flat plates and to fully developed turbulent flow in pipes. More generally, it can be expected to offer a reasonable approximation for h when the pressure drop or friction factor along a particular channel or surface results exclusively from viscous shear. Alternately, when boundary-layer detachment is encountered and form drag is considerable, the analogy between heat and momentum exchange cannot be considered valid.

6.3.2 Turbulent Flow on Flat Plates

The analysis of turbulent flow is far more difficult than the analysis of laminar flow, because in this regime both heat and momentum transfer are dominated by eddy motion. It is thus extremely fortunate that quite a precise expression for the turbulent heat transfer coefficient along a flat plate can be obtained by the use of the Colburn analogy, Eq. (6.41).

In distinction to laminar flow, the turbulent boundary-layer velocity profile follows a $\frac{1}{7}$ power distribution,

$$\frac{u}{U_\infty} = \left(\frac{y}{\delta}\right)^{1/7} \tag{6.42}$$

down to very small values of y. As a consequence of the altered velocity profile and velocity gradient at the wall, skin friction coefficients are not equal to laminar flow values, and in the range $5 \times 10^5 < \text{Re} < 10^7$ are given instead by an empirical relation [11]

$$C_{f,x} = 0.0576 \, \text{Re}^{-1/5} \tag{6.43}$$

Applying the Colburn analogy, Eq. (6.43) can be used to establish the functional form of the heat transfer coefficient relation as

$$\frac{h}{c_p \rho U_\infty} \, \text{Pr}^{2/3} = 0.0288 \, \text{Re}^{-1/5} \tag{6.44}$$

Rearranging terms, the local Nusselt number in turbulent flow is found to be

$$\text{Nu}_x = \frac{hx}{k} = 0.0288 \, \text{Re}^{4/5} \, \text{Pr}^{1/3} \tag{6.45}$$

To find the average value of the Nusselt number along a plate wet by a turbulent boundary layer, it is necessary to combine appropriately the initial laminar boundary-layer region with the predominant turbulent region. This can be done approximately by integrating Eq. (6.36) from the leading edge, $x = 0$, to the point at which transition occurs, $x = x_{tr}$, and Eq. (6.45) from x_{tr} to $x = L$. When this is done for the transition Reynolds number equal to 3×10^5, the average Nu is found to be

$$\text{Nu} = 0.036(\text{Re}_L^{4/5} - 14{,}251) \, \text{Pr}^{1/3} \tag{6.46}$$

For large values of Re_L when the boundary layer becomes turbulent close to the leading edge, the average Nu can be approximated by

$$\text{Nu} = 0.036 \, \text{Re}^{4/5} \, \text{Pr}^{1/3} \tag{6.47}$$

For best results the fluid properties in Eqs. (6.45), (6.46), and (6.47) should be determined at the average boundary-layer temperature.

6.3.3 Laminar Flow in Pipes and Channels

Fluid flow in pipes and channels is distinguished from flow along plates primarily by the limit placed on the growth of the velocity and temperature boundary layers. When these boundary layers, growing on opposite sides of a parellel plate channel or from the internal surface of a pipe, merge, the flow and temperature fields are said to be fully developed as shown in Fig. 6.5.

In this configuration the temperature profile does not vary in the flow direction, that is,

$$\frac{T_0 - T}{T_0 - T_m} = f(y) \quad \text{only} \tag{6.48}$$

and the temperature gradient at the wall

$$\left.\frac{\partial T}{\partial y}\right|_0 = (T_0 - T_m) f'(y)|_0 \tag{6.49}$$

is also independent of x. Consequently, the heat transfer coefficient h,

$$h \equiv \frac{-k(\partial T/\partial y)_0}{T_0 - T_m} = k f'(y)_0 \tag{6.50}$$

is found to be uniform along the pipe or channel wall in fully developed flow, though its actual value is, as before, dependent on the thermal conductivity of the fluid and the wall derivative of the temperature field.

This same conclusion can be reached by assuming that the analytical relation for the heat transfer coefficient along a flat plate, Eq. (6.12), also applies to the present configuration and that the boundary-layer thickness in fully developed flow is constant and equal to the pipe radius. This approach not only yields a uniform h along the pipe but suggests as well that the Nusselt number for fully developed pipe flow, $\text{Nu} \equiv hd/k$, equals approximately 3.

A more rigorous analysis of laminar, fully developed flow shows the Nu for a uniform wall temperature pipe to equal 3.66, whereas for uniform heat flux this value increases to 4.36 [1]. Nusselt number values for other geometries are tabulated in [1].

It is important to note that in actual pipes and channels, the fully developed Nu values are achieved asymptotically as the flow distance increases. In the developing flow region near the pipe entrance, the relatively thin boundary layers result in higher heat transfer coefficients. Consequently, in all but very long pipes, for which L/d is greater than 100, actual Nusselt numbers may be substantially higher than the values cited, especially for gas flow.

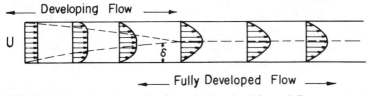

FIG. 6.5 Boundary-layer development in pipe/channel flow.

6.3.4 Turbulent Flow in Pipes and Channels

Much of the complexity in the analysis of turbulent forced convection can be handled by relying on the Reynolds/Colburn analogy between momentum exchange and heat transfer. In turbulent flow through smooth pipes, the friction factor has been correlated by Nikuradse as [12]

$$\frac{1}{\sqrt{f}} = 4.0 \log (\mathrm{Re} \sqrt{f}) - 0.40 \tag{6.51}$$

In the commonly encountered range of $3 \times 10^4 < \mathrm{Re} < 10^6$, this relation can be approximated by

$$f = \frac{0.046}{\mathrm{Re}^{0.2}} \tag{6.52}$$

Applying the Colburn analogy, that is, substituting Eq. (6.52) into Eq. (6.41), and introducing the definition of the Nusselt number, heat transfer in turbulent pipe flow is found to follow

$$\mathrm{Nu} = 0.023 \, \mathrm{Re}^{4/5} \, \mathrm{Pr}^{1/3} \tag{6.53}$$

This relation is identical with Colburn's empirical correlation of data for turbulent forced convection in pipes and is in very good agreement with the widely used McAdams correlation [13], which differs from Eq. (6.53) only in that the Pr exponent is taken to equal 0.4 for heating and 0.3 for cooling rather than the $\frac{1}{3}$ value shown.

As in the laminar regime, the relatively high heat transfer rates in the developing flow region near the pipe entrance can lead to significantly higher Nu values than are predicted by Eq. (6.53). However, in flows that are at least modestly turbulent, that is, where $\mathrm{Re} > 10^4$, this effect is generally limited to pipes with $L/d < 20$ for gases [1].

6.4 FORCED–CONVECTION HEAT TRANSFER COEFFICIENTS–EMPIRICAL CORRELATIONS

The theoretical derivations presented in the preceding section have dealt with relatively simple geometries and well-defined boundary conditions. The resulting analytic relations for the convective heat transfer coefficient are remarkably precise (in their range of applicability) and are a tribute to the ingenuity and persistence of classical as well as modern thermal theoreticians.

The value of these analytic relations is, moreover, not limited to the particular configurations studied. They serve to establish a framework and possible form for the correlation of heat transfer data for far more complex configurations. Based on the derived analytic expressions, it might be anticipated that Nusselt numbers for all modes of convective thermal transport should be correlatable in the form

$$\mathrm{Nu} = C \, \mathrm{Re}^m \, \mathrm{Pr}^n \tag{6.54}$$

where m is near $\frac{1}{2}$ for laminar flow and $\frac{4}{5}$ for turbulent flow and n is approximately $\frac{1}{3}$ for both flow regimes.

In succeeding subsections, correlations for the heat transfer rate from several common, though relatively complex, configurations are presented. Attention focuses on flow around individual and arrays of bluff bodies, such as cylinders and spheres, where the rear detachment of the velocity boundary layer from the heat transfer surface introduces flow and temperature field complexities that are analytically untenable.

6.4.1   External Flows

PLANE SURFACES

The analytical relations developed in the last section offer consistently good to excellent agreement with data for heat transfer from plane or moderately curved surfaces, for Prandtl numbers between 0.6 and 50. For laminar boundary flow on an isothermal plate, the local Nusselt number equals

$$\mathrm{Nu}_x = 0.332\,\mathrm{Re}^{1/2}\,\mathrm{Pr}^{1/3} \tag{6.55}$$

and following integration the average Nu is found to be

$$\mathrm{Nu} = 0.664\,\mathrm{Re}^{1/2}\,\mathrm{Pr}^{1/3} \tag{6.56}$$

where all fluid properties are evaluated at the film temperature, often taken as the arithmetic average of the wall and free stream temperatures.

In turbulent boundary-layer flow the local Nusselt number is given by

$$\mathrm{Nu}_x = 0.0288\,\mathrm{Re}^{4/5}\,\mathrm{Pr}^{1/3} \tag{6.57}$$

and the average value is then expressible as

$$\mathrm{Nu} = 0.036\,\mathrm{Re}^{4/5}\,\mathrm{Pr}^{1/3} \tag{6.58}$$

where properties are again generally evaluated at the film temperature.

In some applications of interest, the surface is not heated over its entire length and an unheated section may be present near the leading edge. For an unheated starting length in laminar flow, the local Nu value can be expressed by

$$\mathrm{Nu}_x = \frac{0.332\,\mathrm{Re}^{1/2}\,\mathrm{Pr}^{1/3}}{[1 - (x_0/x)^{3/4}]^{1/3}} \tag{6.59}$$

where x_0 is the point at which heat transfer begins.

In the turbulent range ($\mathrm{Re} > 3 \times 10^5$), Jakob and Dow [22] proposed the following correlation for air flow:

$$\mathrm{Nu} \equiv \frac{hL}{k_f} = 0.0280\,\mathrm{Re}^{4/5}\left[1 + 0.4\left(\frac{x_0}{x}\right)^{2.75}\right] \tag{6.60}$$

The Nusselt number in the presence of an unheated starting length for other fluids can be estimated by introducing the $Pr^{1/3}$ correction discussed previously.

CYLINDERS IN CROSS FLOW

Flow around cylinders at all but very low Reynolds numbers ($Re \leqslant 1$) results in the detachment of the boundary layer behind the cylinder and the formation of a wake region in which fluid recirculation and vortex formation become more intense with increasing Re. Hilpert was the first to study heat transfer in this configuration; he correlated his results for air flow over a wide range of Re in the form [14]

$$\text{Nu} \equiv \frac{hd}{k_f} = B \left(\frac{\rho U_\infty d}{\mu_f} \right)^n \tag{6.61}$$

where U_∞ is the free stream velocity and k and μ are evaluated at the "film" temperature, that is, the average boundary-layer temperature. Hilpert's values of B, n, and Nu for given ranges of Reynolds numbers are presented in Table 6.1.

Previous analytic results suggest that Eq. (6.61) can be extended to other fluids by dividing this equation by the Pr of air to the $\frac{1}{3}$ power and multiplying by $Pr^{1/3}$ of the fluid of interest. This approach was in fact recommended by McAdams [6]. A more recent detailed treatment of gas and liquid data [15], spanning a range of $1 \leqslant Re \leqslant 10^5$ and $0.67 \leqslant Pr \leqslant 300$, yielded a correlation for the average heat transfer coefficient of

$$\text{Nu} \equiv \frac{hd}{d} = (0.4\,Re^{0.5} + 0.06\,Re^{0.67})\,Pr^{0.4} \left(\frac{\mu_0}{\mu_\infty} \right)^{0.25} \tag{6.62}$$

where all physical properties are evaluated at the free stream temperature except μ_0, the viscosity at the wall temperature. At low values of Re ($Re < 10^2$), where conduction effects begin to compete with thermal transport by forced convection, an additional constant equal to 0.35 should probably be added to the Re terms in parentheses.

It must be noted that Eqs. (6.61) and (6.62) assume that a "natural" turbulence level exists in the oncoming air stream. The presence of grids, fins, and other turbulence promoters can increase the heat transfer coefficient by as much as 50% relative

TABLE 6.1 Constants for Eq. (6.61)[a,b]

Reynolds number range	B	n	Nusselt number range
1–4	0.891	0.330	0.891–1.42
4–40	0.821	0.385	1.42–3.40
40–4,000	0.615	0.466	3.40–29.6
4,000–40,000	0.174	0.618	29.6–121
40,000–400,000	0.0239	0.805	121–840

[a]From McAdams [6], by permission.
[b]The temperature difference defining h is $T_0 - T_w$.

TABLE 6.2   Flow Disturbance Effects on B and n in Eq. (6.61)^a

Disturbance	Range of Re	B	n
Longitudinal fin, 0.1 d thick, on front of tube	1000–4000	0.248	0.603
12 longitudinal grooves, 0.07 d wide	3500–7000	0.082	0.747
Same as above, with burrs	3000–6000	0.0368	0.86

^{a}From Jacob [16], by permission.

to the "natural" condition. Some typical values of the B and n factors in Eq. (6.61) associated with flow turbulence are shown in Table 6.2.

NONCIRCULAR CYLINDERS IN CROSS FLOW
The characteristics of the wake or boundary-layer separation zone, behind a noncircular cylinder, differ from those encountered behind a circular cylinder, and an appropriate correction must be entered in the correlations before they can be applied to these geometries. Jakob [16] found that Eq. (6.61) could be used for noncircular geometries provided that the characteristic dimension used is the diameter of a cylinder of wetted surface equal to that of the geometry of interest and that the B and n values are taken from Table 6.3.

For fluids other than air, introduction of the $Pr^{1/3}$ factor should offer a reasonable first estimate for the prevailing Nu.

SPHERES
Heat transfer from spheres is similar to that encountered in the cylindrical geometry and is strongly affected by the size and behavior of the separated flow region.

For air flow over a single sphere, in the range $17 \leqslant Re \leqslant 7 \times 10^4$, McAdams [6] recommends that the average Nu or h be determined from

$$Nu \equiv \frac{hd}{k_f} = 0.37 \left(\frac{\rho_b \, dU_\infty}{\mu_f} \right)^{0.6} \tag{6.63}$$

where f refers to the film temperature, b to the bulk temperature, and U_∞ is the free stream velocity.

Thermal conduction from a sphere in a nominally stagnant fluid yields an effective Nu value of 2 for Pr near unity. Consequently, it might be anticipated that for low values of Reynolds number, Nu should be expressible as the sum of a constant and a Re-dependent term. For gas flow in the range $1 < Re < 25$, Kreith [17] suggests

$$Nu = \frac{hd}{k} = 2.2 \, Pr + 0.48 \, Pr \, (Re)^{0.5} \tag{6.64}$$

The average heat transfer coefficients for spherical surfaces exposed to the flow of both liquids and gases in the range $3.5 < Re < 7.6 \times 10^4$ and $0.7 < Pr < 380$ were recently correlated [15] in the form

$$Nu \equiv \frac{hd}{k} = 2 + (0.4 \, Re^{0.5} + 0.06 \, Re^{0.67}) \, Pr^{0.4} \left(\frac{\mu_0}{\mu_\infty} \right)^{0.25} \tag{6.65}$$

TABLE 6.3   Values of Band n for Noncircular Cylinders[a]

Flow geometry	B	n	Range of Reynolds number
(ellipse)	0.224	0.612	2,500–15,000
(ellipse)	0.085	0.804	3,000–15,000
(diamond)	0.261	0.624	2,500–7,500
(diamond)	0.222	0.588	5,000–100,000
(square)	0.160	0.699	2,500–8,000
(square)	0.092	0.675	5,000–100,000
(oval)	0.138	0.638	5,000–100,000
(oval)	0.144	0.638	5,000–19,500
(oval)	0.035	0.782	19,500–100,000
(I-shape)	0.205	0.731	4,000–15,000

[a]From Jacob [16], by permission.

FLOW ACROSS TUBE BANKS

The correlation used for the heat transfer coefficient of fluids flowing normal to non-baffled tubes is derived from the work of Colburn [10]:

$$\frac{hd}{k_f} = C \left(\frac{d\rho U_\infty}{\mu_f} \right)^{0.6} \left(\frac{C_p \mu}{k} \right)_f^{0.33} \phi \tag{6.66}$$

This equation is valid in the range $2,000 \leqslant \text{Re} \leqslant 32,000$ and must be corrected by ϕ for the number of tube rows over which the fluid is flowing. Correction factors to be applied are shown in Table 6.4. For in-line tubes the value of C is 0.26, whereas for staggered tubes Colburn recommends $C = 0.33$.

For air, in the parametric range where Pr is nearly constant, Eq. (6.66) can be reduced to

$$\frac{hd}{k_f} = C' \left(\frac{d\rho U_\infty}{\mu_f} \right)^{n'} \tag{6.67}$$

TABLE 6.4   Correlation Factor ϕ for Sparse Tube Banks

Number of rows, N	In line	Staggered
1	0.64	0.68
2	0.80	0.75
3	0.87	0.83
4	0.90	0.89
5	0.92	0.92
6	0.94	0.95
7	0.96	0.97
8	0.98	0.98
9	0.99	0.99
10	1.00	1.00

where the values of C' and n' determined by McAdams [6] are given in Table 6.5 and fluid properties are evaluated at the film temperature. Equation (6.67) is valid in the range $2{,}000 \leqslant \mathrm{Re} \leqslant 40{,}000$, and the corrections for number of rows given in Table 6.4 apply. In Table 6.5, the factors x_L and x_T denote the ratio of longitudinal centerline spacing to tube diameter and transverse centerline spacing to tube diameter, respectively.

For fluids other than air flowing normal to banks of staggered tubes, Fig. 6.6 should be used [6]. In accordance with the procedure recommended by TEMA [19], for $150 \leqslant \mathrm{Re} \leqslant 2000$ the curve is drawn parallel to the curve for air flowing perpendicularly to single cylinders, that is, Eq. (6.61) and Table 6.1. For in-line tubes, TEMA suggests that the data obtained from Fig. 6.6 be reduced by 10% and that the tube number correction given in Table 6.4 be applied.

FLOW OF AIR OVER ELECTRONIC COMPONENTS

Although the correlations for circular and noncircular cylinders and spheres discussed in preceding sections can be used to approximate the heat transfer coefficients along electronic components of similar shape, the finite length and presence of protuberances on the component surface may somewhat alter the fluid flow pattern and thus the value of the heat transfer coefficient.

For the flow of air across an electron tube, Robinson et al. [20] have found that the experimental data can be correlated by an equation of the form of Eq. (6.61) but with n and B factors that are different from those suggested by Hilpert [14]. The recommended values for B and n in the range $10^3 \leqslant \mathrm{Re} \leqslant 10^5$ are shown in Table 6.6. Heat transfer from arrays of electron tubes is discussed in a later chapter.

Heat transfer coefficients for the cooling of prismatic electronic components by the flow of air, either normal or parallel to the sides of a single component in a

TABLE 6.5 Values of Constants C' and n, for Eq. (6.67) [6]

$x_L = \dfrac{S_L}{d_0}$	$x_T = \dfrac{S_T}{d_0} = 1.25$		$x_T = \dfrac{S_T}{d_0} = 1.50$		$x_T = \dfrac{S_T}{d_0} = 2.00$		$x_T = \dfrac{S_T}{d_0} = 3.00$	
	C'	n'	C'	n'	C'	n'	C'	n'
Staggered								
0.600							0.213	0.636
0.900					0.446	0.571	0.401	0.581
1.000			0.497	0.558				
1.125					0.478	0.565	0.518	0.560
1.250	0.518	0.556	0.505	0.554	0.519	0.556	0.522	0.562
1.500	0.451	0.568	0.460	0.562	0.452	0.568	0.488	0.568
2.000	0.404	0.572	0.416	0.568	0.482	0.556	0.449	0.570
3.000	0.310	0.592	0.356	0.580	0.440	0.562	0.421	0.574
In line								
1.250	0.348	0.592	0.275	0.608	0.100	0.704	0.0633	0.752
1.500	0.367	0.586	0.250	0.620	0.101	0.702	0.0678	0.744
2.000	0.418	0.570	0.299	0.602	0.229	0.632	0.198	0.648
3.000	0.290	0.601	0.357	0.584	0.374	0.581	0.286	0.608

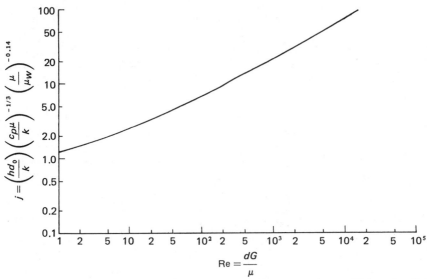

FIG. 6.6   Recommended curve for estimation of heat transfer coefficient for fluids flowing normal to staggered banks of tubes 10 rows deep. (From McAdam [6], by permission.

duct, can be calculated with a correlation of Robinson and Jones [21] for $2.5 \times 10^3 < \mathrm{Re} < 8 \times 10^3$:

$$\mathrm{Nu} = 0.446 \left[\frac{\mathrm{Re}}{(1/6) + (5A_n/6A_0)} \right]^{0.57} \tag{6.68}$$

In Eq. (6.68) the Re is based on the prism-side dimension, A_0 is the gross flow area of the duct, and A_n is the net flow area.

For staggered prismatic components, Robinson and Jones [21] propose multiplying the right side of Eq. (6.68) by a spacing factor $[1 + 0.639(S_T/S_{T,\mathrm{max}}) \times (d/S_L)^{0.172}]$, where S_L is the longitudinal separation, S_T is the transverse separation, and $S_{T,\mathrm{max}}$ is the maximum transverse separation, if different spacings exist.

6.4.2   Internal Flows

PIPES

Heat exchange in fully developed flow within smooth pipes can be determined from the analytic expressions presented in the preceding section. Thus, in laminar flow the Nusselt number is found to be a constant, equal to

Nu = 3.66 for uniform wall temperature

$\quad\ = 4.36$ for uniform surface heat flux   (6.69)

In turbulent, smooth pipe, fully developed flow, the boundary condition is found not to affect the correlation, and the Nu can be expressed by

$$\mathrm{Nu} = 0.023\, \mathrm{Re}^{0.8}\, \mathrm{Pr}^{1/3} \tag{6.70}$$

TABLE 6.6 B and n Factors for Electron Tubes for Use in Eq. (6.61)

Re[a]	B	n
$10^3 - 6 \times 10^3$	0.409	0.531
$6 \times 10^3 - 3 \times 10^4$	0.212	0.606
$3 \times 10^4 - 1 \times 10^5$	0.139	0.806

[a]Re is based on tube diameter.

For large temperature differences between the pipe wall and the fluid, Sieder and Tate [23] recommend evaluating all the properties at the bulk temperature and multiplying the right side of Eq. (6.51) by the ratio of the bulk to wall temperature viscosity, μ_b/μ_0, raised to the 0.14 power.

It must be noted that although Eq. (6.69) offers a most convenient relation for laminar pipe flow, empirical results are generally only in approximate agreement with these values. This discrepancy can be related to the entrance or developing flow effect and the influence of temperature-induced fluid property changes over the long pipe lengths required to achieve fully developed flow. Several empirical correlations, which attempt to address these difficulties, are available in the literature. For short tubes the expression of Seider and Tate [23] may be used with some confidence:

$$\text{Nu} = 1.86 \left(\text{Re Pr} \frac{d}{L} \right)^{1/3} \left(\frac{\mu}{\mu_w} \right)^{0.14} \tag{6.71}$$

where all properties except μ_w are evaluated at the bulk temperature.

NONCIRCULAR GEOMETRIES

Many engineering applications involve heat transfer to fluids flowing within non-circular passages. In fully developed laminar flow the Nusselt number for a particular geometry and wall boundary condition is again found to be a constant [1]. The values for two of the more common configurations, as well as for the previously discussed circular pipe, are shown in Table 6.7. Although an adequate prediction of turbulent heat transfer in noncircular passages can be achieved by use of the equivalent diameter D_e in circular pipe correlations, the values of Nu thus obtained are generally found to lie below the data for rectangular ducts and above the data for other channel configurations [24].

The equivalent diameter, defined by Eq. (6.4), can be shown to correspond to the diameter of a circle inscribed in the cross section of the channel, but no rigorous

TABLE 6.7 Nu Values for Fully Developed Laminar Flow [1]

Configuration	Boundary condition	hD_e/k
Circular pipe	$(q/A)_0$ uniform	4.36
	T_0 uniform	3.66
Parallel plates	$(q/A)_0$ uniform	8.23
	T_0 uniform	7.60
Triangular passage (equilateral)	$(q/A)_0$ uniform	3.00
	T_0 uniform	2.35

justification can be offered for this choice of D_e. Consequently, a more valid geometric translation can perhaps be gained by assuming that in noncircular ducts the so-called *universal velocity distribution* applies along rays that are perpendicular to the isovelocity lines and using twice the average ray length as the corrected equivalent diameter [24]. This technique has been found to offer much better agreement with friction-factor data than reliance on the inscribed circle diameter [24], and values of the correction factor, that is, $2R_{av}/D_e$, are shown in Table 6.8. These correction factors can be applied to D_e calculated via Eq. (6.4) to provide a more precise prediction of turbulent heat transfer rates in noncircular geometries using conventional pipe correlations—for example, Eq. (6.51).

Among the various noncircular channels, rectangular ducts are of considerable importance in electronic cooling applications. The empirical values of the friction factor and heat transfer coefficient for the forced flow of air through rectangular ducts, obtained in an early study by A. L. London [25], are shown in Fig. 6.7.

6.5 FREE OR NATURAL CONVECTION

In free or natural convection, fluid motion is induced by density differences resulting from temperature gradients in the fluid. Under the influence of gravity or other body forces, these density differences give rise to buoyancy forces that circulate the affected fluid and convect heat toward or away from surfaces wetted by the fluid.

The temperature and velocity fields associated with laminar natural convection are typified by the Schmidt and Beckman [26] data given in Fig. 6.8 for a hot, vertical plate immersed in initially quiescent air. At each x location, the vertical air velocity is seen to peak at a modest distance from the plate and decrease toward zero at greater lateral distances. Defining the velocity boundary-layer thickness as the value of the lateral distance at which the velocity has decreased to 1% of the peak velocity, the natural-convection velocity boundary-layer thickness is seen to grow and the peak velocity is seen to shift farther away from the wall as the distance from the leading edge increases. Similarly, the temperature boundary layer is seen to grow along the plate and to lead to progressively lower temperature gradients in the fluid adjacent to the plate, and hence, lower heat transfer rates from the plate to the air. At larger plate-to-air temperature differences and/or greater distances from the leading edge, turbulent flow is encountered and the temperature and velocity profiles attain an asymptotic, that is, x-invariant, form.

TABLE 6.8   Equivalent Diameter Correction for Turbulent Flow [24]

Configuration	$2R_{av}/D_e$
Circular pipe	1
Square cross section	1.156
2:1 elliptical cross section	1.166
Equilateral triangular cross section	1.332
n:1 rectangular cross section	$\dfrac{1.156 + (n-1)}{n}$

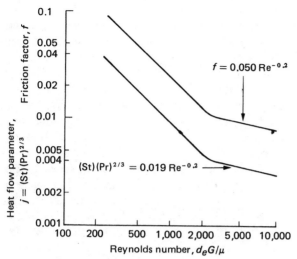

FIG. 6.7 Heat transfer and friction data for forced air through rectangular ducts. St is the Stanton number ($St = h/\rho U_\infty C_p$). (From London [25], by permission.)

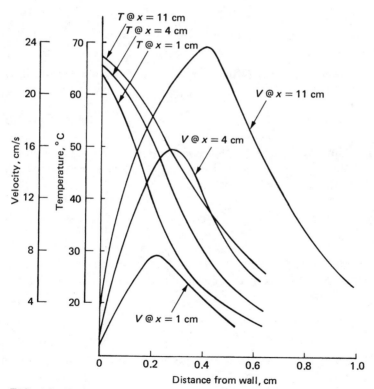

FIG. 6.8 Temperature and velocity fields in air adjacent to vertical plate [26]. The plate is 12.7 cm high.

6.5.1 Governing Parameters

The natural-convection heat transfer rates to be expected for simple geometries in thermally well defined environments can be determined analytically in a manner analogous to that of forced convection. However, the coupling between the velocity and temperature fields inherent in thermally induced circulation substantially complicates both the formulation and solution of the governing equations. In the interest of brevity and in keeping with scope of the present discussion, the parameters governing natural-convection heat transfer will be obtained by the use of a physical analogy; the reader interested in a more rigorous treatment of this subject is referred to [1–3].

In natural-convection flow along a vertical, heated wall, the addition of a frictionless, adiabatic surface at the end of the boundary layer, as shown in Fig. 6.8, does not alter the phenomena but serves to define a channel or chimney for the circulating fluid. By analogy to the flow of gas through a chimney, the pressure "head" responsible for the flow can be characterized by the product of the fluid column height and the difference in density between the hot fluid in the chimney and cold fluid outside, or

$$\Delta p_{\text{avail}} = (\rho_c - \rho_h)gL \tag{6.72}$$

The thermal coefficient of expansion, defined as

$$\beta \equiv \frac{1}{\rho}\frac{d\rho}{dT}$$

can be used to relate the density difference to the prevailing temperature difference and thus to reexpress Eq. (6.72) as

$$\Delta p_{\text{avail}} = \rho \beta \, \Delta T \, gL \tag{6.73}$$

In a naturally circulating fluid this available pressure head must overcome the resistance to flow in the channel, which, as will be shown in the next section, can be related to the velocity of the fluid according to

$$\Delta p_{\text{loss}} \simeq \lambda(\rho V^2) \tag{6.74}$$

where λ is a calculable parameter dependent primarily on the channel geometry.

Equating the available pressure head with the anticipated pressure loss, that is, Eqs. (6.73) and (6.74), it is possible to define a natural-convection reference velocity V_r as

$$V_r \equiv \sqrt{\beta \, \Delta T \, gL} \tag{6.75}$$

where for convenience ΔT may be taken as the temperature difference between the wall and ambient fluid.

Based on the successful prediction and correlation of forced-convection heat transfer by use of the Reynolds number, it might be anticipated that an appropriately defined Re could serve to correlate natural convection data. Inserting the reference velocity into Eq. (6.2), the natural-convection Reynolds number, Re_{nc}, is found to be

$$\text{Re}_{nc} = \frac{\rho L}{\mu} \sqrt{\beta\,\Delta T\,gL} = \sqrt{\frac{\rho^2 \beta\,\Delta T\,gL^3}{\mu^2}} \tag{6.76}$$

The parametric group under the square-root sign is generally referred to as the *Grashof number* Gr, and represents the ratio of the buoyant and viscous forces in the fluid. As suggested by Eq. (6.76), $\sqrt{\text{Gr}}$ can be used to replace Re in the development of natural convection Nusselt number correlations, leading to the form

$$\text{Nu} = C\,\text{Gr}^p\,\text{Pr}^m \tag{6.77}$$

where p may equal approximately $\frac{1}{4}$ in laminar flow and approach $\frac{1}{3}$ in turbulent flow and m is approximately $\frac{1}{3}$.

6.5.2 External Flow Correlations

Because of the relative proximity of m to p, it has been found convenient to relate most natural-convection Nu values to a single parameter, the *Rayleigh number*, defined as $\text{Ra} \equiv \text{Gr}\,\text{Pr}$, and thus to express the correlation in the form

$$\text{Nu} = C(\text{Ra})^n$$

$$= C \left(\frac{\rho^2 \beta g c_p}{\mu k} \right)^n (\Delta T L^3)^n \tag{6.78}$$

When compared with data, n is found to equal approximately $\frac{1}{4}$ for $10^3 < \text{Ra} < 10^9$, representing laminar flow, and approximately $\frac{1}{3}$ for $10^9 < \text{Ra} < 10^{12}$, the region associated with turbulent flow. The value of C is found to vary with the geometry of the heat exchange surface, and empirical values for the more common configurations are shown in Table 6.9 along with other pertinent information.

In using the correlation of Table 6.9 and others in the literature, it must be noted that L is a characteristic dimension of the heat exchange surface. For vertical planes L equals the height, and for horizontal cylinders it represents the diameter. For objects in which horizontal and vertical dimensions are comparable in magnitude, it is clear that both dimensions play a role in determining the value of the characteristic dimension. It has been demonstrated experimentally that for spheres, small, short horizontal or vertical cylinders, and short vertical planes, the expression

$$\frac{1}{L} = \frac{1}{\text{horizontal dimension}} + \frac{1}{\text{vertical dimension}} \tag{6.79}$$

can be used to determine the value of the characteristic dimension.

Returning to Eq. (6.78) it is possible to group all the fluid properties in the Ra into a single term, which, for many fluids, is strongly dependent on the fluid temperature. Definition of a parameter a, which is a function of temperature,

$$a = f(T) = \frac{g \beta \rho^2 c_p}{\mu k}$$

TABLE 6.9 Summary of Natural Convection Data[a]

Geometry and position	Range of Gr Pr	C	n	L	Remarks	References
Horizontal cylinders, vertical plates, vertical cylinders, blocks, spheres	10^4–3.5×10^7 3.5×10^7–10^{12}	0.55 0.13	0.25 0.33	See remarks	For vertical plates and cylinders, L = height for horizontal cylinders, $L = d_0$; for spheres, $L = d_0/2$; for blocks, $1/L = 1/L_{horiz} + 1/L_{vert}$. Data based on tests with air and water, and $L < 12$ in.	14
Short vertical plates	10^4–10^9 10^9–10^{12}	0.59 0.13	0.25 0.33	Height	Based on data for air. Simplified forms for air at moderate temperatures and atmospheric pressure: for Gr Pr = 10^4–10^9, $h = 0.29(\theta/L)^{0.25}$; for Gr Pr = 10^9–10^{12}, $h = 0.19\theta^{0.33}$. For other pressures, use h varying as square root of pressure.	14
Horizontal cylinders	10^3–10^9	0.53	0.25	d_0	Based on data for air and water. As θ approaches zero, Nu approaches 0.45. For spheres with Gr Pr > 10^5, C and n are valid, provided that d_0 is replaced by sphere radius.	12, 14
Vertical cylinders	10^3–10^9 $>10^9$	0.45 0.11	0.25 0.33	d d_0	For $d_0 > 2$ ft, use $d_0 = 2$ ft. For $d_0 > 2$ ft, use $d_0 = 2$ ft. For air at moderate temperatures and atmospheric pressure: For Gr Pr = 10^3–10^9, $h = 0.27(\theta/L)^{0.25}$; for Gr Pr = 10^9–10^{12}, $h = 0.18\theta^{0.33}$.	35 35 14
Spheres	10^3–10^9 $>10^9$	0.65 0.15	0.25 0.33	$d_0/2$ $d_0/2$	For $d_0/2 > 2$ ft, use $d_0/2 = 2$ ft. For $d_0/2 > 2$ ft, use $d_0/2 = 2$ ft.	35 35
Horizontal square plates	10^5–2×10^7 2×10^7–3×10^{10} 3×10^5–3×10^{10}	0.54 0.14 0.27	0.33 0.33 0.25	Side dimension Side dimension Side dimension	Upward-facing heated plates or downward-facing cooled plates. Can be applied to circular disks by setting $L = 0.9d$. Same as above. Downward-facing heated plates or upward-facing cooled plates. Applied to disks as above.	12, 14 12, 14 12, 14

[a]Heat transfer coefficients are in Btu/ft² h °F.

permits Eq. (6.78) to be set down as

$$\frac{hL}{k} = C(ap^2 L^3 \, \Delta T)^n \tag{6.80}$$

where p is the ratio of the actual fluid density to that at standard conditions. This permits one to plot the value of $a = f(T)$ as shown in Fig. 6.9 for air.

6.5.3 Internal Flow Correlations

HORIZONTAL SPACES

In an enclosed horizontal space heated from above, the fluid is stably stratified, no circulation occurs, and heat is transferred by conduction and radiation from one boundary surface to another. When such a space is heated from below, conduction and radiation heat transfer are of importance only at low temperature differences and/or small spacings between the horizontal surfaces, corresponding to Ra (with plate spacing as the characteristic dimension) less than approximately 1700. For Ra $>$ 1700, laminar natural convection obtains; and at values approaching 10^5, transition is made to turbulent flow.

For free convection in horizontal air spaces, Jakob [16] recommends

$$\begin{aligned}
\mathrm{Nu}_s &= 0.195(\mathrm{Gr}_s)^{1/4} \qquad 10^4 < \mathrm{Gr}_s < 4 \times 10^5 \\
\mathrm{Nu}_s &= 0.068(\mathrm{Gr}_s)^{1/3} \qquad \mathrm{Gr}_s > 4 \times 10^5
\end{aligned} \tag{6.81}$$

where s is the plate spacing. Results presented in [27] suggest that in the turbulent-range data for free-convection heat transfer through liquids in horizontal spaces is correlatable by an expression similar to Eq. (6.81), namely,

$$\mathrm{Nu}_s = 0.069(\mathrm{Gr}_s)^{1/3} \, \mathrm{Pr}^{0.407} \qquad 3 \times 10^5 < \mathrm{Gr}_s \, \mathrm{Pr} < 7 \times 10^9 \tag{6.82}$$

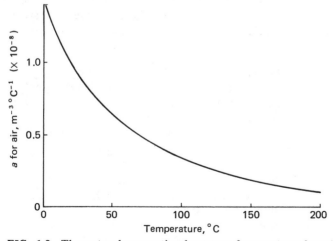

FIG. 6.9 The natural-convection heat transfer property function a for dry air at atmospheric pressure.

VERTICAL SPACES

In vertical spaces formed by two parallel surfaces at different temperatures and two adiabatic end "caps," Ra values (based on plate spacing) below approximately 2×10^3 are again associated with an absence of fluid circulation and the predominance of conduction and radiation effects. For larger Ra, natural convection obtains and progresses from laminar to turbulent flow in the boundary layers along the vertical surfaces.

Because of the importance of the distance separating the two thermally active surfaces of a vertical enclosure in preconvective, steady-state conduction, this dimension, s, appears frequently as the characteristic length in the Nusselt and Rayleigh numbers. Most semiempirical correlations for natural convection in this configuration take the form [28-33]

$$\mathrm{Nu}_s = C'(\mathrm{Ra})^n \left(\frac{H}{s}\right)^r \tag{6.83}$$

and are distinguished from other formulations by the presence of an aspect ratio dependence, where r is typically $-\frac{1}{4}$ for laminar flow and zero for turbulent flow. Examination of the available data reveals, however, that the distance of separation between the plates is generally sufficiently large to ensure that the individual boundary layers on the opposite surfaces do not interfere with each other and thus suggests that this configuration could be more properly analyzed by basing the Ra on the height H [34]. Heat transfer by natural convection in vertical spaces can thus be expected to be governed by the resistance of two boundary layers in series and, when expressed in the form

$$\mathrm{Nu}_H = C''(\mathrm{Ra})^n \tag{6.84}$$

the correlating coefficient C'' can be estimated at one-half the value of C in Eq. (6.77) or approximately 0.28 and 0.065 in laminar and turbulent flow, respectively. This approach is confirmed by the results presented in [35], where the average heat transfer coefficient on each surface was found to differ by not more than 20% from the single plate correlations.

CONFINED COMPONENTS

Heat transfer coefficients for laminar free convection from small components—such as relays, transformers, resistors, and tubes—confined in a larger enclosure may be determined by use of the Masson and Robinson correlation [36]

$$\mathrm{Nu} = 1.45\,\mathrm{Ra}^{0.23} \tag{6.85}$$

Equation (6.85) is based on the average temperature difference between the component and the enclosure. The characteristic dimensions used in correlating the data were as follows: for miniature and subminiature tubes, the tube height; for relays and transformers, the vertical height. For horizontal resistors, the characteristic dimensions are formed as in Eq. (6.79), where the vertical dimension is the diameter and the horizontal dimension is the length of the resistor.

It is to be noted that Eq. (6.85) yields relatively high Nu values. These may be attributable to the plume effect in the center of an enclosure containing a single heat source and/or the irregular geometries involved, but some caution is appropriate in applying Eq. (6.85) to previously untested configurations.

6.5.4 Vertical Channels

Vertical channels formed by parallel plates or fins are a frequently encountered configuration in natural-convection cooling of electronic equipment. In the inlet regions and in relatively short channels, individual boundary layers develop along each surface and heat transfer rates approach those associated with laminar flow along isolated plates in infinite media. For long channels, however, the boundary layers merge near the entrance to the channel and fully developed flow prevails along much of the channel. In this regime, as in forced-convection flow, the heat transfer coefficient is constant, but because the local fluid temperature is not known explicitly, it is far more convenient to express the fully developed heat transfer rate in terms of the ambient or inlet fluid temperature.

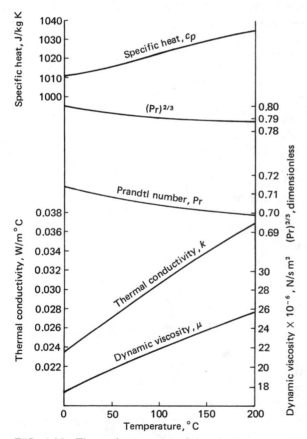

FIG. 6.10 Thermal properties of dry air at atmospheric pressure. The source of these data is Raznjevic [37].

The Nu in laminar flow appropriate to this definition can be determined analytically and has been the subject of detailed experimental investigations and numerical calculations. The results so obtained are presented in detail in Chap. 13.

6.6 THERMAL PROPERTIES OF AIR

Thermal properties of air are displayed in Fig. 6.10.

6.7 NOMENCLATURE

Roman Letter Symbols

A	surface area, m^2
B	coefficient in forced-convection correlation
c	specific heat, $J/kg\,^\circ C$
C	convection constant
C_f	skin friction factor, dimensionless
d	diameter, m
D	equivalent diameter (with subscript e), m (see Eq. 6.5)
f	friction factor, dimensionless
g	acceleration of gravity, m/s^2
Gr	Grashof number, dimensionless
h	heat transfer coefficient, $W/m^2\,^\circ C$
H	height of vertical space, m
I	enthalpy flux, J/m^2
k	thermal conductivity, $W/m\,^\circ C$
L	characteristic length, m
M	momentum flux, $kg/m\,s$
Nu	Nusselt number, dimensionless
P	wetted perimeter, m; or pressure, kg/m^2
Pr	Prandtl number, dimensionless
q	heat flow, W; with double prime superscript, heat flux, W/m^2
Ra	Rayleigh number, dimensionless
Re	Reynolds number, dimensionless
St	Stanton number, dimensionless
T	temperature, $^\circ C$ or K
u	velocity, m/s
U	average pipe or channel velocity, m/s
x	length coordinate, m; with subscript 0, unheated length, m
y	length coordinate, m

Greek Letter Symbols

α	thermal diffusivity, m^2/s
β	coefficient of volumetric expansion, K^{-1}
δ	boundary-layer thickness, m; without subscript, the momentum boundary layer; with subscript T, the thermal boundary layer
Δ	indicates change in variable
ζ	ratio of thermal to momentum boundary-layer thickness, dimensionless

μ dynamic viscosity, kg/m s
ν kinematic viscosity, m^2/s
ρ density, kg/m^3
τ shear stress, kg/m^2

Subscripts

c indicates cold condition
f indicates fluid
h indicates hot condition
m indicates mean condition
p indicates constant-pressure condition
s indicates condition based on spacing
T indicates thermal condition
$w, 0$ indicates wall condition
∞ indicates free stream condition

6.8 REFERENCES

1   Rohsenow, W. M., and Choi, H., *Heat, Mass and Momentum Transfer*, Prentice-Hall, Engle-wood Cliffs, N.J., 1961.

2   Eckert, E. R. G., and Drake, R. M., Jr., *Analysis of Heat and Mass Transfer*, McGraw-Hill, New York, 1972.

3   Chapman, A. J., *Heat Transfer*, 3d ed., Collier Macmillan International, New York, 1974.

4   Prandtl, L., Uber Flussigkeitsbewegung bie sehr bleiner Reibung, *Proc. 3d Int. Math. Congr.*, Heidelberg, pp. 484–491, 1904. Reprinted in NACA TM 452, 1928.

5   Nikuradse, J., Turbulente Strömung in Nichtkreinförmigen Rohren, *Ing. Arch.*, vol. 1, pp. 306–332, 1930.

6   McAdams, W. H., *Heat Transmission*, 3d ed., McGraw-Hill, New York, 1954.

7   Newton, I., Scala Graduum Caloris, *Trans. Roy. Soc. (London)*, vol. 22, p. 824, 1701.

8   Pohlhausen, E., Der Wärmeaustausch zwischen festen Körpern und Flüssigkeiten mit kleiner Reibung und kleiner Wärmeleitung, *Z. Angew. Math. Mech.*, vol. 1, pp. 115–121, 1921.

9   Reynolds, O., An Experimental Investigation of the Circumstances Which Determine Whether the Motion of Water Is Direct or Sinuous, and of The Law of Resistance in Parallel Channels, *Phil. Trans. Roy. Soc. (London)*, vol. 174A, pp. 935–982, 1883.

10 Colburn, A. P., A Method of Correlating Forced Convection Heat Transfer Data and a Comparison of Fluid Friction, *Trans. AIChE*, vol. 29, pp. 174–210, 1933.

11 Schlicting, H., *Boundary Layer Theory*, 6th ed., McGraw-Hill, New York, 1968.

12 Nikuradse, J., Strömungsgesetze in Rauhen Rohren, *Forsch. Arbeiten Ing. Wesen.*, vol. 361, pp. 1–22, 1933.

13 McAdams, W. H., Review and Summary of Developments in Heat Transfer by Conduction and Convection, *Trans. AIChE*, vol. 36, pp. 1–20, 1940.

14 Hilpert, R., Warmeabgue von Geheizten Drähten und Rohren im Lufstrom, *Forsch. Ing.-Wes.*, vol. 4, pp. 215–224, 1933.

15 Whitaker, S., Forced Convection Heat Transfer Correlations for Flow in Pipes, Past Flat Plates, Single Cylinders, Single Spheres and for Flow in Packed Beds and Tube Bundles, *AIChE J.*, vol. 18, pp. 361–371, 1972.

16 Jakob, M., *Heat Transfer*, Wiley, New York, 1949.

17 Kreith, F., *Principles of Heat Transfer*, International Textbook, Scranton, Pa., 1959.

18 Martinelli, R. C., Heat Transfer to Molten Metals, *Trans. ASME*, vol. 69, pp. 947–959, 1947.

19 *Standards of the Tubular Exchanger Manufacturers Association*, New York, 1949.

20 Robinson, W., Han, L. S., Essig, R. H., and Heddleson, C. F., Heat Transfer and Pressure Drop Data for Circular Cylinders in Ducts and Various Arrangements, Ohio State University Research Foundation Report No. 41, Columbus, Oct. 1951.

21   Robinson, W., and Jones, C. D., The Design of Arrangements of Prismatic Components for Crossflow Forced Air Cooling, Ohio State University Research Foundation Report No. 47, Columbus, Oct. 1955.

22   Jakob, M., and Dow, W. M., Heat Transfer from a Cylindrical Surface to Air in Parallel Flow With and Without Unheated Starting Sections, *Trans. ASME*, vol. 68, 1946.

23   Sieder, E. N., and Tate, G. E., Heat Transfer and Pressure Drop of Liquids in Tubes, *Ind. Eng. Chem.*, vol. 28, pp. 1429–1436, 1936.

24   Brundrett, E., Modified Hydraulic Diameter for Turbulent Flow, Ph.D. thesis, University of Waterloo, Waterloo, Ontario, Canada, 1973.

25   London, A. L., Air Coolers for High Power Vacuum Tubes, *Trans. IRE*, ED-1, pp. 9–26, April 1954.

26   Schmidt, E., and Beckman, W., Das Temperatur und Geschwindigkeitsfeld von einer Wärme Abgebenden Senkrechten Platte bei Naturalicher Konvektion, *Forsch. Ing.-Wes.*, vol. 1, pp. 391–404, 1930.

27   Globe, S., and Dropkin, D., Natural Convection Heat Transfer in Liquids Confined by Two Horizontal Plates and Heated from Below, *J. Heat Transfer*, ser. C, vol. 81, pp. 24–28, 1959.

28   Eckert, E. R. G., and Carlson, W. O., Natural Convection in an Air Layer Enclosed Between Two Vertical Plates with Different Temperatures, *Int. J. Heat Mass Transfer*, vol. 2, no. 1, pp. 106–120, 1961.

29   MacGregor, R. K., and Emery, A. F., Free Convection Through Vertical Plane Layers— Moderate and High Prandtl Number Fluids, ASME paper 68-WA/HT-4, 1968.

30   Emery, A., and Chu, N. C., Heat Transfer Across Vertical Layers, *J. Heat Transfer*, vol. 87, pp. 110–115, 1965.

31   Newell, H. E., and Schmidt, F. W., Heat Transfer Laminar Natural Convection Within Rectangular Enclosures, ASME paper 69-HT-42, 1969.

32   Jakob, M., *Heat Transfer*, vol. I, Wiley, New York, 1949.

33   Dropkin, D., and Somerscales, E., Heat Transfer by Natural Convection in Liquids Confined by Two Parallel Plates Which Are Inclined at Various Angles with Respect to the Horizontal, *J. Heat Transfer*, vol. 87, pp. 77–83, 1965.

34   Bar-Cohen, A., Constant Heating Transient Natural Convection in Partially Insulated Vertical Enclosures, ASME paper 77-HT-35, 1977.

35   Eichhorn, R., Natural Convection in a Thermally Stratified Fluid, in *Prog. Heat Mass Transfer*, vol. 2, pp. 41–53, Pergamon Press, Oxford, 1969.

36   Masson, D. J., and Robinson, W., Free Convection Cooling in Air of Confined Small Bodies Similar to Electronic Components, Report no. 43, Contract W33-038ac-14987, U.S. Air Force, Wright Air Development Center, Dayton, Ohio, Jan. 1952.

37   Raznjevic, K., *Handbook of Thermodynamic Tables and Charts*, McGraw-Hill, New York, 1976.

7

■ heat transfer by radiation

7.1 INTRODUCTION

Radiation of thermal energy is believed to be a specific form of radiation within the general phenomenon of electromagnetic radiation. As but one of numerous electromagnetic phenomena, thermal radiant energy travels at a velocity of 3×10^8 m/s within a vacuum.[1]

The existence of radiation as a mode of heat transfer is a matter of everyday experience. Consider a warm body within and without physical contact with a cooler enclosure under complete vacuum. The warm body will eventually attain the temperature of the surrounding enclosure without the aid of conduction or convection. This statement may appear intuitive. However, the configuration of the warm body within a cooler enclosure can be readily approximated to a fair accuracy (e.g., suspension of the body by cords) and the existence of radiation can be easily demonstrated.

All bodies continuously emit radiation. Figure 7.1 shows the electromagnetic spectrum, depicting a range of waves from long radio wavelengths to short wavelengths. The unit of wavelength used in this figure, as well as in the discussion that follows, is the micrometer, denoted μm ($1\ \mu$m $= 10^{-6}$ m). In Fig. 7.1, particular attention should be focused on the infrared region (0.7 to 1400 μm) and on the visible region (0.38 to 0.76 μm). Thermal radiation, which is of concern here, is considered for all practical purposes to be limited to wavelengths ranging from about 0.1 to about 100 μm.

7.2 MONOCHROMATIC EMISSIVE POWER

Radiation from a heat source at a given temperature can be dispersed into a spectrum by a prism. A sensitive measuring device such as a thermopile will indicate that the energy within the spectrum is distributed smoothly among the various wavelengths present.

The relationship between the monochromatic emissive power $E_{b\lambda}$ and wavelength λ was the subject of many investigations (both experimental and mathematical) during the latter part of the nineteenth century. Planck, in formulating the quantum theory, postulated his law, which fits the spectral energy curves. Planck's law is

$$E_{b\lambda} = \frac{2\pi h c^2 \lambda^{-5}}{e^{ch/k\lambda T} - 1} \tag{7.1}$$

where, in addition to c, the velocity of light, h is Planck's constant, 6.624×10^{-27} erg s, and k is Boltzmann's constant, 1.380×10^{-16} erg/K. These constants have been

---

[1] Actually, 2.9979×10^8 m/s.

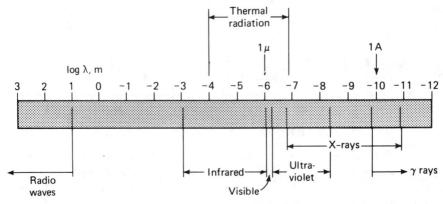

FIG. 7.1 Electromagnetic spectrum. (From Holman [6], by permission.)

taken from Gebhart [1]. This law shows that the emissive power from a perfect radiator is a strong function of both wavelength and absolute temperature. It does not tell
how much total radiation will occur at a given temperature. This is obtained from its
integration over all wavelengths.

Planck's law is plotted in Fig. 7.2 for the temperatures indicated. At a temperature of about 5555 K, corresponding to the effective radiating temperature of the
sun, one can note that a fair portion of the sun's radiation falls into the visible region
(0.38 to 0.76 μm).

7.3 WIEN'S DISPLACEMENT LAW

There will be some value of wavelength for a particular absolute temperature that will
yield a maximum monochromatic emissive power. Indeed, this can be readily observed
in considering Fig. 7.2. A relationship between the wavelength λ and the absolute
temperature T for maximum emissive power can be obtained by differentiating
Planck's law with respect to λ. First, however, it is convenient to change the independent variable from λ to x.

Let

$$x = \frac{A}{\lambda}$$

where $A = ch/kT$, so that

$$\lambda = \frac{A}{x}$$

and

$$\frac{dx}{d\lambda} = \frac{A}{-\lambda^2}$$

Then

$$\frac{dE_{b\lambda}}{d\lambda} = \frac{dE_{b\lambda}}{dx}\frac{dx}{d\lambda} = \frac{d}{dx}\left[\frac{(2\pi hc^2/A^5)x^5}{e^x - 1}\right]\left(-\frac{A}{\lambda^2}\right) = 0$$

and the differentiation and algebraic expansion of terms proceeds to give

$$5(e^x - 1) = xe^x$$

$$5(e^{ch/k\lambda t} - 1) = \frac{ch}{k\lambda T}e^{ch/k\lambda T}$$

which will be satisfied when

$$\lambda T = 2898 \ \mu\text{m K} \tag{7.2}$$

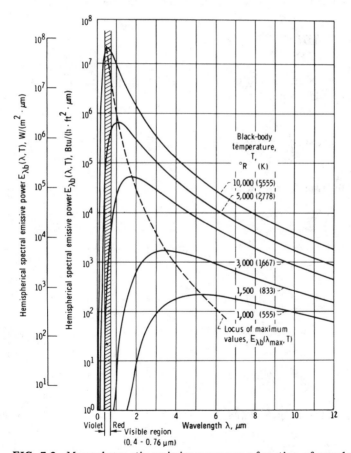

FIG. 7.2 Monochromatic emissive power as a function of wavelength for several temperatures. (From Siegel and Howell [4], by permission.)

Equation (7.2) is the *Wien displacement law*, which gives the wavelength for maximum monochromatic emissive power for any temperature desired.

The glowing of a heating element can be explored. Consider the element at a temperature of 1000 K. The wavelength for maximum emissive power is $\lambda_{max} = 0.2898/1000 = 2.898 \times 10^{-4}$ cm ($\sim$3 μm). For this wavelength, very little of the emission is in the visible spectrum, which has an upper wavelength limit of 0.76 μm.

At higher temperatures the human eye will begin to detect the radiation because the amount of radiation falling into the visible spectrum begins to increase. At a temperature of 3000 K, the wavelength for maximum emissive power is $\lambda_{max} = 0.2898/3000 = 9.66 \times 10^{-5}$ ($\sim$1 μm). Here some of the radiation falls into the visible range (the right-hand skirt, so to speak) and the heating element begins to glow with a dull red color. A further increase in temperature turns the color from red to yellow and then to white. If the surface temperature of the sun, 5555 K, is considered, $\lambda_{max} = 0.2898/5555 = 5.22 \times 10^{-5}$ cm (0.522 μm); it is seen that the wavelength for maximum emissive power falls in the visible spectrum and most of the radiation emitted from the sun falls within the visible spectrum.

7.4 TOTAL RADIATION–THE STEFAN–BOLTZMANN LAW

The total amount of radiation emitted at a particular temperature can be obtained by integrating Planck's law over all wavelengths.

$$E_b = \int_0^\infty E_{b\lambda} \, d\lambda = \int_0^\infty 2\pi hc^2 \, \frac{\lambda^{-5} \, d\lambda}{e^{ch/k\lambda T} - 1}$$

Suppose that the transformation is made:

$$z = \frac{A}{\lambda} \quad \left(A = \frac{ch}{kT} \right)$$

Then

$$\lambda = \frac{A}{z}$$

and

$$d\lambda = -\frac{A}{z^2} \, dz$$

With the substitution of the new variable,

$$E_b = -2\pi hc^2 \int_0^\infty \frac{z^3 \, dz}{A^4(e^z - 1)}$$

However, $1/(e^z - 1)$ can be written as an infinite series, so a term-by-term integration can be performed to yield, after substitution of the limits,

$$E_b = \frac{2\pi h c^2}{A^4} \left(\frac{3!}{1} + \frac{3!}{2^4} + \frac{3!}{3^4} + \frac{3!}{4^4} + \cdots \right)$$

and after evaluating the parenthesized terms one finally obtains

$$E_b = \frac{2\pi h c^2}{A^4} \quad (6.45)$$

But $A = ch/k\lambda T$, so that

$$E_b = 5.67 \times 10^{-12} T^4 \quad (7.3a)$$

the Stefan-Boltzmann law. The constant in Eq. $(7.3a)$ is almost universally designated as σ and is called the *Stefan-Boltzmann constant*. Hence

$$E_b = \sigma T^4 \quad (7.3b)$$

where $\sigma = 5.67 \times 10^{-8}$ W/m^2 K^4.[2]

The Stefan-Boltzmann law gives the total amount of radiant energy flux emitted by an ideal radiator as a function of absolute temperature. The total amount of radiant energy will, of course, be the flux multiplied by the emitting surface area S, so that for an ideal radiator,

$$q = \sigma S T^4 \quad (7.4)$$

7.5 EMISSION AND ABSORPTION— KIRCHHOFF'S LAW

The emission and absorption characteristics of a body may be treated independently of one another but are related under certain conditions.

Let a surface be exposed to a quantity of radiant energy E. The surface will absorb, reflect, and transmit fractions of this radiant energy. Define α, ρ, and τ as the fraction of incident energy absorbed, reflected, and transmitted, respectively. Then if a unit quantity of incident energy is considered,

$$\alpha + \rho + \tau = 1$$

Now define an ideal radiator as a "black" body, one that absorbs all incident radiation and reflects and transmits none of it. The concept of the black body is useful because the laws governing its radiation are simple and many real bodies may be treated approximately as black bodies. Perhaps the term "black" is bothersome. It is

[2] Or 1.713×10^{-9} Btu/h ft^2 °R^4.

used because if a surface actually does absorb all radiant energy falling on it, the surface will appear black to the eye. Some surfaces absorb nearly all incident radiation and yet do not appear black because they do not absorb all visible light rays. Indeed, freshly fallen snow and whitewashed walls, having IR absorptivities of 0.95, are in this category.

Figure 7.3 depicts a small nonblack body totally enclosed by a black body. Let the transmissivity of the small body be zero ($\tau = 0$); that is, let the small body be opaque. Hence

$$\alpha + \rho = 1$$

For the case presented by Fig. 7.3, black-body radiation E_b impinges upon the smaller body and the amount of radiation that is absorbed by the smaller body is equal to $\alpha_1 E_b$. The smaller body, however, will emit radiation, say, E_1, and the net rate of heat interchange between the small body and the enclosure will be

$$q = \alpha_1 E_b - E_1$$

If the temperature of the smaller body is the same as that of the enclosure, there can be no radiant heat interchange and $q = 0$. Hence

$$\alpha_1 E_b - E_1 = 0$$

and

$$E_b = \frac{E_1}{\alpha_1}$$

A similar relationship can be found for any number of nonblack bodies in thermal equilibrium:

$$E_b = \frac{E_1}{\alpha_1} = \frac{E_2}{\alpha_2} = \cdots = \frac{E_n}{\alpha_n} \tag{7.5}$$

a general relationship known as *Kirchhoff's law*.

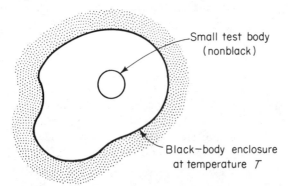

FIG. 7.3 Small test body within large ideal enclosure.

All black bodies at the same temperature will emit radiation at the same rate. If the smaller body in Fig. 7.3 is a black body, then $\alpha_1 = 1.0$ and $E_1 = E_b$. If, for the sake of argument, one proposes that E_1 is greater than E_b, the smaller body must be cooled by a transfer of heat from the lower-temperature smaller body to the higher-temperature black body. This is clearly impossible, because by the second law of thermodynamics, heat cannot flow of its own accord from a low-temperature source to a high-temperature receiver. It is also obvious from the foregoing that, if the smaller body is a black body, at thermal equilibrium it will emit exactly the same amount of radiation as the black body.

The foregoing reasoning leads to three conclusions:

1 For thermal equilibrium, Kirchoff's law shows that the ratio of the emissive power of a surface to its absorptivity is the same for all bodies.
2 Because the absorptivity can never exceed unity at a particular temperature, a black body has the maximum emissive power of any surface.
3 The black body may be considered as a perfect emitter as well as a perfect absorber of radiant energy.

Because a perfect emitter or absorber does not exist, the concept of a black body is an idealized one. Indeed, the emissive characteristics of the so-called nonblack body may be quite different from its absorptive characteristics. The emissive characteristics at a particular temperature are represented by the emissivity ϵ, defined as the ratio of actual rate of energy emission to the rate of energy emission of a black body at the same temperature. Thus

$$E = \epsilon E_b$$

Consider Fig. 7.3 once again and assume that the smaller body is not a black body. For a black-body enclosure, the energy emitted by the enclosure will be E_b. The energy absorbed by the smaller body will be

$$E_1 = \alpha_1 E_b$$

and that emitted by the smaller body will be

$$E_1 = \epsilon_1 E_b$$

as required by the definition of emissivity.

The net rate of heat interchange between the enclosure and the small body is

$$q = \alpha_1 E_b - \epsilon_1 E_b$$

so that for the case of small body and enclosure at the same temperature ($q = 0$),

$$\alpha_1 E_b = \epsilon_1 E_b$$

from which it is seen that the absorptivity equals the emissivity:

$$\alpha_1 = \epsilon_1$$

It can therefore be stated that at thermal equilibrium, the absorptivity and the emissivity are equal. Surfaces that generally exhibit this property are called *gray surfaces*.

7.6 LABORATORY BLACK BODIES

Although there is no perfect emitter or black body, such a body may be approximated by a small hole cut in a hollow cavity. Let the cavity in Fig. 7.4 have its interior wall at a constant temperature; that is, let the interior wall be an isothermal surface. The interior wall, of course, is nonblack, and has an emissivity ϵ.

Now imagine that a small portion of the surface emits energy ϵE_b and that this energy escapes through the hole. Next consider a portion of the surface that emits radiation ϵE_b and let this energy be reflected once before it escapes through the hole. This once-reflected energy will be equal in magnitude to $\rho \epsilon E_b$. For another portion of the surface that emits energy reflected twice, the magnitude of the energy escaping will have a value $\rho^2 \epsilon E_b$. The total energy emitted from the hole will contain energy that has been directly radiated, reradiated once, twice, and so on, and will be equal to

$$E_T = \epsilon E_b + \rho \epsilon E_b + \rho^2 \epsilon E_b + \rho^3 \epsilon E_b + \cdots$$

or $\quad E_T = (1 + \rho + \rho^2 + \rho^3 + \cdots)\epsilon E_b$

However, at any temperature,

$$\epsilon = \alpha = 1 - \rho$$

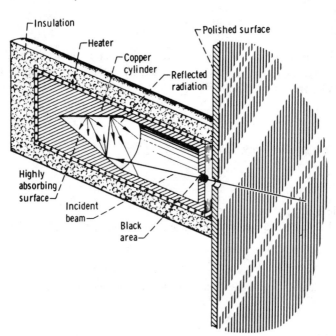

FIG. 7.4 Cavity used to produce black-body radiation. (From Siegel and Howell [4], by permission.)

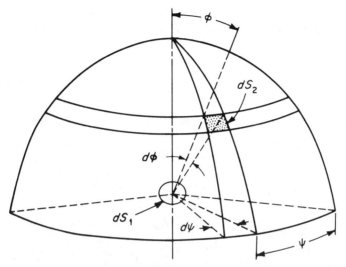

FIG. 7.5   Configuration and nomenclature for the analysis of the intensity of radiation.

and furthermore,

$$1 + \rho + \rho^2 + \rho^3 + \cdots = \frac{1}{1 - \rho}$$

Therefore, the total energy is equal to

$$E_T = \frac{\epsilon E_b}{1 - \rho} = \frac{1 - \rho}{1 - \rho} E_b$$

or $E_T = E_b$

which shows that the energy emitted from the hole is essentially black-body radiation.

7.7   RADIATION INTENSITY AND LAMBERT'S COSINE LAW

So far, only the total radiation has been considered. It becomes necessary at this point to evaluate the radiation intensity I. This takes into account the fact that the radiant intensity varies with the direction of viewing.

Consider Fig. 7.5, which shows a hemispherical surface placed over an element of emitting area dS_1. The hemisphere will intercept all of the radiation emitted, but only from a point directly above will the radiation be "seen" without distortion. From a point on the hemisphere displaced an angle ϕ from the normal to dS_1, the element of area dS_1 will really appear as the projected area $dS_1 \cos \phi$.

Define the intensity of radiation I at some point in space due to emission from area dS_1 as the radiant energy emitted per unit time per unit solid angle subtended at

dS_1 per unit area of emitting surface normal to the direction of the point in space from the emitting source. If dS_2 is considered at an angle ϕ from the normal from dS_1, the heat flow from dS_1 to dS_2 is

$$dq = I \cos \phi \, dS_1 \, \frac{dS_2}{r^2} = I \cos \phi \, dS_1 \, d\omega$$

where $dS_2/r^2 = d\omega$, the solid angle subtended by dS_2 as viewed from dS_1. Note that $dS_1 \cos \phi$ is the effective or projected area of dS_1 as viewed from dS_2.

Hence, the intensity of radiation I has really been defined.

$$I = \frac{dq}{dS_1 \cos \phi \, d\omega} \quad \text{(W/m}^2 \text{ sterad)} \quad \text{(Btu/ft}^2 \text{ h sterad)}$$

The total emissive power will be a summation of the intensity over the surface in question. In the case of Fig. 7.5, the radiation intercepted by the element of area dS_2 is integrated over the half-space represented by the hemisphere

$$E = \int_\omega \int_\phi I \cos \phi \, d\phi \, d\omega = \int_{S_2} \int_\phi I \cos \phi \, \frac{dS_2}{r^2} \, d\phi$$

From Fig. 7.5, it is seen that

$$d\omega = \frac{(r \sin \phi)(d\phi)(r \, d\psi)}{r^2}$$

or $d\omega = \sin \phi \, d\phi \, d\psi$

This may be substituted for $d\omega$ in the integral to obtain

$$E = \int_0^{2\pi} d\psi \int_0^{\pi/2} I \cos \phi \sin \phi \, d\phi$$

First integrate with respect to ψ:

$$E = 2\pi I \int_0^{\pi/2} \cos \phi \sin \phi \, d\phi$$

Then integrate with respect to ϕ:

$$E = 2\pi I (\tfrac{1}{2} \sin^2 \phi)|_0^{\pi/2} = \pi I$$

Thus, the total emissive power is seen to be π multiplied by the radiation intensity. This is consistent with *Lambert's law*, which states that the intensity of radiation I is a constant throughout the hemisphere (half-space) above the emitting element. Indeed, this implies that the intensity of radiation varies inversely as the square of the distance from the source and directly with the cosine of the angle made with the normal to the emitting element.

7.8 HEAT FLOW BETWEEN BLACK BODIES

The Stefan-Boltzmann law applies to emission from a single surface. In this section consideration will be given to the interchange of thermal energy between two surfaces of a physical extent such that not all the emission from one is intercepted by the other. Then, in a subsequent section, the further refinement of considering nonblack surfaces will be made.

Consider Fig. 7.6, which depicts two surfaces, S_1 and S_2, separated by a non-absorbing medium. The distance between the differential elements dS_1, dS_2 is designated r. The rate of radiant energy interchange between dS_1 and dS_2 is

$$dq_{12} = I_1 \cos \phi_1 \, dS_1 \, d\omega_{12}$$

where $d\omega_{12}$ is the solid angle subtended by dS_2 with respect to dS_1. This is equal to the projected area of the receiving surface divided by the square of the distance separating the surfaces:

$$d\omega_{12} = \frac{dS_2 \cos \phi_2}{r^2}$$

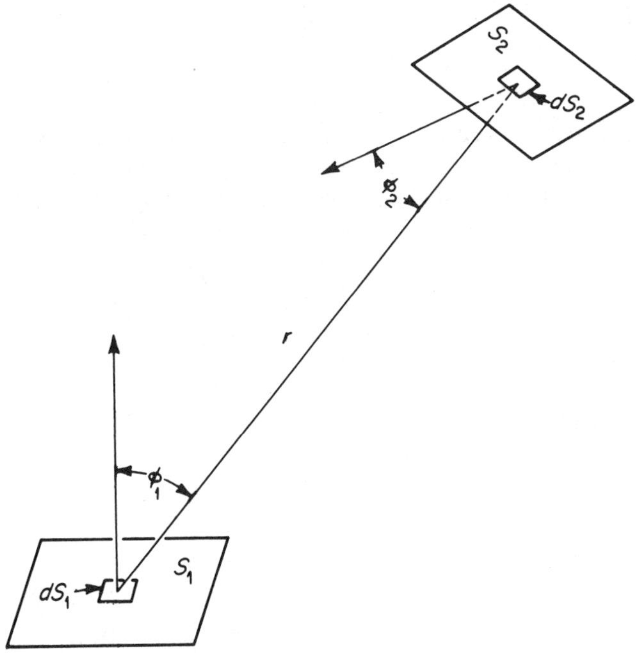

FIG. 7.6 Geometric shape factor notation.

so that

$$dq_{12} = \frac{I_1 \cos \phi_1 \cos \phi_2 \, dS_1 \, dS_2}{r^2}$$

Because $I_1 = E_1 / \pi$,

$$dq_{12} = E_1 \, dS_1 \left(\frac{\cos \phi_1 \cos \phi_2 \, dS_2}{\pi r^2} \right)$$

Consider the term within parentheses as the fraction of the total emission from dS_1 that is intercepted by dS_2.

In a similar manner,

$$dq_{21} = E_2 \, dS_2 \left(\frac{\cos \phi_1 \cos \phi_2 \, dS_1}{\pi r^2} \right)$$

so that the net rate of radiant heat interchange between the differential elements is

$$dq = (E_1 - E_2) \left(\frac{\cos \phi_1 \cos \phi_2 \, dS_1 \, dS_2}{\pi r^2} \right)$$

For the total radiant heat interchange between S_1 and S_2,

$$q = (E_1 - E_2) \int_{S_1} \int_{S_2} \frac{\cos \phi_1 \cos \phi_2 \, dS_1 \, dS_2}{\pi r^2}$$

The double integral may be written $S_1 F_{A12}$, where F_{A12} is called the shape or arrangement factor with respect to area S_1. The value of F_{A12} will depend on the type of configuration under consideration. With the double integral defined in this manner, the heat interchange is

$$q_{12} = F_{A12} S_1 (E_1 - E_2)$$

or, in the case of heat exchange from body 2 to body 1,

$$q_{21} = F_{A21} S_2 (E_2 - E_1)$$

In general,

$$q = S F_A \, \Delta E \tag{7.6}$$

The equality

$$S_1 F_{A12} = S_2 F_{A21} \tag{7.7}$$

is known as the reciprocity theorem and shows that in a two-body system, the heat gained by the receiver is equal to the heat dissipated by the source.

The determination of the arrangement factor by means of the evaluation of the integral is tedious even for the most simple configurations. Fortunately, however, the literature contains many references to specific configurations of practical interest [2-4].

Table 7.1 presents the arrangement factor for several different configurations. This table has been assembled from the work of Hottel [2]. Kreith [3] presents a large glossary of shape factors that are in graphical form. These graphs are easy to read, and for most applications the accuracy is sufficient.

Of interest to the analyst working in the cooling of electronic equipment are the following:[3]

Perpendicular rectangles having a common edge (Fig. 7.7)
Finite, parallel, opposed rectangles (Fig. 7.8)
Differential area and a parallel rectangle (Fig. 7.9)
Differential area and a perpendicular rectangle (Fig. 7.10)

7.9 SHAPE FACTOR ALGEBRA

In the event that a configuration is encountered for which the arrangement factor cannot be obtained, one may frequently resort to some algebra to obtain the desired shape factor. This arrangement or shape factor algebra is based on three properties of the shape factor.

The *reciprocal property* has already been derived and can be expressed without the use of the subscript A:

$$S_1 F_{12} = S_2 F_{21} \tag{7.8}$$

[3] Taken from Gebhart [1].

TABLE 7.1   Radiation Factors[a]

Surfaces	Area, S	F_A	F_ϵ
Infinite parallel planes S_1 and S_2	S_1 or S_2	1	$\dfrac{1}{(1/\epsilon_1) + (1/\epsilon_2) - 1}$
Completely enclosed small body S_1, compared with large enclosing body S_2	S_1	1	ϵ_1
Completely enclosed large body S_1, compared with enclosing body S_2	S_1	1	$\dfrac{1}{(1/\epsilon_1) + (1/\epsilon_2) - 1}$
Concentric spheres, S_1 inside S_2	S_1	1	$\dfrac{1}{(1/\epsilon_1) + (S_1/S_2)[(1/\epsilon_2) - 1]}$
Infinite concentric cylinders, S_1 inside S_2	S_1	1	$\dfrac{1}{(1/\epsilon_1) + (S_1/S_2)[(1/\epsilon_2) - 1]}$

[a]From Hottel [2], by permission of the American Society of Mechanical Engineers.

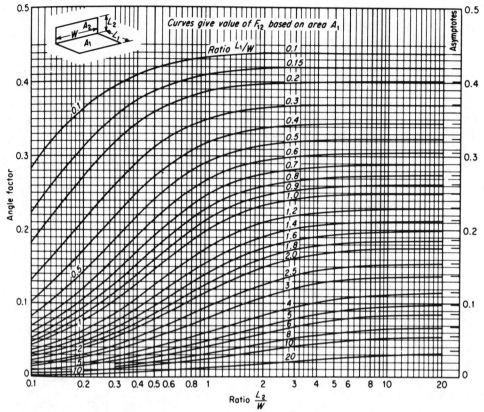

FIG. 7.7 Shape, angle, or arrangement factors between perpendicular rectangular areas with a common edge. (From Gebhart [1], by permission.)

The *additive property* says that if the surface S_1 is subdivided into n subsurfaces, S_i, each with a shape factor $F_{1i,2}$, where $i = 1, 2, 3, \ldots, n$, then, because $S_1 = \Sigma_{i=1}^{n} S_{1i}$, the exchange between surface 1 and surface 2 is completely covered by

$$S_1 F_{12} = \sum_{i=1}^{n} S_{1i} F_{1i,2} \tag{7.9}$$

It is apparent that the reciprocal property of Eq. (7.8) can be employed to show that

$$F_{21} = \sum_{i=1}^{n} F_{2,1i} \tag{7.10}$$

because $S_1 = \Sigma_{i=1}^{n} S_{1i}$. Indeed, this argument can be extended to surface 1 divided into n parts *and* surface 2 divided into m parts, with the result that

$$S_1 F_{12} = \sum_{i=1}^{n} \sum_{j=1}^{m} S_{1i} F_{1i,2j} \tag{7.11}$$

The *enclosure property* pertains to completely enclosed space such as one composed of two printed circuit boards and two sides, the top and the bottom of an electronic package. Indeed, if some surfaces in the enclosure are convex, they may have a "self-to-self" shape factor. Thus for a completely enclosed space where $S = S_1 + S_2 + S_3 + \cdots$, the enclosure property is represented by

$$\sum_{j=1}^{n} F_{i-j} = 1.0 \qquad i = 1, 2, 3, \ldots, n \tag{7.12}$$

EXAMPLE

For the perpendicular wall arrangement shown in Fig. 7.11, determine the shape factor F_{1-4}, where $S_1 = S_2$ and $S_3 = S_4$.

SOLUTION

Call $S_5 = S_1 + S_2$ and $S_6 = S_3 + S_4$. By the additive property,

$$S_5 F_{56} = S_1 F_{13} + S_1 F_{14} + S_2 F_{23} + S_2 F_{24}$$

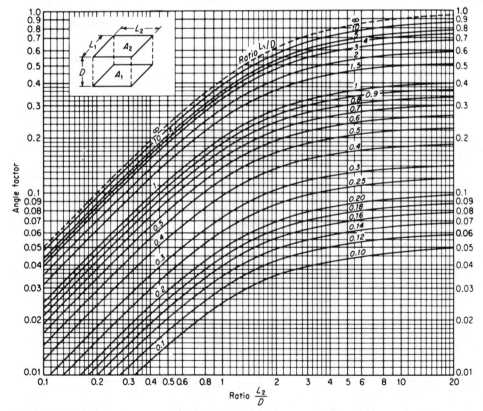

FIG. 7.8 Shape, angle, or arrangement factors between directly opposed rectangular areas. (From Gebhart [1], by permission.)

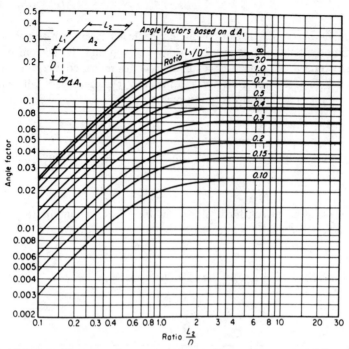

FIG. 7.9 Shape, angle, or arrangement factors between a differential area and a parallel rectangle. (From Gebhart [1], by permission.)

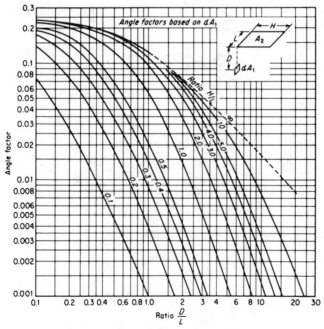

FIG. 7.10 Shape, angle, or arrangement factors between a differential area and a perpendicular rectangle. (From Gebhart [1], by permission.)

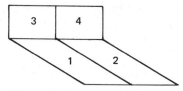

FIG. 7.11 Perpendicular wall arrangement for shape factor example.

Here F_{56} and $F_{13} = F_{24}$ can be found from Fig. 7.7 and are now "knowns." Then, because $S_1 = S_2$, $S_3 = S_4$, and $F_{13} = F_{24}$,

$$2S_1 F_{56} = 2S_1 F_{13} + S_1 (F_{14} + F_{23})$$

But $F_{14} = F_{23}$ as well, so the result is

$$F_{14} = F_{56} - F_{13}$$

7.10 HEAT FLOW BY RADIATION BETWEEN NONBLACK SURFACES

Consider two opaque surfaces, each within full view of the other. As seen previously, if these surfaces are black, the radiant heat interchange between the two is

$$q = S F_A \, \Delta E$$

and if the surfaces are two infinite parallel plates, the arrangement factor F_A will reduce to unity such that

$$q = S \, \Delta E$$

This expression will not hold if the two surfaces are not black. Figure 7.12 shows two parallel plane surfaces that are not black ($\epsilon \neq 1$ and $\alpha \neq 1$). If there is a temperature difference between the surfaces, there will be an interchange of radiant energy.

Begin by considering radiant energy leaving surface 1. Based on black-body radiation, the emission can be represented by

$$E_1 = \epsilon_1 E_{b1}$$

This radiation will strike surface 2, where part of it will be absorbed and part reflected. (Transmission through the surfaces is neglected.) That which is absorbed will be

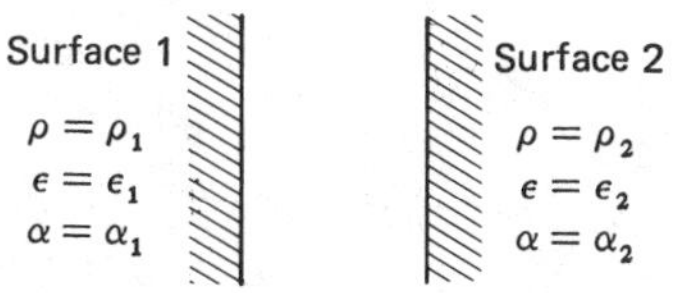

Surface 1

$\rho = \rho_1$
$\epsilon = \epsilon_1$
$\alpha = \alpha_1$

Surface 2

$\rho = \rho_2$
$\epsilon = \epsilon_2$
$\alpha = \alpha_2$

FIG. 7.12 Configuration for study of radiant heat exchange between nonblack parallel planes in full view of one another.

$\alpha_2 \epsilon_1 E_{b1}$ and that reflected will be $\rho_2 \epsilon_1 E_{b1}$. The reflected portion will be intercepted by surface 1, where part is absorbed and part is rereflected (absorbed $= \alpha_1 \rho_2 \epsilon_1 E_{b1}$, reflected $= \rho_1 \rho_2 \epsilon_1 E_{b1}$). This process can be continued indefinitely, as shown by the array of Table 7.2.

In a similar manner, one can consider the emission of surface 2. This emission is

$$E_2 = \epsilon_2 E_{b2}$$

and it will strike surface 1 in a manner such that part is absorbed and part is reflected (absorbed $= \alpha_1 \epsilon_2 E_{b2}$, reflected $= \rho_1 \epsilon_2 E_{b2}$). The reflected portion that is intercepted by surface 2 will be rereflected in part by surface 2 (absorbed $= \alpha_2 \rho_1 \epsilon_2 E_{b2}$, reflected $= \rho_2 \rho_1 \epsilon_2 E_{b2}$). This process can also be continued indefinitely as shown by Table 7.2.

The net heat transferred from surface 1 to surface 2 is the emission $\epsilon_1 E_{b1}$ minus the fraction of $\epsilon_1 E_{b1}$ and of $\epsilon_2 E_{b2}$ that is ultimately absorbed by surface 1 after successive reflections. Hence,

$$q_{12} = S_1 \epsilon_1 E_{b1} - S_2 \alpha_1 \epsilon_2 E_{b2} - S_1 \alpha_1 \rho_2 \epsilon_1 E_{b1} - S_2 \alpha_1 \epsilon_2 \rho_1 \rho_2 E_{b2}$$
$$- S_1 \alpha_1 \rho_2^2 \rho_1 E_{b1} - S_2 \alpha_1 \rho_1^2 \rho_2^2 \epsilon_2 E_{b2}$$
$$- S_1 \alpha_1 \rho_2^3 \rho_1^2 \epsilon_1 E_{b1} - S_2 \alpha_1 \rho_1^3 \rho_2^3 \epsilon_2 E_{b2} - \cdots$$

or
$$q_{12} = S_1 \epsilon_1 E_{b1} (1 - \alpha_1 \rho_2 - \alpha_1 \rho_2^2 \rho_1 - \alpha_1 \rho_2^3 \rho_1^2 - \cdots)$$
$$- S_2 \epsilon_2 E_{b2} (\alpha_1 + \alpha_1 \rho_1 \rho_2 + \alpha_1 \rho_1^2 \rho_2^2 + \alpha_1 \rho_1^3 \rho_2^3 + \cdots)$$

If the difference in temperature between the two bodies is not too great, so that the wavelength of emission is equal for both surfaces, or if the emissivity of either surface is fairly constant with temperature, or if the emissivity, absorptivity, and reflectivity do not vary with wavelength (gray surfaces),

$$\epsilon_1 = \alpha_1$$

$$\epsilon_2 = \alpha_2$$

$$\rho_1 = 1 - \alpha_1 = 1 - \epsilon_1$$

$$\rho_2 = 1 - \alpha_2 = 1 - \epsilon_2$$

Then

$$q_{12} = S_1 \epsilon_1 E_{b1} [1 - \epsilon_1 (1 - \epsilon_2) - \epsilon_1 (1 - \epsilon_1)(1 - \epsilon_2)^2$$
$$- \epsilon_1 (1 - \epsilon_1)^2 (1 - \epsilon_2)^3 + \cdots] - S_2 \epsilon_2 E_{b2} [\epsilon_1 + \epsilon_1 (1 - \epsilon_1)(1 - \epsilon_2)$$
$$+ \epsilon_1 (1 - \epsilon_1)^2 (1 - \epsilon_2)^2 + \cdots]$$

which upon reduction of the infinite series in both cases yields, if $S_1 = S_2$,

$$q_{12} = S(E_{b1} - E_{b2}) \frac{1}{1/\epsilon_1 + 1/\epsilon_2 - 1}$$

TABLE 7.2 Tabulation of Energy Absorbed and Radiated by Two Parallel Surfaces

Pass	Surface 1				Surface 2			
	Beam originating at 1		Beam originating at 2		Beam originating at 1		Beam originating at 2	
	Absorbed	Radiated	Absorbed	Radiated	Absorbed	Radiated	Absorbed	Radiated
1			$\alpha_1\epsilon_2 E_{b2}$	$\rho_1\epsilon_2 E_{b2}$	$\alpha_2\epsilon_1 E_{b1}$	$\rho_2\epsilon_1 E_{b1}$	$\alpha_2\rho_1\epsilon_2 E_{b2}$	$\rho_1\rho_2\epsilon_2 E_{b2}$
2	$\alpha_1\rho_2\epsilon_1 E_{b1}$	$\rho_1\rho_2\epsilon_1 E_{b1}$	$\alpha_1\rho_1\rho_2\epsilon_2 E_{b2}$	$\rho_1^2\rho_2\epsilon_2 E_{b2}$	$\alpha_2\rho_1\rho_2\epsilon_1 E_{b1}$	$\rho_1\rho_2^2\epsilon_1 E_{b1}$	$\alpha_2\rho_1^2\rho_2\epsilon_2 E_{b2}$	$\rho_1^2\rho_2^2\epsilon_2 E_{b2}$
3	$\alpha_1\rho_1\rho_2^2\epsilon_1 E_{b1}$	$\rho_1^2\rho_2^2\epsilon_1 E_{b1}$	$\alpha_1\rho_1^2\rho_2^2\epsilon_2 E_{b2}$	$\rho_1^3\rho_2^2\epsilon_2 E_{b2}$	$\alpha_2\rho_1^2\rho_2^2\epsilon_1 E_{b1}$	$\rho_1^2\rho_2^3\epsilon_1 E_{b1}$	$\alpha_2\rho_1^3\rho_2^2\epsilon_2 E_{b2}$	$\rho_1^3\rho_2^3\epsilon_2 E_{b2}$
4	$\alpha_1\rho_1^2\rho_2^3\epsilon_1 E_{b1}$	$\rho_1^3\rho_2^3\epsilon_1 E_{b1}$	$\alpha_1\rho_1^3\rho_2^3\epsilon_2 E_{b2}$	$\rho_1^4\rho_2^3\epsilon_2 E_{b2}$	$\alpha_2\rho_1^3\rho_2^3\epsilon_1 E_{b1}$	$\rho_1^3\rho_2^4\epsilon_1 E_{b1}$	.	.
⋮	⋮	⋮	⋮	⋮	⋮	⋮		
n	$\alpha_1\rho_1^{n-2}\rho_2^{n-1}\epsilon_1 E_{b1}$	$(\rho_1\rho_2)^{n-1}\epsilon_2 E_{b1}$	$\alpha_1(\rho_1\rho_2)^{n-1}\epsilon_2 E_{b2}$	$\rho_1^n\rho_2^{n-1}\epsilon_2 E_{b2}$	$\alpha_2(\rho_1\rho_2)^{n-1}\epsilon_1 E_{b1}$	$\rho_1^{n-1}\rho_2^n\epsilon_1 E_{b1}$	$\alpha_2\rho_1^{n-1}\rho_2^{n-2}\epsilon_2 E_{b2}$	$(\rho_1\rho_2)^{n-1}\epsilon_2 E_{b2}$

or $q_{12} = SF_e(E_{b1} - E_{b2})$ (7.13)

with F_e defined as the emissivity factor.

The emissivity factor, like the arrangement factor, will depend on the configuration. Unlike the arrangement factor, F_e will depend on the emissivities of the surfaces. Emissivity factors have been worked out for a number of other cases. A compilation of some of these due to Hottel [2] is given in Table 7.1.

7.11   THE COMPLETE RADIATION EQUATION

The reader may note that the discussion of the shape factor was based on a consideration of perfect emitters or black bodies. In addition, the introduction to the subject of the emissivity factor considered a configuration that had a unity arrangement factor. Table 7.1 presents the arrangement and emissivity factors for several configurations. It is obvious that the complete relationship for radiant heat transfer will be a function of both the arrangement and emissivity factors:

$$q_{12} = SF_A F_e (E_1 - E_2)$$

which can be written, using the Stefan-Boltzmann law, as

$$q_{12} = SF_A F_e (oT_1^4 - oT_2^4)$$

or $q = oSF_A F_e (T_S^4 - T_R^4)$ (7.14)

where the subscripts refer respectively to the source and receiver.

7.12   RADIATION WITHIN ENCLOSURES BY A NETWORK METHOD

For surfaces within an enclosure that obey Lambert's cosine law and reflect diffusely, a network method is quite useful to evaluate the multiple heat flow paths from surface to surface.

Define the radiosity J as the rate at which radiation leaves a surface per unit surface area and the irradiation G as the total radiation incident on a surface per unit area. Then, for the ith surface in an enclosure,

$$J_i = \rho_i G_i + \epsilon_i E_{bi}   (\text{W/m}^2)$$ (7.15)

where ρ_i and ϵ_i are the surface reflectivity and emissivity, respectively, and E_{bi} is the black-body emissive power.

If the surface temperature and irradiation over the surface are uniform, then the net rate of heat loss (or gain) is

$$q_i = S_i (J_i - G_i)$$ (7.16)

If the surfaces are gray (see Sec. 7.5), then

$$\epsilon_i = \alpha_i$$

an important limiting assumption, and

$$\rho_i = 1 - \epsilon_i$$

so that Eqs. (7.15) and (7.16) can be combined by noting from Eq. (7.15) that

$$G_i = \frac{1}{\rho_i}(J_i - \epsilon_i E_{bi})$$

and hence

$$q_i = S_i \left[J_i - \frac{1}{\rho_i}(J_i - \epsilon_i E_{bi}) \right]$$

or $q_i = \dfrac{S_i}{\rho_i}[\epsilon_i E_{bi} + (\rho_i - 1)J_i] = \dfrac{S_i \epsilon_i}{1 - \epsilon_i}(E_{bi} - J_i)$ (7.17)

If Eq. (7.17) is rewritten in the form

$$q_i = \frac{E_{bi} - J_i}{(1 - \epsilon_i)/S_i \epsilon_i}$$ (7.18)

an analogy can be drawn that shows $E_{bi} - J_i$ as a difference of potential and $(1 - \epsilon_i)/S_i \epsilon_i$ as a radiative thermal resistance. This resistance governs the flow of heat between a node point representing $E_{bi} = \sigma T_i^4$ and another node point representing J_i.

Now the direct radiative interchange between surface i and surface j is governed by the difference of the two radiosities, the surface S_i and the shape factor F_{ij}. It is easy to see that

$$q_{ij} = \frac{J_i - J_j}{S_i F_{ij}}$$

and that use of the reciprocity relationship of Eq. (7.8) will allow this to be enhanced to

$$q_{ij} = \frac{J_i - J_j}{S_i F_{ij}} = \frac{J_i - J_j}{S_j F_{ji}}$$ (7.19)

Here the radiative thermal resistance is $1/S_i F_{ij} = 1/S_j F_{ji}$. This resistance governs the flow of heat between two node points, each represented by a radiosity.

Equations (7.18) and (7.19) form the basis for the electrical network. The nodes marked E_{bi} represent heat sources, one of which must be known. Each of these may be related to a temperature, $E_{bi} = \sigma T_i^4$. The resistances $(1 - \epsilon_i)/S_i \epsilon_i$ transfer the E_{bi}'s to surface radiosities J_i, thereby accounting for the fact that the surfaces are not black but gray. The resistances $1/S_i F_{ij} = 1/S_j F_{ji}$ link the radiosities and account for the actual radiative interchange between the ith and jth surfaces.

The setup of the actual circuit will now be shown in an example. It should be borne in mind that the evaluation of the radiosities is only rarely required. The name of the game is "obtain the surface temperatures." The example treats the radiative interchange between two surfaces. The extension of the method (a computer will be required) to many surfaces, such as arrays of printed circuit boards, is obvious.

■   **EXAMPLE**
Table 7.1 shows that, for an infinitely long cylinder of surface area S_1 and having emissivity ϵ_1 within and concentric to an infinitely long cylinder of surface area S_2 having emissivity ϵ_2, the emissivity factor is

$$F_e = \frac{1}{(1/\epsilon_1) + (S_1/S_2)[(1/\epsilon_2) - 1]}$$

If $T_1 > T_2$, verify this by using a radiative network.

SOLUTION
Observe first that because every bit of the radiation from the inner cylinder is intercepted by the outer cylinder, $F_{12} = 1.0$.

Figure 7.13 shows a series circuit that represents the radiative heat transport. The resistances have values as follows:

$$R_1 = \frac{1 - \epsilon_1}{S_1 \epsilon_1}$$

$$R_2 = \frac{1}{S_1 F_{12}}$$

$$R_3 = \frac{1 - \epsilon_2}{S_2 \epsilon_2}$$

The total resistance is therefore

$$R_T = R_1 + R_2 + R_3$$

or $$R_T = \frac{1 - \epsilon_1}{S_1 \epsilon_1} + \frac{1}{S_1 F_{12}} + \frac{1 - \epsilon_2}{S_2 \epsilon_2}$$

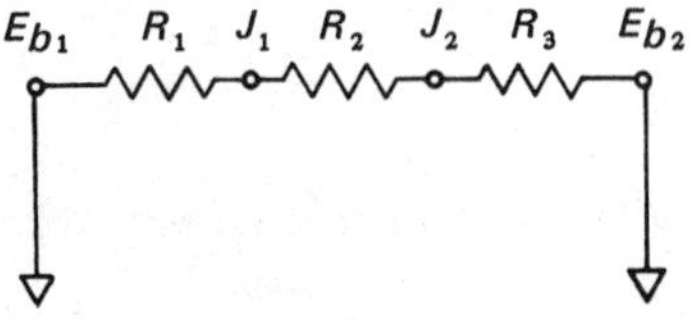

FIG. 7.13 Electrothermal analog (radiation network) for concentric cylinder example.

Algebraic adjustment of the last term permits writing

$$R_T = \frac{1}{S_1}\left(\frac{1-\epsilon_1}{\epsilon_1} + 1 + \frac{S_1}{S_2}\frac{1-\epsilon_2}{\epsilon_2}\right)$$

because $F_{12} = 1.0$. Further simplification gives

$$R_T = \frac{1}{S_1}\left[\frac{1}{\epsilon_1} - 1 + 1 + \frac{S_1}{S_2}\left(\frac{1}{\epsilon_2} - 1\right)\right]$$

or $R_T = \frac{1}{S_1}\left[\frac{1}{\epsilon_1} + \frac{S_1}{S_2}\left(\frac{1}{\epsilon_2} - 1\right)\right]$

Now, because

$$q = F_A \frac{E_{b1} - E_{b2}}{R_T}$$

it is seen that with $E_{b1} = \sigma T_1^4$ and $E_{b2} = \sigma T_2^4$,

$$q = \sigma S_1 F_A \left\{\frac{1}{(1/\epsilon_1) + (S_1/S_2)[(1/\epsilon_2) - 1]}\right\}(T_1^4 - T_2^4)$$

With $F_A = 1$, comparison with Eq. (7.14),

$$q = \sigma S F_A F_\epsilon (T_S^4 - T_R^4)$$

shows that

$$F_\epsilon = \frac{1}{(1/\epsilon_1) + (S_1/S_2)[(1/\epsilon_2) - 1]}$$

which is the required result.

7.13 NOMENCLATURE

Roman Letter Symbols

A a combination of terms, defined where used

C velocity of light, 3×10^8 m/s

E emissive power, W/m^2; with subscript $b\lambda$, monochromatic emissive power, W/m^2 μm; with subscript b, black-body emissive power, W/m^2

F with subscript A, the arrangement or shape factor, dimensionless; with subscript ϵ, the emissivity factor, dimensionless

G irradiation, W/m^2

h Planck's constant, 6.624×10^{-27} erg s

i	an index
I	radiation intensity, W/m^2 steradian
j	an index
J	radiosity, W/m^2
k	Boltzmann's constant, 1.380×10^{-16} erg/K
m	designates limit of an index
n	designates limit of an index
q	heat flow, W
r	radius or radial distance, m
S	surface area, m^2
T	absolute temperature, K
x	a transformed variable, defined where used
z	a transformed variable, defined where used

Greek Letter Symbols

α	absorptivity, dimensionless
Δ	indicates change in variable
ϵ	emissivity, dimensionless
λ	wavelength, μm
μ	unit of wavelength, μm
ρ	reflectivity, dimensionless
σ	Stefan-Boltzmann constant, 5.67×10^{-8} W/m^2 K^4
τ	transmissivity, dimensionless
ϕ	an angle, degrees or radians
ψ	an angle, degrees or radians
ω	solid angle, steradians

Subscripts

A	indicates arrangement or shape factor
b	indicates black or ideal surface·
$b\lambda$	indicates monochromatic emission
ϵ	indicates emissivity factor

7.14   REFERENCES

1   Gebhart, B., *Heat Transfer*, McGraw-Hill, New York, 1971.
2   Hottel, H. C., Radiant Heat Transmission, *Mech. Eng.*, vol. 52, pp. 699–704, 1930.
3   Kreith, F., *Radiation Heat Transfer*, International Text Book, Scranton, Pa., 1962.
4   Siegel, R., and Howell, J. R., *Thermal Radiation Heat Transfer*, 2d ed., Hemisphere, Washington, D.C., 1981.
5   Holman, J. P., *Heat Transfer*, 4th ed., McGraw-Hill, New York, 1976.

8

■ evaporation, boiling, and condensation

8.1 INTRODUCTION

Previous chapters have dealt exclusively with heat transfer to fluids in a single liquid or gaseous phase, but in many applications of interest, heat exchange is accompanied by evaporation of a liquid or condensation of a vapor. The resulting flow of vapor toward or away from the heat transfer surface and the high rates of thermal transport associated with change-of-phase processes make it incorrect to apply single-phase formulations in these situations and necessitate the use of the appropriate two-phase relations.

This chapter is devoted to exploring the mechanisms active in liquid-vapor phase changes and presents the basic theory and empirical correlations needed to analyze evaporation, boiling, and condensation.

8.2 EVAPORATION

8.2.1 Thermodynamic Basis

When liquid in a constant-pressure vessel is uniformly heated at a slow rate, the resulting changes in temperature and volume are related in a characteristic way as shown in Fig. 8.1. Starting at point A, the addition of heat results in a temperature rise and a modest increase in volume until point 1 is reached. Along this all-liquid locus, the amount of heat required to raise the temperature of a unit mass one degree is nearly constant and is referred to as the liquid constant-pressure specific heat, c_p. In the region between points 1 and 2, vapor and liquid are present as coexisting phases in equilibrium. The addition of heat increases the vapor fraction and, hence, the volume without altering the temperature of the two-phase mixture. The amount of heat needed to convert a unit mass of liquid into vapor is termed the latent heat of vaporization, h_{fg}. At point 2 all the liquid has vaporized, and further heating results in a temperature rise of the vapor along a locus appropriate to the *vapor* constant-pressure specific heat. The termination of the all-liquid state, point 1, and the beginning of the all-vapor state, point 2, are generally referred to as the liquid and vapor saturation points, respectively.

This description pertains to addition of heat to a quantity of liquid at a single, constant pressure. If the process is repeated for many different pressures, the saturation curve shown in Fig. 8.2 is obtained. The saturated-liquid line shown in the figure represents the locus of all points 1, and the saturated-vapor line represents the locus of all points 2. It is of interest to note that a "critical" temperature and pressure, at which liquid and vapor molecules are energetically identical and at which no difference

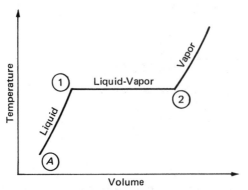

FIG. 8.1 Constant-pressure heat addition.

exists between the specific volume of the liquid and vapor, can be defined for each fluid. This point corresponds to the top of the dome in Fig. 8.2.

For convenience, the relation between the saturation temperature (line 1-2 in Fig. 8.1) and pressure implicit in Fig. 8.2 is often plotted directly, to reflect the dependence of the vapor pressure on the liquid temperature. The appropriate curves for several fluorocarbon fluids, of importance in liquid cooling of electronic components, are shown in Fig. 8.3.

8.2.2 Thermal Transport Mechanism

Because of the higher internal energy of a vapor molecule relative to a liquid molecule of the same species and at the same temperature, evaporation of a fluid, under conditions far removed from the critical point, generally results in a very substantial heat flux. The migration or diffusion of vapor away from a free liquid-vapor interface is governed by the difference in vapor pressure or vapor concentration between the interface and the ambient. Along the interface it is generally assumed that the vapor and liquid are in thermodynamic equilibrium, and therefore that the vapor pressure at the interface corresponds to the saturation value at the local liquid temperature. Because, as seen in Fig. 8.3, the saturation pressure increases with temperature, the rate of evaporation from a free surface in an environment of fixed vapor pressure

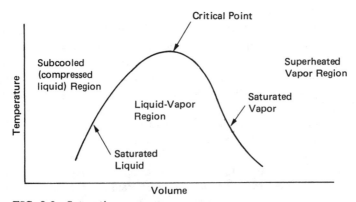

FIG. 8.2 Saturation curve.

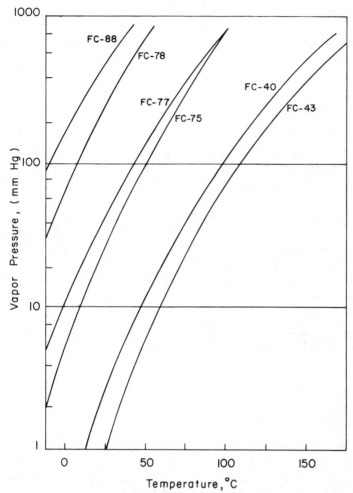

FIG. 8.3 Vapor pressure of several fluorocarbon fluids [25].

increases as the liquid temperature approaches the boiling point. At a given total ambient pressure, that is, vapor pressure plus pressure of other gas components, the liquid temperature cannot exceed the boiling point and, consequently, the maximum interface vapor pressure and evaporation rate are attained at this point.

It must be noted, however, that when the heat of vaporization needed for this process is transferred to the liquid by boiling along surfaces submerged in the liquid, the bursting of vapor bubbles at the liquid-vapor interface may enhance the evaporative process and result in heat transfer rates much above those associated with diffusive vapor transport from the interface.

8.2.3 Predicting Evaporation Rates

By analogy to convective heat transfer as discussed in Chap. 6, the convection of vapor from wetted surfaces can be related to a mass transfer coefficient h_D and can be described by an equation of the form [1]

$$w_{vapor} = h_D A(C_w - C_\infty)$$ (8.1)

Values of h_D for a variety of liquids evaporating into air in wetted-wall columns were compiled by Gilliland and Sherwood [2] and correlated in the form

$$h_D = 0.023 \left(\frac{D}{d}\right) (\text{Re})^{0.83} (\text{Sc})^{0.44}$$ (8.2)

where Sc is the Schmidt number, $\text{Sc} = \mu/\rho D$.

The diffusivity D of the vapor in air is analogous to thermal conductivity of solids and, like conductivity, is a tabulated quantity. Equation (8.2) is recommended for Reynolds numbers in the range 2,000–35,000 and Schmidt numbers between 0.6 and 2.5 [1].

The similarity between Eqs. (8.2) and (6.70) for heat transfer in forced convection inside passages is most noteworthy and suggests that mass transfer coefficients can be obtained by replacing the thermal conductivity by diffusivity and the Prandtl number by the Schmidt number in the standard heat transfer correlations. This analogy is generally stated in terms of the Chilton-Colburn j factors [3] and for evaporation from flat-wetted surfaces yields

$$h_D = 0.664 \frac{V}{\sqrt{\text{Re}_L}(\text{Sc})^{0.67}}$$ (8.3)

for laminar flow and

$$h_D = 0.037 \frac{V}{(\text{Re}_L)^{0.2} (\text{Sc})^{0.67}}$$ (8.4)

for turbulent flow. Agreement between Eqs. (8.3) and (8.4) and the available data is generally good, though not excellent [1].

Evaporation in a quiescent medium is more nearly analogous to heat conduction than heat convection, and the calculation of the associated heat transfer rate must be suitably modified. Assuming that the vapor and ambient gas behave as an isothermal mixture of perfect gases and diffuse past each other, the mass flux of vapor can be shown to equal [1]

$$\frac{w_v}{A} = \frac{M_v D P}{RTl} \ln\left(\frac{P - p_l}{P - p_w}\right)$$ (8.5)

Here M_v is the molecular weight of the vapor, P is the total pressure, and R is the universal gas constant. Multiplication of the mass flux w_v/A by the latent heat of vaporization h_{fg} yields the desired surface heat flux or cooling effect.

8.3 BOILING

Although boiling, like evaporation, involves a change from the liquid to the vapor phase, in distinction to evaporation, boiling generally refers to a process that occurs along surfaces submerged in the fluid and is characterized by the dominant influence of vapor bubbles. Boiling typically commences when the surface temperature exceeds the liquid saturation temperature (the boiling point) by 3 to 10°C, and vapor bubbles are then found to grow and issue from minute cavities in the surface.

In the analysis and correlation of boiling phenomena, it is convenient to distinguish between *pool boiling*, referring to boiling in an initially quiescent liquid, and *flow boiling*, referring to boiling in the presence of a strong velocity field, as may occur, for example, in pipe flow. Furthermore, since the boiling process depends primarily on the temperature of the heated surface, both *subcooled boiling*, during which the bulk fluid is below the saturation temperature, and *saturated boiling*, during which the bulk fluid is uniformly at the saturation temperature, may be observed.

In boiling the variation of heat flux along a heated element or surface results in a characteristic temperature response at the surface, reflecting a progression through particular regimes of ebullient heat transfer. This behavior is generally represented by a log-log plot of the heat flux q'' versus the surface superheat, $T_w - T_s$. A typical "boiling curve" for the saturated pool boiling of Refrigerant 113 (R-113) at atmospheric pressure is shown in Fig. 8.4.

The initial, preboiling part of the curve, labeled I, represents thermal transport by natural convection, resulting from the temperature difference between the heated surface and the liquid. At point a, sufficient superheat is available to initiate the growth of vapor bubbles at nucleation sites on the surface and boiling commences. In region II, modest increases in surface superheat result in the activation of many more nucleation sites and a rapid increase in the frequency of bubble departure at each site. The violent disturbance of the hot boundary layer along the heated surface

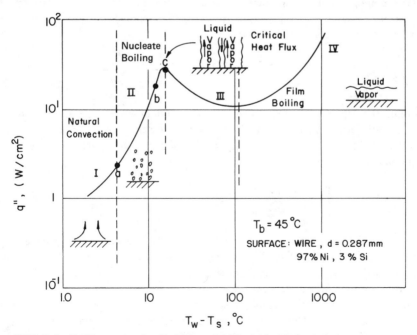

FIG. 8.4 Boiling curve for Refrigerant 113 at atmospheric pressure.

and gross fluid circulation induced by the motion of the vapor bubbles leads to a very steep rise in the heat flux from the heater in this nucleate boiling regime. At point b, bubble departure frequency at each nucleation site is so high that the trailing bubble merges with the leading bubble and vapor columns rooted at particular nucleation sites appear. Beyond this inflection point, bubble interference constrains the incremental increase in heat flux resulting from higher wall superheat, and at point c, bubble interference is so severe that the flow of liquid to the surface is halted and *dryout* occurs. This peak or critical heat flux represents a local maximum in the boiling curve, and higher heat fluxes can be obtained only in the film boiling regime, labeled IV, at much higher wall superheats than are encountered in nucleate boiling. In region IV, a vapor layer blankets the heater, and relatively ineffective thermal conduction through the vapor (augmented by radiation) must be relied upon to transfer the heat from the heated surface to the liquid. Region III on the boiling curve corresponds to unstable film boiling and is characterized by rapid, local oscillations between nucleate and film boiling.

Because of the complex interactions of the distinct heat transfer mechanisms active in ebullient thermal transport, it is not yet possible to represent the entire boiling curve by a single analytical function. Rather, it is common practice to provide separate analytical or semiempirical relations for each inflection point and boiling regime. Predictive expressions for the boiling curve and salient features of the individual boiling regimes are discussed in succeeding sections.

8.3.1 Nucleate Boiling

Mechanical stability of a spherical void or bubble of radius r, such as formed during nucleate boiling, requires that the internal pressure exceed the external pressure. The relationship between the vapor pressure within the bubble and the ambient liquid pressure can be expressed according to

$$p_v - p_l = \frac{2\sigma}{r} \tag{8.6}$$

Assuming the vapor to be close to the saturation condition and the liquid and vapor to be at a uniform temperature, the required pressure difference can be translated via the Clausius-Clapeyron relation into the degree of liquid superheat needed to sustain the bubble, or [4]

$$T_v = T_l = T_s \left[1 + \frac{2\sigma}{r} \frac{v_{fg}}{h_{fg}} \frac{\rho}{\rho - \rho_v} \right] \tag{8.7}$$

Thus, bubbles of increasing radius can be supported by progressively lower liquid superheat and on rough surfaces boiling can be expected to commence at temperatures only modestly above the saturation temperature.

Whereas Eq. (8.7) pertains to isothermal fluids, it must be noted that a bubble growing at a heated surface encounters decreasing temperature as it protrudes into the liquid boundary layer. Under such circumstances, it is the liquid superheat at the bubble "roof" that must follow Eq. (8.7), and the wall temperature must exceed this value. Following Bergles and Rohsenow [5], the wall superheat required for the formation of the first bubble or boiling incipience at a heated surface can be estimated by

$$(T_w - T_s)_{\text{incip}} = \left(\frac{8q''T\sigma}{\rho_v h_{fg} k_f} \right)^{1/2}$$

(8.8)

The preboiling or incipience heat flux is itself a function of the wall-to-liquid temperature difference and a convection heat transfer coefficient h. Consequently, in a saturated liquid the estimated incipience superheat can also be expressed as

$$(T_w - T_s)_{\text{incip}} = \frac{8hT\sigma}{\rho_v h_{fg} k_f}$$

(8.9)

It must be noted that Eqs. (8.8) and (8.9) presume that cavities of all radii exist on the surface and that the liquid temperature decreases linearly from the wall. As a result, when applied to real surfaces with low heat fluxes or heat transfer coefficients, these equations are found to underpredict the superheat required for boiling incipience.

DEPARTURE DIAMETER
A vapor bubble growing in a surface cavity will remain attached to the surface until the appropriate combination of buoyancy, fluid drag, and inertia forces overcome the retarding force of surface tension. For the commonly encountered slow-growing bubbles, prime attention can be focused on buoyancy and surface tension forces and the bubble departure diameter for saturated pools correlated by [6]:

$$E_0^{1/2} = (1.5 \times 10^{-4})(\text{Ja}^*)^{5/4} \quad \text{for water} \tag{8.10a}$$

$$E_0^{1/2} = (4.65 \times 10^{-4})(\text{Ja}^*)^{5/4} \quad \text{for other fluids} \tag{8.10b}$$

where $E_0 = g(\rho - \rho_v)D_b^2/g_0\sigma$ and $\text{Ja}^* = \rho c T_s/\rho_v h_{fg}$.

DEPARTURE FREQUENCY
The frequency with which bubbles depart a specified nucleation site is dependent on the heat flux and thermofluid properties and has been found to be related as well to the bubble departure diameter. Much of the data are inconsistent, but for slowly growing bubbles (thermally controlled) the frequency can be estimated by [4]

$$f = \frac{\frac{9}{16}\pi\alpha(\text{Ja})^2}{D_b^2} \quad \pm 10\%$$

(8.11)

with $\text{Ja} = \rho c(T_w - T_s)/h_{fg}\rho_v$. The departure frequency of rapidly growing bubbles (dynamically controlled) may be expressed as [7]

$$f = \left[\frac{\frac{4}{3}g(\rho - \rho_v)}{C_D \rho D_b} \right]^{1/2}$$

(8.12)

Caution must be used in attempting to predict this particular parameter.

The foregoing bubble frequency data can be used to determine the bubble flux or total number of bubbles departing a heated, boiling surface when the density and size of bubble nucleation sites are known. This information, however, is rarely available. Alternatively, the bubble flux can be determined from the vapor flux and bubble departure diameter. Unfortunately, the vapor flux Q_v is itself difficult to predict, and although Q_v can be expressed as

$$Q_v = \eta \frac{q''}{h_{fg}\rho_v} \tag{8.13}$$

the value of η varies in a complex fashion, from 0.01 to 0.02 at low heat fluxes to $\eta > 0.5$ at 20% of the critical heat [8], and to near unity as the critical heat flux is approached.

NUCLEATE BOILING CURVE

Since increased heat flux in the nucleate boiling regime is generally associated with the appearance of more vapor bubbles, and the heat content of the vapor in the bubbles typically accounts for only a small percentage of the total heat transfer [8]. it appears possible to relate the observed heat flux to "bubble pumping" of hot liquid away from the heated surface. Based on this postulate and the definition of a bubble Reynolds number and bubble Nusselt number, Rohsenow was able to obtain one of the more successful correlations of nucleate pool boiling [1]:

$$\frac{c_l(T_w - T_s)}{h_{fg}} = C_{sf}\left[\frac{q/A}{\mu h_{fg}}\sqrt{\frac{g_0\sigma}{g(\rho - \rho_v)}}\right]^r \Pr^s \tag{8.14}$$

The C_{sf} coefficient is interpreted as reflecting the influence of the cavity size distribution of the surface on vapor bubble generation, and hence ebullient heat transfer, and has been tabulated for various fluid-surface combinations [1, 4]. The surface quality can also effect the exponents r and s in Eq. (8.14) [4], but nevertheless, r is generally taken to equal $\frac{1}{3}$ and $s = 1.0$ for water, but 1.7 for all other fluids. With these exponents a C_{sf} value of 0.013 correlates a wide spectrum of data to within ±20% [4].

To obtain a precise relation between heat flux and wall superheat in the nucleate boiling regime, it is generally advisable to perform a limited number of experiments with the surface-fluid combination of interest and to use the results to establish C_{sf}, r, and s empirically.

The wall superheat encountered during boiling in flowing liquids is influenced not only by the heat flux, fluid properties, and surface characteristics, but reflects, as well, the liquid flow rate and the vapor fraction in the liquid. Accurate prediction of flow-boiling heat transfer rates is thus extremely difficult and requires, in addition to precise property values, a detailed knowledge of the geometry and orientation of the heated element and the flow regime and thermal history of the working fluid.

In the absence of such detail, it is possible to obtain a first estimate of flow boiling behavior by superimposing the single-phase convective heat transfer rate on the *bubble pumping effect* [1] by

$$\frac{q}{A}\bigg|_{fb} = \frac{q}{A}\bigg|_{c} + \frac{q}{A}\bigg|_{b} \tag{8.15}$$

In using Eq. (8.15), $q/A|_c$ can be calculated from the standard natural and forced convection heat transfer coefficient correlations presented in Chap. 6 and the surface-to-liquid (actual, not saturation) temperature difference. The *pure* boiling heat flux can be obtained from the Rohsenow correlation, given by Eq. (8.14).

Under most circumstances this same relation can also be used to account approximately for the effect of significant liquid subcooling on the pool boiling curve.

PARAMETRIC EFFECTS

Because of the "bubble pumping" nature of nucleate pool boiling, the heat transfer rates and wall superheats attained in this regime are sensitive to both bubble and thermal conduction parameters. Some of the important dependencies are embodied in Eq. (8.14), and this relation can generally be used to obtain at least a first estimate of the effects of variations in the fluid properties. Thus it is found that, with increasing pressure, the boiling curve shifts toward the left or toward lower wall superheats, as shown, for example, in Fig. 8.5.

As noted earlier, the character of the surface can influence the location and slope of the boiling curve and is one of the primary variables in ebullient behavior. Rougher surfaces generally contain more and larger cavities than polished surfaces of the same material and, as a consequence, nucleate boiling from rough surfaces commences and continues at lower wall superheats than from smooth surfaces [4] and is thus characterized by lower values of C_{sf}.

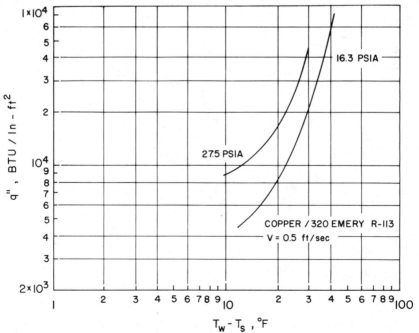

FIG. 8.5 Effect of pressure on nucleate boiling of R-113 [9].

Although not shown explicitly in the previous equations, the presence of non-condensable gases dissolved in the working fluid can have a profound influence on boiling behavior and is especially significant in dielectric fluids (used for immersion cooling of electronics) that possess a very high solubility for air. In such fluids, bubble formation and growth are aided by the presence of gas in the cavities along the heated surface, which reduces the vapor partial pressure and wall superheat needed for nuclea-tion and shifts the nucleate boiling curve to lower wall superheats.

In this boiling regime, vapor and possibly noncondensable gas occupying minute surface cavities serve as nuclei for vapor bubbles. Evaporation at the bubble-liquid interface provides the vapor needed for bubble growth, and the available superheat determines the size of the activated cavities. The available boiling data suggest that, for most working fluid/surface combinations, the wetting angle, shown in Fig. 8.6, is sufficiently large that the inflow of liquid in the wake of a departing bubble traps some vapor in the cavity and provides a nucleus for a subsequent bubble. The proper wetting characteristics are thus central to the nucleate boiling process, and extremely wetting or nonwetting surface/fluid combinations can result in anomolous behavior.

The boiling of nonwetting fluids is accompanied by the generation of very large vapor bubbles, which blanket wide areas of the heated surface and result in rather substantial wall superheats [4]. Alternately, highly wetting fluids penetrate and fill even small cavities and thus eliminate many potential nucleation sites. Boiling in-cipience, via the small-diameter cavities remaining active in such fluids, may necessitate very high wall superheats and be followed by a sudden decrease in surface temperature as the growing bubble activates the previously dormant, larger-diameter cavities. Such behavior may be encountered in the boiling of highly wetting dielectric fluids and may help explain the thermal hysteresis observed in the boiling of Refrigerant 113 (see Fig. 8.7 [9, 10]).

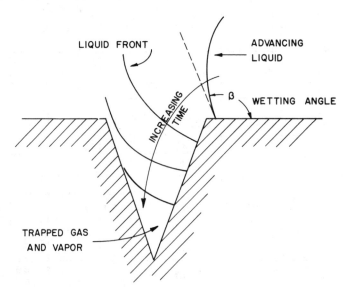

FIG. 8.6 Vapor trapping in surface cavities by an advancing liquid front (time increasing).

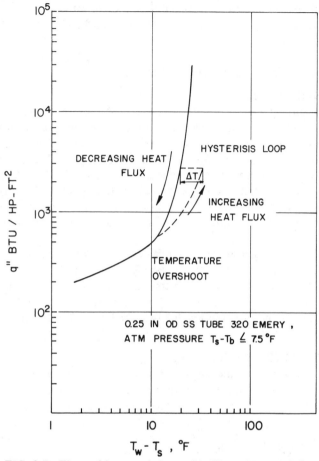

FIG. 8.7 Thermal hysteresis in pool boiling of R-113 [9].

8.3.2 Critical Heat Flux

CORRELATION

The peak, critical, or burn-out heat flux marks the end of the nucleate boiling regime, and heat fluxes in excess of this value result in wall superheats that are often one to two orders of magnitude above those prevailing in nucleate boiling. Although some controversy still exists in the literature over the physical mechanism responsible for the critical heat flux, it appears possible to explain much of the available data within the framework of Zuber's hydrodynamic instability model [11, 12]. This model asserts that the critical heat flux results when sufficient liquid can no longer flow past the vapor columns to the heated surface as a result of the breakdown of the vertical liquid-vapor interface. For boiling on large, horizontal surfaces, q''_{crit} is found to equal [4]

$$q''_{crit} = 0.18 \rho_v h_{fg} \left[\frac{g g_0 \sigma(\rho - \rho_v)}{\rho_v^2} \right]^{1/4} \left(\frac{\rho + \rho_v}{\rho} \right)^{-1/2} \tag{8.16}$$

where 0.18 is determined empirically. Experimental results for saturated pool boiling of many fluids are close to the values of q''_{crit} calculated via Eq. (8.16), but heater size, fluid velocity, and liquid subcooling exert a very profound influence on the critical heat flux. In distinction to nucleate boiling, surface characteristics are generally found not to affect the magnitude of the critical heat flux.

HEATER CONFIGURATION

A series of investigations by Lienhard and co-workers [13, 14] produced a unified approach to geometric effects on the pool-boiling critical heat flux and showed that the variation between experimental and predicted values could be related to a characteristic length L', equal to $L \sqrt{g(\rho - \rho_v)/\sigma g_0}$. The results, shown in Fig. 8.8, reveal that for low values of L', corresponding to cylindrical, spherical, and vertical plate heater dimensions that are small relative to the characteristic vapor jet diameter $[D_j \approx \sigma g_0/g(\rho - \rho_v)]$, the critical heat flux is considerably higher than predicted by Eq. (8.16) but the values of q''_{crit} approach the Zuber relation asymptotically as L' increases toward 10.

It is interesting to note that at very small values of L', as may be encountered in boiling from surfaces of integrated circuit packages, the first bubble nucleated may blanket the heater and lead to an abrupt transition from natural convection to film boiling. Under these circumstances, it is difficult to define the critical heat flux [15], and extrapolation of the curves in Fig. 8.8 below values of $L' = 0.1$ must therefore be done with caution [12].

LIQUID SUBCOOLING AND VELOCITY

Experimental results suggest that the degree of subcooling in the bulk fluid can markedly affect the critical heat flux. This parametric variation has been correlated by several investigators, including Ivey and Morris, who found the simple expression [16]

$$\frac{q''_{crit,sub}}{q''_{crit,sat}} = 1 + 0.1 \left(\frac{\rho}{\rho_v}\right)^{3/4} \frac{c_l(T_s - T_l)}{h_{fg}} \qquad (8.17)$$

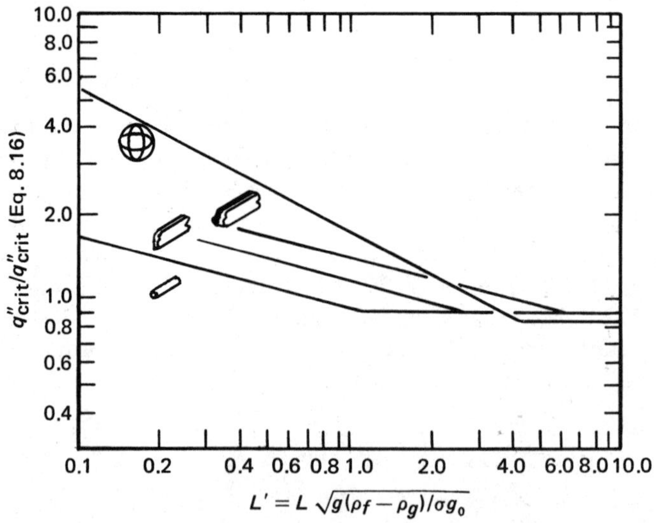

FIG. 8.8 Predictions of critical heat flux for various heater configurations [12].

The flow of liquid toward and along the heated element provides an alternative path for liquid seeking to reach the boiling surface and eases the removal or collapse of vapor columns generated at the surface. These two mechanisms can be expected to increase the critical heat flux, and even a modest natural circulation velocity in a nominally quiescent liquid can substantially increase the critical heat flux [17].

In the presence of forced convection, the critical heat flux is often as much as an order of magnitude higher than in pool boiling, but precise prediction is made difficult by the dependence of q''_{crit} on the two-phase flow regime, system dynamics, and thermal boundary conditions. Distinct correlations, many of which are presented in [4], are available for particular parametric combinations but are generally limited to high-pressure water flow, of interest to the nuclear industry.

For other fluids and as a general tool for obtaining a first estimate of q''_{crit} in subcooled flow boiling, it may be convenient to follow Gambill's [18] recommendation in modified form and set the critical heat flux in flow boiling equal to the sum of the pool boiling q''_{crit} given by Eqs. (8.16) and (8.17) and a convection component calculated as the product of the all-liquid convective coefficient and the wall-to-liquid temperature difference,

$$q''_{crit} = 0.18\rho_v h_{fg} \left[\frac{\sigma g_0 g(\rho - \rho_v)}{\rho_v^2} \right]^{1/4} \left(\frac{\rho + \rho_v}{\rho} \right)^{-1/2} \left[1 + 0.1 \left(\frac{\rho}{\rho_v} \right)^{3/4} \frac{c_l(T_s - T)}{h_{fg}} \right]$$
$$+ h(T_{w,crit} - T_l) \tag{8.18}$$

8.3.3 Film Boiling

The large temperature excursion resulting from heat dissipation in excess of the critical heat flux has minimized interest in the film boiling regime, and detailed experimental data are generally available only for cryogenic fluids and low-molecular-weight fuels.

The minimum film boiling heat flux has been shown by Zuber to be derivable from a stability analysis of the liquid-vapor boundary that yields [19]

$$q''_{min} = 0.09 \rho_{v,f} h_{fg} \left[\frac{g(\rho - \rho_v)}{\rho + \rho_v} \right]^{1/2} \left[\frac{g_0 \sigma}{g(\rho - \rho_v)} \right]^{1/4} \tag{8.19}$$

where $\rho_{v,f}$ is the vapor density evaluated at the average film temperature and all other properties are at the saturation temperature.

The minimum wall superheat required to sustain film boiling was found by Berenson to be expressible as [20]

$$\Delta T_{min} = 0.127 \frac{\rho_{v,f} h_{fg}}{k_{v,f}} \left[\frac{g(\rho - \rho_v)}{\rho + \rho_v} \right]^{2/3} \left[\frac{g_0 \sigma}{g(\rho \rho_v)} \right]^{1/2} \left[\frac{\mu_f}{g_0 \sigma(\rho - \rho_v)} \right]^{1/3} \tag{8.20}$$

The effective heat transfer coefficient associated with thermal conduction through the vapor film blanketing the heated surface in stable film boiling can be determined by analogy to the Nusselt analysis commonly employed in film condensation (and presented in the next section). A balance between buoyant and frictional forces and consideration of evaporation at the liquid-vapor interface yields the film thickness and subsequently the heat transfer coefficient. For horizontal tubes in pool boiling, Bromley et al. found [21]

$$h_c = 0.62 \left\{ \frac{k_v^3 \rho_v (\rho - \rho_v) g [h_{fg} + 0.4 c_{p,v}(T_w - T_s)]}{d \mu_v (T_w - T_s)} \right\}^{1/4}$$

(8.21)

After the addition of thermal radiation, the total heat transfer coefficient is expressed as

$$h = h_c \left(\frac{h_c}{h} \right)^{1/3} + h_r$$

(8.22)

where h_r is calculated for radiation between parallel plates [4]. Following Berenson [20], h_c in film boiling on horizontal surfaces can be obtained by replacing the 0.62 coefficient by 0.425 and the pipe diameter by $\sqrt{g_0 \sigma / g(\rho - \rho_v)}$. It may be observed that the diameter of vapor bubbles released during film boiling is proportional to this factor and ranges from [21]

$$3.14 \sqrt{\frac{g_0 \sigma}{g(\rho - \rho_v)}} \leqslant D_b \leqslant 5.45 \sqrt{\frac{g_0 \sigma}{g(\rho - \rho_v)}}$$

(8.23)

For flow film boiling along horizontal cylinders with significant velocity ($V_\infty > 2\sqrt{g d_0}$), Bromley [21] suggests

$$h_c = 2.7 \sqrt{\frac{V_\infty k_v \rho_v [h_{fg} + 0.4 c_p (T_w - T_s)]}{d(T_w - T_s)}}$$

(8.24)

and $h = h_c + \frac{7}{8} h_r$

8.4 CONDENSATION

When vapor is cooled below the saturation temperature, that is, when it traverses the temperature-pressure curve of Fig. 8.1 from right to left, the vapor molecules undergo a phase change and liquid is formed. This process is generally accompanied by a very substantial decrease in volume and is thus referred to as *condensation*. In engineering applications, condensation most often occurs along surfaces whose temperature is below the local saturation value and commences with the nucleation of liquid drops on minute surface discontinuities. For most liquid-surface combinations, these liquid drops then spread across the condenser surface and merge with neighboring drops to establish a stable liquid film. Subsequent vapor condensation, in this *film condensation mode*, occurs along the surface of the liquid film and the latent heat of condensation, released at this liquid-vapor interface, must then be conducted through the liquid film to the condenser surface. Under most circumstances, the rate of condensation is controlled by the thermal resistance of the liquid film, and analysis of film condensation thus centers on the prediction of the condensate film thickness.

When the condensed liquid does not wet the surface, a continuous liquid film cannot form, and the dropwise mode of condensation persists beyond the initial period. The absence of the thermal resistance associated with the liquid film leads to dropwise condensation rates that are typically an order of magnitude higher than those encountered in film condensation at identical values of condenser surface and vapor temperature, respectively. To ensure continuous dropwise condensation, however, it is generally necessary to coat or treat the surface with so-called promoters, which prevent wetting of the surface by the condensate. The limited availability and/or high cost of such promoters has discouraged the commercial utilization of dropwise condensation.

In some thermal systems, vapor, generated by boiling along heated surfaces, remains contained within vapor bubbles rather than forming a vapor space. Upon encountering subcooled liquid, the bubbles begin to collapse as vapor condenses on the bubble surface (the liquid-vapor interface). The rate of vapor condensation in bubbles varies with both bubble characteristics and thermodynamic properties and is far more difficult to predict than in film condensation.

There are thus two important categories of condensive heat transfer: film condensation and vapor bubble collapse. The equations governing these two modes are presented and discussed in detail in the following sections.

8.4.1 Film Condensation

Film condensation is distinguished by the formation of a continuous liquid film on a subcooled surface immersed in vapor. In this mode, thermal conduction through the condensate film generally governs the rate of heat transfer and the analysis of this phenomenon, including the pioneering effort by Nusselt [22], thus focuses on prediction of the condensate film thickness.

CLASSICAL ANALYSIS

Following [1], a force balance on an element of the condensate film, shown in Fig. 8.9, accounting for gravitational acceleration, pressure forces, and shear forces along the wall, but neglecting liquid-vapor shear and momentum effects, yields

$$\mu \left(\frac{dV_z}{dy} \right) = g\rho(\delta - y) - g\rho_v(\delta - y) \tag{8.25}$$

By noting that the liquid velocity at the wall is zero, Eq. (8.25) can be integrated to yield the velocity profile

$$V_z = \frac{g(\rho - \rho_v)(\delta y - y^2/2)}{\mu} \tag{8.26}$$

and a second integration yields

$$\Gamma = \int_0^\delta \rho V_z \, dy = \frac{g\rho(\rho - \rho_v)\delta^3}{3\mu} \tag{8.27}$$

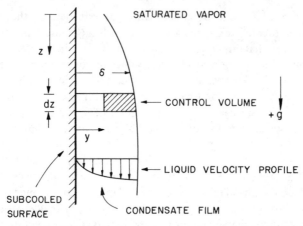

FIG. 8.9   Schematic drawing and parameter definitions for film condensation analysis.

which provides the liquid flow rate (per unit width) in the film. Differentiation of this relation with respect to δ, followed by algebraic manipulation, yields an expression for the film thickness in differential form:

$$\delta^2 \, d\delta = \frac{\mu \, d\Gamma}{g\rho(\rho - \rho_v)} \tag{8.28}$$

For negligible subcooling in the condensate film, the heat removal rate at the wall necessary to sustain condensation equals

$$q'' = h_{fg} \frac{d\Gamma}{dz} \tag{8.29}$$

However, for an assumed linear temperature profile in the liquid film, with the interface temperature equal to the saturation value, the heat flux at the wall can be expressed as

$$q'' = \frac{k(T_s - T_w)}{\delta} \tag{8.30}$$

Combining Eqs. (8.29) and (8.30) to solve for $d\Gamma$ and substituting the resulting expression for $d\Gamma$ in Eq. (8.28), the variation of film thickness along the plate is found to be governed by

$$\delta^3 \, d\delta = \frac{k\mu(T_s - T_w)}{g\rho(\rho - \rho_v)h_{fg}} \, dz \tag{8.31}$$

Integration of Eq. (8.31) along the condenser surface with $\delta = 0$ at the top then yields

$$\delta = \left[\frac{4k\mu z(T_s - T_w)}{g\rho(\rho - p_v)h_{fg}} \right]^{1/4} \tag{8.32}$$

Finally, since by virtue of its definition the heat transfer coefficient equals $q''/(T_s - T_w)$, which then must equal k/δ via Eq. (8.30), the sought-after relation for the local transfer coefficient is found to be

$$h_z = \left[\frac{g\rho(\rho - \rho_v)k^3 h_{fg}}{4z\mu(T_s - T_w)} \right]^{1/4} \tag{8.33}$$

and the average coefficient is then given by

$$\bar{h} = \frac{1}{L} \int_0^L h_z \, dz = 0.943 \left[\frac{g\rho(\rho - \rho_v)k^3 h_{fg}}{4L\mu(T_s - T_w)} \right]^{1/4} \tag{8.34}$$

Despite the many assumptions inherent in this classical analysis, Eq. (8.34) is generally found to offer acceptable to excellent agreement with the data for film condensation along subcooled, vertical flat plates and large tubes ($d > 1/8$ in) immersed in saturated vapor, free of noncondensable gas. Nevertheless, for low-h_{fg} fluids and/or condensers operating at relatively high surface subcooling, it is desirable to modify the above analysis to reflect the actual (nonlinear) temperature profile and the associated subcooling of the liquid film. As shown in [1], these effects can be incorporated into Eq. (8.34) by replacing h_{fg} by h'_{fg} set equal to

$$h'_{fg} = h_{fg} \left[1 + \frac{0.68c_l(T_s - T_w)}{h_{fg}} \right] \tag{8.35}$$

This modification has been found valid for the parametric range defined by $\mathrm{Pr} > 0.5$ and $c_l(T_s - T_w)/h_{fg} < 1.0$, when fluid properties are evaluated at $T_f = T_w + 0.25 \times (T_s - T_w)$ [1].

GEOMETRIC VARIATIONS
The preceding analysis of film condensation on vertical surfaces can be applied to other condenser configurations by replacing g with the proper gravitational force acting on the liquid film. Thus, for condensation on a flat plate inclined at an angle of ϕ to the horizontal plane, Eq. (8.34) can be modified to

$$\bar{h} = 0.943 \left[\frac{g \sin \phi \rho(\rho - \rho_v)k^3 h_{fg}}{4L\mu(T_s - T_w)} \right]^{1/4} \tag{8.36}$$

Similarly, for condensation on a large-diameter horizontal tube, Eq. (8.34) must be modified to reflect the variation in $g \sin \phi$ as the angle varies from 0 to $180°$. The classical film condensation analysis followed by integration around the tube yields

$$\bar{h} = 0.728 \left[\frac{g\rho(\rho - \rho_v)k^3 h_{fg}}{d\mu(T_s - T_w)} \right]^{1/4} \tag{8.37}$$

For a vertical bank of n closely spaced, horizontal tubes, d in Eq. (8.37) is replaced by nd. However, when, as is often the case, the tubes are widely spaced, account must be taken of condensation on the liquid film as it "drips" down to the next lowest tube. Following [23], the average heat transfer coefficient for this configuration is found to be

$$\bar{h} = 0.728 \left[1 + \frac{0.2c_l(T_s - T_w)(n-1)}{h_{fg}}\right]\left[\frac{g\rho(\rho - \rho_v)k^3 h_{fg}}{nd\mu(T_s - T_w)}\right]^{1/4}$$ (8.38)

Condensation inside vertical large-diameter tubes is analogous to the vertical flat-plate configuration, and Eq. (8.34) can be expected to apply. However, when the tube diameter is small and the entering vapor velocity is high, vapor shear stress on the liquid film may be substantial. Consideration of this phenomenon as well as the influence of turbulence and vapor superheat on film condensation rates is beyond the scope of this discussion but is explored in detail by Rohsenow and Choi [1].

NONCONDENSABLE GAS

The condensation of vapor in the presence of noncondensable gas or liquefaction of a single component of a gas mixture is generally controlled by the rate at which vapor can diffuse to the surface rather than by thermal conduction through the liquid film. Vapor diffusion, by its nature, must require a pressure gradient (see Sec. 8.2) and, as a consequence, for a steady condensation rate, the vapor pressure at the liquid interface must be lower than in the bulk mixture. As a consequence of this lower vapor pressure, the interface saturation temperature must fall below the *dew point* of the vapor in the mixture and can thus be substantially below the mixture temperature. The driving force for condensation heat transfer at the surface $(T_s - T_w)$ is, therefore, also reduced substantially. This effect, shown in Fig. 8.10, is quite pronounced for even modest bulk concentrations of noncondensable gas, and an air mass fraction of 0.5% has been shown to result in a 50% reduction in the condensation of stagnant steam at atmospheric pressure [24]. The heat transfer ratio in forced convection condensation is significantly less sensitive to the presence of noncondensables (approximately 10% reduction for 0.005 air fraction), but the effect on both stagnant and forced convection condensation becomes more pronounced as the total mixture pressure is reduced.

   This phenomenon is of special significance in the design of condensers for electronic immersion cooling systems. The dielectric fluids used in this application often display an extremely high solubility for air (typically 0.3 g/1000 g of liquid [25]), and air coming out of solution as the liquid is heated may occupy much of the vapor space and blanket the condenser surface.

   Following Rohsenow [4], a first approximation for the performance of a vertical condensing surface in the presence of noncondensable gas can be obtained by defining an effective heat transfer coefficient for vapor diffusion up to the liquid interface,

$$h_{nc,d} \equiv \left(\frac{w_v}{A}\right)\frac{h_{fg} + c_{pv}(T_s - T_i)}{T_s - T_i}$$ (8.39)

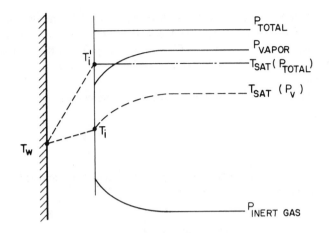

$T_i - T_w$ = ACTUAL CONDENSATION DRIVING FORCE

$T_i' - T_w$ = INCORRECT CONDENSATION DRIVING FORCE

FIG. 8.10 The effect of noncondensable gas on the interface temperature in film condensation [1].

and summing this resistance with the liquid film resistance to yield an overall heat transfer coefficient

$$\frac{1}{h_{nc,T}} = \frac{1}{h_{nc,d}} + \frac{1}{\bar{h}} \tag{8.40}$$

with $\bar{h}$ from Eq. (8.34).

As discussed in Sec. 8.2.2, the vapor flow rate appearing in Eq. (8.39) is dependent on the vapor pressure gradient, thermodynamic properties, and the diffusion coefficient. For vapor-gas flow parallel to the condenser surface, the vapor flow rate can be determined by use of Eq. (8.1) with Eqs. (8.2), (8.3), and (8.4). In a stagnant vapor-gas mixture, [31] recommends using

$$\frac{w_v}{A} = M_v \left[\frac{(p - p_{vi})}{(W/A)} \right] \left(\frac{Dp_v}{R_v R_g T_s^2} \right) \tag{8.41}$$

8.4.2 Vapor Bubble Collapse

Vapor condensation along the internal surface of a bubble rising through subcooled liquid reduces the vapor bubble volume and, in the absence of noncondensable gas, ultimately results in bubble collapse. The distance required for the collapse or complete condensation of vapor bubbles is related to the departure diameter, rise velocity, liquid subcooling, and the prevalent collapse mechanism.

BUBBLE PARAMETERS

The departure diameter of isolated vapor bubbles in saturated pool boiling has been successfully correlated by Eq. (8.10). In the absence of a distinct relation for departure diameter in subcooled pool boiling and in recognition of the modest subcooling generally encountered in electronic cooling applications, this correlation must suffice for the present analysis. Similarly, when film boiling prevails on the heated surface and for vigorous nucleate pool boiling, during which vertical vapor columns may appear on the surface, Eq. (8.23) can be used.

The rise velocity of an undisturbed bubble in the diameter range of interest can be approximated [26] as

$$V_b = \frac{1}{3}\sqrt{4gr} \tag{8.42}$$

However, the presence of neighboring bubbles, fluid circulation, and side walls can exert a significant influence on the rise velocity.

BUBBLE COLLAPSE

The mechanics of vapor bubble collapse under spherically symmetrical conditions have received extensive attention, and it has been shown [27–29] that the collapse of vapor bubbles in moderately subcooled pools is governed by heat transfer considerations. An approximate expression for the collapse rate can be obtained by neglecting translational velocity effects and assuming that a thin thermal boundary later exists at the bubble surface. Utilizing the Plesset-Zwick temperature integral [30], the collapse rate is expressible [27] as

$$\tau_H = \frac{4}{\pi}\,\mathrm{Ja}^2\,\frac{\alpha}{r_0^2}\,t = \frac{1}{3}\left(\frac{2}{\gamma}+\gamma^2-3\right) \tag{8.43}$$

where $\gamma = r/r_0$.

Because of the asymptotic nature of heat transfer-controlled collapse, no precise criterion exists for the definition of a collapse period. However, because it is the bubble volume that is of prime interest, and at $\gamma = 0.2$, the bubble volume is only 1% of its departure volume, the collapse period τ_c can be defined as

$$\tau_c = \frac{1}{3}\left(\frac{2}{\gamma}+\gamma^2-3\right)\Bigg|_{\gamma=0.2} \approx 2.32 \tag{8.44}$$

whereupon

$$t_c = \frac{2.32}{4}\,\frac{\pi r_0^2}{\mathrm{Ja}^2\,\alpha} = 0.580\,\frac{\pi r_0^2}{\mathrm{Ja}^2\,\alpha}$$

Utilizing the relations for bubble departure diameter, rise velocity, and rate of collapse, it is now possible to determine an upper bound on the collapse length of the vapor bubbles in a moderately subcooled, liquid-filled enclosure free of noncondensable gas. The collapse length L_c can be set as

$$L_c = \int_0^{t_c} V_b\,dt \tag{8.45}$$

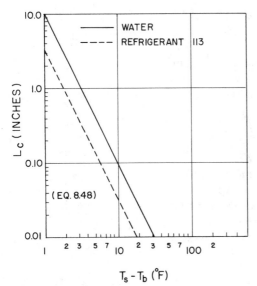

FIG. 8.11 Dependence of bubble collapse distance on bulk subcooling [32].

With the substitution of

$$\tau = \frac{4}{\pi} \, \text{Ja}^2 \, \frac{\alpha}{r_0^2} \, t \tag{8.46}$$

and introduction of Eq. (8.42),

$$L_c = \frac{\pi r_0^2}{4 \, \text{Ja}^2 \, \alpha} \int_0^{\tau_c} \frac{1}{3} \sqrt{4gr} \, d\tau \tag{8.47}$$

Simplification, integration and letting $\gamma = D/D_0$ yields [32]

$$L_c = 1.4 \, \frac{D_b^{5/2} \sqrt{2g}}{\text{Ja}^2 \, \alpha} \, \frac{\pi}{48} = 0.0292 \, \frac{\pi D_b^{5/2} \sqrt{2g}}{\text{Ja}^2 \, \alpha} \tag{8.48}$$

In the isolated bubble regime D_b is calculated from Eq. (8.10), whereas in the vapor column and film boiling regimes, Eq. (8.23) can be utilized. The collapse length L_c, as a function of bulk subcooling, is plotted in Fig. 8.11 for the isolated bubble regime.

8.5 NOMENCLATURE

Roman Letter Symbols

A area, m^2

c_p specific heat at constant pressure, J/kg°C

C mass coefficient, kg/m^3

C_D	drag coefficient, dimensionless
C_{sf}	boiling coefficient, dimensionless
d	pipe diameter, m
D	diffusivity, m^2/s; or bubble diameter, m
D_b	bubble departure diameter, m
Eo	Eustis number, dimensionless
f	bubble departure frequency, s^{-1}
g	acceleration of gravity, m/s^2
g_0	gravitational constant, $1 \ kg \, m/n \, s^2$
h	heat transfer coefficient, $W/m^2 \, °C$
h_D	mass transfer coefficient, m/s
h_{fg}	latent heat of vaporization, J/kg
h'_{fg}	modified latent heat of vaporization, J/kg
Ja	Jakob number, dimensionless
Ja*	modified Jakob number, dimensionless
k	thermal conductivity, $W/m \, °C$
l	diffusion path, m
L	length, m
L'	characteristic length, m
M	molecular weight
p	partial pressure of vapor, n/m^2
Pr	Prandtl number, dimensionless
n	number of pipes or tubes
q	heat flow, W
q''	heat flux, W/m^2
Q_v	vapor flux, m/s
r	bubble radius, m; or exponent, dimensionless
r_0	initial bubble radius, m
R	universal gas constant, $8315 \ J/kg \, mole \, K$
Re	Reynolds number, dimensionless
s	an exponent, dimensionless
Sc	Schmidt number, dimensionless
t	time, s
T	temperature, $°C$
V	velocity, m/s
w	mass flow, kg/s
W	total mass of noncondensable gas, kg
y	length coordinate, m
z	length coordinate, m

Greek Letter Symbols

α	thermal diffusivity, m^2/s
β	wetting angle, deg
γ	bubble radius ratio, dimensionless
Γ	condensate flow rate per unit width, $kg/s \, m$
δ	film thickness, m

η	fraction of heat flux resulting in net vapor generation, dimensionless
μ	dynamic viscosity, kg/s m
ρ	density, kg/m^3
σ	surface tension, N/m
τ	bubble collapse period, dimensionless
ϕ	angle, measured from vertical, degrees or radians

Subscripts

b	indicates boiling
c	indicates convection or collapse
crit	indicates critical boiling condition
f	indicates film condition
fb	indicates flow boiling
fg	indicates latent heat or liquid to vapor transition
H	indicates a heat transfer-controlled condition
i	indicates an interface
incip	indicates incipience
j	indicates a vapor jet
l	indicates a liquid condition or $x = l$
L	indicates length at $x = L$
min	indicates a minimum condition
nc, d	indicates diffusion in the presence of a noncondensable gas
r	indicates radiation
s	indicates a saturation condition
sub	indicates a subcooled condition
T	indicates a total condition
v	indicates vapor
w	indicates a wall condition
z	indicates a local value
∞	indicates a free stream condition

Superscripts

$^-$	indicates the average value
$'$	indicates a characteristic length
$''$	indicates a heat flux

8.6 REFERENCES

1 Rohsenow, W. M., and Choi, H., *Heat, Mass and Momentum Transfer*, Prentice-Hall, Englewood Cliffs, N.J., 1961.
2 Gilliland, E. R., and Sherwood, T. K., Diffusion of Vapors into Air Streams, *Ind. Eng. Chem.*, vol. 26, pp. 516–523, 761, 1934.
3 Colburn, A. P., A Method of Correlating Forced Convection Heat Transfer Data and a Comparison with Fluid Friction, *Trans. AIChE*, vol. 29, pp. 174–210, 1933.
4 Hetsroni, G. (ed.), *Handbook of Multiphase Systems*, Hemisphere, Washington, D.C., 1982.
5 Bergles, A. E., and Rohsenow, W. M., The Determination of Forced Convection Surface-Boiling Heat Transfer, *J. Heat Transfer*, vol. 86, pp. 365–372, 1964.

6 Cole, R., and Rohsenow, W. M., Correlation of Bubble Diameters of Saturated Liquids, *Chem. Eng. Prog. Symp. Ser.*, vol. 65, pp. 92, 211–213, 1969.

7 Cole, R., Photographic Study of Boiling in Region of Critical Heat Flux, *AIChE J.*, vol. 6, pp. 533–538, 1960.

8 Graham, R. W., and Hendricks, R. C., Assessment of Convection, Conduction and Evaporation in Nucleate Boiling, NASA TND-3943, National Technical Information Service, Springfield, Va., May 1967.

9 Bergles, A. E., Bakhru, N., and Shirer, J. W., Jr., Cooling High-Power-Density Computer Components, Rept. DSR 70712-60, Department of Mechanical Engineering, Massachusetts Institute of Technology, Cambridge, Mass., 1968.

10 Seely, J., and Chu, R., *Heat Transfer in Microelectronic Equipment*, Marcel Dekker, New York, 1972.

11 Zuber, N., On the Stability of Boiling Heat Transfer, *J. Heat Transfer*, vol. 80, no. 3, pp. 711–720, 1958.

12 Bergles, A. E., Burnout in Boiling Heat Transfer. Part I: Pool Boiling Systems, *Nuclear Safety*, vol. 16, no. 1, pp. 29–42, 1975.

13 Sun, K. H., and Lienhard, J. H., The Peak Boiling Heat Flux on Horizontal Cylinders, *Int. J. Heat Mass Transfer*, vol. 13, no. 9, pp. 1425–1439, 1970.

14 Ded, J. S., and Lienhard, J. H., The Peak Boiling Heat Flux from a Sphere, *AIChE J.*, vol. 18, no. 2, pp. 337–342, 1972.

15 Bakhru, N., and Lienhard, J. H., Boiling from Small Cylinders, *Int. J. Heat Mass Transfer*, vol. 15, no. 11, pp. 2011–2025, 1972.

16 Ivey, H. J., and Morris, D. J., On the Relevance of the Vapor-Liquid Exchange Mechanism for Subcooled Boiling Heat Transfer at High Pressure, UK Rept. AEEW-R-137, 1962.

17 Lienhard, J. H., and Keeling, K. B., Jr., An Induced Convection Effect upon the Peak-Boiling Heat Flux, *J. Heat Transfer*, vol. 92, no. 1, pp. 1–5, 1970.

18 Gambill, W. R., Generalized Prediction of Burnout Heat Flux for Flowing Subcooled, Wetting Liquids, AIChE preprint 17, 5th Natl. Heat Transfer Conf., Houston, Texas, August 1962.

19 Zuber, N., and Tribus, M., Further Remarks on the Stability of Boiling Heat Transfer, UCLA Rept. No. 58-5, 1958.

20 Berenson, P., Transition Boiling Heat Transfer from a Horizontal Surface, AIChE Paper no. 18, Fourth ASME-AIChE National Heat Transfer Conf., Buffalo, N.Y., August 1960.

21 Bromley, L. A., Leroy, N. R., and Robbers, A., Heat Transfer in Forced Convection Film Boiling, *Ind. Eng. Chem.*, vol. 45, pp. 2639–2646, 1953.

22 Nusselt, W. Z., Die Oberflächencondensation der Wasserdamfes, *Z. Ver. Deut. Ing.*, vol. 60, pp. 541, 569, 1916.

23 Chen, M. M., An Analytical Study of Laminar Film Condensation: Part 1–Flat Plates and Part 2–Single and Multiple Horizontal Tubes, *J. Heat Transfer*, Ser. C, vol. 83, pp. 48–60, 1961.

24 Minkowycz, W. J., and Sparrow, E. M., Condensation Heat Transfer in the Presence of Non-Condensables, Interfacial Resistance, Superheating, Variable Properties and Diffusion, *Int. J. Heat Mass Transfer*, vol. 9, pp. 1125–1144, 1966.

25 Fluorinert Technical Information, Chemical Division 3M Company, St. Paul, Minn., 1965.

26 Zuber, N., Tribus, M., and Westwater, J. W., The Hydrodynamic Crisis in Pool Boiling of Saturated and Subcooled Liquids, Paper 27, *International Developments in Heat Transfer*, ASME, New York, 1961.

27 Florschuetz, L. W., and Chao, B. T., On the Mechanics of Vapor Bubble Collapse, *J. Heat Transfer*, vol. 87, pp. 209–220, 1965.

28 Wittke, D. D., and Chao, B. T., Collapse of Vapor Bubbles with Translatory Motion, *J. Heat Transfer*, vol. 89, pp. 17–24, 1967.

29 Tokuda, N., Yang, W. J., and Clark, J. A., Dynamics of Moving Gas Bubbles in Injection Cooling, *J. Heat Transfer*, vol. 90, pp. 371–378, 1968.

30 Plesset, M. S., and Zwick, S. A., A Non Steady Heat Diffusion Problem with Spherical Symmetry, *J. Appl. Phys.*, vol. 23, pp. 95–98, 1952.
31 Rohsenow, W. M., and Hartnett, J. P., *Handbook of Heat Transfer*, McGraw-Hill, New York, 1973.
32 Markowitz, A., and Bergles, A. E., Operational Limits of a Submerged Condenser, *Progress in Heat and Mass Transfer*, vol. 6, pp. 701–716, Pergamon, Oxford, 1972.

9

■ contact resistance

9.1 INTRODUCTION

Equations governing the flow of heat through composite walls and cylinders have been presented in Chap. 4. These equations cover ideal cases, where there is no resistance to the flow of heat at the interfaces between the walls and cylinders in contact. However, heat transfer across an interface formed by two contacting bodies is usually accompanied by a measurable temperature difference across the interface, because a *thermal contact resistance* to heat flow exists at the region of the interface.

In modern electronic equipment, where high reliability is of such importance, many physical contacts exist between the "buried" components and the enclosure. These contacts, each contributing to a thermal contact resistance, can cause excessive component temperature. It is clear that an ability to evaluate and then minimize thermal contact resistance is a tool the electronic equipment designer and analyst must possess. To ignore this important aspect of heat transfer is to make assessment of reliability meaningless or to make overdesign of heat conducting paths necessary.

9.2 SIMPLIFIED MODEL

Suppose that two solid metal bars are brought into intimate contact as shown in Fig. 9.1a. Imagine that the lateral faces of the bars are insulated so that heat is forced to flow in the axial direction and that each bar possesses its own thermal conductivity. If the actual temperature profile is as shown in Fig. 9.1b, with a temperature drop at the interface ΔT_c, there is said to be a *thermal contact resistance* at the interface.

A steady-state energy balance dictates that the same heat flow q must proceed, in turn, through bar 1, the contact or interface between the two metals, and through bar 2,

$$q = k_1 A_a \frac{T_1 - T_a}{L_1} = \frac{T_a - T_b}{1/(h_c A_a)} = k_2 A_a \frac{T_b - T_2}{L_2} \qquad (9.1)$$

where A_a is the *apparent contact area* and h_c is called the *contact coefficient* or *contact conductance*.

The apparent contact area is composed of an area where metal-to-metal contacts exist, A_c, and a void area where they do not, A_v, so that

$$A_a = A_c + A_v \qquad (9.2)$$

This says that the heat flow across the interface is by conduction through the spots of contact and by conduction and possibly radiation through the fluid trapped in the void spaces created by the contact. Convection in the entrapped fluid is not really a factor, because the effective void space thickness δ_v is presumed to be very small.

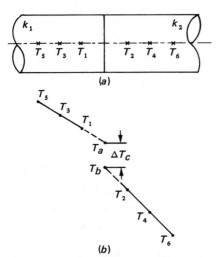

FIG. 9.1 Two solid metal bars in contact. (*a*) Temperature designations for simulated probe. (*b*) Linear temperature profile for this real interface condition.

If metal 1 and metal 2 each have surface irregularities of height $\delta_v/2$, then the heat flow across the interface is composed of two paths

$$q = \frac{T_a - T_b}{\delta_v/(2k_1 A_c) + \delta_v/(2k_2 A_c)} + \frac{k_f A_v (T_a - T_b)}{\delta_v} = h_c A_a (T_a - T_b)$$

and an evaluation of h_c then follows:

$$h_c = \frac{1}{\delta_v} \left[2 \frac{A_c}{A_a} \frac{k_1 k_2}{(k_1 + k_2)} + \frac{A_v}{A_a} k_f \right] \tag{9.3}$$

The use of Eq. (9.3) is dangerous. If one assumes that air, or a similar fluid, fills the void space and that $k_f \ll k_1$, $k_f \ll k_2$, there is a tendency to neglect the second term within the brackets in Eq. (9.3).

For example, for two polished steel surfaces in contact with $k_1 = k_2 = 20$ W/m °C and $A_c/A_a = 0.005$. Then from Eq. (9.2),

$$A_v = A_a - A_c = 1.000 - 0.005 = 0.995$$

and $A_v/A_a = 0.995$. Here

$$2 \frac{A_c}{A_a} \frac{k_1 k_2}{k_1 + k_2} = 2(0.005) \frac{(20)^2}{2(20)} = 0.100$$

and with $k_f = 0.030$ W/m °C,

$$\frac{A_v}{A_a} k_f = 0.995(0.030) \approx 0.030$$

It would appear that the metal-to-metal path controls, but this is quite incorrect because the air does contribute to the heat flow across the interface. However, A_c is usually very much smaller than A_v, and the two terms within the brackets in Eq. (9.3) may be of comparable magnitude.

Because it is exceedingly difficult to determine the proper values of A_c, A_v, and δ_v, the foregoing theory is almost useless. A better theory is required, and this is the purpose of the balance of this chapter.

9.3 THE GEOMETRY OF CONTACTING SURFACES

An actual surface is not necessarily smooth; it may contain peaks and valleys super-imposed on waves as shown in Fig. 9.2. The surface irregularities with large wavelength are termed *waviness*, the length of these waves depending on a number of conditions vary from 0.10 to 1.00 cm. The superimposed irregularities are termed *roughness*, and the actual depth of the interface can range from smooth surfaces ($\simeq 1.5~\mu$m) to very rough surfaces ($\simeq 25~\mu$m).

The peaks shown in Fig. 9.3 are called *asperities*, and whenever reference is made to surface asperities, surface roughness is implied. These individual peaks, which are projections on an actual surface, are separated by *valleys*, which are individual recesses in an actual surface.

The ratio of the peak line Y to the root-mean-square (RMS) roughness σ is a function of the finishing operation. Typical values of this ratio are given in Table 9.1.

The length of a section of the surface, such as L in Fig. 9.3, which is chosen to measure the roughness, is called the *base length*. It neglects any type of irregularity whose spacing is greater than L. Also shown in Fig. 9.3 are the *peak* and *valley lines*, which pass respectively through the top of the highest peak and the lowest point of the deepest valley, and the *mean line*, which is positioned so that the sum of the squares of equally spaced ordinates measured from it is a minimum. Both peaks and valleys are called *irregularities*, and the *maximum height of irregularities* is the distance between the peak and valley lines.

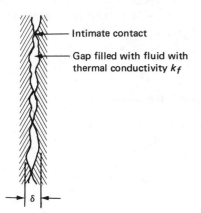

Intimate contact

Gap filled with fluid with thermal conductivity k_f

δ

FIG. 9.2 A schematic drawing of a real interface between two solids. The dimension δ may range between 1.5×10^{-6} and 2.5×10^{-5} m.

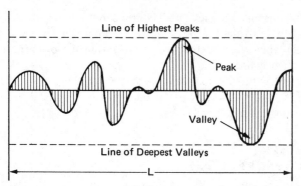

FIG. 9.3 A section of an interface between two surfaces. The length L is termed the base length.

Figure 9.4 shows the *slope*, the *vertex angle* 2γ, and the *radius of curvature* ρ of an asperity. In this figure it is observed that the vertex angle is obtained by extrapolating the positive and negative asperity slopes.

The *lay* of a surface is the principal direction of the surface pattern resulting from production methods. The *transverse roughness* is determined for a section normal to the lay, and the *longitudinal roughness* is in the direction of the lay.

9.4 FACTORS INFLUENCING CONTACT RESISTANCE

The heat transfer across an interface of two materials in contact is a very complex phenomenon and is a function of many parameters. The following appear to be of the greatest importance:

1 The number of contact spots
2 The shape of the contact spots: circular, elliptic, band, or rectangular
3 The size of the contact spots
4 The disposition or arrangement of the contact spots
5 The geometry of the contacting surfaces with regard to roughness and waviness
6 The average thickness of the void space (the noncontact region)
7 The fluid in the void space: gas, liquid, grease, vacuum
8 The pressure of the fluid in the void space

TABLE 9.1 Typical Values of the Ratio Y/σ as a Function of Type of Surface Finishing Operation

Finishing operation	Y/σ
Grinding	4.5
Hyperlap	6.5
Sandpaper	7.0
Superfinish	7.0
Lap with loose abrasive	10.0

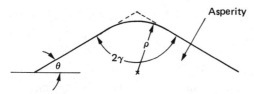

FIG. 9.4 Nomenclature for an asperity. The slope of the asperity is the tangent of the angle θ. The radius of curvature of the asperity is ρ. Half of the vertex angle is the angle γ.

9 Thermal conductivities of the contacting solids and of the fluid in the void space
10 The hardness of the contacting asperities (Will the asperities undergo plastic deformation?)
11 The moduli of elasticity of the contacting asperities (Will the asperities undergo elastic deformation?)
12 The average temperature of the interface (Will radiation be a factor?)
13 The past history of the contact with regard to the number of previous compressions and decompressions
14 The contact pressure
15 The duration of the contact with regard to relaxation effects
16 Vibrational effects
17 Directional effects
18 Contact cleanliness (Oxides?)

In this list, it should be recognized that the geometry of the contacting surfaces (see item 5) has a predominating influence on the other items (such as those given in items 1 through 4 and 6). Furthermore, other factors may well need to be considered in special cases, such as those that arise when some unusual interstitial fluid is employed.

9.5 CONTACT BETWEEN NOMINALLY FLAT, ROUGH SOLIDS

It is shown in this section that the hardness of the solids in contact plays a significant role and that the contact area is a result of plastic deformation of the asperities at least when the surfaces are first put into contact.

The ensuing discussion is based on the following assumptions:

1 The surfaces are rough but have no waviness (microscopically rough and macroscopically flat).
2 The apparent contact area does not change during the contact.
3 The surfaces are isotropic and remain isotropic during the contact.
4 Plastic deformation is related to the hardness of the softer solid.
5 The hardness remains constant and is equal to three times the elastic limit under tensile stress.
6 During the plastic deformation of the highest peaks, the sublayers and the bulk of the solids undergo elastic deformation.

7   During the deformation, the profile does not change; that is, the profile retains the same analytic form.

8   The analysis considers only the first compression of newly machined solids.

Assume that the asperities are hemispheres having identical radii of curvature ρ, and that the materials have identical moduli of elasticity E and identical Poisson ratios ν. Then, with F the load and a the radius of the contact, according to the theory of elasticity, the maximum contact pressure is

$$P_{max} = \frac{1.5F}{\pi a^2} = 0.62 \left(\frac{FE^2}{\rho^2} \right)^{1/3} \qquad (9.4)$$

If, for example, the two contacting surfaces are of mild steel with $E = 21 \times 10^{10}$ N/m^2 and a tensile yield stress of 35×10^7 N/m^2, plastic deformation will occur when the pressure is 105×10^7 N/m^2. For $\rho = 10 \ \mu$m, which is a typical radius of curvature for relatively smooth surfaces, a calculation for F that will be the load for plastic deformation using Eq. (9.4) yields $F = 3.2 \times 10^{-5}$ N. Because actual contact loads are orders of magnitude greater than this value, it is seen that, upon first contact, all asperities experience plastic deformation and the hardness of the material is an important parameter.

Greenwood and Williamson [1] have defined a *plasticity index* for two solids in contact:

$$\frac{E'}{H} = \left(\frac{\sigma}{\rho} \right)^{1/2} \qquad (9.5)$$

where ρ is the radius of curvature of a typical asperity and H is the hardness of the softer solid,

$$E' = \frac{E_1 E_2}{(1 - \nu_1^2)E_2 + (1 - \nu_2^2)E_1} \qquad (9.6)$$

and σ is the RMS roughness of the contact, approximated by

$$\sigma = (\sigma_1^2 + \sigma_2^2)^{1/2} \qquad (9.7)$$

where σ_1 and σ_2 are taken as the RMS heights from mean line to peak and valley lines, respectively (see Fig. 9.3).

Materials with values greater than unity, which are obtainable for most machined surfaces, will undergo plastic contact at the lightest loads. Values below 0.7, which can be obtained with careful polishing, experience elastic contact, although Greenwood and Williamson have also concluded that, at first contact, all asperities experience some plastic deformation.

Fenech, in his Sc.D. dissertation at the Massachusetts Institute of Technology [2], generated Table 9.2 for a particular set of aluminum surfaces. This table shows that the contact spot density depends strongly on the apparent contact pressure, defined as the actual load divided by the area of contact if the surface had no rough-

TABLE 9.2   Summary of Fenech [2] Studies[a]

P_a	n	a	$F = P_a/n$
7	59	16.4	0.119
35	244	17.8	0.144
70	390	20.0	0.180
350	1260	25.0	0.267
700	1730	30.0	0.405

[a]P_a = apparent pressure, N/cm²
n = number of contacts per square centimeter
a = contact radius, μm
F = force, N

ness or waviness. The table also shows that the average contact radius is not markedly increased by load. Because the load per asperity is considerably larger than the critical load [F in Eq. (9.4)], the asperities must surely undergo plastic deformation.

Bowden and Tabor [3] obtained the results displayed in Table 9.3, which show that the ratio of the real contact area to the apparent area, A_r/A_a, is always less than unity. Observe that the apparent area does not vary, whereas the real contact area is directly proportional to the load, and if the hardness is taken as three times the yield tensile stress, the area ratio is proportional to the ratio of the apparent pressure to the hardness.

From these studies and other empirical work, one may conclude that for the first compressive cycle of nominally flat but rough surfaces:

1 The total real contact area is always a small fraction of the apparent contact area.
2 The number of contact spots increases rapidly during the first stages of compression and then very slowly as the apparent pressure continues to increase.
3 The size or radius of the average contact spot increases slowly with increase in apparent pressure.
4 The size distribution of contact spot radii appears to be Gaussian.
5 The contact spots are uniformly distributed over the apparent contact area.
6 The contact spots appear to be elliptical.

TABLE 9.3   Summary of Bowden and Tabor [3] Studies[a]

F	A_a	A_r	A_r/A_a
2	21	0.0002	0.00001
5	21	0.0005	0.00005
100	21	0.01	0.0005
500	21	0.05	0.0025
1000	21	0.10	0.0050

[a]F = force, N
A_a = apparent area, cm²
A_r = real area, cm²

9.6 CORRELATIONS FOR THE CONTACT COEFFICIENT

It has been shown that in order to understand how heat flows across an interface be-tween two contacting solids, an understanding of two additional and prerequisite problems is necessary. The first problem is the *metrology problem*, which is the problem of surface description. This problem deals with the types of surfaces—whether they are smooth, nominally flat but rough, or wavy and rough—and defines such items as contact area, contour area, and apparent area.

The second problem is the mechanical problem, which deals with the manner in which contacting surfaces deform. Here it is observed that elasticity is important when smooth, wavy surfaces contact and plasticity is important in the contact of nominally flat, rough surfaces. Elasticity and plasticity play equally important roles in the deformation of rough, wavy surfaces in contact.

With the background afforded by the metrology and mechanical problems, several investigators have tackled the thermal problem and have proposed working correlations for either the contact coefficient h or the contact resistance R_c. Because the mathematical development, in most cases, is rather tedious, only the results of these investigations are presented in this section.

9.7 SMOOTH, WAVY SURFACES IN VACUUM

Clausing and Chao [4] have considered the special case of two circular cylinders of radius b, having smooth hemispherical caps coming into contact (Fig. 9.5). They assumed negligible resistance due to asperities and took equal Poisson ratios for the two materials ($\nu_1^2 = \nu_2^2 = 0.1$) and proposed

$$\frac{hb}{k} = \frac{2}{\pi} \frac{\epsilon}{\psi(\epsilon)} \quad (\epsilon < 0.65) \tag{9.8}$$

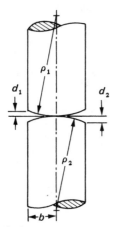

FIG. 9.5 Two cylinders with hemispherical caps in contact. Each cylinder has a modulus of elasticity E, a Poisson ratio ν, and a thermal conductivity k.

where

$$k = \frac{2k_1 k_2}{k_1 + k_2}$$

$$\epsilon = 1.285 \left(\frac{P_a b}{Ed} \right)^{1/3}$$

$$P_a = \frac{F}{\pi b^2}$$

where F is the load,

$$E = \frac{2E_1 E_2}{E_1 + E_2}$$

$$d = d_1 + d_2$$

and $\quad \psi(\epsilon) = 1 - 1.40925\epsilon + 0.29591\epsilon^3 + 0.05254\epsilon^5 + \cdots$

9.8 NOMINALLY FLAT, ROUGH SURFACES IN VACUUM

The problem of nominally flat, rough surfaces in vacuum has received considerable attention by many investigators. All of their investigations are similar in that they assume:

1 The existence of n identical circular contact spots uniformly distributed over the apparent contact area
2 Isothermal contact spots
3 That the total constriction resistance of one contact spot is the sum of two constriction resistances because two asperities serve to make a contact spot
4 That the total constriction resistance for n contact spots is $1/n$ of the constriction resistance for one spot
5 That the formation of the contact spots is due to plastic deformation

The differences in the investigations are due to additional assumptions:

1 Some assume no interaction between adjacent contact spots.
2 Some assume a particular value for the average contact spot radius and assume it to be independent of surface roughness and applied load.
3 Some assume a variation of the contact spot radius with surface roughness but not with applied load.
4 Some assume a variation of contact spot radius with surface roughness and applied load.
5 Some assume that a relationship exists between the contact spot density and the real contact area due to plastic theory.

Holm [5], with the plasticity assumption in mind, defined the real contact area as

$$A_r = \frac{P_a A_a}{H} = \frac{F}{H} \tag{9.9}$$

where P_a is the apparent contact pressure, A_a is the contact area ($P_a A_a$ is merely the force F), and H is the hardness of the softer material. He then correlates h:

$$h = \frac{1/A_a}{(1/\pi kna) \arctan [(l/a)^2 - 1]^{1/2} - (0.3/knl)[1 - (a/l)^2]^{1/2}} \tag{9.10}$$

where

$$l = \left(\frac{A_r}{4n}\right)^{1/2}$$

a, the contact spot radius, is

$$a = \left(\frac{A_r}{n\pi}\right)^{1/2}$$

and $k = \dfrac{2k_1 k_2}{k_1 + k_2}$

The second term in the denominator of Eq. (9.10) is a measure of the interaction between adjacent contact spots. For small, isolated contact spots, $a/l \to 0$ and Eq. (9.10) reduces to

$$h = \frac{2kna}{A_a} \tag{9.11}$$

In [6], Kraus discusses the correlation of Fenech and Rohsenow [7], who proposed a correlation based on a contact model that subsequently proved not to be a good representation of the real situation. Realizing this, Yovanovich and Fenech [8] proposed a correlation that requires detailed knowledge of the surface profiles in order to determine n, the number of contacts. With the ratio to apparent contact area,

$$\epsilon^2 = \frac{A_r}{A_a} = \frac{P_a}{H} = \left(\frac{a}{b}\right)^2$$

because

$$A_r = n\pi a^2$$
$$A_a = n\pi b^2$$

With $k = 2k_1 k_2/(k_1 + k_2)$,

$$h_c = \frac{k\epsilon\sqrt{\pi n}}{\arctan (1/\epsilon) - 2\epsilon} \tag{9.12}$$

Cooper et al. [9], in considering nominally flat surfaces, based their work on a statistical analysis of contacting surfaces possessing Gaussian distributions of asperity heights and a Gaussian distribution of asperities over the apparent area. Assuming plastic deformation of asperities,

$$h_c = 1.45 \frac{k(P_a/H)^{0.985}}{\sigma} |\overline{\tan \theta}| \tag{9.13}$$

where again $k = 2k_1 k_2 /(k_1 + k_2)$, P_a is the apparent contact pressure, and H is the hardness of the softer material. In Eq. (9.13), σ is the RMS roughness given by Eq. (9.7) and $|\overline{\tan \theta}|$ is the average absolute asperity angle (see Fig. 9.4).

$$|\overline{\tan \theta}|^2 = |\overline{\tan \theta_1}|^2 + |\overline{\tan \theta_2}|^2 \tag{9.14}$$

The agreement between Eq. (9.13) and test data is very good for

$$0.00035 < \frac{P_a}{H} < 0.01$$

$$8 < k < 77 \,(\text{Btu/ft h} \,°F) \quad \text{or} \quad 13.8 < k < 133.2 \,(\text{W/m} \,°C)$$

$$1.0 \,\mu\text{m} < \sigma < 85 \,\mu\text{m}$$

$$0.08 < |\overline{\tan \theta}| < 0.160$$

However, the correlation overpredicts h when $P_a/H > 0.01$ by as much as 100%, especially at very high contact pressures, because the analysis on which the correlation was based does not consider the effects of contact maldistribution.

Tien [10] proposed a correlation

$$h = 0.55 \frac{k}{\sigma} m \left(\frac{P_a}{H}\right)^{0.85} \tag{9.15}$$

where k, σ, P_a, and H are as listed previously and m is the RMS slope of the contacting asperities. Tien has determined that m ranges from 0.01 for very smooth surfaces to 0.13 for rough surfaces.

9.9 ROUGH, WAVY SURFACES IN VACUUM

Mikic et al. [11] provide the correlation

$$h = \frac{1}{[\delta \psi(\epsilon)/\sqrt{\pi n} k\epsilon] + [\delta \psi(\lambda_e)/k\lambda_e]} \tag{9.16}$$

where k, P_a, n, and H take on previously cited values and

$$\epsilon^2 \lambda_e^2 = \frac{P_a}{H}$$

$$\lambda_e = 1.285 \left(\frac{P_a b}{2Ed} \right)^{1/3}$$

where b is the apparent contact radius,

$$E = \frac{2E_1 E_2}{E_1 + E_2}$$

and d is the *out-of-flatness* of solids 1 and 2,

$$d = d_1 + d_2$$

In Eq. (9.16), Yovanovich [12] gives $\psi(x)$, where x can take on either ϵ or λ_e:

$$\psi(x) = 1 - \frac{4}{\pi} x - x^2 \tag{9.17}$$

9.10   EFFECT OF INTERSTITIAL FLUIDS

Correlations that handle both contact spot conductance and fluid layer conductance are based on the simple assumption that the two heat flow paths are independent of each other but that their combined effect is a parallel combination.

Shlykov [13] has developed the correlation

$$h = \frac{k_f Y}{7(\sigma_1 + \sigma_2)} + 8000k \left(\frac{CP_a}{H} \right)^{0.86} \quad \text{W/m}^2 \, ^\circ\text{C} \tag{9.18}$$

where k, P_a, and H are as cited previously, k_f is the fluid thermal conductivity, σ_1 and σ_2 are RMS roughnesses of the two surfaces, and

$$C = 1 \qquad\qquad\qquad \sigma_1 + \sigma_2 > 8.5 \, \mu\text{m}$$

$$C = \left(\frac{8.5}{\sigma_1 + \sigma_2} \right)^{1/3} \qquad 2.9 \, \mu\text{m} < (\sigma_1 + \sigma_2) < 8.5 \, \mu\text{m}$$

$$C = \frac{4.3}{\sigma_1 + \sigma_2} \qquad\qquad \sigma_1 + \sigma_2 < 2.9 \, \mu\text{m}$$

$$Y = \frac{10}{3} + \frac{10}{x} + \frac{4}{x^2} - 4 \left(\frac{1}{x^3} + \frac{3}{x^2} + \frac{2}{x} \right) \ln (1 + x)$$

$$x = \frac{7(\sigma_1 + \sigma_2)}{2l} \tag{9.19}$$

where $2l$ and $2l/\Lambda$ are obtained for various gases at 20°C in Table 9.4. Conversions for conditions at temperatures other than 20°C are obtained from

$$\Lambda(T) = \Lambda(293) \left(\frac{T + 273}{293} \right)$$

TABLE 9.4   Values of Λ (293 K) and $2l/\Lambda$ (293 K) for Use in Eq. (9.19)

Gas	Λ (293 K)	$2l/\Lambda$ (293 K)
Air	9.6×10^{-2}	4.6
Hydrogen	16.0×10^{-2}	22.1
Helium	28.5×10^{-2}	14.8
Argon	10.0×10^{-2}	5.1

EXAMPLE

A molybdenum (Mo) and copper (Cu) contact that is formed after a grinding operation on both materials is to be evaluated under four conditions at room temperature (25°C):

1 Contact pressure 6.895×10^4 N/m² (10 psi)
   (a) Dry contact–air only
   (b) Grease contact
2 Contact pressure 1.379×10^6 N/m² (200 psi)
   (a) Dry contact–air only
   (b) Grease contact

Thermal properties and physical data are as follows:

	Mo	Cu
Thermal conductivity k (W/m°C)	140	380
Hardness H (N/m²) $\times 10^{-8}$	18	5
RMS roughness σ (μm)	0.625	0.625
Asperity slope, tan θ (dimensionless)	0.08	0.08

Consider thermal conductivities of air and grease as follows at 25°C:

Air, $k = 0.0252$ (W/m°C)

Grease, $k = 0.213$ (W/m°C)

The contact is one of nominally flat, rough surfaces. Evaluate the contact coefficient.

SOLUTION

The total contact coefficient is composed of a metal-to-metal contact in parallel with a fluid conduction effect due to the fluid (air or grease) in the gap. Thus

$$h = h_c + h_g$$

where h_c may be obtained from Eq. (9.13) due to Cooper et al. [9], and where

$$h_g = \frac{k}{Y}$$

where Y is the nominal gap height in meters. First,

$$o = (o_1^2 + o_2^2)^{1/2} = [(0.625)^2 + (0.625)^2]^{1/2}$$
$$= 0.625\sqrt{2} \times 10^{-6} \text{ m}$$
$$|\overline{\tan \theta}| = (\tan^2 \theta_1 + \tan^2 \theta_2)^{1/2} = [(0.08)^2 + (0.08)^2]^{1/2} = 0.08\sqrt{2}$$
$$k = 2\frac{k_1 k_2}{k_1 + k_2} = 2\left[\frac{(140)(380)}{140 + 380}\right]$$
$$= 204.62 \text{ W/m}^2 \text{ °C}$$

For case 1(a),

$$\frac{P_a}{H} = \frac{6.895 \times 10^4}{5 \times 10^8} = 1.379 \times 10^{-4}$$

where the hardness of the softer of the two materials is used.
By Eq. (9.13),

$$h_c = 1.45\frac{k|\overline{\tan \theta}|}{o}\left(\frac{P_a}{H}\right)^{0.985}$$
$$= 1.45\frac{(204.62)0.08\sqrt{2}}{0.625\sqrt{2} \times 10^{-6}}\left(\frac{P_a}{H}\right)^{0.985}$$
$$= 3.7977 \times 10^7\left(\frac{P_a}{H}\right)^{0.985}$$

For cases 1(a) and 1(b),

$$h_c = 3.7977 \times 10^7(1.379 \times 10^{-4})^{0.985} = 5,984 \text{ W/m}^2 \text{ °C}$$

For cases 2(a) and 2(b), where

$$\frac{P_a}{H} = \frac{1.379 \times 10^6}{5 \times 10^8} = 2.758 \times 10^{-3}$$
$$h_c = 3.7977 \times 10^7(2.758 \times 10^{-3})^{0.985} = 114,422 \text{ W/m}^2 \text{ °C}$$

For conduction across the gap fluid,

$$h_g = \frac{k}{Y} = \frac{k}{(\gamma/o)o}$$

where $Y/o = 4.5$ for grinding (Table 9.1). Thus for cases 1(a) and 2(a) for air,

$$h_g = \frac{0.0252}{4.5(0.625\sqrt{2} \times 10^{-6})} = 6,336 \text{ W/m}^2 \text{ °C}$$

and for cases 1(b) and 2(b) for grease,

$$h_g = \frac{0.213}{4.5(0.625\sqrt{2}\times 10^{-6})} = 53{,}552 \text{ W/m}^2 \, ^{\circ}\text{C}$$

The results can now be established:

Case 1(a) $h = 5{,}984 + 6{,}336 = 12{,}320 \text{ W/m}^2 \, ^{\circ}\text{C}$

Case 1(b) $h = 5{,}984 + 53{,}552 = 59{,}536 \text{ W/m}^2 \, ^{\circ}\text{C}$

Case 2(a) $h = 114{,}422 + 6{,}336 = 120{,}758 \text{ W/m}^2 \, ^{\circ}\text{C}$

Case 2(b) $h = 114{,}422 + 53{,}552 = 167{,}974 \text{ W/m}^2 \, ^{\circ}\text{C}$

and conclusions can be readily drawn as to the effects of increasing the contact pressure and the employment of grease.

9.11 NOMENCLATURE

Roman Letter Symbols

a radius of contact, m

A area, m^2

b radius of cylinder, m; apparent contact radius, m

C a constant used in Eq. (9.18), dimensionless

d height of hemispherical cap, m

E modulus of elasticity, kg/m^2

F load, kg

h heat transfer coefficient, W/m^2 $^{\circ}$C, with subscript c, contact coefficient, W/m^2 $^{\circ}$C

H hardness, kg/m^2

k thermal conductivity, W/m^2 $^{\circ}$C

L length, m

l an adjusted length used in Eqs. (9.10) and (9.19), m

m root-mean-square slope of contacting asperities, dimensionless

n indicates number of contacts

P pressure, kg/m^2

q heat flow, W

R thermal resistance, $^{\circ}$C/W; with subscript c, contact resistance, $^{\circ}$C/W

T temperature, $^{\circ}$C

x an argument of a functional relationship given by Eq. (9.17); a function defined by Eq. (9.19)

Y indicates peak line, m; a function used in Eq. (9.19)

Greek Letter Symbols

γ vertex angle, degrees or radians

Δ indicates change in variable

δ	void space thickness, m
ϵ	a parameter used in Eqs. (9.8) and (9.16); square root of the ratio of real to contact area, dimensionless
λ	with subscript e, a parameter used in Eq. (9.16), dimensionless
Λ	a functional relationship used in Eq. (9.19); see Table 9.4
ν	Poisson's ratio, dimensionless
ρ	radius of curvature, m
σ	root-mean-square roughness, m
ψ	indicates functional relationship

Subscripts

a	indicates an apparent value or a point in an interface
b	indicates a point in an interface
c	indicates contact
f	indicates fluid
r	indicates real value
v	indicates void

9.12 REFERENCES

1 Greenwood, A. J., and Williamson, P. B. J., Contact of Nominally Flat Surfaces, *Proc. R. Soc. London Ser. A*, vol. 295, pp. 300–319, 1966.

2 Fenech, H., The Thermal Conductance of Metallic Surfaces in Contact, Sc.D. dissertation, Massachusetts Institute of Technology, Cambridge, Mass., 1959.

3 Bowden, F. P., and Tabor, D., *The Friction and Lubrication of Solids, Part II*, Oxford University Press, London, 1966.

4 Clausing, A. M., and Chao, B. T., Thermal Contact Resistance in a Vacuum Environment, NASA Rept. ME-TN-242-1, University of Illinois, Champaign-Urbana, 1963.

5 Holm, R., *Electric Contacts Handbook*, Springer-Verlag, Berlin, 1958.

6 Kraus, A. D., *Cooling Electronic Equipment*, Prentice-Hall, Englewood Cliffs, N.J., 1965.

7 Fenech, H., and Rohsenow, W. M., Prediction of Thermal Conductance of Metallic Surfaces in Contact, *J. Heat Transfer*, vol. 82, pp. 15–24, 1962.

8 Yovanovich, M. M., and Fenech, H., Thermal Contact Conductance of Nominally Flat, Rough Surfaces in a Vacuum Environment, AIAA Paper 66-42, 1966.

9 Cooper, M. G., Mikic, B. B., and Yovanovich, M. M., Thermal Contact Resistance, *Int. J. Heat Mass Transfer*, vol. 12, pp. 279–300, 1969.

10 Tien, C. L., A Correlation for Thermal Contact Conductance of Nominally Flat Surfaces in a Vacuum, *Proc. 7th Conf. Thermal Conductivity, National Bureau of Standards, Gaithersburg, Md.*, pp. 755–759, November 1967. Published in National Bureau of Standards Special Publication 302, September 1968.

11 Mikic, B. B., Yovanovich, M. M., and Rohsenow, W. M., The Effect of Surface Roughness and Waviness upon the Overall Thermal Contact Resistance, EPL Rept. no. 79361-43, Massachusetts Institute of Technology, Cambridge, Mass., 1966.

12 Yovanovich, M. M., Thermal Contact Conductance in a Vacuum, Sc.D. dissertation, Massachusetts Institute of Technology, Cambridge, Mass., 1967.

13 Shlykov, Y. L., Calculating Thermal Contact Resistance of Machined Metal Surfaces, *Teploenergetika*, vol. 12, no. 10, pp. 79–83, 1965.

10

■ heat exchangers

10.0 A NOTE ON UNITS

Because the material in this chapter is based heavily on the dimensional data for heat exchanger surfaces investigated by Kays and London [1], use is made of the units in the English engineering system.

10.1 INTRODUCTION

A heat exchanger is a device that permits the transfer of heat from a warm fluid to a cooler fluid through an intermediate surface and without mixing of the fluids. In electronic equipment cooling applications, the heat exchanger is most often used to remove heat from a recirculating liquid coolant that has been applied directly to the electronic components. Sometimes an exchanger is used to regulate temperature levels within a package containing electronic equipment or to control the temperature of a series of electronic packages. In all cases of heat exchanger application, the heat exchanger itself may be regarded as a closed system within which the heat exchange occurs.

Common tubular exchangers employ a limited range of tube diameters arrayed in cylindrical shells with about 20 to 40 ft² of surface per cubic foot of exchanger when employing plain tubing $\frac{5}{8}$ to 1 in OD. Compact heat exchangers, through the development of unusual stamping and brazing techniques, contain as much as 1350 ft² of surface per cubic foot of exchanger, and the number continues to rise. Compact exchangers are offered in a large variety of external shapes with an even larger geometric variation of the internals whereby fluids are kept separate as they pass through the exchanger.

With so many shapes from which to choose, it is not unusual that there may be some overlapping of nomenclature. To clarify terms as they are used in this chapter, one variation of the fundamental compact exchanger element, the core, is shown in Fig. 10.1. The core consists of a pair of parallel plates with connecting metal members that are bonded to the plates. The arrangement of plates and bonded members provides both a fluid flow channel and prime and extended surface. It will be shown in Chap. 14 that if a plane were drawn midway between the two plates, each half of the connecting metal members could be considered as longitudinal fins.

Indeed, if two or more identical cores are connected by separation plates, the arrangement may be called a *stack* or *sandwich*. Heat can enter a stack through either or both end plates. However, the heat is removed from the successive separating plates and fins by a fluid flowing in parallel through the entire network with a single average convection heat transfer coefficient. For this reason the stack is treated as a finned passage rather than as a fluid-fluid heat exchanger. The stack is particularly useful

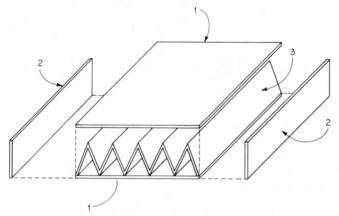

FIG. 10.1 Exploded view of a compact heat exchanger core: 1, plates; 2, side bars; 3, corrugated fins stamped from a continuous strip of metal. By spraying braze powder on the plates, the entire assembly of plates, fins, and bars can be thermally bonded in a single furnace operation.

where the convection heat transfer coefficient for the fluid is small compared with the quantity of heat that may enter the stack by conduction and the amount of heat transfer surface incorporated in it. Of course, due consideration must be given to the fact that as more and more fins are placed in a core, the hydraulic radius of the core will decrease. The heat transfer coefficient of the fluid may be lowered while the fluid friction loss is increased significantly.

Next, consider a pair of cores arranged as components of a two-fluid heat exchanger in cross flow as shown in Fig. 10.2. Fluids enter alternate cores from separate headers at right angles to each other and leave through separate headers at the opposite

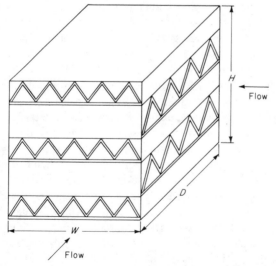

FIG. 10.2 A two-fluid compact heat exchanger with headers removed.

ends of the exchanger. The separation plate spacing need not be the same for both fluids, nor need the cores for both fluids contain the same numbers or kinds of fins. These are dictated by the allowable pressure drops for both fluids and the resulting heat transfer coefficients. Where one coefficient is quite large compared with the other, it is entirely permissible to have no extended surface in the alternate cores through which the fluid with the higher coefficient travels. Very often the body of an exchanger built up of cores as shown in Fig. 10.2, and without headers, is also called a core. An exchanger built up of plates and fins as in Fig. 10.2 is a *plate-fin* exchanger.

10.2   CLASSIFICATION OF COMPACT HEAT EXCHANGERS

Compact heat exchangers may be classified by the kinds of compact elements they employ. The compact elements generally fall into four classes: (1) *Tubular surfaces* are arrays of tubes of small diameter from $\frac{1}{2}$ in down to perhaps $\frac{1}{4}$ in. These tubes are used in service where the ruggedness and cleanability of the conventional shell-and-tube exchanger are not required. Usually tubesheets are comparatively thin, and soldering or brazing a tube to a tubesheet provides an adequate seal against inter-leakage and differential thermal expansion. (2) *Surfaces with flow normal to finned banks of smooth tubes* are smooth, round tubes that have been expanded into fins that can accept a number of tube rows as shown in Fig. 10.3a. Holes may be stamped in the fin with a drawn hub or foot to improve contact resistance or as a spacer be-tween successive fins as shown, or brazed directly to the fin with or without a hub. Other types reduce flow resistance outside the tubes by using flattened tubes and brazing as shown in Figs. 10.3b and 10.3c. Flat tubing is made from strip similar to the manufacture of welded circular tubing but is much thinner and is joined by solder-ing or brazing rather than welding. (3) *Plate-fin surfaces* are shown in Figs. 10.3d through 10.3i. (4) *Matrix surfaces* are surfaces that are used in rotating, regenerative equipment such as combustion flue gas-air preheaters for conventional fossil fuel furnaces. In this application, metal is deployed for its ability to absorb heat with minimal fluid friction while exposed to hot flue gas and to give up the heat to in-coming cold combustion air with minimal fluid friction when it is rotated into the incoming cold air stream. Each of these classes and their various subdivisions are de-fined in Table 10.1.

Compact heat exchanger surfaces are described in the literature by geometric factors that have been standardized largely through the extensive work of Kays and London [1]. These factors and the relationships between them are essential for the application of the basic heat transfer and flow friction data to a particular design problem. They are listed and defined in Table 10.2. Physical data for a number of compact heat exchanger surfaces are given in Table 10.3. The relationships between the geometric factors in Table 10.2 will now be established.

Consider an exchanger composed of n_1 layers of one type of plate-fin surface and n_2 layers of a second type, as shown in Fig. 10.2. The separation plate thickness is established by the pressure differential to which it is exposed or through discretion. Retaining the subscripts 1 and 2 for the respective types of surface, the overall ex-changer height H is

$$H = n_1(b_1 + a) + n_2(b_2 + a) \tag{10.1}$$

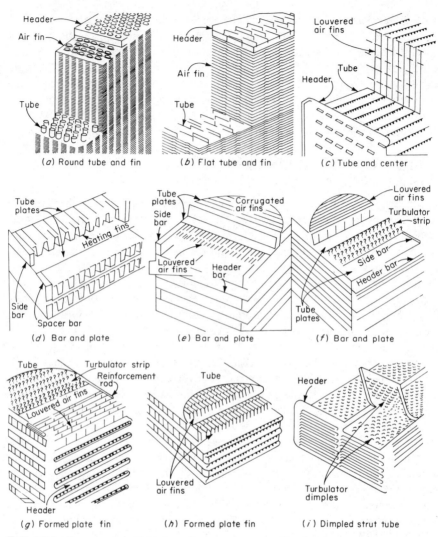

FIG. 10.3 Some compact heat exchanger elements. (Sketched from illustrations provided by the Harrison Radiator Division, Lockport, N.Y.)

where b_1 and b_2 are the separation distances between the plates for the two kinds of surface. With the width W and depth D selected, the overall volume V is

$$V = W \cdot D \cdot H \tag{10.2}$$

In Fig. 10.2, the length L_1 is along the depth of the exchanger, where $L_1 = D$, and the length L_2 is along the width, $L_2 = W$.

The frontal areas are also established. Again referring to Fig. 10.2,

$$A_{f1} = HW \tag{10.3a}$$

TABLE 10.1 Compact Heat Exchanger Classifications

Exchanger type	How designated	Remarks
Flow inside circular and flattened circular tubes	ST indicates flow inside straight tubes. Example: ST-1. FT indicates flow inside straight flattened tubes. Example: FT-1. FTD indicates flow inside straight flattened dimpled tubes.	Simplest form of compact heat exchanger surface. Dimpling interrupts boundary layer, which tends to increase heat transfer coefficients without increasing flow velocity.

Tubular surfaces

Flow normal to banks of smooth tubes	Designation considers staggered (S) and in-line (I) arrangements of tubes and identifies transverse and longitudinal pitch ratios. Suffix (s) indicates data correlation by steady-state tests. All others correlated by a transient technique. Examples:	

Designation	Arrange-ment	Pitch ratio in diameters	
		Trans-verse	Longi-tudinal
S1.50–1.25(s)	Staggered	1.50	1.25
S1.50–1.25	Staggered	1.50	1.25
I1.25–1.25	In-line	1.25	1.25

Plate-fin surfaces

Exchanger type	How designated	Remarks
Plain fins	Designated by a numeral that indicates number of fins per inch. The suffix T is appended when the passages are of definite triangular shape. Examples: 19.86, 15.08, 46.45T.	Characterized by long, uninterrupted flow passages.
Louvered fins	Designated by a fraction that indicates length of fin in flow direction followed by a numeral that indicates fins per inch. Example: $\frac{1}{2}$–6.06 indicates 6.06 $\frac{1}{2}$-in-long fins per inch.	Characterized by fins that are cut and bent into the flow stream at frequent intervals.
Strip fins	Designated in the same manner as louvered fins. The suffixes (D) and (T) indicate double and triple stacks or sandwiches.	Often called *offset* fins. Fins are offset at frequent intervals and exchanger is essentially a series of plain plate fins with alternate lengths offset.
Wavy fins	Designated by two figures. The first gives the number of fins per inch. The second gives the length of the wave. Wavy-fin designations are always followed by the letter W. Example: 11.44–$\frac{3}{8}$ W is a wavy fin with 11.44 fins per inch and a $\frac{3}{8}$-in length of wave.	Characterized by a continuous curvature. The change in flow direction introduced by the waves in the surface tends to interrupt the boundary layer as in the case of louvered and strip fins.

TABLE 10.1   Compact Heat Exchanger Classifications (*Continued*)

Exchanger type	How designated	Remarks
Plate fin surfaces		
Pin fins	Designation method is nondescriptive	Fins are constructed from small-diameter wire. They yield very high heat transfer coefficients because the effective flow length is very small.
Perforated fins	Designated by the number of fins per inch followed by the letter P.	Holes cut in the fins provide boundary layer interruption.
Finned-tube surfaces		
Circular tubes with spiral radial fins	Designated by the letters CF followed by one or two numerals. The first numeral designates the number of fins per inch. The second numeral, if one is used, refers to the nominal tube size.	When CF does not appear in the designation, the surface may be presumed to have continuous fins.
Circular tubes with continuous fins	No letter prefix is employed. The first numeral indicates the number of fins per inch. The second numeral, if one is used, refers to the nominal tube size.	
Finned flat tubes	No letter prefix is employed. The first numeral indicates the fins per inch. The second numeral indicates the largest tube dimension.	
Matrix surfaces		
All types	No designation employed.	

and $A_{f2} = HD$ $\hfill (10.3b)$

If the entire exchanger consisted of a single exchanger surface, surface 1 or surface 2, the total surface area would be the product of the ratio of the total surface to total volume, β ft^2/ft^3, and the total exchanger volume V. However, where there are two different surfaces, it is necessary to employ the factor α, which is the ratio of the total surface on *one* side to the total volume on *both* sides of the exchanger. By taking simple proportions,

$$\alpha_1 = \frac{b_1}{b_1 + b_2 + 2a} \beta_1 \hfill (10.4a)$$

$$\alpha_2 = \frac{b_2}{b_1 + b_2 + 2a} \beta_2 \hfill (10.4b)$$

and the total heat transfer surfaces are

$$S_1 = \alpha_1 V \hfill (10.5a)$$

$$S_2 = \alpha_2 V \hfill (10.5b)$$

The hydraulic radius is defined by

$$r_h = \frac{A}{P} = \frac{AL}{S} \tag{10.6}$$

TABLE 10.2 Compact Heat Exchanger Geometric Factors

Factor and symbol	Descriptive comment
A	The free flow area on one side of the exchanger. To distinguish between hot and cold sides, the free flow areas are often designated A_h and A_c.
A_f	The frontal area of one side of the exchanger. This is merely the product of the overall exchanger width and height or depth and height.
a	The separation plate thickness. Applies to plate-fin surfaces only.
b	The separation plate spacing. This dimension is an approximation of the fin height. Applies to plate-fin surfaces only.
D_e	The equivalent diameter is also used to correlate heat transfer and flow friction and is four times the hydraulic radius.
L	The flow length on one side of the exchanger. Note that this factor *always* concerns the flow length of a *single* side of the exchanger, although two sides may be present, and that ambiguity is avoided with the overall exchanger dimensions, which are designated *width, depth,* and *height.* It is therefore reasonable to have the overall exchanger depth be the length of one side of the exchanger and the overall width the length of the other side.
P	Perimeter of passage.
p	The porosity, which is the ratio of the exchanger void volume to the total exchanger volume. Applies to matrix surfaces only.
r_h	The hydraulic radius. Heat transfer and flow friction are correlated on the basis of the hydraulic radius, which is the ratio of the passage flow area to the passage wetted perimeter.
S	The heat transfer surface on one side of the exchanger. Subscripts are often appended to distinguish between hot- and cold-side surfaces. The surface designated by S refers to *both* prime and finned surfaces.
S_f	The surface of the fins, only, on one side of the exchanger. Applies to finned surfaces only.
V	The total exchanger volume. This applies to both sides of the heat exchanger and is merely the product of the overall heat exchanger core width, depth, and height.
α	The ratio of total surface area on *one* side of the exchanger to the total volume on *both* sides of the exchanger. Applies to tubular, plate-fin surfaces, and crossed-rod matrices only.
β	Ratio of total surface to the total volume on one side of the exchanger. The surface alone is S. The total volume includes the overall exchanger dimensions. Applies to plate-fin surfaces only.
δ	Fin width.
η	Fin efficiency.
η_w	Weighted fin efficiency for a compact exchanger, $\eta_w = [\eta S_f + 1(S - S_f)]/S = 1 - (1 - \eta)S_f/S$. Also see Eq. (10.15).
σ	Ratio of free flow area to frontal area on one side of the exchanger.

TABLE 10.3 Surface Geometry of Some Plate-Fin Surfaces

Plain plate fins

Designation	b (in)	Fins/in	δ_0 (in)	r_h (ft)	L (in)	β (ft²/ft³)	S_f/S (ft²/ft²)
11.1	0.250	11.10	0.006	0.002530	2.50	367	0.756
15.08	0.418	15.08	0.006	0.002190	6.84	414	0.870
19.86	0.250	19.86	0.006	0.001538	2.51	561	0.849
46.45T	0.100	46.45	0.002	0.000661	2.63	1332.5	0.837

Louvered fins

Designation	b (in)	Fins/in	δ_0 (in)	r_h (ft)	Louver spacing (in)	Louver gap (in)	β (ft²/ft³)	S_f/S (ft²/ft²)
$\frac{3}{8}$-6.06	0.250	6.06	0.006	0.003650	0.375	0.055	256	0.640
$\frac{1}{2}$-6.06	0.250	6.06	0.006	0.003650	0.500	0.055	256	0.640
$\frac{3}{16}$-11.1	0.250	11.1	0.006	0.002530	0.1875	0.055	367	0.756
$\frac{3}{4}$-11.1	0.250	11.1	0.006	0.002530	0.750	0.040	367	0.756

Strip fins

Designation	b (in)	Fins/in	δ_0 (in)	r_h (ft)	L (in)	β (ft²/ft³)	S_f/S (ft²/ft²)
$\frac{1}{8}$-13.95	0.375	13.95	0.010	0.002198	0.125	381	0.840
$\frac{1}{8}$-16.00(D)	0.255	16.00	0.006	0.001528	0.125	550	0.845
$\frac{1}{8}$-19.82(D)	0.205	19.82	0.004	0.001262	0.125	680	0.841
$\frac{1}{8}$-20.06(D)	0.201	20.06	0.004	0.001223	0.125	698	0.843

Wavy and pin fins

Designation	b (in)	Fins/in	δ_0 or d (in)	r_h (ft)	β (ft²/ft³)	S_f/S (ft²/ft²)	Remarks
11.5-$\frac{3}{8}$W	0.375	11.50	0.010	0.002482	347	0.822	Length of waves 0.375 in
17.8-$\frac{3}{8}$W	0.413	17.8	0.006	0.001740	514	0.892	Length of waves 0.375 in
AP-1	0.240		0.040	0.003610	188	0.512	In-line pins
PF-3	0.750		0.031	0.001340	339	0.834	In-line pins

and from the definition of porosity, $p = A/A_f$,

$$r_h = \frac{pA_fL}{S}$$

Because $V = A_fL$, α may be defined for a matrix surface as

$$\alpha = \frac{S}{V} = \frac{p}{r_h} \tag{10.7}$$

The factor σ for all but matrix surfaces is determined in a similar manner. It is defined by

$$\sigma = \frac{A}{A_f}$$

Substituting A from Eq. (10.6),

$$\sigma = \frac{Sr_h}{A_fL} = \frac{Sr_h}{V} = \alpha r_h \tag{10.8}$$

Thus the flow areas are given by

$$A_1 = \sigma_1 A_{f1} \tag{10.9a}$$
$$A_2 = \sigma_2 A_{f2} \tag{10.9b}$$

For matrix surfaces, the flow areas are obtained directly from the porosity p.

10.3 HEAT TRANSFER AND FLOW FRICTION DATA

Heat transfer data for compact heat exchangers are correlated on an individual surface basis using a Colburn-type of representation. This representation plots the heat transfer factor

$$j_h = \left(\frac{h}{cG}\right)\left(\frac{c\mu}{k}\right)^{2/3} \tag{10.10}$$

as a function of the Reynolds number, which is obtained by employing the equivalent diameter, $4r_h$:

$$\text{Re} = \frac{4r_hG}{\mu} = \frac{D_eG}{\mu} \tag{10.11}$$

The term h/cG in Eq. (10.10) is the Stanton number St, the ratio of the Nusselt number Nu to the product of the Reynolds number Re and the Prandtl number Pr.[1]

$$St = \frac{Nu}{Re\,Pr} = \frac{hD_e/k}{(D_e G/\mu)(c\mu/k)} = \frac{h}{Gc} \tag{10.12}$$

The fluid properties in Eqs. (10.10) and (10.11) are evaluated at the bulk temperature. Employing the nomenclature T_1 and T_2 for the hot fluid inlet and outlet temperatures, respectively, the hot-side bulk temperature is given by

$$T_b = \tfrac{1}{2}(T_1 + T_2) \tag{10.13a}$$

In a similar manner, the cold fluid bulk temperature is

$$t_b = \tfrac{1}{2}(t_2 + t_1) \tag{10.13b}$$

The pressure drop ΔP in a compact heat exchanger was computed by Kays and London [1] from the equation

$$\frac{\Delta P}{P_1} = \frac{(G')^2 v_1}{2g'_c P_1}\left[(1 + K_c - \sigma^2) + 2\left(\frac{v_2}{v_1} - 1\right) + f\frac{S}{A}\frac{v_m}{v_1} - (1 - \sigma^2 - K_e)\frac{v_2}{v_1}\right] \tag{10.14}$$

where $G' =$ mass velocity, lb/ft^2 s
  $v_1 =$ fluid specific volume at inlet conditions, ft^3/lb
  $v_2 =$ fluid specific volume at outlet conditions, ft^3/lb
  $v_m =$ fluid specific volume at mean conditions, ft^3/lb
  $S =$ surface area, ft^2
  $A =$ free flow area, ft^2
  $g'_c =$ local gravitational acceleration, ft/s^2
  $P_1 =$ pressure at inlet conditions
  $\Delta P =$ pressure drop in consistent units with P_1
  $K_c =$ entrance contraction loss coefficient, dimensionless
  $K_e =$ exit loss coefficient, dimensionless

Friction factors are correlated on an individual surface basis and are usually plotted as a function of the Reynolds number. The entrance and exit loss coefficients differ for the various types of passages and are plotted as a function of the parameter σ and the Reynolds number.

Four terms may be noted within the brackets in Eq. (10.14). These terms denote, respectively, entrance or contraction loss as the fluid approaches the exchanger at line velocity and changes to the initial exchanger velocity, acceleration loss or gain as the fluid expands or contracts during its passage through the exchanger, flow-friction loss, and exit loss.

---

[1] Because compact heat exchangers find their greatest utility in gas-to-gas applications, in Eqs. (10.10) and (10.12) the specific heat is the value for constant pressure.

Kays and London have presented heat transfer and flow friction data for 120 surfaces in the form described above. Some typical examples are shown in Figs. 10.4 through 10.7. Entrance and exit loss coefficients for plate-fin cores and rectangular passages are plotted in Figs. 10.8 and 10.9.

10.4 FIN AND OVERALL PASSAGE EFFICIENCY

The factor S_f/S that is available for each surface indicates the ratio of finned surface to total surface on one side (hot or cold) of the exchanger. For example, it may be observed in Table 10.3 that values of S_f/S range from 0.640 to 0.870. This shows that a considerable amount of surface in the exchanger will be in the form of fins and cognizance must be taken of the fin and overall passage efficiency, which are discussed in detail in Chap. 14.

The fin efficiency accounts for the fact that as heat flows from the base to the tip of a fin, this heat flow must be due to a falling temperature gradient.[2] Hence, there is decrease in fin operating temperature from base to tip and not all portions of the fin are equally *efficient* in dissipating the heat. In order to account for this, the *fin efficiency*, which is defined as the ratio of the actual heat dissipated to the ideal heat that would be dissipated if the entire fin were to operate at the base temperature, is used.

For the longitudinal fin of rectangular profile that is employed in all of the surfaces listed in Table 10.3 (except for the pin-fin surfaces), the fin efficiency is given by

$$\eta_f = \frac{\tanh mb}{mb} \tag{14.5}$$

where $m = (2h/k\delta_0)^{1/2}$. For pin fins, Eq. (14.5) applies but, in this case $m = (4h/kd)^{1/2}$. For single stacks with equal dissipations from both sides, b is taken as one-half of the Table 10.3 value.

The weighted passage efficiency is, as the name implies, a weighting of the finned surface S_f operating at its efficiency η_f and the base or prime surface S_b operating at an efficiency of 100%. The total surface is, of course,

$$S = S_b + S_f$$

and the weighing process yields the weighted passage efficiency η_w for the total surface S. Hence

$$\eta_w S = 1.00 S_b + \eta_f S_f$$

But the base surface is $S_b = S - S_f$, so

$$\eta_w S = 1.00(S - S_f) + \eta_f S_f$$

---

[2] If the fin is dissipating heat to a flowing fluid.

$L/4r_h = 20.6$

0.25"

0.18"

Fin pitch = 11.1 per in.
Plate spacing, b = 0.250 in.
Flow passage hydraulic diameter, $4r_h$ = 0.01012 ft.
Fin metal thickness = 0.006 in., aluminum
Total transfer area/volume between plate, β = 367 ft²/ft³
Fin area/total area = 0.756

(b)

$L/4r_h = 35.0$

0.25"

0.1006"

Fin pitch = 19.86 per in.
Plate spacing, b = 0.250 in.
Flow passage hydraulic diameter, $4r_h$ = 0.00615 ft
Fin metal thickness = 0.006 in., aluminum
Total transfer area/volume between plates, β = 561 ft²/ft³
Fin area/total area = 0.849

(d)

$L/4r_h = 83.0$

0.100"

0.0431"

Fin pitch = 46.45 per in.
Plate spacing, b = 0.100 in.
Fin length flow direction = 2.63 in.
Flow passage hydraulic diameter, $4r_h$ = 0.002643 ft
Fin metal thickness = 0.002 in., stainless steel
Total heat transfer area/volume between plates, β = 1332.5 ft²/ft³
Fin area/total area = 0.837

(a)

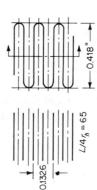

$L/4r_h = 65$

0.418"

0.1326

Fin pitch = 15.08 per in.
Plate spacing, b = 0.418 in.
Flow passage hydraulic diameter, $4r_h$ = 0.00876 ft
Fin metal thickness = 0.006 in., aluminum
Total transfer area/volume between plates, β = 414 ft²/ft³
Fin area/total area = 0.870

(c)

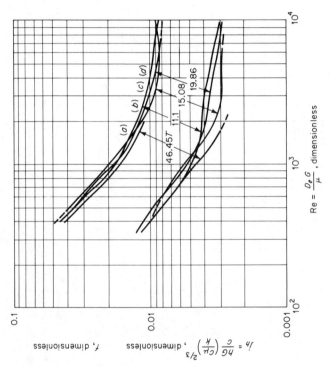

FIG. 10.4 Heat transfer and flow friction characteristics of some plain plate-fin heat exchanger surfaces. (From Kays and London [1], by permission.)

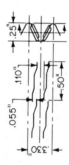

Fin pitch 11.1 per in.
Plate spacing, b = 0.250 in.
Louver spacing = 0.75 in.
Fin gap = 0.05 in.
Louver gap = 0.04 in.
Flow passage hydraulic diameter, $4r_h$ = 0.01012 ft
Fin metal thickness = 0.006 in., aluminum
Total heat transfer area / volume between plates, β = 367 ft^2/ft^3
Fin area / total area = 0.756

(b)

Fin pitch = 6.06 per in.
Plate spacing, b = 0.250 in.
Louver spacing = 0.50 in.
Fin gap = 0.110 in.
Louver gap = 0.055 in.
Flow passage hydraulic diameter, $4r_h$ = 0.01460 ft
Fin metal thickness = 0.006 in., aluminum
Total heat transfer area/volume between plates, β = 256 ft^2/ft^3
Fin area/total area = 0.640

(d)

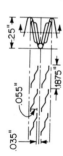

Fin pitch = 11.1 per in.
Plate spacing, b = 0.250 in.
Louver spacing = 0.1875 in.
Fin gap = 0.035 in.
Louver gap = 0.055 in.
Flow passage hydraulic diameter, $4r_h$ = 0.01012 ft
Fin metal thickness = 0.006 in., aluminum
Total heat transfer area / volume between plates, β = 367 ft^2/ft^3
Fin area / total area = 0.756

(a)

Fin pitch = 6.06 per in.
Plate spacing, b = 0.250 in.
Louver spacing = 0.375 in.
Fin gap = 0.110 in.
Louver gap = 0.055
Flow passage hydraulic diameter, $4r_h$ = 0.01460 ft
Fin metal thickness = 0.006 in., aluminum
Total transfer area/volume between plates, β = 256 ft^2/ft^3
Fin area/total area = 0.640

(c)

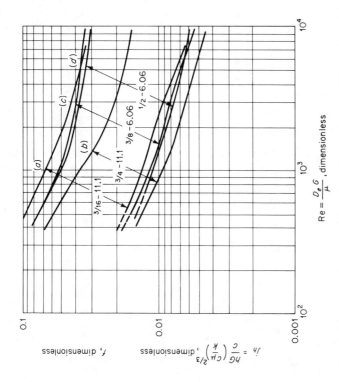

FIG. 10.5 Heat transfer and flow friction characteristics of some louvered-fin heat exchanger surfaces. (From Kays and London [1], by permission.)

229

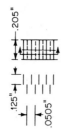

.125" | .201"

.0499"

Fin pitch = 20.06 per in.
Plate spacing, b = 0.201 in.
Splitter symmetrically located
Fin length flow direction = 0.125 in.
Flow passage hydraulic diameter, $4r_h$ = 0.004892 ft.
Fin metal thickness = 0.004 in., aluminum
Splitter metal thickness = 0.006 in.
Total heat transfer area / volume between plates, β = 698 ft²/ft³
Fin area (including splitter) / total area = 0.843

(a)

.125" | .205"

.0505"

Fin pitch = 19.82 per in.
Plate spacing, b = 0.205 in.
Splitter symmetrically located
Fin length flow direction = 0.125 in.
Flow passage hydraulic diameter, $4r_h$ = 0.005049 ft
Fin metal thickness = 0.004 in., nickel
Splitter metal thickness = 0.006 in.
Total heat transfer area / volume between plates, β = 680 ft²/ft³
Fin area (including splitter) / total area = 0.841

(b)

.125" | .255"

.0625"

Fin pitch = 16.00 per in.
Plate spacing, b = 0.255 in.
Splitter symmetrically located
Fin length flow direction = 0.125 in.
Flow passage hydraulic diameter, $4r_h$ = 0.00612 ft.
Fin metal thickness = 0.006 in., aluminum
Splitter metal thickness = 0.006 in.
Total heat transfer area / volume between plates, β = 550 ft²/ft³
Fin area (including splitter) / total area = 0.845

(c)

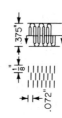

$1\frac{1}{8}$" | .375"

.072"

Fin pitch = 13.95 per in.
Plate spacing, b = 0.375 in.
Fin length = 0.125 in.
Flow passage hydraulic diameter, $4r_h$ = 0.00879 ft.
Fin metal thickness = 0.010 in., aluminum
Total heat transfer area / volume between plates, β = 381 ft²/ft³
Fin area / total area = 0.840
Note: The fin surface area on the leading and trailing edges of the fins have not been included in area computations.

(d)

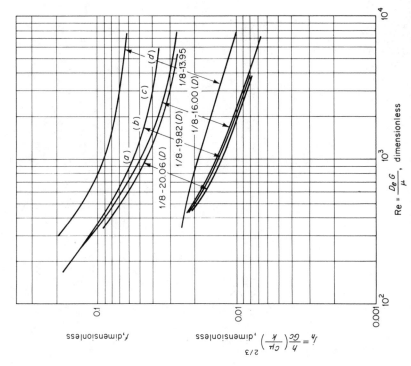

FIG. 10.6 Heat transfer and flow friction characteristics of some strip-fin heat exchanger surfaces. (From Kays and London [1], by permission.)

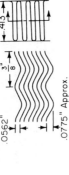

Pin diameter = 0.031 in., aluminum
Pin pitch parallel to flow = 0.062 in.
Pin pitch perpendicular to flow = 0.062 in.
Plate spacing, b = 0.750 in.
Flow passage hydraulic diameter, $4r_h$ = 0.00536 ft
Total heat transfer area/volume between plates, β = 339 ft²/ft³
Fin area/total area = 0.834

(a)

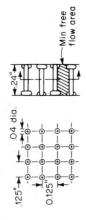

Fin pitch = 17.8 per in.
Plate spacing, b = 0.413 in.
Flow passage hydraulic diameter, $4r_h$ = 0.00696 ft
Fin metal thickness = 0.006 in., aluminum
Total heat transfer area/volume between plates, β = 514 ft²/ft³
Fin area/total area = 0.892
Note: Hydraulic diameter based on free-flow area
normal to mean flow direction.

(b)

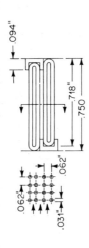

Fin pitch = 11.5 per in
Plate spacing, b = 0.375 in.
Flow passage hydraulic diameter, $4r_h$ = 0.00993 ft.
Fin metal thickness = 0.010 in., aluminum
Total heat transfer area/volume between plates, β = 347 ft²/ft³
Fin area/total area = 0.822
Note: Hydraulic diameter based on free-flow area
normal to mean flow direction.

(c)

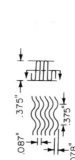

Pin diameter = 0.04 in., copper
Pin pitch parallel to flow = 0.125 in.
Pin pitch perpendicular to flow = 0.125 in.
Plate spacing, b = 0.24 in.
Flow passage hydraulic diameter, $4r_h$ = 0.01444 ft
Total heat transfer area/volume between plates, β = 188 ft²/ft³
Fin area/total area = 0.512

(d)

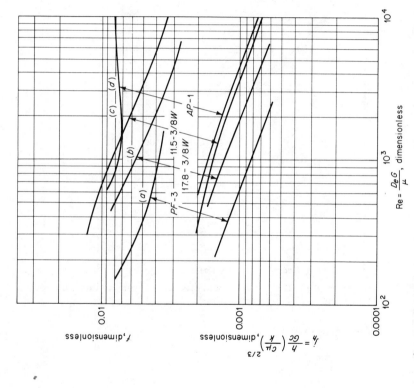

FIG. 10.7 Heat transfer and flow friction characteristics of some wavy and pin-fin heat exchanger surfaces. (From Kays and London [11], by permission.)

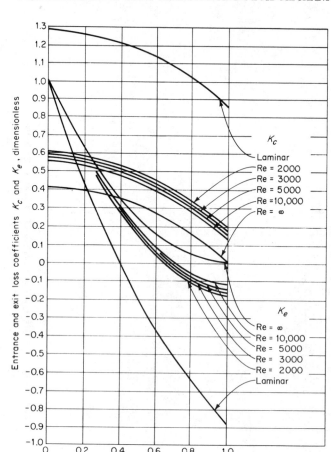

FIG. 10.8 Entrance and exit loss coefficients for flow-through plate-fin cores. (From Kays and London [1], by permission.)

It is now only a matter of algebra:

$$\eta_w = 1.00 \frac{S}{S} - 1.00 \frac{S_f}{S} + \eta_f \frac{S_f}{S}$$

or $$\eta_w = 1 - (1 - \eta_f) \frac{S_f}{S} \qquad (10.15)$$

10.5 THE OVERALL HEAT TRANSFER COEFFICIENT

The overall heat transfer coefficient to be used in the rate equation

$$q = US(\text{LMTD}) \qquad (10.16)$$

may be based on the hot- or the cold-side surface. With k_m and S_m, respectively, the thermal conductivity and surface of the metal separating plates, U is either

$$U_h = \frac{1}{(1/\eta_{wh}h_h) + (a/k_m)(S_h/S_m) + (1/\eta_{wc}h_c)(S_h/S_c)} \qquad (10.17a)$$

or

$$U_c = \frac{1}{(1/\eta_{wh}h_h)(S_c/S_h) + (a/k_m)(S_c/S_m) + (1/\eta_{wc}h_c)} \qquad (10.17b)$$

10.6 THE ϵ-N_{tu} METHOD

The ϵ-N_{tu} method of heat exchanger design (or performance estimation) is based on the rate equation as given by Eq. (10.16), with LMTD defined as the logarithmic mean temperature difference. The ϵ-N_{tu} method makes use of three dimensionless parameters, ϵ, R, and N_{tu}.

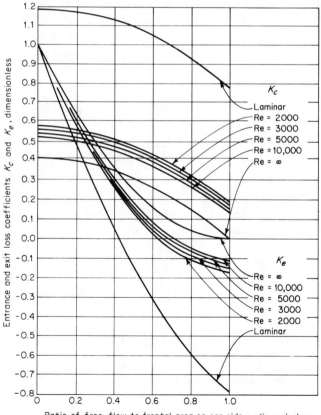

FIG. 10.9 Entrance and exit loss coefficients for flow-through rectangular passages. (From Kays and London [1], by permission.)

The exchanger heat transfer effectiveness, or simply the effectiveness ϵ, is designated in the nomenclature of the ϵ-N_{tu} method by

$$\epsilon = \frac{C_h(T_1 - T_2)}{C_{min}(T_1 - t_1)} = \frac{C_c(t_2 - t_1)}{C_{min}(T_1 - t_1)} \tag{10.18}$$

where $C_h = WC$ is the product of the weight flow W and specific heat C of the hot fluid, $C_c = wc$ corresponds to the cold fluid, and C_{min} is the smaller of the C_h and C_c values. Here ϵ is a measure of the amount of heat transferred by the exchanger to the maximum amount of heat that could be transferred with infinite surface.

The *number of transfer units* N_{tu} is defined by

$$N_{tu} = \frac{US}{C_{min}} \tag{10.19}$$

where U is the overall heat transfer coefficient and S is the surface to which it refers. Let Δt be the true temperature difference. Consider $t_2 - t_1$, the duty to be accomplished per pound of fluid, and Δt, the potential for accomplishing it. Then N_{tu} has the significance of measuring the size of the heat transfer task, namely, the duty to be accomplished divided by the temperature potential available for accomplishing it.

The third dimensionless parameter, the *capacity rate ratio*, is uniquely expressed by

$$R = \frac{C_{min}}{C_{max}} \tag{10.20}$$

Observe that, by definition, $R \leqslant 1.00$.

There is a correspondence between the ϵ-N_{tu} method dimensionless parameters and the conventional rate-energy equation approach. Writing the heat balance,

$$q = US\,\Delta t = C_h(T_1 - T_2) = C_c(t_2 - t_1)$$

Applying the definitions of ϵ, R, and N_{tu} given by Eqs. (10.18) through (10.20), the heat transferred or the duty of the exchanger when $C_c = C_{min}$ is

$$q = C_c(t_2 - t_1) = \epsilon C_c \frac{C_{min}}{C_c}(T_1 - t_1) = \epsilon C_{min}(T_1 - t_1)$$

and when $C_h = C_{min}$,

$$q = C_h(T_1 - T_2) = \epsilon C_h \frac{C_{min}}{C_h}(T_1 - t_1) = \epsilon C_{min}(T_1 - t_1)$$

10.6.1   The Counter-Flow Compact Heat Exchanger

In the counter-flow heat exchanger shown in Fig. 10.10, the logarithmic mean temperature difference (LMTD) is such that the rate-energy equation balance, using the cold fluid as C_{min} ($C_c = C_{min} = wc$), is

$$q = US \frac{(T_1 - t_2) - (T_2 - t_1)}{\ln\left[(T_1 - t_2)/(T_2 - t_1)\right]} = C_c(t_2 - t_1)$$

Rearrangement gives

$$N_{tu} = \frac{US}{C_c} = \left[\frac{t_2 - t_1}{(T_1 - T_2) - (t_2 - t_1)}\right] \ln \frac{T_1 - t_2}{T_2 - t_1}$$

or, with further adjustment and substitution of Eq. (10.20),

$$\frac{T_1 - t_2}{T_2 - t_1} = e^{(N_{tu})(R-1)} \tag{10.21}$$

Adding and subtracting t_1 to the numerator and T_1 to the denominator of Eq. (10.21) and applying Eq. (10.20) yields

$$\frac{T_1 + t_1 - t_2 - t_1}{T_2 - t_1 + T_1 - T_1} = \frac{T_1 - t_1 - (t_2 - t_1)}{T_1 - t_1 - R(t_2 - t_1)} = e^{(N_{tu})(R-1)}$$

With further expansion, factoring, and rearrangement employing Eq. (10.18),

$$\frac{1 - (t_2 - t_1)/(T_1 - t_1)}{1 - R[t_2 - t_1)/(T_1 - t_1)]} = \frac{1 - \epsilon}{1 - R\epsilon} = e^{(N_{tu})(R-1)} \tag{10.22}$$

Solving for ϵ yields

$$\epsilon = \frac{t_2 - t_1}{T_1 - t_1} = \frac{1 - e^{(N_{tu})(R-1)}}{1 - Re^{(N_{tu})(R-1)}}$$

Because R is always equal to or less than unity,

$$\epsilon = \frac{t_2 - t_1}{T_1 - t_1} = \frac{1 - e^{-(N_{tu})(1-R)}}{1 - Re^{-(N_{tu})(1-R)}} \tag{10.23}$$

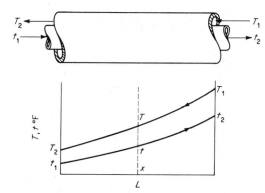

FIG. 10.10 Simple double-pipe heat exchanger arranged for counter flow. Arrows designate fluid flow paths.

Using $C_h = C_{min}$ leads to an identical result but with ϵ having the hot-fluid temperature range in the numerator:

$$\epsilon = \frac{T_1 - T_2}{T_1 - t_1} = \frac{1 - e^{-(N_{tu})(1-R)}}{1 - Re^{-(N_{tu})(1-R)}} \tag{10.24}$$

Both Eqs. (10.23) and (10.24) become indeterminate when R is unity. In this case the logarithmic mean temperature difference is $\Delta t = T_1 - t_2 = T_2 - t_1$, and the heat balance may be written

$$US(T_1 - t_2) = C_c(t_2 - t_1) = C_h(T_1 - T_2)$$

or, using $C_c = C_{min}$,

$$N_{tu} = \frac{US}{C_c} = \frac{t_2 - t_1}{T_1 - t_2} = \frac{t_2 - t_1}{T_1 - t_1 - (t_2 - t_1)}$$

Proceeding algebraically,

$$N_{tu} = \frac{1}{(T_1 - t_1)/(t_2 - t_1) - 1} = \frac{\epsilon}{1 - \epsilon}$$

and solving for ϵ,

$$\epsilon = \frac{t_2 - t_1}{T_1 - t_1} = \frac{N_{tu}}{N_{tu} + 1} \tag{10.25}$$

10.6.2 The Cross-Flow Heat Exchanger with Both Fluids Unmixed

For cross flow with both fluids unmixed (Fig. 10.2 is typical), effectiveness values were obtained by Nusselt [2, 3], whose work was extended by Smith [4]. Mason [5] obtained a solution in the form of an infinite series by employing the Laplace transformation. Mason's solution converges more rapidly than those of the previous investigators and is more readily adaptable to the computer. London [6] has shown that Mason's work can be incorporated into the single relationship

$$\epsilon = \frac{1}{R(N_{tu})} \sum_{n=0}^{\infty} \left[1 - e^{-N_{tu}} \sum_{m=0}^{n} \frac{(N_{tu})^m}{m!} \right] \left\{ 1 - e^{-R(N_{tu})} \sum_{m=0}^{n} \frac{[R(N_{tu})]^m}{m!} \right\} \tag{10.26}$$

which reduces for the case of $R = 0$ to

$$\epsilon = 1 - e^{-N_{tu}} \tag{10.27}$$

These equations are tabulated in Table 10.5, where effectiveness values may be obtained as a function of R and N_{tu}.

10.7 DESIGN PROCEDURE

The design of a heat exchanger by the ϵ-N_{tu} method requires a trial-and-error procedure and is accomplished by a logical progression through the following steps:

1. Establish the heat balance. Thermal properties of air are obtained from Fig. 6.10 with appropriate conversion.
2. Assume an overall exchanger size (length, width, depth) and select the types of surface to be used. Using the geometric data for the assumed surfaces, compute the surface area, free flow area, and other physical parameters.
3. Obtain the thermal properties of the fluids at their bulk temperatures.
4. Compute the film coefficients.
5. Compute the fin and overall surface efficiencies.
6. Determine the overall heat transfer coefficient using Eqs. (10.17a) or (10.17b).
7. Obtain R and N_{tu} and determine the actual exchanger effectiveness from Table 10.5 if cross flow with both fluids unmixed.
8. Compare the actual exchanger effectiveness with the required effectiveness determined from Eq. (10.18). If actual is less than required, return to step 2.
9. Determine the pressure drops.
10. Compare the pressure drops with those specified. If excessive, return to step 2.

10.8 EXAMPLE–LIQUID–TO–AIR EXCHANGER

In Chap. 22, an example is presented that illustrates how Coolanol-45 can be used in a four-pass cooling passage for the anode of a traveling wave tube dissipating 1.5 kW. It is observed that the anode can be held at 302°F (150°C) if 2030 lb/h of Coolanol-45 enters the cooling passages at 170°F.

In this example an air-liquid exchanger will be designed to bring the Coolanol-45, which leaves the anode of the tube at approximately 175°F, back to 170°F so that it can be recirculated. The exchanger will be designed to support this cooling task for 10 traveling wave tubes.

The specifications are as follows:

Heat duty: 51,195 Btu/h (15 kW)
Hot fluid: Coolanol-45
Inlet: 175.1°F
Flow: 20,300 lb/h
Cold fluid: air at atmospheric pressure
Inlet: 80°F
Flow: 932 ft^3/min at 80°F
Arrangement: cross flow
Envelope: maximum 18 in X 18 in, depth maximum 3 in
Pressure drops: air maximum, 1.10 in H$_2$O; Coolanol-45 maximum, 0.75 psi

Table 10.4 presents a worksheet for use in the design and performance calculations for a compact heat exchanger regardless of the flow arrangement. Its only restriction is that single-layered compact surfaces must be used for each fluid. Departures

TABLE 10.4 Compact Heat Exchanger Work Sheet

Hot Fluid: Coolanol-45
Cold Fluid: Air

$T_1 = 175.1$

$T_2 = 170$

$t_2 = 131.9$

$t_1 = 80$

Heat Balance

	Hot	Cold
T_1 or t_2	175.1	131.9
T_2 or t_1	170.0	80.0
ΔT or Δt	5.1	51.9
$\frac{1}{2}\Delta T$ or $\frac{1}{2}\Delta t$	2.55	25.95
T_a or t_a	172.6	105.95
c_p @ t_a or T_a (1)	0.495	0.241
W (2)	20,300	4,096
$q = Wc_p\,\Delta T$	51,195	51,195
$C = Wc_p$	10,048.5	987
C_c, C_h, C_{min}	C_h	$C_c = C_{min}$

Surface Characteristics

Cold side

Type (3)	PLAIN 46.45T
b (4)	0.100
δ (4)	0.002
r_h (4)	0.000661
D_e (5)	0.00264
S_f/S (4)	0.837
β (4)	1,332.5
a (6)	0.01
k_{met} (7)	100
$b/2$	0.050

Hot side

Type	LOUV 3/8-606
b	0.250
δ	0.006
r_h	0.00365
D_e	0.0146
S_f/S	0.640
β	256
a	0.01
k_{met}	100
$b/2$	0.125

Exchanger Physical Data

	Cold side		Hot side	
L (6)	16	16	16	16
H (6)	18	18	18	18
D (6)	1	3/4	1	3/4
Flow?	D		L	
V (8)	0.1481	0.1406	0.1481	0.1406
α (9)	360.14		172.97	
σ (9)	0.238		0.631	
S (10)	53.35	50.65	25.63	24.32
A_f (10)	1.778	2.25	0.111	0.094
A (10)	0.423	0.534	0.0702	0.0592
Heat Transfer Coefficient				
t_a	106.0		172.6	
μ (1)	0.0450		10.65	
k (1)			0.0777	
c_p (1)	0.241		0.495	
$\mathrm{Pr} = c_p \mu / k$	0.793		67.85	
$\mathrm{Pr}^{2/3}$ (1)			16.63	
w	4,096		20,300	
G (11)	9,679	7,648	289,174	342,968
Re (11)	568	449	396	470
j (12)	0.0074	0.00915	0.0185	0.0165
h (13)	21.78	21.52	159.24	168.44

TABLE 10.4 Compact Heat Exchanger Work Sheet (*Continued*)

Efficiencies

	Cold side			Hot side	
$m^2 = 2h/k\delta_0$ (14)	2,613.6	2,582.4	m^2	6,369.6	6,737.6
m	51.12	50.82	m	79.81	82.08
$mb/2$ (14)	0.213	0.212	$mb/2$	0.831	0.855
η_f (15)	0.985	0.985	η_f	0.819	0.811
$\psi_1 = 1 - \eta_f$	0.015	0.015	ψ_1	0.181	0.189
$\psi_2 = \psi_1(S_f/S)$	0.012	0.012	ψ_2	0.116	0.121
η_w (16)	0.988	0.988	η_w	0.884	0.879

Overall coefficient of heat transfer

			Required effectiveness
$1/[(S_h/S_c)\eta_{wh}h_h]$ (17)	0.0148	0.0141	$C_c = C_{min}$
$1/N_{wc}h_c$	0.0465	0.0470	$\epsilon = \dfrac{\Delta t}{T_1 - t_1} = \dfrac{51.9}{(175.1)-(80.0)}$
$R_h/(S_h/S_c)$ (18)			$= \dfrac{51.9}{95.1} = 0.546$
$R_c/1$ (18)			$C_h = C_{min}$
$1/U_c$ (19)	0.0613	0.0611	$\epsilon = \dfrac{\Delta T}{T_1 - t_1} = \dfrac{(\ \ )-(\ \ )}{\rule{2cm}{0.4pt}}$
U_c (19)	16.32	16.37	$=\rule{2cm}{0.4pt}=$

Effectiveness

$R = C_{min}/C_{max}$	0.098	
$N_{tu} = U_c S_c/C_{min}$	0.882	0.840
ϵ (20)	0.570	0.553

Pressure drop

	Cold side			Hot side	
t_1	80.0		T_1	175.1	
t_2	131.9		T_2	170.0	
v_1 (21)	13.65		v_1	0.0184	
v_2	14.97		v_2	0.0184	
v_m	14.31		v_m	0.0184	
v_m/v_1	1.048		v_m/v_1	1.00	
v_2/v_1	1.096		v_2/v_1	1.00	
G' (22)	2.689	2.124	G'	80.33	95.27
$(G')^2$	7.229	4.513	$(G')^2$	6,452.3	9,076.2
σ^2	0.057		σ^2	0.399	
K_c (23)	1.24		K_c	1.11	
K_e (23)	0.36		K_e	-0.41	
f (12)	0.0290	0.0375	f	0.0870	0.0800
$\psi_3 = 1 + K_c - \sigma^2$	2.183		ψ_3	1.711	
$\psi_4 = f(S/A)(v_m/v_1)$	3.832	3.716	ψ_4	31.764	32.865
$\psi_5 = 2[(v_2/v_1) - 1]$	0.192		ψ_5	0.00	
$\psi_6 = \psi_3 + \psi_4 + \psi_5$	6.207	5.991	ψ_6	33.475	34.576
$\psi_7 = 1 - \sigma^2 - K_e$	0.583		ψ_7	1.011	
$\psi_8 = \psi_7(v_2/v_1)$	0.639		ψ_8	1.011	
$\psi_9 = \psi_6 - \psi_8$	5.568	5.352	ψ_9	32.464	33.565
$\psi_{10} = (G')^2 v_1/2g$	1.534	0.958	ψ_{10}	1.845	2.596
$\Delta P = (\psi_9)(\psi_{10})/144$	0.0593	0.0356	ΔP (26)	0.42	0.61
ΔP, in H_2O (25) (26)	1.64	0.99			

from this requirement can, of course, be handled by skipping the proper slots in the worksheet.

The worksheet follows the procedure outlined. To assist the reader, numbers in parentheses have been placed at various locations on the worksheet. These numbers refer to the dialogue that follows.

1 Coolanol-45 properties are from Fig. 22.3. Air properties are from Fig. 6.10. Note that the c_p in the worksheet is for gas-to-gas applications. For a liquid, c_p is meaningless; just C will suffice.

2 W is in pounds per hour. For air at 80°F and at atmospheric pressure, $\rho = 0.0732$ lb/ft^3, and 932 ft^3/min amounts to

$$(932)(60)(0.0732) = 4096 \text{ lb/h}$$

3 Just a selection for better or for worse. This is usually at the designer's discretion.
4 See Table 10.3.
5 $D_e = 4r_h$ in feet.
6 Designer's discretion and in inches.
7 Aluminum ($k = 100$ Btu/ft h °F).
8 In cubic feet.
9 See Eqs. (10.4a) and (10.4b). Using Eq. (10.4a) for the air side,

$$\alpha_1 = \frac{b_1}{b_1 + b_2 + 2a} \quad \beta_1 = \frac{0.10}{0.25 + 0.25 + 2(0.01)} \, 1332.5$$

or $\alpha_1 = 360.14$ ft^2/ft^3. In similar fashion, from Eq. (10.4b) using $\beta_2 = 256$, $\alpha_2 = 172.97$ ft^2/ft^3. The value of σ is obtained from Eq. (10.8).

10 S, A_f, and A are in square feet. To obtain the values, use Eqs. (10.3a), (10.3b), (10.5a), (10.5b), (10.9a), and (10.9b).
11 $G = W/A$ lb/ft^2 h and Re $= D_e G/\mu$.
12 Figure 10.4 or Fig. 10.5 with extrapolation or interpolation.
13 $h = jGc_p/\text{Pr}^{2/3}$ Btu/ft^2 h °F.
14 δ_0 and $b/2$ must be in feet.
15 $\eta_f = \tanh mb/mb$.
16 See Eq. (10.15).
17 See Eq. (10.17b). No effect of the wall is included in the worksheet.
18 Not required.
19 Eq. (10.17b).
20 The effectiveness of the assumed exchanger ($\epsilon = 0.570$) is slightly greater than the required effectiveness ($\epsilon = 0.546$). If the pressure drops are not excessive, this exchanger is acceptable for the specified performance.
21 For air at atmospheric pressure,

$$v = \frac{460 + t}{39.55} \text{ ft}^3/\text{lb}$$

TABLE 10.5 Cross-Flow Heat Exchanger Effectiveness—Both Fluids Unmixed

N_{tu}	R = 0.05	R = 0.10	R = 0.15	R = 0.20	R = 0.25	R = 0.30	R = 0.35	R = 0.40	R = 0.45	R = 0.50
0.01	0.010	0.010	0.010	0.010	0.010	0.010	0.010	0.010	0.010	0.010
0.02	0.020	0.020	0.020	0.020	0.020	0.020	0.020	0.020	0.020	0.020
0.03	0.030	0.030	0.029	0.029	0.029	0.029	0.029	0.029	0.029	0.029
0.04	0.039	0.039	0.039	0.039	0.039	0.039	0.039	0.039	0.039	0.039
0.05	0.049	0.049	0.049	0.049	0.048	0.048	0.048	0.048	0.048	0.048
0.06	0.058	0.058	0.058	0.058	0.058	0.058	0.058	0.058	0.057	0.057
0.07	0.067	0.067	0.067	0.067	0.067	0.067	0.067	0.067	0.067	0.066
0.08	0.077	0.077	0.076	0.076	0.076	0.076	0.076	0.076	0.076	0.075
0.09	0.086	0.086	0.085	0.085	0.085	0.085	0.085	0.085	0.084	0.084
0.10	0.095	0.095	0.094	0.094	0.094	0.094	0.094	0.093	0.093	0.093
0.11	0.104	0.104	0.103	0.103	0.103	0.102	0.102	0.102	0.102	0.101
0.12	0.113	0.112	0.112	0.112	0.111	0.111	0.111	0.110	0.110	0.110
0.13	0.122	0.121	0.121	0.120	0.120	0.120	0.119	0.119	0.118	0.118
0.14	0.130	0.130	0.129	0.129	0.128	0.128	0.127	0.127	0.127	0.126
0.15	0.139	0.138	0.138	0.137	0.137	0.136	0.136	0.135	0.135	0.134
0.16	0.147	0.147	0.146	0.146	0.145	0.144	0.144	0.143	0.143	0.143
0.17	0.156	0.155	0.154	0.154	0.153	0.152	0.152	0.152	0.151	0.150
0.18	0.164	0.163	0.163	0.162	0.161	0.160	0.160	0.159	0.159	0.158
0.19	0.172	0.171	0.171	0.170	0.169	0.169	0.168	0.167	0.166	0.166
0.20	0.180	0.179	0.179	0.178	0.177	0.176	0.176	0.175	0.174	0.173
0.21	0.188	0.187	0.186	0.185	0.185	0.184	0.183	0.182	0.182	0.181
0.22	0.196	0.195	0.194	0.194	0.193	0.192	0.191	0.190	0.189	0.188
0.23	0.204	0.203	0.202	0.201	0.200	0.199	0.198	0.197	0.196	0.195
0.24	0.212	0.211	0.210	0.209	0.208	0.207	0.206	0.205	0.203	0.202
0.25	0.220	0.218	0.218	0.216	0.215	0.214	0.213	0.212	0.211	0.209
0.26	0.227	0.226	0.225	0.224	0.223	0.221	0.220	0.219	0.218	0.216
0.27	0.235	0.234	0.232	0.231	0.230	0.228	0.227	0.226	0.225	0.223
0.28	0.243	0.241	0.240	0.238	0.237	0.236	0.234	0.233	0.231	0.230
0.29	0.250	0.249	0.247	0.246	0.244	0.243	0.241	0.240	0.238	0.237
0.30	0.257	0.256	0.254	0.253	0.251	0.249	0.248	0.246	0.245	0.243
0.31	0.264	0.263	0.261	0.260	0.258	0.256	0.255	0.253	0.251	0.250
0.32	0.272	0.270	0.268	0.267	0.265	0.263	0.261	0.259	0.258	0.256

TABLE 10.5 Cross-Flow Heat Exchanger Effectiveness—Both Fluids Unmixed (*Continued*)

N_{tu}	R = 0.05	R = 0.10	R = 0.15	R = 0.20	R = 0.25	R = 0.30	R = 0.35	R = 0.40	R = 0.45	R = 0.50
0.33	0.279	0.277	0.275	0.273	0.272	0.270	0.268	0.266	0.264	0.262
0.34	0.286	0.284	0.282	0.280	0.278	0.276	0.274	0.272	0.270	0.269
0.35	0.293	0.291	0.289	0.287	0.285	0.283	0.281	0.279	0.277	0.275
0.36	0.300	0.298	0.296	0.293	0.291	0.289	0.287	0.285	0.283	0.281
0.37	0.307	0.305	0.302	0.300	0.298	0.295	0.293	0.291	0.289	0.287
0.38	0.314	0.311	0.309	0.306	0.304	0.302	0.299	0.297	0.295	0.293
0.39	0.320	0.318	0.315	0.313	0.310	0.308	0.306	0.303	0.301	0.298
0.40	0.327	0.324	0.322	0.319	0.317	0.314	0.312	0.309	0.307	0.304
0.41	0.334	0.331	0.328	0.325	0.323	0.320	0.318	0.315	0.312	0.310
0.42	0.340	0.337	0.334	0.332	0.329	0.326	0.323	0.321	0.318	0.315
0.43	0.347	0.344	0.341	0.338	0.335	0.332	0.329	0.326	0.324	0.321
0.44	0.353	0.350	0.347	0.344	0.341	0.338	0.335	0.332	0.329	0.326
0.45	0.359	0.356	0.353	0.350	0.347	0.344	0.341	0.338	0.335	0.332
0.46	0.365	0.362	0.359	0.356	0.352	0.349	0.346	0.343	0.340	0.337
0.47	0.372	0.368	0.365	0.361	0.358	0.355	0.352	0.349	0.345	0.342
0.48	0.378	0.374	0.371	0.367	0.364	0.361	0.357	0.354	0.351	0.348
0.49	0.384	0.380	0.377	0.373	0.370	0.366	0.363	0.359	0.356	0.353
0.50	0.390	0.386	0.382	0.379	0.375	0.372	0.368	0.365	0.361	0.358
0.51	0.396	0.392	0.388	0.384	0.381	0.377	0.373	0.370	0.366	0.363
0.52	0.401	0.398	0.394	0.390	0.386	0.382	0.378	0.375	0.371	0.368
0.53	0.407	0.403	0.399	0.395	0.391	0.387	0.384	0.380	0.376	0.372
0.54	0.413	0.409	0.405	0.401	0.397	0.393	0.389	0.385	0.381	0.377
0.55	0.419	0.414	0.410	0.406	0.402	0.398	0.394	0.390	0.386	0.382
0.56	0.424	0.420	0.416	0.411	0.407	0.403	0.399	0.395	0.391	0.387
0.57	0.430	0.425	0.421	0.417	0.412	0.408	0.404	0.399	0.395	0.391
0.58	0.435	0.431	0.426	0.422	0.417	0.413	0.409	0.404	0.400	0.396
0.59	0.441	0.436	0.431	0.427	0.422	0.418	0.413	0.409	0.405	0.400
0.60	0.446	0.441	0.437	0.432	0.427	0.423	0.418	0.414	0.409	0.405
0.61	0.452	0.447	0.442	0.437	0.432	0.427	0.423	0.418	0.414	0.409
0.62	0.457	0.452	0.447	0.442	0.437	0.432	0.427	0.423	0.418	0.414
0.63	0.462	0.457	0.452	0.447	0.442	0.437	0.432	0.427	0.423	0.418

0.64	0.422	0.427	0.432	0.437	0.442	0.447	0.452	0.457	0.462	0.467
0.65	0.426	0.431	0.436	0.441	0.446	0.451	0.456	0.462	0.467	0.472
0.66	0.431	0.435	0.440	0.445	0.451	0.456	0.461	0.467	0.472	0.478
0.67	0.435	0.440	0.445	0.450	0.455	0.460	0.466	0.471	0.477	0.483
0.68	0.439	0.444	0.449	0.454	0.460	0.465	0.471	0.476	0.482	0.488
0.69	0.443	0.448	0.453	0.458	0.464	0.470	0.475	0.481	0.487	0.492
0.70	0.447	0.452	0.457	0.463	0.468	0.474	0.480	0.486	0.491	0.497
0.71	0.451	0.456	0.461	0.467	0.473	0.478	0.484	0.490	0.496	0.502
0.72	0.455	0.460	0.466	0.471	0.477	0.483	0.489	0.495	0.501	0.507
0.73	0.459	0.464	0.470	0.475	0.481	0.487	0.493	0.499	0.505	0.512
0.74	0.462	0.468	0.474	0.479	0.485	0.491	0.497	0.504	0.510	0.516
0.75	0.466	0.472	0.478	0.483	0.489	0.495	0.502	0.508	0.515	0.521
0.76	0.470	0.476	0.482	0.488	0.493	0.500	0.506	0.512	0.519	0.526
0.77	0.473	0.479	0.485	0.492	0.497	0.504	0.510	0.517	0.523	0.530
0.78	0.477	0.483	0.489	0.495	0.501	0.508	0.514	0.521	0.528	0.535
0.79	0.481	0.487	0.493	0.499	0.505	0.512	0.519	0.525	0.532	0.539
0.80	0.484	0.490	0.497	0.503	0.509	0.516	0.523	0.530	0.536	0.544
0.81	0.488	0.494	0.500	0.507	0.513	0.520	0.527	0.534	0.541	0.548
0.82	0.491	0.498	0.504	0.511	0.517	0.524	0.531	0.538	0.545	0.552
0.83	0.495	0.501	0.508	0.514	0.521	0.528	0.535	0.542	0.549	0.556
0.84	0.498	0.505	0.511	0.518	0.525	0.531	0.539	0.546	0.553	0.561
0.85	0.501	0.508	0.515	0.522	0.528	0.535	0.542	0.550	0.557	0.565
0.86	0.505	0.511	0.518	0.525	0.532	0.539	0.546	0.554	0.561	0.569
0.87	0.508	0.515	0.522	0.529	0.536	0.543	0.550	0.558	0.565	0.573
0.88	0.511	0.518	0.525	0.532	0.539	0.547	0.554	0.562	0.569	0.577
0.89	0.514	0.521	0.528	0.536	0.543	0.550	0.558	0.565	0.573	0.581
0.90	0.518	0.525	0.532	0.539	0.546	0.554	0.561	0.569	0.577	0.585
0.91	0.521	0.528	0.535	0.542	0.550	0.557	0.565	0.573	0.581	0.589
0.92	0.524	0.531	0.538	0.546	0.553	0.561	0.569	0.577	0.585	0.593
0.93	0.527	0.534	0.542	0.549	0.557	0.565	0.573	0.580	0.589	0.597
0.94	0.530	0.537	0.545	0.552	0.560	0.568	0.576	0.584	0.592	0.601
0.95	0.533	0.540	0.548	0.556	0.563	0.571	0.579	0.588	0.596	0.605
0.96	0.536	0.543	0.551	0.559	0.567	0.575	0.583	0.591	0.600	0.608
0.97	0.539	0.546	0.554	0.562	0.570	0.578	0.586	0.595	0.603	0.612
0.98	0.542	0.549	0.557	0.565	0.573	0.581	0.590	0.598	0.607	0.616

TABLE 10.5 Cross-Flow Heat Exchanger Effectiveness—Both Fluids Unmixed (*Continued*)

N_{tu}	R = 0.05	R = 0.10	R = 0.15	R = 0.20	R = 0.25	R = 0.30	R = 0.35	R = 0.40	R = 0.45	R = 0.50
0.99	0.619	0.610	0.602	0.593	0.585	0.576	0.568	0.560	0.552	0.545
1.00	0.623	0.614	0.605	0.597	0.588	0.580	0.571	0.563	0.555	0.547
1.01	0.627	0.617	0.608	0.600	0.591	0.583	0.574	0.566	0.558	0.550
1.02	0.630	0.621	0.612	0.603	0.594	0.586	0.577	0.569	0.561	0.553
1.03	0.634	0.624	0.615	0.606	0.598	0.589	0.580	0.572	0.564	0.556
1.04	0.637	0.628	0.618	0.610	0.601	0.592	0.583	0.575	0.567	0.559
1.05	0.640	0.631	0.622	0.613	0.604	0.595	0.586	0.578	0.569	0.561
1.06	0.644	0.634	0.625	0.616	0.607	0.598	0.589	0.581	0.572	0.564
1.07	0.647	0.638	0.628	0.619	0.610	0.601	0.592	0.583	0.575	0.567
1.08	0.651	0.641	0.631	0.622	0.613	0.604	0.595	0.586	0.578	0.569
1.09	0.654	0.644	0.634	0.625	0.616	0.607	0.598	0.589	0.580	0.572
1.10	0.657	0.647	0.637	0.628	0.619	0.610	0.601	0.592	0.583	0.574
1.11	0.660	0.650	0.641	0.631	0.622	0.612	0.603	0.594	0.586	0.577
1.12	0.664	0.653	0.644	0.634	0.625	0.615	0.606	0.597	0.588	0.579
1.13	0.667	0.656	0.647	0.637	0.627	0.618	0.609	0.600	0.591	0.582
1.14	0.670	0.660	0.650	0.640	0.630	0.621	0.611	0.602	0.593	0.584
1.15	0.673	0.663	0.653	0.643	0.633	0.624	0.614	0.605	0.596	0.587
1.16	0.676	0.666	0.656	0.646	0.636	0.626	0.617	0.607	0.598	0.589
1.17	0.679	0.668	0.659	0.648	0.639	0.629	0.619	0.610	0.601	0.592
1.18	0.682	0.671	0.661	0.651	0.641	0.632	0.622	0.612	0.603	0.594
1.19	0.685	0.674	0.664	0.654	0.644	0.634	0.624	0.615	0.606	0.596
1.20	0.688	0.677	0.667	0.657	0.647	0.637	0.627	0.617	0.608	0.599
1.21	0.691	0.680	0.670	0.660	0.649	0.639	0.630	0.620	0.610	0.601
1.22	0.694	0.683	0.673	0.662	0.652	0.642	0.632	0.622	0.613	0.603
1.23	0.697	0.686	0.675	0.665	0.655	0.644	0.634	0.625	0.615	0.606
1.24	0.699	0.688	0.678	0.667	0.657	0.647	0.637	0.627	0.617	0.608
1.25	0.702	0.691	0.681	0.670	0.660	0.649	0.639	0.629	0.620	0.610
1.26	0.705	0.694	0.683	0.673	0.662	0.652	0.642	0.632	0.622	0.612
1.27	0.708	0.697	0.686	0.675	0.665	0.654	0.644	0.634	0.624	0.614
1.28	0.711	0.699	0.689	0.678	0.667	0.657	0.646	0.636	0.626	0.616
1.29	0.713	0.702	0.691	0.680	0.670	0.659	0.649	0.639	0.628	0.619

1.30	0.621	0.631	0.641	0.651	0.661	0.672	0.683	0.694	0.704	0.716
1.31	0.623	0.633	0.643	0.653	0.664	0.674	0.685	0.696	0.707	0.719
1.32	0.625	0.635	0.645	0.656	0.666	0.677	0.688	0.699	0.710	0.721
1.33	0.627	0.637	0.647	0.658	0.668	0.679	0.690	0.701	0.712	0.724
1.34	0.629	0.639	0.650	0.660	0.671	0.682	0.693	0.704	0.715	0.726
1.35	0.631	0.641	0.652	0.662	0.673	0.684	0.695	0.706	0.717	0.729
1.36	0.633	0.643	0.654	0.664	0.675	0.686	0.697	0.709	0.720	0.731
1.37	0.635	0.645	0.656	0.667	0.677	0.688	0.700	0.711	0.722	0.734
1.38	0.637	0.647	0.658	0.669	0.680	0.691	0.702	0.713	0.725	0.736
1.39	0.639	0.649	0.660	0.671	0.682	0.693	0.704	0.716	0.727	0.739
1.40	0.641	0.651	0.662	0.673	0.684	0.695	0.706	0.718	0.730	0.741
1.41	0.643	0.653	0.664	0.675	0.686	0.697	0.709	0.720	0.732	0.744
1.42	0.645	0.655	0.666	0.677	0.688	0.699	0.711	0.722	0.734	0.746
1.43	0.647	0.657	0.668	0.679	0.690	0.702	0.713	0.725	0.737	0.748
1.44	0.649	0.659	0.670	0.681	0.692	0.704	0.715	0.727	0.739	0.751
1.45	0.651	0.661	0.672	0.683	0.694	0.706	0.717	0.729	0.741	0.753
1.46	0.653	0.663	0.674	0.685	0.696	0.708	0.720	0.731	0.743	0.755
1.47	0.654	0.665	0.676	0.687	0.698	0.710	0.722	0.734	0.746	0.758
1.48	0.656	0.667	0.678	0.689	0.700	0.712	0.724	0.736	0.748	0.760
1.49	0.658	0.669	0.680	0.691	0.702	0.714	0.726	0.738	0.750	0.762
1.50	0.660	0.671	0.681	0.693	0.704	0.716	0.728	0.740	0.752	0.764
1.51	0.661	0.672	0.683	0.695	0.706	0.718	0.730	0.742	0.754	0.766
1.52	0.663	0.674	0.685	0.697	0.708	0.720	0.732	0.744	0.756	0.769
1.53	0.665	0.676	0.687	0.698	0.710	0.722	0.734	0.746	0.758	0.771
1.54	0.667	0.678	0.689	0.700	0.712	0.724	0.736	0.748	0.760	0.773
1.55	0.668	0.680	0.691	0.702	0.714	0.726	0.738	0.750	0.763	0.775
1.56	0.670	0.681	0.693	0.704	0.716	0.728	0.740	0.752	0.765	0.777
1.57	0.672	0.683	0.694	0.706	0.718	0.730	0.742	0.754	0.767	0.779
1.58	0.673	0.685	0.696	0.707	0.719	0.732	0.744	0.756	0.769	0.781
1.59	0.675	0.686	0.698	0.709	0.721	0.733	0.746	0.758	0.771	0.783
1.60	0.677	0.688	0.700	0.711	0.723	0.735	0.748	0.760	0.773	0.785
1.61	0.678	0.690	0.701	0.713	0.725	0.737	0.749	0.762	0.774	0.787
1.62	0.680	0.691	0.703	0.714	0.727	0.739	0.751	0.764	0.776	0.789
1.63	0.681	0.693	0.705	0.717	0.728	0.741	0.753	0.766	0.778	0.791
1.64	0.683	0.695	0.706	0.718	0.730	0.742	0.755	0.767	0.780	0.793

TABLE 10.5 Cross-Flow Heat Exchanger Effectiveness—Both Fluids Unmixed (*Continued*)

N_{tu}	R = 0.05	R = 0.10	R = 0.15	R = 0.20	R = 0.25	R = 0.30	R = 0.35	R = 0.40	R = 0.45	R = 0.50
1.65	0.795	0.782	0.769	0.757	0.744	0.732	0.720	0.708	0.696	0.685
1.66	0.797	0.784	0.771	0.758	0.746	0.733	0.722	0.710	0.698	0.686
1.67	0.798	0.786	0.773	0.760	0.748	0.735	0.723	0.711	0.699	0.688
1.68	0.800	0.788	0.775	0.762	0.749	0.737	0.725	0.713	0.701	0.689
1.69	0.802	0.789	0.776	0.764	0.751	0.738	0.727	0.714	0.703	0.691
1.70	0.804	0.791	0.778	0.765	0.753	0.740	0.728	0.716	0.704	0.692
1.71	0.806	0.793	0.780	0.767	0.754	0.742	0.730	0.718	0.706	0.694
1.72	0.807	0.795	0.782	0.769	0.756	0.743	0.731	0.719	0.707	0.695
1.73	0.809	0.796	0.783	0.770	0.758	0.745	0.733	0.721	0.709	0.697
1.74	0.811	0.798	0.785	0.772	0.759	0.747	0.734	0.722	0.710	0.698
1.75	0.813	0.800	0.787	0.774	0.761	0.749	0.736	0.724	0.712	0.700
1.76	0.814	0.801	0.788	0.775	0.762	0.750	0.738	0.725	0.713	0.701
1.77	0.816	0.803	0.790	0.777	0.764	0.752	0.739	0.727	0.714	0.702
1.78	0.818	0.805	0.792	0.779	0.766	0.753	0.741	0.728	0.716	0.704
1.79	0.819	0.806	0.793	0.780	0.767	0.755	0.742	0.730	0.717	0.705
1.80	0.821	0.808	0.795	0.782	0.769	0.756	0.744	0.731	0.719	0.707
1.81	0.823	0.810	0.796	0.783	0.770	0.758	0.745	0.733	0.720	0.708
1.82	0.824	0.811	0.798	0.785	0.772	0.759	0.747	0.734	0.722	0.709
1.83	0.826	0.813	0.800	0.786	0.773	0.761	0.748	0.735	0.723	0.711
1.84	0.827	0.814	0.801	0.788	0.775	0.762	0.749	0.737	0.724	0.712
1.85	0.829	0.816	0.803	0.789	0.776	0.764	0.751	0.738	0.726	0.713
1.86	0.831	0.818	0.804	0.791	0.778	0.765	0.752	0.740	0.727	0.715
1.87	0.832	0.819	0.806	0.792	0.779	0.767	0.754	0.741	0.728	0.716
1.88	0.834	0.821	0.807	0.794	0.781	0.768	0.755	0.742	0.730	0.717
1.89	0.835	0.822	0.809	0.795	0.782	0.769	0.756	0.744	0.731	0.719
1.90	0.837	0.824	0.810	0.797	0.783	0.771	0.758	0.745	0.732	0.720
1.91	0.838	0.825	0.812	0.798	0.785	0.772	0.759	0.746	0.734	0.721
1.92	0.840	0.826	0.813	0.800	0.787	0.774	0.761	0.748	0.735	0.722
1.93	0.841	0.828	0.814	0.801	0.788	0.775	0.762	0.749	0.736	0.724
1.94	0.843	0.829	0.816	0.802	0.789	0.776	0.763	0.750	0.738	0.725
1.95	0.844	0.831	0.817	0.804	0.791	0.778	0.765	0.752	0.739	0.726

1.96	0.727	0.740	0.753	0.766	0.779	0.792	0.805	0.819	0.832	0.846
1.97	0.729	0.741	0.754	0.767	0.780	0.793	0.806	0.820	0.833	0.847
1.98	0.730	0.743	0.755	0.768	0.782	0.795	0.808	0.821	0.835	0.848
1.99	0.731	0.744	0.757	0.770	0.783	0.796	0.809	0.823	0.836	0.850
2.00	0.732	0.745	0.758	0.771	0.784	0.797	0.810	0.824	0.838	0.851
2.01	0.733	0.746	0.759	0.772	0.785	0.799	0.812	0.825	0.839	0.852
2.02	0.735	0.747	0.760	0.773	0.787	0.800	0.813	0.827	0.840	0.854
2.03	0.736	0.749	0.762	0.775	0.788	0.801	0.814	0.828	0.842	0.855
2.04	0.737	0.750	0.763	0.776	0.789	0.803	0.816	0.829	0.843	0.856
2.05	0.738	0.751	0.764	0.777	0.790	0.804	0.817	0.831	0.844	0.858
2.06	0.739	0.752	0.765	0.778	0.792	0.805	0.818	0.832	0.845	0.859
2.07	0.740	0.753	0.766	0.780	0.793	0.806	0.819	0.833	0.847	0.860
2.08	0.741	0.754	0.768	0.781	0.794	0.808	0.821	0.834	0.848	0.862
2.09	0.743	0.756	0.769	0.782	0.795	0.809	0.822	0.836	0.849	0.863
2.10	0.744	0.757	0.770	0.783	0.797	0.810	0.823	0.837	0.850	0.864
2.11	0.745	0.758	0.771	0.784	0.798	0.811	0.824	0.838	0.852	0.865
2.12	0.746	0.759	0.772	0.785	0.799	0.812	0.825	0.839	0.853	0.866
2.13	0.747	0.760	0.773	0.787	0.800	0.814	0.827	0.840	0.854	0.868
2.14	0.748	0.761	0.774	0.788	0.801	0.815	0.828	0.842	0.855	0.869
2.15	0.749	0.762	0.776	0.789	0.802	0.816	0.829	0.843	0.856	0.870
2.16	0.750	0.763	0.777	0.790	0.803	0.817	0.831	0.844	0.858	0.871
2.17	0.751	0.764	0.778	0.791	0.805	0.818	0.832	0.845	0.859	0.872
2.18	0.752	0.765	0.779	0.792	0.806	0.819	0.833	0.846	0.860	0.873
2.19	0.753	0.767	0.780	0.793	0.807	0.820	0.834	0.847	0.861	0.875
2.20	0.754	0.768	0.781	0.794	0.808	0.822	0.835	0.848	0.862	0.876
2.21	0.755	0.769	0.782	0.796	0.809	0.823	0.836	0.850	0.863	0.877
2.22	0.756	0.770	0.783	0.797	0.810	0.824	0.837	0.851	0.864	0.878
2.23	0.757	0.771	0.784	0.798	0.811	0.825	0.838	0.852	0.865	0.879
2.24	0.759	0.772	0.785	0.799	0.812	0.826	0.839	0.853	0.867	0.880
2.25	0.760	0.773	0.786	0.800	0.813	0.827	0.841	0.854	0.868	0.881
2.26	0.761	0.774	0.787	0.801	0.814	0.828	0.842	0.855	0.869	0.882
2.27	0.762	0.775	0.788	0.802	0.815	0.829	0.843	0.856	0.870	0.883
2.28	0.763	0.776	0.789	0.803	0.816	0.830	0.844	0.857	0.871	0.884
2.29	0.764	0.777	0.790	0.804	0.818	0.831	0.845	0.858	0.872	0.885
2.30	0.765	0.778	0.791	0.805	0.819	0.832	0.846	0.859	0.873	0.886

TABLE 10.5 Cross-Flow Heat Exchanger Effectiveness—Both Fluids Unmixed (*Continued*)

N_{tu}	$R = 0.05$	$R = 0.10$	$R = 0.15$	$R = 0.20$	$R = 0.25$	$R = 0.30$	$R = 0.35$	$R = 0.40$	$R = 0.45$	$R = 0.50$
2.31	0.887	0.874	0.860	0.847	0.833	0.820	0.806	0.792	0.779	0.766
2.32	0.888	0.875	0.861	0.848	0.834	0.821	0.807	0.793	0.780	0.767
2.33	0.889	0.876	0.862	0.849	0.835	0.822	0.808	0.794	0.781	0.768
2.34	0.890	0.877	0.863	0.850	0.836	0.823	0.809	0.795	0.782	0.768
2.35	0.891	0.878	0.864	0.851	0.837	0.824	0.810	0.796	0.782	0.769
2.36	0.892	0.879	0.865	0.852	0.838	0.825	0.811	0.797	0.784	0.770
2.37	0.893	0.880	0.866	0.853	0.839	0.825	0.812	0.798	0.785	0.771
2.38	0.894	0.881	0.867	0.854	0.840	0.826	0.813	0.799	0.786	0.772
2.39	0.895	0.882	0.868	0.855	0.841	0.827	0.814	0.800	0.787	0.773
2.40	0.896	0.883	0.869	0.856	0.842	0.828	0.815	0.800	0.788	0.774
2.41	0.897	0.884	0.870	0.857	0.843	0.829	0.816	0.802	0.788	0.775
2.42	0.898	0.885	0.871	0.858	0.844	0.830	0.816	0.803	0.789	0.776
2.43	0.899	0.885	0.872	0.859	0.845	0.831	0.817	0.804	0.790	0.777
2.44	0.900	0.886	0.873	0.859	0.846	0.832	0.818	0.805	0.791	0.778
2.45	0.901	0.887	0.874	0.860	0.847	0.833	0.819	0.805	0.792	0.778
2.46	0.902	0.888	0.875	0.861	0.848	0.834	0.820	0.806	0.792	0.779
2.47	0.902	0.889	0.875	0.862	0.849	0.835	0.821	0.807	0.793	0.780
2.48	0.903	0.890	0.876	0.863	0.849	0.836	0.822	0.808	0.794	0.781
2.49	0.904	0.891	0.877	0.864	0.850	0.837	0.823	0.809	0.795	0.782
2.50	0.905	0.892	0.878	0.865	0.851	0.837	0.824	0.810	0.796	0.783

N_{tu}	$R = 0.55$	$R = 0.60$	$R = 0.65$	$R = 0.70$	$R = 0.75$	$R = 0.80$	$R = 0.85$	$R = 0.90$	$R = 0.95$	$R = 1.00$
0.01	0.010	0.010	0.010	0.010	0.010	0.010	0.010	0.010	0.010	0.010
0.02	0.020	0.020	0.020	0.020	0.020	0.020	0.020	0.020	0.020	0.020
0.03	0.029	0.029	0.029	0.029	0.029	0.029	0.029	0.029	0.029	0.029
0.04	0.039	0.039	0.039	0.039	0.039	0.039	0.039	0.039	0.038	0.038
0.05	0.048	0.048	0.048	0.048	0.048	0.048	0.048	0.048	0.048	0.048
0.06	0.057	0.057	0.057	0.057	0.057	0.057	0.057	0.057	0.057	0.057
0.07	0.066	0.066	0.066	0.066	0.066	0.066	0.066	0.066	0.065	0.065
0.08	0.075	0.075	0.075	0.075	0.075	0.074	0.074	0.074	0.074	0.074

0.09	0.084	0.084	0.084	0.083	0.083	0.083	0.083	0.083	0.082	0.082
0.10	0.093	0.092	0.092	0.092	0.092	0.091	0.091	0.091	0.091	0.091
0.11	0.101	0.101	0.101	0.100	0.100	0.100	0.099	0.099	0.099	0.099
0.12	0.109	0.109	0.109	0.108	0.108	0.108	0.108	0.107	0.107	0.107
0.13	0.118	0.117	0.117	0.117	0.116	0.116	0.115	0.115	0.115	0.115
0.14	0.126	0.125	0.125	0.125	0.124	0.124	0.124	0.123	0.123	0.122
0.15	0.134	0.134	0.133	0.133	0.132	0.132	0.131	0.131	0.130	0.130
0.16	0.142	0.141	0.141	0.140	0.140	0.139	0.139	0.138	0.138	0.137
0.17	0.150	0.149	0.149	0.148	0.148	0.147	0.146	0.146	0.145	0.145
0.18	0.158	0.157	0.156	0.156	0.155	0.154	0.154	0.153	0.153	0.152
0.19	0.165	0.164	0.164	0.163	0.162	0.162	0.161	0.160	0.160	0.159
0.20	0.173	0.172	0.171	0.170	0.170	0.169	0.168	0.167	0.167	0.166
0.21	0.180	0.179	0.178	0.177	0.177	0.176	0.175	0.174	0.173	0.173
0.22	0.187	0.186	0.185	0.184	0.184	0.183	0.182	0.181	0.180	0.179
0.23	0.194	0.193	0.192	0.191	0.190	0.190	0.189	0.188	0.187	0.186
0.24	0.201	0.200	0.199	0.198	0.197	0.196	0.195	0.194	0.193	0.192
0.25	0.208	0.207	0.206	0.205	0.204	0.203	0.202	0.201	0.200	0.199
0.26	0.215	0.214	0.213	0.212	0.210	0.209	0.208	0.207	0.206	0.205
0.27	0.222	0.221	0.219	0.218	0.217	0.216	0.214	0.213	0.212	0.211
0.28	0.229	0.227	0.226	0.225	0.223	0.222	0.221	0.219	0.218	0.217
0.29	0.235	0.234	0.232	0.231	0.230	0.228	0.227	0.225	0.224	0.223
0.30	0.242	0.240	0.239	0.237	0.236	0.234	0.233	0.231	0.230	0.228
0.31	0.248	0.246	0.245	0.243	0.242	0.240	0.239	0.237	0.236	0.234
0.32	0.254	0.253	0.251	0.249	0.248	0.246	0.244	0.243	0.241	0.240
0.33	0.261	0.259	0.257	0.255	0.254	0.252	0.250	0.249	0.247	0.245
0.34	0.267	0.265	0.263	0.261	0.259	0.258	0.256	0.254	0.252	0.251
0.35	0.273	0.271	0.269	0.267	0.265	0.263	0.261	0.260	0.258	0.256
0.36	0.279	0.277	0.275	0.273	0.271	0.269	0.267	0.265	0.263	0.261
0.37	0.285	0.282	0.280	0.278	0.276	0.274	0.272	0.270	0.268	0.266
0.38	0.290	0.288	0.286	0.284	0.282	0.280	0.278	0.275	0.273	0.271
0.39	0.296	0.294	0.292	0.289	0.287	0.285	0.283	0.281	0.278	0.276
0.40	0.302	0.299	0.297	0.295	0.292	0.290	0.288	0.286	0.283	0.281
0.41	0.307	0.305	0.302	0.300	0.298	0.295	0.293	0.291	0.288	0.286
0.42	0.313	0.310	0.308	0.305	0.303	0.300	0.298	0.295	0.293	0.291
0.43	0.318	0.316	0.313	0.310	0.308	0.305	0.303	0.300	0.298	0.295

TABLE 10.5 Cross-Flow Heat Exchanger Effectiveness—Both Fluids Unmixed (*Continued*)

N_{tu}	$R = 0.55$	$R = 0.60$	$R = 0.65$	$R = 0.70$	$R = 0.75$	$R = 0.80$	$R = 0.85$	$R = 0.90$	$R = 0.95$	$R = 1.00$
0.44	0.324	0.321	0.318	0.315	0.313	0.310	0.308	0.305	0.302	0.300
0.45	0.329	0.326	0.323	0.321	0.318	0.315	0.312	0.310	0.307	0.304
0.46	0.334	0.331	0.328	0.325	0.323	0.320	0.317	0.314	0.312	0.309
0.47	0.339	0.336	0.333	0.330	0.327	0.324	0.322	0.319	0.316	0.313
0.48	0.344	0.341	0.338	0.335	0.332	0.329	0.326	0.323	0.320	0.318
0.49	0.349	0.346	0.343	0.340	0.337	0.334	0.331	0.328	0.325	0.322
0.50	0.354	0.351	0.348	0.345	0.341	0.338	0.335	0.332	0.329	0.326
0.51	0.359	0.356	0.352	0.349	0.346	0.343	0.339	0.336	0.334	0.330
0.52	0.364	0.361	0.357	0.354	0.350	0.347	0.344	0.341	0.338	0.335
0.53	0.369	0.365	0.362	0.358	0.355	0.351	0.348	0.345	0.342	0.339
0.54	0.374	0.370	0.366	0.363	0.359	0.356	0.352	0.349	0.346	0.342
0.55	0.378	0.374	0.371	0.367	0.363	0.360	0.357	0.353	0.350	0.346
0.56	0.383	0.379	0.375	0.371	0.368	0.364	0.361	0.357	0.354	0.350
0.57	0.387	0.383	0.379	0.375	0.372	0.368	0.365	0.361	0.357	0.354
0.58	0.392	0.388	0.384	0.380	0.376	0.372	0.369	0.365	0.361	0.358
0.59	0.396	0.392	0.388	0.384	0.380	0.376	0.373	0.369	0.365	0.361
0.60	0.400	0.396	0.392	0.388	0.384	0.380	0.376	0.373	0.369	0.365
0.61	0.405	0.401	0.397	0.392	0.388	0.384	0.380	0.376	0.372	0.369
0.62	0.409	0.405	0.401	0.396	0.392	0.388	0.384	0.380	0.376	0.372
0.63	0.414	0.409	0.405	0.400	0.396	0.392	0.388	0.384	0.379	0.375
0.64	0.418	0.413	0.409	0.404	0.400	0.396	0.391	0.387	0.383	0.379
0.65	0.422	0.417	0.413	0.408	0.404	0.399	0.395	0.391	0.386	0.382
0.66	0.426	0.421	0.417	0.412	0.407	0.403	0.398	0.394	0.390	0.386
0.67	0.430	0.425	0.420	0.416	0.411	0.406	0.402	0.398	0.393	0.389
0.68	0.434	0.429	0.424	0.419	0.415	0.410	0.405	0.401	0.396	0.392
0.69	0.438	0.433	0.428	0.423	0.418	0.413	0.409	0.404	0.400	0.395
0.70	0.442	0.437	0.432	0.427	0.422	0.417	0.412	0.408	0.403	0.398
0.71	0.446	0.440	0.435	0.430	0.425	0.420	0.416	0.411	0.406	0.401
0.72	0.449	0.444	0.439	0.434	0.429	0.424	0.419	0.414	0.409	0.405
0.73	0.453	0.448	0.442	0.437	0.432	0.427	0.422	0.417	0.412	0.408
0.74	0.457	0.451	0.446	0.441	0.435	0.430	0.425	0.420	0.415	0.411

0.75	0.460	0.455	0.449	0.444	0.439	0.434	0.428	0.423	0.418	0.413
0.76	0.464	0.458	0.453	0.447	0.442	0.437	0.432	0.426	0.421	0.416
0.77	0.468	0.462	0.456	0.451	0.445	0.440	0.435	0.429	0.424	0.419
0.78	0.471	0.465	0.460	0.454	0.448	0.443	0.438	0.432	0.427	0.422
0.79	0.475	0.469	0.463	0.457	0.452	0.446	0.441	0.435	0.430	0.425
0.80	0.478	0.472	0.466	0.460	0.455	0.449	0.444	0.438	0.433	0.428
0.81	0.482	0.475	0.470	0.464	0.458	0.452	0.447	0.441	0.436	0.430
0.82	0.485	0.479	0.473	0.467	0.461	0.455	0.449	0.444	0.438	0.433
0.83	0.488	0.482	0.476	0.470	0.464	0.458	0.452	0.447	0.441	0.436
0.84	0.492	0.485	0.479	0.473	0.467	0.461	0.455	0.449	0.444	0.438
0.85	0.495	0.488	0.482	0.476	0.470	0.464	0.458	0.452	0.446	0.441
0.86	0.498	0.492	0.485	0.479	0.473	0.467	0.461	0.455	0.449	0.443
0.87	0.501	0.495	0.488	0.482	0.476	0.469	0.463	0.457	0.452	0.446
0.88	0.504	0.498	0.491	0.485	0.478	0.472	0.466	0.460	0.454	0.448
0.89	0.508	0.501	0.494	0.488	0.481	0.475	0.469	0.463	0.457	0.451
0.90	0.511	0.504	0.497	0.491	0.484	0.478	0.471	0.465	0.459	0.453
0.91	0.514	0.507	0.500	0.493	0.487	0.480	0.474	0.468	0.462	0.456
0.92	0.517	0.510	0.503	0.496	0.489	0.483	0.477	0.470	0.464	0.458
0.93	0.520	0.513	0.506	0.499	0.492	0.486	0.479	0.473	0.466	0.460
0.94	0.523	0.515	0.508	0.502	0.495	0.488	0.482	0.475	0.469	0.463
0.95	0.526	0.518	0.511	0.504	0.497	0.491	0.484	0.478	0.471	0.465
0.96	0.528	0.521	0.514	0.507	0.500	0.493	0.487	0.480	0.473	0.467
0.97	0.531	0.524	0.517	0.510	0.503	0.496	0.489	0.482	0.476	0.469
0.98	0.534	0.527	0.519	0.512	0.505	0.498	0.491	0.485	0.478	0.472
0.99	0.537	0.529	0.522	0.515	0.508	0.501	0.494	0.487	0.480	0.474
1.00	0.540	0.532	0.525	0.517	0.510	0.503	0.496	0.489	0.482	0.476
1.01	0.542	0.535	0.527	0.520	0.513	0.505	0.498	0.491	0.485	0.478
1.02	0.545	0.537	0.530	0.522	0.515	0.508	0.501	0.494	0.487	0.480
1.03	0.548	0.540	0.532	0.525	0.517	0.510	0.503	0.496	0.489	0.483
1.04	0.551	0.543	0.535	0.527	0.520	0.512	0.505	0.498	0.491	0.485
1.05	0.553	0.545	0.537	0.530	0.522	0.515	0.507	0.500	0.494	0.487
1.06	0.556	0.548	0.540	0.532	0.524	0.517	0.510	0.502	0.496	0.489
1.07	0.558	0.550	0.542	0.534	0.527	0.519	0.512	0.505	0.498	0.491
1.08	0.561	0.553	0.545	0.537	0.529	0.521	0.514	0.507	0.500	0.493
1.09	0.563	0.555	0.547	0.539	0.531	0.524	0.516	0.509	0.502	0.495

TABLE 10.5 Cross-Flow Heat Exchanger Effectiveness—Both Fluids Unmixed (*Continued*)

N_{tu}	R = 0.55	R = 0.60	R = 0.65	R = 0.70	R = 0.75	R = 0.80	R = 0.85	R = 0.90	R = 0.95	R = 1.00
1.10	0.566	0.558	0.549	0.541	0.533	0.526	0.519	0.511	0.504	0.497
1.11	0.568	0.560	0.552	0.544	0.536	0.528	0.521	0.513	0.506	0.498
1.12	0.571	0.562	0.554	0.546	0.538	0.530	0.523	0.515	0.508	0.500
1.13	0.573	0.565	0.556	0.548	0.540	0.532	0.525	0.517	0.510	0.502
1.14	0.576	0.567	0.559	0.550	0.542	0.535	0.527	0.519	0.512	0.504
1.15	0.578	0.569	0.561	0.552	0.545	0.537	0.529	0.521	0.513	0.506
1.16	0.580	0.572	0.563	0.555	0.547	0.539	0.531	0.523	0.515	0.508
1.17	0.583	0.574	0.565	0.557	0.549	0.541	0.533	0.525	0.517	0.510
1.18	0.585	0.576	0.567	0.559	0.551	0.543	0.535	0.527	0.519	0.511
1.19	0.587	0.578	0.570	0.561	0.553	0.545	0.537	0.529	0.521	0.513
1.20	0.590	0.581	0.572	0.563	0.555	0.547	0.538	0.530	0.523	0.515
1.21	0.592	0.583	0.574	0.566	0.557	0.549	0.540	0.532	0.524	0.517
1.22	0.594	0.585	0.576	0.568	0.559	0.551	0.542	0.534	0.526	0.518
1.23	0.596	0.587	0.578	0.570	0.561	0.552	0.544	0.536	0.528	0.520
1.24	0.598	0.589	0.580	0.572	0.563	0.554	0.546	0.538	0.530	0.522
1.25	0.601	0.591	0.582	0.574	0.565	0.556	0.548	0.539	0.531	0.523
1.26	0.603	0.593	0.585	0.576	0.567	0.558	0.550	0.541	0.533	0.525
1.27	0.605	0.596	0.587	0.577	0.569	0.560	0.551	0.543	0.535	0.527
1.28	0.607	0.598	0.589	0.579	0.570	0.562	0.553	0.545	0.536	0.528
1.29	0.609	0.600	0.590	0.581	0.572	0.564	0.555	0.546	0.538	0.530
1.30	0.611	0.602	0.592	0.583	0.574	0.565	0.557	0.548	0.540	0.531
1.31	0.613	0.604	0.594	0.585	0.576	0.567	0.558	0.550	0.541	0.533
1.32	0.615	0.606	0.596	0.587	0.578	0.569	0.560	0.551	0.543	0.534
1.33	0.617	0.608	0.598	0.589	0.580	0.571	0.562	0.553	0.544	0.536
1.34	0.619	0.610	0.600	0.591	0.581	0.572	0.563	0.555	0.546	0.537
1.35	0.621	0.612	0.602	0.592	0.583	0.574	0.565	0.556	0.548	0.539
1.36	0.623	0.613	0.604	0.594	0.585	0.576	0.567	0.558	0.549	0.540
1.37	0.625	0.615	0.606	0.596	0.587	0.577	0.568	0.559	0.551	0.542
1.38	0.627	0.617	0.607	0.598	0.588	0.579	0.570	0.561	0.552	0.543
1.39	0.629	0.619	0.609	0.600	0.590	0.581	0.571	0.562	0.554	0.545
1.40	0.631	0.621	0.611	0.601	0.592	0.582	0.573	0.564	0.555	0.546

1.41	0.548	0.557	0.565	0.575	0.584	0.593	0.603	0.613	0.623	0.633
1.42	0.549	0.558	0.567	0.576	0.585	0.595	0.605	0.614	0.625	0.635
1.43	0.551	0.559	0.568	0.578	0.587	0.597	0.606	0.616	0.626	0.637
1.44	0.552	0.561	0.570	0.579	0.589	0.598	0.608	0.618	0.628	0.638
1.45	0.553	0.562	0.571	0.581	0.590	0.600	0.610	0.620	0.630	0.640
1.46	0.555	0.564	0.573	0.582	0.592	0.601	0.611	0.621	0.632	0.642
1.47	0.556	0.565	0.574	0.584	0.593	0.603	0.613	0.623	0.633	0.644
1.48	0.557	0.566	0.576	0.585	0.595	0.605	0.615	0.625	0.635	0.645
1.49	0.559	0.568	0.577	0.587	0.596	0.606	0.616	0.626	0.637	0.647
1.50	0.560	0.569	0.579	0.588	0.598	0.608	0.618	0.628	0.638	0.649
1.51	0.561	0.570	0.580	0.589	0.599	0.609	0.619	0.630	0.640	0.651
1.52	0.563	0.572	0.581	0.591	0.601	0.611	0.621	0.631	0.642	0.652
1.53	0.564	0.573	0.583	0.592	0.602	0.612	0.622	0.633	0.643	0.654
1.54	0.565	0.574	0.584	0.594	0.604	0.614	0.624	0.634	0.645	0.656
1.55	0.566	0.576	0.585	0.595	0.605	0.615	0.625	0.636	0.647	0.657
1.56	0.568	0.577	0.587	0.596	0.606	0.617	0.627	0.637	0.648	0.659
1.57	0.569	0.578	0.588	0.598	0.608	0.618	0.628	0.639	0.650	0.661
1.58	0.570	0.580	0.589	0.599	0.609	0.619	0.630	0.641	0.651	0.662
1.59	0.571	0.581	0.591	0.600	0.611	0.621	0.631	0.642	0.653	0.664
1.60	0.572	0.582	0.592	0.602	0.612	0.622	0.633	0.644	0.654	0.665
1.61	0.574	0.583	0.593	0.603	0.613	0.624	0.634	0.645	0.656	0.667
1.62	0.575	0.584	0.594	0.604	0.615	0.625	0.636	0.646	0.657	0.669
1.63	0.576	0.586	0.596	0.606	0.616	0.626	0.637	0.648	0.659	0.670
1.64	0.577	0.587	0.597	0.607	0.617	0.628	0.639	0.649	0.660	0.672
1.65	0.579	0.588	0.598	0.608	0.619	0.629	0.640	0.651	0.662	0.673
1.66	0.580	0.589	0.599	0.610	0.620	0.631	0.641	0.652	0.663	0.675
1.67	0.581	0.591	0.600	0.611	0.621	0.632	0.643	0.654	0.665	0.676
1.68	0.582	0.592	0.602	0.612	0.623	0.633	0.644	0.655	0.666	0.678
1.69	0.583	0.593	0.603	0.613	0.624	0.634	0.645	0.656	0.668	0.679
1.70	0.584	0.594	0.604	0.614	0.625	0.636	0.647	0.658	0.669	0.681
1.71	0.585	0.595	0.606	0.616	0.626	0.637	0.648	0.659	0.671	0.682
1.72	0.587	0.597	0.607	0.617	0.628	0.638	0.649	0.661	0.672	0.683
1.73	0.588	0.598	0.608	0.618	0.629	0.640	0.651	0.662	0.673	0.685
1.74	0.589	0.599	0.609	0.619	0.630	0.641	0.652	0.663	0.675	0.686
1.75	0.590	0.600	0.610	0.620	0.631	0.642	0.653	0.665	0.676	0.688

TABLE 10.5 Cross-Flow Heat Exchanger Effectiveness—Both Fluids Unmixed (*Continued*)

N_{tu}	$R = 0.55$	$R = 0.60$	$R = 0.65$	$R = 0.70$	$R = 0.75$	$R = 0.80$	$R = 0.85$	$R = 0.90$	$R = 0.95$	$R = 1.00$
1.76	0.689	0.677	0.666	0.655	0.643	0.632	0.622	0.611	0.601	0.591
1.77	0.691	0.679	0.667	0.656	0.645	0.634	0.623	0.613	0.602	0.592
1.78	0.692	0.680	0.669	0.657	0.646	0.635	0.624	0.614	0.603	0.593
1.79	0.693	0.681	0.670	0.658	0.647	0.636	0.625	0.615	0.604	0.594
1.80	0.695	0.683	0.671	0.660	0.648	0.637	0.627	0.616	0.605	0.595
1.81	0.696	0.684	0.672	0.661	0.649	0.639	0.628	0.617	0.606	0.596
1.82	0.697	0.685	0.674	0.662	0.651	0.640	0.629	0.618	0.607	0.597
1.83	0.699	0.687	0.675	0.663	0.652	0.641	0.630	0.619	0.608	0.598
1.84	0.700	0.688	0.676	0.664	0.653	0.642	0.631	0.620	0.610	0.599
1.85	0.701	0.689	0.677	0.666	0.654	0.643	0.632	0.621	0.611	0.600
1.86	0.702	0.690	0.679	0.667	0.656	0.644	0.633	0.622	0.612	0.601
1.87	0.704	0.692	0.680	0.668	0.657	0.645	0.634	0.623	0.613	0.602
1.88	0.705	0.693	0.681	0.669	0.658	0.647	0.635	0.624	0.614	0.603
1.89	0.706	0.694	0.682	0.670	0.659	0.648	0.636	0.625	0.615	0.604
1.90	0.708	0.695	0.683	0.671	0.660	0.649	0.638	0.626	0.616	0.605
1.91	0.709	0.697	0.684	0.673	0.661	0.650	0.639	0.627	0.617	0.606
1.92	0.710	0.698	0.686	0.674	0.662	0.651	0.640	0.628	0.618	0.607
1.93	0.711	0.699	0.687	0.675	0.664	0.652	0.641	0.629	0.619	0.608
1.94	0.712	0.700	0.688	0.676	0.665	0.653	0.642	0.630	0.619	0.609
1.95	0.714	0.701	0.689	0.677	0.666	0.654	0.643	0.631	0.620	0.610
1.96	0.715	0.702	0.690	0.679	0.667	0.655	0.644	0.632	0.621	0.610
1.97	0.716	0.704	0.691	0.680	0.668	0.656	0.645	0.633	0.622	0.611
1.98	0.717	0.705	0.693	0.681	0.669	0.657	0.646	0.634	0.623	0.612
1.99	0.718	0.706	0.694	0.682	0.670	0.658	0.647	0.635	0.624	0.613
2.00	0.720	0.707	0.695	0.683	0.671	0.659	0.648	0.636	0.625	0.614
2.01	0.721	0.708	0.696	0.684	0.672	0.660	0.649	0.637	0.626	0.615
2.02	0.722	0.709	0.697	0.685	0.673	0.661	0.650	0.638	0.627	0.616
2.03	0.723	0.710	0.698	0.686	0.674	0.662	0.651	0.639	0.628	0.617
2.04	0.724	0.711	0.699	0.687	0.675	0.663	0.652	0.640	0.629	0.617
2.05	0.725	0.713	0.701	0.688	0.676	0.664	0.653	0.641	0.630	0.618
2.06	0.726	0.714	0.702	0.689	0.677	0.665	0.653	0.642	0.631	0.619

2.07	0.620	0.631	0.643	0.654	0.666	0.678	0.690	0.703	0.715	0.727
2.08	0.621	0.632	0.644	0.655	0.667	0.679	0.691	0.704	0.716	0.729
2.09	0.622	0.633	0.645	0.656	0.668	0.680	0.692	0.705	0.717	0.730
2.10	0.623	0.634	0.645	0.657	0.669	0.681	0.693	0.706	0.718	0.731
2.11	0.623	0.635	0.646	0.658	0.670	0.682	0.694	0.707	0.719	0.732
2.12	0.624	0.636	0.647	0.659	0.671	0.683	0.695	0.708	0.720	0.733
2.13	0.625	0.637	0.648	0.660	0.672	0.684	0.696	0.709	0.721	0.734
2.14	0.626	0.637	0.649	0.661	0.673	0.685	0.697	0.710	0.723	0.735
2.15	0.627	0.638	0.650	0.662	0.674	0.686	0.698	0.711	0.724	0.736
2.16	0.628	0.639	0.651	0.663	0.675	0.687	0.699	0.712	0.725	0.737
2.17	0.628	0.640	0.652	0.663	0.676	0.688	0.700	0.713	0.726	0.738
2.18	0.629	0.641	0.652	0.664	0.676	0.689	0.701	0.714	0.727	0.739
2.19	0.630	0.641	0.653	0.665	0.677	0.690	0.702	0.715	0.728	0.741
2.20	0.631	0.642	0.654	0.666	0.678	0.691	0.703	0.716	0.729	0.742
2.21	0.631	0.643	0.655	0.667	0.679	0.691	0.704	0.717	0.730	0.743
2.22	0.632	0.644	0.656	0.668	0.680	0.692	0.705	0.718	0.731	0.744
2.23	0.633	0.645	0.657	0.669	0.681	0.693	0.706	0.719	0.732	0.745
2.24	0.634	0.645	0.657	0.669	0.682	0.694	0.707	0.720	0.732	0.746
2.25	0.634	0.646	0.658	0.670	0.683	0.695	0.708	0.721	0.733	0.747
2.26	0.635	0.647	0.659	0.671	0.683	0.696	0.709	0.721	0.734	0.747
2.27	0.636	0.648	0.660	0.672	0.684	0.697	0.710	0.722	0.735	0.748
2.28	0.637	0.649	0.661	0.673	0.685	0.698	0.710	0.723	0.736	0.749
2.29	0.637	0.649	0.661	0.674	0.686	0.699	0.711	0.724	0.737	0.750
2.30	0.638	0.650	0.662	0.674	0.687	0.699	0.712	0.725	0.738	0.751
2.31	0.639	0.651	0.663	0.675	0.688	0.700	0.713	0.726	0.739	0.752
2.32	0.640	0.652	0.664	0.676	0.688	0.701	0.714	0.727	0.740	0.753
2.33	0.641	0.652	0.664	0.677	0.689	0.702	0.715	0.728	0.741	0.754
2.34	0.641	0.653	0.665	0.678	0.690	0.703	0.716	0.729	0.742	0.755
2.35	0.642	0.654	0.666	0.678	0.691	0.704	0.717	0.730	0.743	0.756
2.36	0.643	0.655	0.667	0.679	0.692	0.704	0.717	0.730	0.744	0.757
2.37	0.644	0.656	0.667	0.680	0.693	0.705	0.718	0.731	0.745	0.758
2.38	0.644	0.656	0.668	0.681	0.693	0.706	0.719	0.732	0.745	0.759
2.39	0.645	0.657	0.669	0.681	0.694	0.707	0.720	0.733	0.746	0.760
2.40	0.646	0.658	0.670	0.682	0.695	0.708	0.721	0.734	0.747	0.761
2.41	0.646	0.658	0.670	0.683	0.696	0.709	0.722	0.735	0.748	0.761

TABLE 10.5 Cross-Flow Heat Exchanger Effectiveness—Both Fluids Unmixed (*Continued*)

N_{tu}	R = 0.55	R = 0.60	R = 0.65	R = 0.70	R = 0.75	R = 0.80	R = 0.85	R = 0.90	R = 0.95	R = 1.00
2.42	0.762	0.749	0.736	0.722	0.709	0.696	0.684	0.672	0.659	0.647
2.43	0.763	0.750	0.736	0.723	0.710	0.697	0.684	0.672	0.660	0.648
2.44	0.764	0.751	0.737	0.724	0.711	0.698	0.685	0.673	0.661	0.648
2.45	0.765	0.751	0.738	0.725	0.712	0.699	0.686	0.674	0.661	0.649
2.46	0.766	0.752	0.739	0.726	0.712	0.699	0.687	0.674	0.662	0.650
2.47	0.767	0.753	0.740	0.726	0.713	0.700	0.687	0.675	0.663	0.650
2.48	0.767	0.754	0.741	0.727	0.714	0.701	0.688	0.676	0.663	0.651
2.49	0.768	0.755	0.741	0.728	0.715	0.702	0.689	0.677	0.664	0.652
2.50	0.769	0.756	0.742	0.729	0.716	0.702	0.690	0.677	0.665	0.652

for Coolanol-45, Fig. 22.3 shows a specific gravity of approximately 0.87 at both 170°F and 175.1°F. Thus

$$v = \frac{1}{0.87(62.4)} = 0.0184 \text{ ft}^3/\text{lb}$$

22 $G' = G/3600 \text{ lb/ft}^2 \text{ s}$.
23 Figure 10.8.
24 All ψ terms here are steps in the computation of ΔP using Eq. (10.14).
25. 1 psi $= 27.7$ in H_2O.
26 The pressure drop is excessive on the air side and well within the specified level on the Coolanol-45 side. A rerun using $L = 18$ in, $W = 18$ in, and $D = \frac{3}{4}$ in is to be made. The worksheet is adjusted accordingly using column II in each case. Only changes need be shown.

The second run shows that the design is quite satisfactory. The required effectiveness is $\epsilon = 0.546$ and the effectiveness provided is $\epsilon = 0.553$. The air-side pressure drop is 0.99 in $H_2O < 1.10$ in H_2O allowed, and the Coolanol-side pressure drop is 0.61 psi < 0.75 psi allowed.

10.9 NOMENCLATURE

Roman Letter Symbols

a	separation plate thickness, ft
A	area, ft^2; with subscripts h and c, free flow area on hot and cold side, respectively; with subscript f, frontal area
b	fin height, ft
c	specific heat in general, Btu/lb °F; or specific heat of cold fluid, Btu/lb °F
C	capacity rate, Btu/h °F; or specific heat of hot fluid, Btu/lb °F
D	exchanger depth, ft; with subscript e, equivalent diameter, ft
f	friction factor, dimensionless
g_c'	local acceleration of gravity, 32.17 ft/s^2
G	mass velocity, lb/ft^2 h; G' is mass velocity in lb/ft^2 s
h	heat transfer coefficient, Btu/ft^2 h °F
H	exchanger height, ft
j	with subscript h, a heat transfer factor, dimensionless
k	thermal conductivity, Btu/ft h °F
K	loss coefficient, K_c and K_e are entrance (contraction) and exit loss coefficients, respectively, dimensionless
L	length, ft
LMTD	logarithmic mean temperature difference, °F
m	fin performance factor, ft
n	number of layers
N_{tu}	number of transfer units defined by Eq. (10.19), dimensionless
Nu	Nusselt number, dimensionless
p	porosity of passage, dimensionless

P	perimeter of passage, ft; or pressure, lb/ft^2
Pr	Prandtl number, dimensionless
q	heat flow, Btu/h
r	with subscript h, hydraulic radius, ft
R	capacity rate ratio, dimensionless
Re	Reynolds number, dimensionless
S	surface, ft^2; with subscript f, finned surface, ft^2
St	Stanton number, dimensionless
t	temperature, °F
T	temperature, °F
U	overall heat transfer coefficient, Btu/ft^2 h °F
v	specific volume, ft^3/lb
V	volume, ft^3
w	flow rate, lb/min
W	exchanger width, ft or flow rate, lb/h

Greek Letter Symbols

α	ratio of total surface area on *one* side of the exchanger to total volume on *both* sides of the exchanger, dimensionless
β	ratio of total surface on *one* side of the exchanger to total volume on *one* side of the exchanger, dimensionless
δ	fin width, ft
Δ	indicates change in variable
ϵ	exchanger effectiveness, dimensionless
η	efficiency, dimensionless; without subscript, fin efficiency; with subscript w, weighted fin efficiency
μ	viscosity, lb/h ft
ρ	density, lb/ft^3
σ	ratio of free flow area to frontal area, dimensionless
ψ	indicates a grouping of terms in pressure loss calculations, dimensions vary

Subscripts

b	indicates bulk condition
c	indicates cold fluid or contraction coefficient
e	indicates equivalent or exit condition
f	indicates frontal area or finned surface
h	indicates hydraulic radius or hot fluid
m	indicates mean condition or metal interface
max	indicates maximum condition
min	indicates minimum condition
p	indicates constant-pressure condition
w	indicates weighted condition
0	indicates fin base condition

10.10 REFERENCES

1 Kays, W. M., and London, A. L., *Compact Heat Exchangers*, 2d ed., McGraw-Hill, New York, 1964.

2 Nusselt, W., Der Wärmeubergang im Kreuzstrom, *Z. Ver. Dtsch. Ing.*, vol. 55, pp. 2021–2024, 1911.

3 Nusselt, W., Eine Neue Formel für den Wärmedurchgang im Kreuzstrom, *Tech. Mech. Thermodynam.*, vol. 1, pp. 417–422, 1930.

4 Smith, D. M., Mean Temperature Difference in Cross Flow, Parts I and II, *Engineering*, vol. 38, pp. 479–481, 606–607, 1934.

5 Mason, J. L., Heat Transfer in Cross Flow, *Proc. 2d U.S. Natl. Congr. Applied Mechanics*, ASME, Ann Arbor, Mich., pp. 801–808, 1955.

6 London, A. L., personal communication to A. D. Kraus, 1964.

11

■ air handling

11.0 A NOTE ON UNITS

As in the previous chapter, use is made, in part, of units in the English engineering system because some of the literature referred to in this chapter is based on this system. For example, Welsh's data [1] provide the equivalent length of a straight duct for a turn of a given radius (in inches) of a circular duct (with diameter in inches). Furthermore, fan manufacturer's catalogs still show air flow in ft^3/min and static pressure in in H_2O and even in an attempt to use SI units, some literature still shows pressure in Kgf/m^2 rather than N/m^2.

Whenever possible, use is made of the Standard International (SI) system of units.

11.1 INTRODUCTION

The abundance of air in the atmosphere makes it the prime coolant for ground and airborne electronic equipment. It is available, costs nothing, and has heat transfer properties that permit its use in a wide variety of applications. Indeed, utilization of air as a coolant is possible with little system penalty even in high-altitude applications.

The thermal engineer is concerned not only with the heat transfer properties of a coolant, but also with the problems of coolant distribution. For this reason, knowledge of air handling (air movers, air distribution systems, frictional resistance, and altitude effects) is of paramount importance.

In this chapter it is presumed that an air handling system is part of the electronic equipment. Frictional and other resistances to air flow are discussed, and equations for the calculation of these resistances are given. The establishment of a system resistance curve and the matching of a system to a fan (blower) are illustrated in a typical example. Because fans are the most frequently used air movers, fan dynamics are treated in detail, and it is shown through the use of the fan laws how fan performance is influenced by departures from design conditions.

11.2 FRICTION LOSS AND SYSTEM RESISTANCE

An air handling system is composed of two basic items: (1) an air mover such as a fan or blower and (2) all other items through which the air will flow (plenum chamber, ducts or pipes, orifices and flow limiters, fittings such as valves, tees, and elbows, and the electronic equipment itself). Frequently, air is delivered directly to the equipment by an environmental control system. When this is the case, judicious air handling system design is still required, even though the design may be less complicated and tedious than when a fan or blower is used as the air mover.

The air mover may be considered as the source of energy in the air handling system. All other items are considered as impedances of the system; the sum of the frictional and other resistances establishes the total system resistance or system impedance to which the air mover is matched. Hence the design of any air handling system requires generating the total system resistance characteristic curve.

There is a vast literature that can be used to determine losses in an air ducting system. In Secs. 11.2.1 through 11.2.6, pertinent extracts from the literature (with references) are presented.

11.2.1   Friction in Pipes or Ducts

As will be shown in Chap. 12, the head loss in friction for flow through a pipe, duct, or channel (when the pressure loss is less than one-tenth of the inlet pressure) can be represented by the relationship

$$\Delta p = \frac{fL\rho V^2}{2dg} \quad \text{lb/ft}^2 \tag{11.1}$$

where f is the friction factor, a function of the Reynolds number and the relative roughness of the confining duct, pipe, or channel. Values of the friction factor were established by Moody [2], and a graphical representation known as a *Moody chart* is shown in Fig. 12.2.

Many authors, such as Rohsenow and Choi [3], give the pressure loss as

$$\Delta p = \frac{4fL\rho V^2}{2dg}$$

It can be seen that this relationship differs from Eq. (11.1) by a factor of 4. For equal L, ρ, V, and d, either relationship must yield the same value of pressure loss, Δp. Hence, it is the friction factors in the two equations that differ by a factor of 4.

This difference has been a source of considerable confusion, and care must be exercised to assure that the correct friction factor is used with its corresponding equation. Perhaps it will be helpful to note that in the laminar flow region (Re $\leqslant$ 2100), the friction factor associated with Eq. (11.1) is given by $f = 64/\text{Re}$. It is this friction factor that is plotted in Fig. 12.3.

If the confining channel is a duct rather than a pipe of circular cross section, the value of d is the equivalent diameter, given by

$$d_e = \frac{4A}{P} \tag{11.2}$$

This equivalent diameter may not be equal to the equivalent diameter used for determining the heat transfer coefficient. This is because of the value of P used in Eq. (11.2). All surfaces are wetted by the flowing air, and the frictional equivalent diameter is determined directly from Eq. (11.2). However, only a portion of the surfaces may transfer heat. Only the perimeter of the heat transfer surfaces is used in computing the heat transfer equivalent diameter.

For example, consider a rectangular duct a ft wide and b ft high. The equivalent diameter for friction can be computed from Eq. (11.2):

$$d_e = \frac{4A}{P} = \frac{4(ab)}{2(a+b)} = \frac{2ab}{a+b} \tag{11.3}$$

If the heat is transferred from one side of the duct, say, dimension a, the equivalent diameter for heat transfer, also from Eq. (11.2), is computed as

$$d_e = \frac{4A}{P} = \frac{4ab}{a} = 4b \tag{11.4}$$

Many references [1, 4–6] give charts of frictional resistance of air flowing through straight circular pipe. It is a simple matter to locate the air flow velocity in cubic feet per minute or cubic meters per minute and the pipe diameter in inches or meters. At the intersection of these two values one can read, on the appropriate scale, the value of the friction loss, usually given in inches or centimeters of water per 100 ft or 100 m of circular pipe or round duct. Use of these frictional resistance charts is not recommended, because they are based on air at a particular density, usually 0.075 lb/ft^3, and a pipe material of specific roughness. They are also based on a given number of pipe joints. Hence, any departure from the specified conditions should be treated with extreme care when using the air friction charts.

Air handling systems for electronic equipment often utilize rectangular ducting. The friction loss in these ducts can be calculated from Eq. (11.1) using the equivalent diameter given by Eq. (11.3). The resultant friction loss is the same for identical velocities in the rectangular duct of diameter d_e and the circular duct or pipe of diameter d. For identical quantity of flow, the same friction loss values will be obtained when the equivalent diameter in the rectangular duct is given by

$$d_e = \frac{1.30(ab)^{0.625}}{(a+b)^{0.25}} \tag{11.5}$$

11.2.2 Turn and Elbow Losses

When a duct or piping system changes direction, an additional pressure loss is encountered. This pressure loss will be over and above the frictional pressure loss calculated from Eq. (11.1).

Two methods of calculating turn losses are currently in use. One method adds a loss based on a fraction of the velocity head in the turn. The velocity head can be calculated in feet or meters of flowing fluid from

$$h_v = \frac{V^2}{2g} \tag{11.6}$$

The turn loss may be obtained by assigning some correction factor C, which is based on the angle, radius, length, and diameter of the turn. Hence, the value of the turn loss is

$$\Delta p = Ch_v = C \frac{V^2}{2g} \tag{11.7}$$

where the value of C can be determined from a number of references.

Madison and Parker [7] give values of C for $90°$ round and square elbows as a function of the r/d ratio. Here r is the turn radius and d is the circular pipe (duct) diameter, or the side of the square duct as the case may be.

The Society of Automotive Engineers (SAE) *Aerospace Applied Thermodynamics Manual* [8] gives charts of the correction factor C for $90°$-radius circular pipe or duct bends. These values are based on the work of Henry [9] and Kittredge and Rowley [10], and C is given as a function of the Reynolds number in the turn. Similar data are given for square ducts based on the work of Henry [9] and Wirt [11].

Loss coefficients for rectangular ducts are given as a function of aspect ratio (width to height) for several values of Reynolds number. These data are also based on Henry's work [9]. Data for elbows or bends less than $90°$ are in the form of an additional correction factor and are given [8] on the basis of other SAE data [12] and the experiments of Madison and Parker [7].

Madison and Parker [7] also show the effect of an exit duct on the turn loss in a $90°$ elbow. A graph of the results of this work [8] shows an additional correction as a function of radius ratio (turn radius to duct height or diameter) for several values of aspect ratio (duct width to height). This graph indicates that the pressure loss in the elbow can be as much as three times higher when no exit duct from the elbow is provided.

These data are for turns and bends only. They do not account for frictional resistance in the turn or bend. To evaluate the total frictional loss, one must find the appropriate coefficient to be applied to Eq. (11.7), compute the loss by using Eq. (11.7), and then add a loss computed from Eq. (11.1) to account for the length of the turn. The total loss for the turn or bend is therefore

$$\Delta p = C \frac{V^2}{2g} + \frac{fLV^2}{2dg} \tag{11.8}$$

in feet or meters of flowing fluid.

The second method of computing the turn loss is to add an equivalent length of straight pipe or duct to the actual length of the pipe or duct. This method is based on the fact that a turn will yield a pressure loss equal to an equivalent length of straight run.

Many sources [1, 5, 13] list equivalent lengths for a variety of fittings. Welsh [1] presents a very inclusive table, which is not confined to turns or bends but considers tees, contractions, expansions, transitions, and entrance and exit losses as well.

Use of the equivalent-length-of-straight-pipe method requires the evaluation of the frictional loss, using Eq. (11.1) with a fictitious length substituted for the length L. As an example, consider 6 ft of 2-in-diameter round duct containing a turn with radius 4 in. From Welsh's data [1], it is determined that the equivalent length of straight duct for a radius-diameter ratio of 2 is 9 diameters. The equivalent length of straight duct for the turn is therefore $9(2) = 18$ in or 1.5 ft, and the length to be used in Eq. (11.1) is $6 + 1.5 = 7.5$ ft. Note that in this example the turn accounts for 20% of the total frictional pressure loss.

11.2.3   Use of Turning Vanes and Splitters

The magnitude of the turn or bend loss in a ducting system may be substantially reduced by using turning vanes or flow splitters. A *turning vane* is a curved vane of small radius inserted in a *square* turn in a duct. The vanes usually cover a $90°$ arc and are spaced uniformly across the turn. *Flow splitters* are turning vanes that are applied to circular elbows, cover a $90°$ arc, and are spaced, not uniformly, but in accordance with the relationship [4, 8]

$$R' = R^{1/(n+1)} \tag{11.9}$$

A nomograph that quickly determines the location of n flow splitters within the duct [4] may be derived from Eq. (11.9).

As an example, consider a rectangular duct having an elbow with outer radius $\frac{1}{4}$ m and inner radius $\frac{1}{8}$ m. If it is desired to use one flow splitter,

$$R = \frac{1/4}{1/8} = 2$$

$$n = 1$$

and, by Eq. (11.9),

$$R' = (R)^{1/n+1} = (2)^{1/2} = \sqrt{2}$$

The radius of curvature of the splitter is therefore

$$r_o = r_i R' = \tfrac{1}{8}\sqrt{2} = 0.177 \text{ m}$$

The effect of using flow splitters is to reduce the turn loss in the elbows. To determine the turn loss, one merely computes the value of C (see Sec. 11.2.2) using the dimensions of one passage between the flow splitters.

11.2.4   Duct Area Changes

Patterson [14] recommends the use of Eq. (11.7) for the loss in an abrupt expansion and gives data for the choice of the value of C. SAE data [12] enable one to choose the proper value of C for an abrupt contraction.

The pressure loss in a gradual contraction can be calculated from an equation recommended by Madison [4].

11.2.5   Diffusers

A diffuser is a gradual transition between ducts of smaller to larger size. It may be straight or it may be curved, but whenever possible the expansion angle between the diffuser walls should be limited to about $12°$ [8]. When the included angle exceeds $15°$, as required by space limitations, guide vanes should be employed [14–16].

Although a resistance unit such as a heat exchanger is not a downstream duct, such a unit tends to decrease the pressure loss in the diffuser because the tendency for flow separation is reduced. McClellan and Nichols [17] recommend a conical diffuser and give data for the calculation of the pressure loss.

11.2.6 Manifolds

When an electronic system is composed of several packages of equipment fed by a single, central source of air such as a fan, proper manifold design is exceedingly important. The equation governing the pressure loss between ports is that recommended by the SAE [8]:

$$\Delta p = \left(C + \frac{f \, \Delta L}{d} \frac{V^2}{2g} \right) \tag{11.10}$$

In Eq. (11.10), ΔL is the length between exhaust ports, f is the friction factor for the manifold based on the velocity V, and C is a loss coefficient that has been reported by Vazsonyi [18].

The pressure loss across the ports is determined using Eq. (11.7), employing the loss coefficients of Allen and Albinson [19]. When the flow of air at any point in the manifold is considered, a trial-and-error solution for the area of the ports can be evolved. It should be realized, however, that manifold sizing based on this procedure will give approximate results. One should bear in mind that the final requirement will be satisfied only after test and suitable adjustment.

11.3 THE SYSTEM RESISTANCE CURVE

The system resistance curve is determined from the individual pressure losses in the system. It cannot be accurately established until a test is run on the equipment and its air distribution system. For purposes of analysis, however, one can construct the curve by calculating the pressure loss in all component parts of the system at particular air flow values. The sum of these pressure losses, the total pressure loss, is then plotted as the ordinate against the air flow, usually in cubic feet per minute or liters per minute, as the abscissa.

The curve so obtained is usually parabolic in shape. However, contrary to the positive statement made in many references, this curve is not necessarily a parabola of the form $y = ax^2$. Because the friction factor decreases as velocity increases, the friction loss is not a function of the square of the velocity. More probably, one will find that the system resistance curve is of the approximate form $y = ax^{1.85}$. Neglect of this important consideration may quite possibly lead to the selection of a fan that is too large for a given application.

11.4 FAN DYNAMICS

The core of the air handling system design problem lies in matching the air handling system to a suitable fan or blower. In applications where an environmental control system is used, this means that the duct and distribution system resistance characteristic falls within limits that are specified by the environmental control system. In cases where fans or blowers are employed, the designer has more control over the ultimate performance of the system but must cope with a more difficult problem.

The effects of several of the fittings and appurtenances that comprise an air distribution system are discussed in Sec. 11.2. A discussion of the fan and blower, the available types, the relationships between variables that influence air mover performance, and equipment selection now follows.

11.4.1 Types of Fans

In the material that follows, the word *impeller* is used as an inclusive term pertaining to both fans and blowers. Impellers may be of two types: centrifugal and axial.

The *centrifugal* impeller receives air at its axis of rotation and exhausts it at its periphery through an outlet opening that is often called a scroll, housing, or shroud. The centrifugal impeller is essentially a high head, moderate flow device. It is further classified by the configuration of its blades.

The *axial* impeller or propeller exhibits a high flow, moderate head characteristic and is classified by its mounting. A free air impeller has no housing or casing and is used to circulate air within the space it occupies and within which it operates. The partially cased impeller, which is truly the propeller, is mounted within an orifice and is designed to move air from one enclosed space to another. The ducted impeller is completely enclosed by a duct and may be used to move air from one enclosed space to another or may be used merely to keep air moving in an enclosed system.

Control is accomplished in a centrifugal impeller by inlet vanes or outlet louvers. In an axial impeller, straightening vanes or the duct itself may be used to keep radial motion, and attendant energy loss, to a minimum.

11.4.2 Centrifugal Impellers—Blade Configurations

The blades of a centrifugal impeller may be forward-curved, radial, or backward-curved. These blade configurations are illustrated in Fig. 11.1. Note the shape of the vector diagrams. In all cases, the vector designated by **u** represents the tip speed of the impeller, and the vector designated by **r** represents the relative velocity of the discharging air with respect to the impeller. The resultant air velocity is the vector sum of relative velocity and the tip speed, and is designated by **a**. It is interesting to note that in the case of the forward-curved and radial blades, the air discharges at a velocity higher than the impeller tip speed.

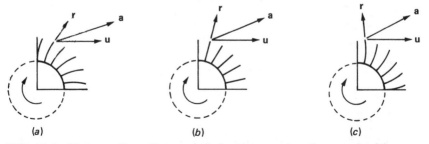

FIG. 11.1 Blade configurations and air departure vector diagrams for (*a*) forward-curved impeller, (*b*) radial impeller, and (*c*) backward-curved impeller.

The total pressure developed by a centrifugal impeller is proportional to the tangential component of the resultant velocity vector. Refer to Fig. 11.1 and imagine that an increase in flow can be represented by a suitable lengthening of the relative velocity vector **r**. Note that the tangential component of the vector **a** will decrease for the backward-curved blades and increase for the case of the forward-curved blades. The tangential component will remain constant only for the case of the radial blades.

This explains why a backward-curved impeller will exhibit a falling pressure characteristic with increased flow. Furthermore, the radial impeller shows an essentially constant characteristic with flow rate. Finally, the forward-curved impeller, for a certain range of flows, possesses a rising pressure versus flow rate characteristic. The static pressure flow rate curves for these three blade types are shown in Fig. 11.2. In considering Fig. 11.2, one must imagine that all three impellers are of equal diameters and rotate at the same speed.

11.4.3 Characteristic Fan Curves

The performance of a fan is adequately represented by a group of curves known as characteristic curves. It is this set of curves that is provided by fan manufacturers to assist engineers in fan selection. The curves usually provided are the static pressure and static efficiency characteristics. These, along with the total pressure and total efficiency curves are shown, for a typical fan, in Fig. 11.3.

The difference between the static and total pressure can be seen from inspection of the steady-state energy equation where there is no heat flow and negligible change in internal energy:

$$z_1 + p_1 v_1 + \frac{V_1^2}{2g} + W = z_2 + p_2 v_2 + \frac{V_2^2}{2g} \tag{11.11}$$

The units of Eq. (11.11) are feet or meters. The $V^2/2g$ term is due to the kinetic energy and is often called the velocity pressure. The pv term is called the static pressure. One may therefore define

$$p_T \equiv p_S + p_V \equiv \frac{p}{\rho} + \frac{V^2}{2g} \tag{11.12}$$

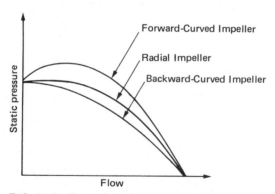

FIG. 11.2 Typical static pressure-air flow characteristics for three impeller blade types.

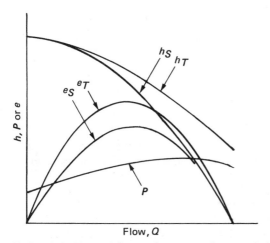

FIG. 11.3   Typical fan performance characteristics.

where

$$p_S \equiv \frac{p}{\rho} \tag{11.13}$$

and $p_V \equiv \dfrac{V^2}{2g}$ (11.14)

With reference to Fig. 11.3, observe that the static and total pressures are identical at zero flow rate. This is the "cutoff" condition where the velocity is zero. Hence $V^2/2g = 0$ and $p_T = p_S$. On the other hand, at the "wide open" condition where the flow rate is maximum, there is an air flow at a velocity ($V^2/2g \neq 0$), and there will be some velocity pressure even though a free discharge occurs with no static pressure developed. In this case, $p_T = p_v$.

Consider a moving fluid with volumetric flow rate designated by Q (say, ft^3/min). Then the weight rate of flow per unit time will be ρQ (lb/min or kg/min). If this weight of fluid is moved a distance h (ft or m), then the work done will be $\rho Q h$ (ft lb/min or kg m/min) or $\rho Q h/33,000$ (HP).

The input horsepower will be this developed horsepower divided by the efficiency:

$$P = \frac{\rho Q h}{33,000e} \tag{11.15}$$

For an air mover one may use standard density and the head developed in inches of water to determine $\rho Q h/33,000$, the air horsepower, AHP, as

$$AHP = 0.0001575 Q h \tag{11.16}$$

and depending on whether total pressure h_T or static pressure h_S is used, the total efficiency

$$e_T = \frac{0.0001575Qh_T}{P} \tag{11.17}$$

or the static efficiency

$$e_S = \frac{0.0001575Qh_S}{P} \tag{11.18}$$

may be evaluated. In Eqs. (11.17) and (11.18), P is the input horsepower to the fan shaft; and in Eqs. (11.16), (11.17), and (11.18), h is expressed in inches of water.

11.4.4 Relationships among Fan Performance Variables

The characteristic curves discussed in Sec. 11.4.3 indicate that static pressure h_S, flow rate Q, and input power P can be considered as three of the variables influencing fan performance. Three others are the fan or impeller size, which can be related to the impeller diameter d, rotation n, and air density ρ. Dimensional analysis, treated in more detail in the next chapter, can be employed to determine the relationships among these six variables.

Let the fundamental dimensions be M for mass, L for length, and T for time. The six variables to be considered are listed with their dimensions in Table 11.1.

The first step is the determination of the variation of flow rate as a function of fan size, speed of rotation, and air density. Write a functional relationship

$$Q = f_1(d, n, \rho)$$

or, with a constant of proportionality C,

$$Q = C_1 d^a n^b \rho^c$$

with the exponents to be determined. Then, with dimensions substituted,

$$L^3 T^{-1} = L^a (T^{-1})^b (ML^{-3})^c = L^a T^{-b} M^c L^{-3c}$$

TABLE 11.1 Variables and Dimensions for Fan Analysis

| Quantity | Primary dimensions | | Dimensions |
	English	SI	
Flow rate Q	ft³/min	m³/min	L^3/T
Static pressure h_S	lbf/ft²	N/m²	M/LT^2
Size (diameter) d	ft	m	L
Power P	lbf ft/min	J/min	$ML^2 T^3$
Speed n	r/min	r/min	$1/T$
Density ρ	lbm/ft³	kg/m³	M/L^3

the exponents of like terms on either side of the equation may be equated with the result:

$M:$ $\quad 0 = c$

$L:$ $\quad 3 = a - 3c$

$T:$ $\quad -1 = -b$

from which $a = 3, b = 1$, and $c = 0$. Thus

$$Q = C_1 d^3 n \tag{11.19}$$

A similar procedure for $h_S = f_2(d, n, \rho)$ and $P = f_3(d, n, p)$ yields

$$h_S = C_2 d^2 n^2 \rho \tag{11.20}$$

and $\quad P = C_3 d^5 n^3 \rho \tag{11.21}$

11.4.5 The Fan Laws

Equations (11.19), (11.20), and (11.21) lead to the *fan laws*. These are concise statements that relate the flow rate, static pressure, and power requirements with any one of the variables, size, speed, or density, when the other two are held constant. The fan laws are also useful in providing the means for determining the departure of fan performance from a given operating point.

The first set of fan laws considers the case of a given fan (fixed d) operating under constant air density (fixed ρ) at variable speed n.

1 (a) Q is directly proportional to n.
 (b) h_S is directly proportional to n^2.
 (c) P is directly proportional to n^3.

The second set of fan laws revolves about the consideration of a variable air density with fixed size d and constant speed n.

2 (a) Q is invariant with ρ.
 (b) h_S is directly proportional to ρ.
 (c) P is directly proportional to ρ.

Finally, for a fan operating at constant speed (fixed n) and handling air at the same density (fixed ρ), for variable size of impeller,

3 (a) Q is directly proportional to d^3.
 (b) h_S is directly proportional to d^2.
 (c) P is directly proportional to d^5.

EXAMPLE

Consider a fan with diameter d operating at a nominal speed n_0, with air at a nominal density ρ_0. Consider a point on the characteristic fan curve designated by Q_0, h_{S0}, and ρ_0 and determine the values of Q, h_S, and P when the fan is operated at twice the speed of rotation ($n = 2n_0$) with air at one-tenth the density ($\rho = \rho_0/10$).

SOLUTION

The problem reduces to the determination of Q, h_S, and P given $n, \rho, Q_0, h_{S0}, P_0, n_0$, and ρ_0. From fan law 1a, Q varies directly with n, and from fan law 2a, Q is invariant with density. Hence

$$Q = \frac{n}{n_0} Q_0 = 2Q_0$$

Next, from fan laws 1b and 2b, it is observed that static pressure varies directly as the square of the speed and directly as the air density. Hence

$$h_S = \left(\frac{n}{n_0}\right)^2 \left(\frac{\rho}{\rho_0}\right) h_{S0} = (2)^2 (0.1) h_{S0} = 0.4 h_{S0}$$

Finally, the power P is evaluated from fan laws 1c and 2c. From law 1c, it is seen that the power required to drive the fan varies as the cube of the speed; from law 2c, it is observed that the power is directly proportional to the density. Thus

$$P = \left(\frac{n}{n_0}\right)^3 \left(\frac{\rho}{\rho_0}\right) P_0 = (2)^3 (0.1) P_0 = 0.8 P_0$$

11.4.6 Selection of a Fan: Specific Speed

All types of fans have a point of maximum efficiency, and it is wise to operate the fan and system at this point. At the point of maximum efficiency for a given speed, there will be particular values of flow rate and static pressure.

The specific speed of a fan or blower is defined as

$$n_S = \frac{nQ^{0.5}}{(h_S)^{0.75}} \tag{11.22}$$

Because the maximum efficiency and values of n, Q, and h_S are related for a particular fan, it is possible to select fans on the basis of specific speed. Indeed, when it is established that a fan is to possess a specific speed of say, 40,000, a backward-curved centrifugal impeller is called for (as Table 11.2 clearly shows), because it is known that

TABLE 11.2 Ranges of Specific Speeds for Several Impeller Types

Type of impeller	Specific speed n_S
Centrifugal	
Forward curved	3,000–40,000
Radial	12,500–60,000
Backward curved	22,000–60,000
Axial	
Vane axial	45,000–125,000
Tube axial	70,000–200,000
Propeller	120,000–400,000

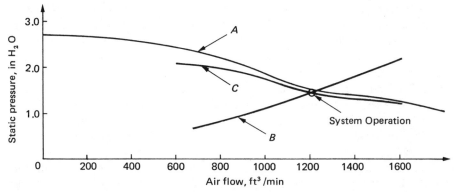

FIG. 11.4 Fan-exchanger performance curves. Curve A is the fan characteristic at 77°F (25°C) at sea level. Curve B is the exchanger (system) resistance curve. Curve C is the fan characteristic curve adjusted for the air density at the fan. The circled point indicates the operating point of this induced-draft system.

this type of impeller operates near its maximum efficiency at the values of n, Q, and h_S that have established this particular value of specific speed.

Ranges of specific speeds for the various types of impeller are given in Table 11.2.

11.5 MATCHING THE FAN TO THE SYSTEM

Previous sections have considered the establishment of the system resistance characteristic and the types of fan available and their performance. The next step is to consider the matching of the fan to the system. This is best done through the use of an example.

EXAMPLE

In Chap. 10, a heat exchanger was designed to handle 51,195 Btu/h in a liquid cooling application. Air entered at 80°F in a cross-flow arrangement at a flow rate of 932 ft³/min (4096 lb/h), and the pressure loss on the air side was 0.99 in H_2O, where 1.10 in H_2O was permitted.

Suppose that this exchanger is to be used in an application where the air flow, again at an 80°F inlet, is *induced* by an axial flow fan (or a group of axial-flow fans operating in parallel) that exhibit a standard (77°F, sea level) characteristic shown in curve A of Fig. 11.4. It is required to determine the air flow rate for this condition.

SOLUTION
The problem is solved in two steps.

1   Assume four air flows and determine the air-side pressure loss for the exchanger. The exchanger heat duty remains at 51,195 Btu/h, and the air inlet remains at 80°F. Plot a curve of static pressure loss as a function of air flow.

2   For each air flow selected in step 1, adjust the fan characteristic to reflect the fact that the induced-draft fan operates at the outlet of the exchanger and handles

air at a temperature higher than the standard 77°F (25°C). The adjustment is easily made using fan laws 2a and 2b, which show that the only consideration is that fan static pressure is directly proportional to air density.

The exchanger calculations are simply effected by using a form displayed with all of the calculations necessary in Table 11.3. Here, air flows of 750, 1200, 1400, and 1600 ft³/min have been assumed, and the inlet air temperature is held at 80°F. The results, as well as the result of 0.99 in H_2O at 932 ft³/min from Table 10.2 are plotted as curve B in Fig. 11.4. This completes step 1 in the outlined solution strategy.

The fan, operating at the outlet of the exchanger, must have its characteristic curve, which is provided by the fan manufacturer as curve A in Fig. 11.4, adjusted to reflect the fact that the fan does not handle air at the so-called standard condition of 77°F (25°C). Step 2 of the solution strategy makes this adjustment as shown in the very simple step-by-step procedure presented in Table 11.4. According to fan law 2b, the static pressure at each air flow is lowered in direct proportion to the density ratio of the air at the exchanger outlet to the air at sea level and 77°F, namely, $\rho = 0.0736$ lb/ft³. With data for the second column of Table 11.4 taken from Table 10.4, the result is plotted as curve C of Fig. 11.4. This completes step 2 in the solution strategy. The operating point of the system is found at the intersection of curves B and C: 1195 ft³/min at 1.43 in water.

TABLE 11.3 Calculations for Air-Side Pressure Loss for Cross-Flow Heat Exchanger Design in Table 10.2[a]

Flow, ft³/min	750	1,200	1,400	1,600
$W = $ lb/h at 80°F in	3,293	5,269	6,147	7,025
$t_2 = 80 + (51,195/0.241\,W)$	144.5	120.3	114.6	110.2
$t_a = (80 + t_2)/2$	112.3	100.2	97.3	95.1
μ at t_a	0.0455	0.0447	0.0446	0.0445
$G = W/A = W/0.534$	6,166.7	9,867.0	11,511.2	13,155.4
$Re = 0.00264\,G/\mu$	357.8	582.8	681.4	780.5
f at Re (Fig. 10.4)	0.047	0.0304	0.0265	0.0234
$v_1 = (460 + 80)/39.55$	13.65	13.65	13.65	13.65
$v_2 = (460 + t_2)/39.55$	15.28	14.67	14.53	14.42
$v_m = (v_1 + v_2)/2$	14.467	14.161	14.089	14.034
v_2/v_1	1.120	1.075	1.064	1.056
v_m/v_1	1.060	1.037	1.032	1.028
$\psi_3 = 1 + K_c - \sigma^2$	2.183	2.183	2.183	2.183
$\psi_4 = f(S/A)v_m/v_1$	4.725	2.972	2.594	2.282
$\psi_5 = 2[(v_2/v_1) - 1]$	0.240	0.150	0.128	0.112
$\psi_6 = \psi_3 + \psi_4 + \psi_5$	7.148	5.305	4.905	4.577
$\psi_7 = (1 - \sigma^2 - K_e)$	0.583	0.583	0.583	0.583
$\psi_8 = \psi_7(v_2/v_1)$	0.653	0.627	0.620	0.616
$\psi_9 = \psi_6 - \psi_8$	6.495	4.678	4.285	3.961
$\psi_{10} = (G')^2 v_1/2g$	0.623	1.594	2.169	2.833
$\Delta P = \psi_9 \psi_{10}/144$	0.0281	0.0518	0.0645	0.0779
ΔP, in H_2O	0.77	1.43	1.79	2.16

[a]$A = 0.534$ ft², $S = 50.65$ ft², $\sigma = 0.238$, $d_e = 0.00264$ ft, $K_c = 1.24$ (laminar flow), $K_e = 0.36$ (laminar flow).

TABLE 11.4 Calculations for Adjustment of Fan Curve for Operation at Higher Temperatures

Air flow, ft³/min	750	932	1200	1400	1600
Static pressure, in H_2O	2.26	2.01	1.55	1.38	1.28
Air temperature, °F	114.5	131.9	120.3	114.6	110.2
Air density ρ, lb/ft³	0.0654	0.0668	0.0682	0.0688	0.0694
Density ratio, $\rho/0.0736$	0.888	0.907	0.925	0.935	0.942
Corrected static pressure, in H_2O	2.01	1.82	1.43	1.29	1.21

11.6 OPERATION AT ALTITUDE CONDITIONS

Operation of an air cooling system at other than standard conditions is a prevalent condition. Even the term "sea level" is a misnomer; one can argue that even at sea level, operation of equipment can occur over a wide range of environmental temperatures.

Even the word "standard" may refer to a multitude of conditions. It has been observed that some fan manufacturers most often use 25°C (77°F) and 760 mm Hg (14.696 psi) as their standard. A military standard at 15°C (59°F) and a commercial standard at 20.56°C (69°F) have also been proposed.

Table 11.5 presents values for the *model atmosphere* proposed by the Air Research and Development Command of the U.S. Air Force in 1960 [20].

Consider an electronic system that requires a certain air flow at a given inlet temperature. When this air flow is provided at sea level, a certain air pressure loss across the electronic system can be measured. Suppose that it is desired to operate the electronic system at some altitude at the same inlet temperature. For a precisely designed cooling system, the same mass flow rate of air will be required. But, because at higher altitudes the air density is lower, a higher air velocity will be required. This higher velocity, as an inspection of the pressure loss relationships provided here clearly show, will result in a higher pressure loss than that exhibited at sea level for the same mass flow rate of air.

In order to compare pressure losses at various altitudes, use is made of a density ratio σ, defined as the ratio of air density at any condition to that at some standard condition. Designating ρ_0 as the density at standard conditions, σ becomes

$$\sigma = \frac{\rho}{\rho_0} \tag{11.23}$$

If the model atmosphere presented in Table 11.5 is used and standard conditions are taken at sea level and 59°F, then $\rho_0 = 0.0762$ lb/ft³ and

$$\sigma = \frac{\rho}{0.0762} = 13.123\rho \tag{11.24}$$

The ratio σ can be written as a function of pressure and temperature. Assuming that air is a perfect gas, $\rho = p/RT$ and hence

$$\sigma = \frac{\rho}{\rho_0} = \frac{p/RT}{p_0/RT_0} = \frac{pT_0}{p_0 T}$$

and $\sigma = \dfrac{p(460 + 59)}{14.7T} = 35.31 \dfrac{p}{T}$ (11.25a)

where p must be expressed in pounds per square inch and T in degrees Rankine. In SI units this relationship becomes

$$\sigma = \frac{p(273 + 15)}{1033.2T} = 0.279 \frac{p}{T}$$ (11.25b)

where p must be expressed in grams force per square centimeter and T in kelvins.

At sea level, little error results if σ is calculated at system inlet conditions, because the pressure drop is quite low. At altitude, however, σ should be based on the

TABLE 11.5   Air Research and Development Command Model Atmosphere

Altitude (ft)	Pressure (psi)	Temperature ($^\circ$F)
0	14.70	59.0
5,000	12.23	41.2
10,000	10.11	23.4
15,000	8.30	5.5
20,000	6.76	−12.3
25,000	5.46	−30.1
30,000	4.37	−47.8
35,000	3.47	−65.6
40,000	2.73	−69.7
45,000	2.15	−69.7
50,000	1.69	−69.7
55,000	1.33	−69.7
60,000	1.05	−69.7
65,000	0.83	−69.7
70,000	0.65	−69.7
75,000	0.51	−69.7

Altitude (m)	Pressure (kg/m^2)	Temperature ($^\circ$C)
0	10,332	15.0
2,000	8,106.9	2.0
4,000	6,287.6	−11.0
6,000	4,814.8	−24.0
8,000	3,635.4	−36.9
10,000	2,702.2	−49.9
12,000	1,978.2	−56.5
14,000	1,445.0	−56.5
16,000	1,055.7	−56.5
18,000	771.4	−56.5
20,000	563.8	−56.5
22,000	412.2	−56.5
24,000	301.4	−56.5

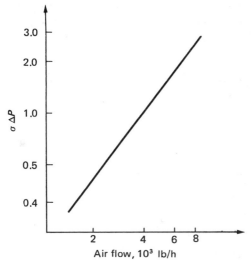

FIG. 11.5 $\sigma \Delta P$ curve for heat exchanger example.

average between system inlet and outlet conditions, and it is easy to see that in the English engineering system,

$$\sigma = 17.655 \left(\frac{p_1}{T_1} + \frac{p_2}{T_2} \right) \tag{11.26a}$$

and in the SI system,

$$\sigma = 0.1395 \left(\frac{p_1}{T_1} + \frac{p_2}{T_2} \right) \tag{11.26b}$$

where in Eqs. (11.26) the subscripts 1 and 2 are used to designate system inlet and outlet conditions, respectively.

It is customary to describe the pressure loss characteristics of an air cooling system using the product of σ and the pressure loss. This is known as the $\sigma \Delta P$ characteristic. The system is fully described by plotting $\sigma \Delta P$ against air mass flow on a log-log plot. The result, as shown in Fig. 11.5 for the heat exchanger discussed in this chapter and in Chap. 10, is a straight line with slope 1.4. The value of the slope will vary between about 1.2 for laminar flow with restrictions in the air flow path to about 2.0 for fully turbulent flow.

When given a pressure drop at a specified condition, one can find the pressure drop at a different condition. It is this fact, which is easily gleaned from an inspection of the relationships for air pressure loss, that enables the analyst quickly and accurately to determine pressure loss for "off-standard" conditions.

EXAMPLE

Consider the heat exchanger discussed previously in this chapter and observe that when a static pressure of 1.17 in water is provided, 1040 ft^3/min at an inlet temperature of 80°F will flow through the exchanger. It is desired to determine the static pressure loss when the exchanger is operated with an identical mass flow rate, again at an inlet air temperature of 80°F but at an altitude of 50,000 ft.

SOLUTION

At 80°F and 14.7 psia, $\rho = 0.0732$ lb/ft^3. Using the model atmosphere and Eq. (11.25a),

$$\sigma = 35.31 \left(\frac{14.7}{540} \right) = 0.961$$

and hence, for the sea level conditions,

$$\sigma \Delta P = 0.961(1.17) = 1.124$$

The mass flow rate at sea level and with an 80°F inlet is

$$W = 0.0732(1040) = 76.13 \text{ lb/min}$$

For a 51,195 Btu/h heat duty, regardless of the altitude, the exit air temperature will be

$$t_2 = 80 + \frac{51,195}{(0.241)(60)(76.13)} = 80 + 46.5 = 126.5°F$$

where $60(76.13) = 4567.8$ lb of air per hour.

The model atmosphere (Table 11.5) shows that the ambient pressure at 50,000 ft is 1.69 psia. Assuming that this value will be p_2,

$$p_2 = 1.69 \text{ psia}$$

then p_1 will be $1.69 + \Delta P$, where ΔP must be expressed in pounds per square inch (recall that 1 psi = 27.7 in water). Hence

$$p_1 = 1.69 + \frac{\Delta P}{27.7}$$

where ΔP is in inches of water.

The value of σ at 50,000 ft must be obtained from a trial-and-error solution employing Eq. (11.26a):

$$\sigma = 17.655 \left[\frac{p_1 + (\Delta P/27.7)}{T_1} + \frac{p_2}{T_2} \right]$$

With the values obtained thus far, this becomes

$$\sigma = 17.655 \left[\frac{1.69 + (\Delta P/27.7)}{540} + \frac{1.69}{586.5} \right]$$

or $\sigma = 0.1061 + 1.1803 \times 10^{-3} \, \Delta P$

When σ is calculated from this equation at an assumed ΔP, the product of σ and ΔP must equal $\sigma \Delta P = 1.124$, which is the value obtained from the sea level conditions *at an identical air mass flow rate and inlet temperature.*

Thus, assume that $\Delta P = 40$ in water. Then $\sigma = 0.1061 + (1.1803 \times 10^{-3}) \times (40) = 0.1533$ and $\sigma \Delta P = (0.1533)(40) = 6.133$, which shows that $\Delta P = 40$ in water is much too high.

With $\Delta P = 15$ in water, $\sigma = 0.1061 + (1.1803 \times 10^{-3})(15) = 0.1238$ and $\sigma \Delta P = (0.1238)(15) = 1.857$. With $\Delta P = 8$ in water, $\sigma = 0.1061 + (1.1803 \times 10^{-3}) \times (8) = 0.1155$ and $\sigma \Delta P = (0.1155)(8) = 0.9243$.

Repeated calculations made in this manner will show that the static pressure loss at 50,000 ft is 9.58 in water.

EXAMPLE
Suppose that 3600 lb/h at 40°F are provided to the heat exchanger. What will be the static pressure loss at sea level and 50,000 ft?

SOLUTION
At 40°F (4.6°C) and sea level,

$$\sigma = 35.31 \left(\frac{14.7}{460 + 40} \right) = 1.038$$

From the $\sigma \Delta P$ versus air flow curve of Fig. 11.5 at 3600 lb/h, $\sigma \Delta P = 0.838$. Hence at sea level,

$$\Delta P = \frac{0.838}{\sigma} = \frac{0.838}{1.038} = 0.807 \text{ in } H_2O$$

For a heat duty of 51,195 Btu/h and an air mass flow of 3600 lb/h,

$$\Delta T = \frac{51,195}{0.241(3,600)} = 59°F$$

and $t_2 = 40 + 59 = 99°F$. Hence at 50,000 ft, where $p = 1.69$ psia (see Table 11.5),

$$\sigma = 17.655 \left[\frac{1.69 + (\Delta P/27.7)}{460 + 40} + \frac{1.69}{460 + 99} \right]$$

or $\sigma = 0.1130 + (1.275 \times 10^{-3}) \Delta P$

At 50,000 ft, $\sigma \Delta P$ must also equal 0.838. Thus

$$\sigma \Delta P = [0.1130 + (1.275 \times 10^{-3} \, \Delta P)] \, \Delta P = 0.838$$

This is a quadratic in ΔP,

$$1.275 \times 10^{-3} \, \Delta P^2 + 0.1130 \, \Delta P - 0.838 = 0$$

which yields $\Delta P = 6.88$ in water.

11.7   THE FAN–DRIVEN SYSTEM AT ALTITUDE

If a fan is the air mover, the device, component, or system to be cooled must be matched, in terms of pressure loss, to the fan at all altitudes. The analyst is therefore concerned with the determination of the operating point, not only at sea level as discussed in Sec. 11.5, but at all altitudes. This involves the plotting of curves exactly as in Fig. 11.4.

The heat exchanger ΔP as a function of air flow is easy to determine once the $\sigma \, \Delta P$ as a function of air flow is known (see Fig. 11.5). The fan characteristic is a little harder to determine and involves application of the fan laws.

■   *EXAMPLE*

Determine the operating point of the heat exchanger-fan combination of the example in Sec. 11.5 if the system is placed in a compartment where the pressure is 4.37 psi (30,000 ft) and the temperature is 40°F (4.6°C). The fan employed is equipped with a motor that doubles its speed of rotation when the air density falls below 0.0245 lb/ft^3.

SOLUTION

Determination of the operating point requires plotting two curves: (1) the pressure loss of the exchanger as a function of the air flow in cubic feet per minute; and (2) the fan static pressure versus ft^3/min characteristic at 30,000 ft and the temperature at which it handles the air.

Table 11.6 displays the pertinent calculation steps for developing the two curves. Six values of air flow in cubic feet per minute are assumed. Using the air density at 40°F and 30,000 ft (4.37 psia from Table 11.5), air flow rates in pounds per minute and pounds per hour are easily found. The outlet air temperature from the exchanger, which is the temperature at the fan, is determined from the heat duty of the exchanger and the specific heat of the air by the relationship

$$t_2 = 40 + \frac{51,195}{0.241W}$$

where W is the air flow rate in pounds per hour. The $\sigma \, \Delta P$ for the exchanger is found from Fig. 11.5, and the quadratic relationship of the previous example is used to establish the static pressure loss. The resulting ΔP as a function of cubic feet per minute is plotted as curve A in Fig. 11.6.

From fan laws 1a and 2a, it is known that Q is directly proportional to fan speed and that Q does not vary with air density. Thus, for each air flow in cubic feet per minute listed in Table 11.6, half of this value is taken to establish the base cubic feet

TABLE 11.6 Pertinent Calculations for Establishing the Fan Heat Exchanger Operating Point at 30,000 ft (4.37 psia)

Assumed ft³/min	1000	1250	1500	1750	2000	2250
ρ at 40°F and 4.37 psia	0.0235	0.0235	0.0235	0.0235	0.0235	0.0235
lb/min	23.51	29.39	35.27	41.15	47.03	52.91
lb/h	1410.9	1763.6	2116.3	2469.1	2821.8	3174.5
Δt	150.6	120.5	100.4	86.0	75.3	66.9
t_2	190.6	160.5	140.4	126.0	115.3	106.9
$\sigma \Delta P$ (from Fig. 11.5)	0.24	0.32	0.41	0.50	0.60	0.70
ΔP, in H_2O	0.88	1.14	1.44	1.74	2.06	2.39
Base ft³/min	500	625	750	875	1000	1125
Base h_S	2.49	2.38	2.26	2.10	1.88	1.63
ρ at t_2 and 4.37 psia	0.0181	0.0189	0.0196	0.0201	0.0204	0.0207
ρ at 77°F and sea level	0.0736	0.0736	0.0736	0.0736	0.0736	0.0736
Density ratio	0.2454	0.2573	0.2659	0.2724	0.2775	0.2816
New ft³/min	1000	1250	1500	1750	2000	2250
New h_S	2.44	2.45	2.40	2.29	2.09	1.84

per minute value because the fan speed doubles at 30,000 ft. The static pressure loss is then read from curve *A* of Fig. 11.4 [at 77°F (25°C)] at the base cubic feet per minute value, and this is called the base static pressure h_S.

From fan laws 1b and 2b, it is seen that fan static pressure is directly proportional to the square of the speed and directly proportional to the air density. The air density is determined at 30,000 ft and t_2 and the result is divided by the density at sea level and 77°F. This is called the density ratio and, when the density ratio is multiplied by 4 (fan speed has doubled) and the base static pressure, the static pressure for 30,000 ft and t_2 is established. This is plotted as a function of the cubic feet per minute at 30,000 ft at 40°F inlet as curve *B* in Fig. 11.6.

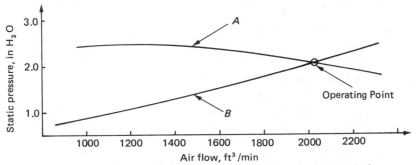

FIG. 11.6 Heat exchanger (system) resistance and fan characteristic for operation at 30,000 ft (9144 m) and 40°F (25°C).

The operating point is at the intersection of curves *A* and *B* in Fig. 11.6. It is seen to be at 2010 ft³/min, where the static pressure is 2.07 in water.

11.8 GOOD FAN UTILIZATION PRACTICE

This chapter now concludes with several keys to proper fan utilization.

1 For altitude requirements where the fan forces air through the electronic components, be sure to account for the temperature rise of the air in passing across the fan motor. When operating at altitude, the fan will speed up considerably. The fan motor power may increase accordingly and the additional heat dissipated by the motor will have to be carried away.

2 For altitude requirements where the fan delivers against a high static resistance, the static pressure is likely to be a significant percentage of the environmental pressure. This means that the fan is acting as an air compressor and, in an ideal sense, the air delivered by the fan will undergo an isentropic temperature increase that must be accounted for. This isentropic temperature rise can be calculated from the relationship

$$\frac{T_2}{T_1} = \left(\frac{p_2}{p_1}\right)^{0.285}$$

where the subscripts 1 and 2 indicate the fan inlet and outlet, respectively, and the temperatures *T* are in degrees Rankine.

3 The fan motor should be located so that the coolest air passes over it; it will be more reliable this way.

4 Be sure to select the proper fan for the job. Pay attention to the recommendations based on specific speed. Try to come as close as possible on flow rate and static pressure. If, on selection, you must depart from the exact requirements, depart on the conservative side.

5 Make the equipment discharge port or grill as large as possible. Smaller openings will tend to choke and restrict the flow. Check the effect of the discharge opening on the routing of air across the equipment.

11.9 NOMENCLATURE

Roman Letter Symbols

a	side dimension of duct, m or ft
a	resultant air velocity from impeller, m/s or ft/s
A	flow area, m² or ft²
b	side dimension of duct, m or ft
C	a correction factor defined where used
d	diameter, m or ft; with subscript e, equivalent diameter, m or ft
e	fan or blower efficiency, dimensionless
f	friction factor, dimensionless
g	local acceleration of gravity, m/s² or ft/s²
G	mass velocity, kg/m² h or lb/ft² h

h	head, m or ft; may also be given in cm or in water
L	length, m or ft; also used to designate a fundamental dimension of length
M	designates a fundamental dimension of mass
n	rotational speed, s^{-1} or min^{-1}; with subscript S, specific speed
p	pressure, kg/m^2 or lb/ft^2
P	perimeter, m or ft; or power, in W or HP (1 HP = 33,000 ft lb/s)
Q	volumetric flow rate, liters/min, m^3/min, or ft^3/min
r	turn radius, m or ft
$\mathbf{r}$	relative velocity of discharging air with respect to impeller, m/s or ft/s
R	radius ratio, dimensionless
Re	Reynolds number, dimensionless
T	temperature, °C, K, °F, or °R; also used to designate a fundamental dimension of temperature
$\mathbf{u}$	tip speed of impeller, m/s or ft/s
v	specific volume, m^3/kg or ft^3/lb
V	velocity, m/s or ft/s
W	work, J or ft lb; or flow rate, kg/s or lb/min
z	elevation, m or ft

Greek Letter Symbols

Δ	indicates change in variable
ρ	density, kg/m^3 or lb/ft^3
σ	air density ratio, dimensionless

Subscripts

e	indicates equivalent
i	indicates inner radius of impeller
o	indicates outer radius of impeller
S	indicates static pressure or specific speed
T	indicates total pressure
t	indicates blade tip
V	indicates velocity or velocity head
0	indicates nominal condition

11.10 REFERENCES

1 Welsh, J. P., *Handbook of Methods of Cooling Air Force Ground Electronic Equipment*, RADC-TR-58-126, Astia No. AD-148907, Rome Air Development Center-Griffis Air Force Base, Rome, N.Y., June 1959.

2 Moody, L. F., Friction Factors for Pipe Flow, *J. Heat Transfer*, vol. 66, pp. 671–684, 1944.

3 Rohsenow, W. M., and Choi, H. Y., *Heat, Mass, and Momentum Transfer*, Prentice-Hall, Englewood Cliffs, N.J., 1961.

4 Madison, R. D., *Fan Engineering*, Buffalo Forge Co., Buffalo, N.Y., 1949.

5 Salisbury, J. K., *Kent's Mechanical Engineer's Handbook*, 12th ed., Wiley, New York, 1960.

6 Baumeister, T., *Mark's Mechanical Engineer's Handbook*, McGraw-Hill, New York, 1958.

7 Madison, R. D., and Parker, J. R., Pressure Losses in Rectangular Elbows, *Trans. ASME*, vol. 58, pp. 167–176, April 1936.

8 *Aerospace Applied Thermodynamics Manual*, Society of Automotive Engineers, New York, 1962.

9 Henry, J. R., Design of Power Plant Installations; Pressure Loss Characteristics of Duct Components, NACA Rept. L-208, National Technical Information Service, Springfield, Va., 1944.

10 Kittredge, C. P., and Rowley, D. S., Resistance Coefficients for Laminar and Turbulent Flow through One-half Inch Valves and Fittings, *Trans. ASME*, vol. 79, pp. 1759–1766, 1957.

11 Wirt, L., New Data for the Design of Elbows in Duct Systems, *General Electric Review*, vol. 30, pp. 286–296, June 1927.

12 Society of Automotive Engineers, Airplane Heating and Ventilating Equipment Engineering Data, SAE Rept. no. 23, 1951.

13 National Valve and Manufacturing Company Catalog, Pittsburgh, Pa., 1950.

14 Patterson, G. N., Modern Diffuser Design, *Aircraft Eng.*, vol. 10, pp. 267–273, September 1938.

15 Reid, E. G., Performance Characteristics of Plane Wall Two Dimensional Diffusers, NACA TN-2888, National Technical Information Service, Springfield, Va., February 1953.

16 Moore, C. A., and Kline, S. J., Some Effects of Vanes and Turbulence in Two Dimensional Wide Angle Subsonic Diffusers, NACA TN-4080, National Technical Information Service, Springfield, Va., 1958.

17 McClellan, C. H., and Nichols, M. R., Investigation of Diffuser Resistance Combinations in Duct Systems, NACA WR-L-324, National Technical Information Service, Springfield, Va., February 1942.

18 Vazsonyi, A., Pressure Losses in Elbows and Duct Branches, *Trans. ASME*, vol. 66, pp. 177–183, 1949.

19 Allen, J., and Albinson, B., Investigation of Manifold Problems for Incompressible Fluids with Special Reference to Use of Manifolds for Canal Locks, *Proc. ICE*, vol. 4, pp. 114–138, 1955.

20 U.S. Air Force, *Handbook of Geophysics*, MacMillan, New York, 1960.

12

■ dimensional analysis and the correlation of test data

12.1 INTRODUCTION

In Chap. 6 it was pointed out that the coefficient of heat transfer for laminar and turbulent forced convection for flow over a flat plate could be obtained, approximately, through the solution of the momentum integral equation of the boundary layer. It was also pointed out that the solution of the momentum integral equation, as well as analog methods for heat, mass, and momentum transfer, and an exact solution of the boundary layer equations, all have limitations in their scope of application. Indeed, because of these limitations, no one method can be used to solve all convection problems.

The method of dimensional analysis [1, 2], which is treated in some detail in this chapter, has found a wide range of applications and is quite simple and easy to apply. It is a method of correlating a number of independent variables into a single equation for the dependent variable.

When a mathematical relationship governing the system cannot be found, or when such a relationship is too complex for ready solution, dimensional analysis may be used to indicate, in a semiempirical manner, the form of the solution. Indeed, in considering friction loss and the heat transfer coefficient, one uses dimensional analysis to reduce the number of variables that require investigation, to suggest logical groupings for presentation of results, and to pave the way for a proper experimental program.

In this chapter, the method of dimensional analysis is discussed and two ways are shown in which it can be used: in the determination of (1) a convection heat transfer coefficient and (2) a pressure loss relationship. Through the use of some fictitious test data, it is shown how a working correlation for the heat transfer coefficient can be formed.

Dimensional analysis can also play a role in the development of new electronic thermal control techniques, including air jet impingement and liquid spray cooling as well as electrostatic and acoustic convection, where the controlling mechanisms may not be sufficiently well understood to provide an analytical base for the interpretation and evaluation of available data. Furthermore, a working knowledge of such a fundamental, yet simple technique, is essential for the thermal specialist who must bridge the gap between available heat transfer and friction factor equations and the thermal fluid complexity of the configurations often encountered in air and liquid cooling of electronic components. Successful correlation of the effects of component geometry and distribution patterns on convective heat transfer coefficients in direct air cooling or bulk liquid and component temperature in immersion cooling modules may not be possible without the use of scaling laws derived from dimensional analysis.

12.2 DIMENSIONAL ANALYSIS–HEAT TRANSFER COEFFICIENT

Dimensional analysis can be applied to a quantitative determination of the heat transfer coefficient h in terms of the variables having an influence on h. The method of dimensional analysis is used when insufficient information is available for the formulation of a mathematical relationship or a physical law. In considering only the unit dimensions of the many variables involved, dimensional analysis does not produce a numerical solution but instead yields an empirical relationship that expresses the effect of each of the independent variables on the desired dependent parameter. Observed experimental data can then be correlated to yield an expression that is usable in further studies.

In the many data correlations given in Chap. 6, it was observed that the heat transfer coefficient in forced convection is dependent on the fluid velocity, certain fluid properties (viscosity, thermal conductivity, density, and specific heat), and the dimensions of the limiting surface. A dimensional analysis using these variables, which are tabulated with both actual and abbreviated dimensions in Table 12.1, is performed.

Note that, in addition to the "customary" three basic dimensions of mass, length, and time, those of heat and temperature have been added. Because it is known that the heat transfer coefficient is likely to be a function of all of the variables listed in Table 12.1, one writes the functional relationship

$$h = f(V, \rho, k, \mu, C_p, d) \tag{12.1}$$

TABLE 12.1 Dimensions and Variables for Dimensional Analysis of a Heat Transfer Relationship

Quantity	Primary dimensions		Dimensions
	English	SI	
Mass	lb	kg	M
Length	ft	m	L
Time	h	s	T
Heat	Btu	J	Q
Temperature	°F	°C	θ
		Variables	
Velocity	ft/h	m/s	L/T
Diameter	ft	m	L
Density	lb/ft³	kg/m³	M/L^3
Thermal conductivity	Btu/ft h °F	W/m °C	$Q/LT\theta$
Specific heat	Btu/lb °F	J/kg °C	$Q/M\theta$
Viscosity	lb/h ft	kg/s m	M/TL
Heat transfer coefficient	Btu/ft² h °F	W/m² °C	$Q/L^2 T\theta$

Equation (12.1) may be expressed more exactly in exponential form as

$$h = A V^a \rho^b k^c \mu^d C_p^e d^f \qquad (12.2)$$

and if dimensions of Table 12.1 are substituted into Eq. (12.2),

$$\frac{Q}{L^2 \theta T} = A \left(\frac{L}{T}\right)^a \left(\frac{M}{L^3}\right)^b \left(\frac{Q}{L\theta T}\right)^c \left(\frac{M}{TL}\right)^d \left(\frac{Q}{M\theta}\right)^e (L)^f \qquad (12.3)$$

Now equate the exponents of each of the dimensions on both sides of Eq. (12.3):

Q: $1 = c + e$

L: $-2 = a - 3b - c - d + f$

T: $-1 = -a - c - d$ $\qquad (12.4)$

θ: $-1 = -c - e$

M: $0 = b + d - e$

which forms a system of five simultaneous algebraic equations in six unknowns. These may be solved in terms of two of the unknowns.

Let $a = a$ and $c = c$. Then the solution to Eqs. (12.4) is

$a = a$

$b = a$

$c = c$

$d = 1 - a - c$

$e = 1 - c$

$f = a - 1$

so that Eq. (12.2) becomes

$$h = A V^a \rho^a k^c \mu^{1-a-c} C_p^{1-c} d^{a-1} \qquad (12.5)$$

Equation (12.5) may be adjusted by multiplying both sides by k^{-1} so that

$$\frac{hd}{k} = A \left(\frac{\rho V d}{\mu}\right)^a \left(\frac{C_p \mu}{k}\right)^{1-c} \qquad (12.6)$$

becomes the final expression for the forced convection relationship. Equation (12.6) is known as *Nusselt's equation*; note that it relates the Nusselt number, a heat transfer grouping, with the Reynolds number and the Prandtl number, which are, respectively, fluid flow and fluid thermal property groupings. The completely dimensionless relationship of Eq. (12.6) may now be used to correlate experimental data. As soon as three reliable sets of data are obtained, the constants A, a, and $(1-c)$ can be determined. Usually, however, many more than three sets of data are taken to assure that no poor sample of data is used.

From the assumed experimental data given in Table 12.2,[1] one can demonstrate how an actual forced convection correlation can be obtained from Eq. (12.6), which was determined merely from the dimensions of the variables thought to have an effect on the process. Ten runs are assumed with air flowing in pipes of different sizes. In an actual test, air inlet and outlet temperature can be measured and the quantity of air flowing can be obtained by using a precision floating bob-type flow meter. The pipe wall temperature can be maintained at an essentially constant value through the use of condensing steam, and this temperature can be varied by adjusting the steam pressure. The pipes are assumed to be 5 ft 4 in long. The fictitious values in Table 12.2 have been chosen purposely to yield a reasonable expression for the heat transfer coefficient.

The computations necessary for the correlation of data are given in Table 12.3. Note that the exponent, $1-c$, associated with the dimensionless parameter $(C_p\mu/k)$ in Eq. (12.6) has been assumed to be equal to $\frac{1}{3}$. This assumption is permissible because the other constants A and a will numerically adjust themselves accordingly.

After the computations listed in Table 12.3 are performed, a plot of $(hd/k) \times (C_p\mu/k)^{-1/3}$ is made against $(\rho Vd/\mu)$. As shown in Fig. 12.1, the data lie on a straight line connecting the points. Point scatter is almost nonexistent.. Because it appears that no runs are obviously out of line, the constants A and a can be computed from any pair of points. Take, for example, runs 4 and 6 (Table 12.3) using the equation

$$\frac{hd}{k} = A \left(\frac{\rho Vd}{\mu}\right)^a \left(\frac{C_p\mu}{k}\right)^{1/3}$$

converted to logarithmic form to the base 10:

$$\log\left(\frac{hd}{k}\right) = \log A + a \log\left(\frac{\rho Vd}{\mu}\right) + \log\left(\frac{C_p\mu}{k}\right)^{1/3} \tag{12.7}$$

[1] Presumably taken in the United States in the English engineering system of units.

TABLE 12.2   Heat Transfer Data Obtained from an Assumed Test

Run no.	Pipe diam. (in)	Flow (ft³/min)	Temperatures (°F)		
			T_s	T_1	T_2
1	1	9.45	225	65	181
2	1	12.32	225	80	180
3	1	15.08	225	80	179
4	1	15.08	240	105	199
5	2	40.10	225	70	138
6	2	48.25	225	95	149
7	2	48.25	240	120	170
8	2	56.30	240	145	185
9	$2\frac{1}{2}$	54.65	225	90	137
10	$2\frac{1}{2}$	60.35	240	115	159

TABLE 12.3 Computations for Establishing Correlation of Test Data from Table 12.2[a]

Run no.	T_a (°F)	T_f (°F)	ρ (lb/ft³)	d (ft)	A (ft²)	V (ft/h)	k	μ	C_p
1	123	174	0.0678	0.0833	0.00545	104,000	0.01735	0.0485	0.241
2	131	178	0.0671	0.0833	0.00545	135,900	0.01745	0.0488	0.241
3	129.5	177.3	0.0672	0.0833	0.00545	166,000	0.0174	0.0487	0.241
4	152	196	0.0644	0.0833	0.00545	166,000	0.0179	0.0498	0.241
5	104	164.5	0.0700	0.1667	0.0218	110,200	0.0171	0.0480	0.241
6	122	173.5	0.0679	0.1667	0.0218	132,800	0.01735	0.0484	0.241
7	145	192.5	0.0654	0.1667	0.0218	132,800	0.0178	0.0495	0.241
8	165	202.5	0.0633	0.1667	0.0218	155,000	0.01805	0.0502	0.241
9	113.5	169.8	0.0689	0.2083	0.0342	96,000	0.0172	0.0485	0.241
10	137	188.5	0.0664	0.2083	0.0342	106,000	0.0177	0.0494	0.241

Run no.	$\rho V d/\mu$	$C_p\mu/k$	$(C_p\mu/k)^{1/3}$	S (ft²)	θ_m (°F)	q (Btu/h)	h	hd/k	$(hd/k)(C_p\mu/k)^{-1/3}$
1	12,110	0.674	0.876	1.392	89.8	1,072	8.58	41.2	47.0
2	15,600	0.675	0.877	1.392	85.5	1,198	10.08	48.0	54.8
3	19,150	0.674	0.876	1.392	86.2	1,459	12.14	58.0	66.2
4	17,800	0.673	0.876	1.392	78.6	1,320	12.05	56.0	64.0
5	26,800	0.680	0.879	2.782	117.2	2,740	8.39	81.6	93.0
6	31,000	0.674	0.876	2.782	100.8	2,555	9.09	87.5	99.9
7	29,050	0.673	0.876	2.782	98.6	2,275	8.81	82.6	94.4
8	32,600	0.670	0.875	2.782	73.1	2,065	10.15	94.6	107.0
9	28,400	0.680	0.879	3.48	110.0	2,560	6.69	81.0	92.3
10	29,700	0.672	0.875	3.48	101.9	2,400	6.79	80.0	91.5

[a]Derivation of computed variables: T_a = average fluid temperature = $\frac{1}{2}(T_1 + T_2)$; T_f = mean film temperature = $\frac{1}{2}(T_s + T_a)$; T_f = mean fluid temperature at T_f; ρ = air density at T_a; k = air thermal conductivity at T_f; μ = air viscosity at T_f; C_p = specific heat of air, constant at 0.241; θ_m = long mean temperature difference = $(T_2 - T_1)/\log_e [T_s - T_1]/(T_s - T_2)]$; q = rate of heat transferred = $WC_p(T_2 - T_1)$; h = heat transfer coefficient = $q/S\theta_m$.

293

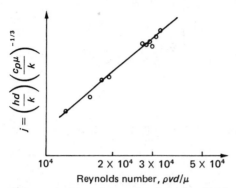

FIG. 12.1   Plot of data developed in Table 12.3.

By substitution of the corresponding numbers computed from the assumed test data, two simultaneous equations are obtained:

$$\log 64 = \log A + a \log 17{,}800 + \log 0.876$$
$$\log 99.9 = \log A + a \log 31{,}000 + \log 0.876$$

(12.8)

Simultaneous solution of Eqs. (12.8) yields

$$A = 0.0283 \quad \text{and} \quad a = 0.803$$

so that the final expression for the forced convection heat transfer relationship for the assumed data is

$$\frac{hd}{k} = 0.0283 \left(\frac{\rho V d}{\mu}\right)^{0.803} \left(\frac{C_p \mu}{k}\right)^{0.33}$$

(12.9)

12.3   DIMENSIONAL ANALYSIS BY THE BUCKINGHAM PI THEOREM

The relationship among those variables that have a bearing on the friction loss of a fluid flowing within a pipe or duct may also be obtained through dimensionless analysis.

Another way of performing a dimensional analysis is through the use of the *Buckingham pi theorem*: "If r physical quantities having s fundamental dimensions are considered, then there exists a maximum number q of the r quantities that themselves cannot form a dimensionless group. This maximum number of quantities q may never exceed the number of s fundamental dimensions ($q \leqslant s$). By combining each one of the remaining quantities, one at a time, with the q quantities, n dimensionless groups can be formed, where $n = r - q$. These dimensionless groups are called 'π terms,' and are represented by $\pi_1, \pi_2, \pi_3, \ldots$."

The Buckingham pi theorem may be illustrated quite simply. Suppose that there is a list of eight variables known or assumed to have a bearing on a certain problem;

then $r = 8$. If it is desired to express these variables in terms of four physical dimensions such as length, mass, temperature, and time, then $s = 4$. Thus, there are possibly $q = r - s = 8 - 4 = 4$ physical quantities, which by themselves cannot form a dimensionless group. The usual practice is to make $q = s$ to minimize labor. In addition, one should strive to select the q quantities so that each of the fundamental physical dimensions occurs at least once. This, of course, is not always possible.

Suppose that $q = 4$. Then there will be $n = r - q = 8 - 4 = 4$ different π terms, and the equation that relates the eight variables will be

$$\phi(\pi_1, \pi_2, \pi_3, \pi_4) = 0$$

Consider, for example, the determination of a relationship that will yield the pressure loss per unit length of pipe. It is expected that this pressure loss will be a function of fluid velocity (V), pipe diameter (d), pipe roughness (e), and the fluid properties of density (ρ) and dynamic viscosity (μ). These variables are assumed to be the only ones having a bearing on $\Delta P/L$ and may be related symbolically by

$$\frac{\Delta P}{L} = f(V, d, e, \rho, \mu) \tag{12.10}$$

Noting that $r = 6$, select the fundamental dimensions of mass (M), length (L), and time (T) so that $s = 3$. This means that the maximum number of variables that cannot, by themselves, form a dimensionless group will be $q = r - s = 3$.

The variables themselves, along with their dimensions, are given in Table 12.4. Note that because mass is a fundamental dimension, pressure in the English engineering system is represented by pounds force per unit area, not pounds mass per unit area. Because $W = mg$, pressure is represented as $W/A = mg/A$ and dimensionally by $MLT^{-2}/L^2 = ML^{-1}T^{-2}$.

Suppose that V, ρ, and d are selected as the three primary quantities ($q = 3$). These clearly contain all three of the fundamental dimensions. Thus there will be $n = r - q = 6 - 3 = 3$ dimensionless π groups, namely:

$$\pi_1 = \frac{\Delta P}{L} V^a \rho^b d^c \tag{12.11}$$

TABLE 12.4   Variables and Dimensions for Pipe Friction Analysis

Variable	Dimension
Pressure loss ΔP	$M/(LT^2)$
Length L	L
Velocity V	L/T
Diameter d	L
Roughness e	L
Density ρ	M/L^3
Viscosity μ	$M/(LT)$
Pressure loss per unit length $\Delta P/L$	$M/(L^2 T^2)$

$$\pi_2 = eV^a\rho^b d^c$$

$$\pi_3 = \mu V^a \rho^b d^c$$

(12.11 *Continued*)

The exponents a, b, and c must be determined for each group such that each group is completely dimensionless. Each π group is written in terms of its dimensions:

$$\pi_1 = (ML^{-2}T^{-2})(LT^{-1})^a(ML^{-3})^b(L)^c \qquad (12.12)$$

$$\pi_2 = (L)(LT^{-1})^a(ML^{-3})^b(L)^c \qquad (12.13)$$

$$\pi_3 = (ML^{-1}T^{-1})(LT^{-1})^a(ML^{-3})^b(L)^c \qquad (12.14)$$

In each of the π groups, the exponents are collected and equated to zero. The equations are then solved simultaneously for the exponents. Take the first case, π_1, and collect exponents:

M:　　$0 = 1 + b$

L:　　$0 = -2 + a - 3b + c$

T:　　$0 = -2 - a$

A simultaneous solution yields $a = -2$, $b = -1$, and $c = +1$, so that

$$\pi_1 = \frac{\Delta P}{L} V^{-2}\rho^{-1}d = \frac{\Delta Pd}{\rho L V^2} = \frac{\Delta P}{(L/d)\rho V^2} \qquad (12.15)$$

For the second case, π_2, the same procedure follows:

M:　　$0 = -b$

L:　　$0 = 1 + a - 3b + c$

T:　　$0 = -a$

Here, simultaneous solution yields $a = 0$, $b = 0$, and $c = -1$, so that

$$\pi_2 = ed^{-1} = \frac{e}{d} \qquad (12.16)$$

Finally, for the third case, π_3:

M:　　$0 = 1 + b$

L:　　$0 = -1 + a - 3b + c$

T:　　$0 = -1 - a$

from which $a = -1$, $b = -1$, and $c = -1$, so that

$$\pi_3 = \mu V^{-1}\rho^{-1}d^{-1} = \frac{\mu}{\rho V d} \qquad (12.17)$$

the reciprocal of the Reynolds number.

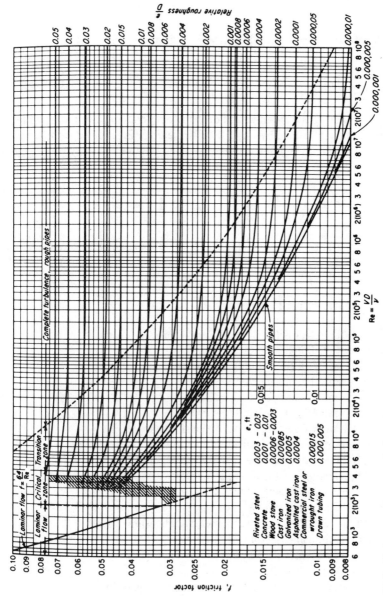

FIG. 12.2 Moody chart [3] for friction loss characteristic for laminar and turbulent flow in pipes. (From Gebhart [4], by permission.)

297

Let a friction factor f be defined as

$$f = \frac{2g(\Delta P/L)}{\rho V^2 d}$$

(12.18)

such that the pressure loss per unit length will be given by

$$\frac{\Delta P}{L} = \frac{f\rho V^2}{2dg}$$

(12.19)

Equation (12.19) is a modification of the D'Arcy-Fanning head loss relationship, and the friction factor defined by Eq. (12.18) is clearly a function (due to dimensional analysis) of the Reynolds number and the so-called relative roughness of the containing pipe. Hence,

$$f = \frac{2g(\Delta P/L)}{\rho V^2 d} = \phi \left(\frac{\rho V d}{\mu}, \frac{e}{L} \right)$$

(12.20)

The representation given by Eq. (12.20) has been graphed by Moody [3] and is shown in Fig. 12.2.

12.4 NOMENCLATURE

Roman Letter Symbols

a	an exponent, dimensionless
A	area, m^2 or ft^2; a constant, dimensionless
b	an exponent, dimensionless
c	an exponent, dimensionless
C	specific heat, J/kg °C or Btu/lb °F
d	diameter, m or ft; an exponent, dimensionless
e	roughness, m or ft; an exponent, dimensionless
f	friction factor, dimensionless; an exponent, dimensionless
g	local acceleration of gravity, m/s^2 or ft/h^2
h	heat transfer coefficient, W/m^2 °C or Btu/ft^2 h °F
k	thermal conductivity, W/m °C or Btu/ft h °F
L	length, m or ft; designates fundamental dimension of length
M	designates fundamental dimension of mass
n	number of dimensionless groups
P	pressure, kg/m^2 or lb/ft^2
q	heat flow, W or Btu/h; a maximum number of quantities
Q	designates fundamental dimension of heat
r	number of physical quantities
Re	Reynolds number, dimensionless
s	number of fundamental dimensions
S	surface, m^2 or ft^2

T	temperature, °C or °F; designates fundamental dimension of time (not temperature)
V	velocity, m/s or ft/h
W	mass flow, kg/s or lb/h

Greek Letter Symbols

Δ	indicates change in variable
θ	temperature difference, °F or °C; designates fundamental dimension of temperature
μ	dynamic viscosity, kg/h m or lb/h ft
ρ	density, kg/m^3 or lb/ft^3
ϕ	indicates a function

Subscripts

a	indicates average value
f	indicates value at film
m	indicates mean value
p	indicates value at constant pressure

12.5 REFERENCES

1 Bridgman, P. W., *Dimensional Analysis*, Yale University Press, New Haven, Conn., 1931.
2 Langhaar, H. L., *Dimensional Analysis and Theory of Models*, Wiley, New York, 1951.
3 Moody, L. F., Friction Factors for Pipe Flow, *Trans. ASME*, vol. 66, pp. 671–684, 1944.
4 Gebhart, B., *Heat Transfer*, 2d ed., McGraw-Hill, New York, 1971.

III

■ thermal control techniques

13

■ modeling direct air cooling of components

13.1 INTRODUCTION

Despite changing demands and the availability of new heat transfer technology, direct air cooling of electronic components continues to command substantial attention. The vast body of empirical results and published analyses, as well as the undoubted convenience of direct air cooling, make this technique a natural first choice for many thermal packaging tasks and an effective baseline design for more sophisticated thermal control techniques.

The increasingly stringent constraints imposed on electronic thermal control systems require that the prediction of local heat transfer coefficients and the determination of configurations and component distribution patterns achieve a degree of precision unknown in earlier decades. Accurate correlations for free and forced convection heat transfer coefficients along the surface of *isolated* components are readily available (see for example, Chap. 6), but the precise thermal behavior of component *arrays* forming vertical or horizontal air channels cannot be easily deduced from the available literature. Numerical solutions of the laminar flow field around individual protuberances on a flat surface indicate that, in the flow direction, the disturbance generally diminishes to negligible proportions after a distance three to five times the length scale of the protuberance. Similarly, when adjacent protuberances are separated in the streamwise direction by less than one-fifth to one-tenth of the length scale, the flow is essentially undisturbed. These limiting approximations can be applied to arrays of electronic components, but greater rigor must be applied to determine inter-component influences in the intermediate range of separation distances. Although such a level of analytical sophistication is beyond the scope of this text, succeeding sections do provide the thermal analyst with the tools needed to assess and quantify the channel influence on heat transfer from individual components and to identify the primary parametric trends as well as opportunities for design optimization.

The chapter is divided into two major sections, the first dealing with natural convection and the second with forced convection. To determine the prevalent flow regime in specific applications, it is possible to rely on the ratio of the Grashof number (representing the natural convection potential) to the square of the Reynolds number (embodying forced convection behavior). For Gr/Re^2 substantially greater than unity, natural convection effects dominate; whereas for values significantly below unity, acceptable accuracy can generally be obtained from predictions based exclusively on forced convection formulations. For intermediate values, natural convection may augment or retard the forced convection flow (or vice versa), and caution must be exercised in attempting to predict heat transfer coefficients in this mixed convection regime. Advanced heat transfer texts and the current literature contain various analytical and empirical treatments of this complex thermal phenomenon.

13.2   NATURAL CONVECTION IN CHANNELS

Vertical channels formed by parallel plates or fins are a frequently encountered configuration in natural convection cooling in air of electronic equipment, ranging from transformers to mainframe computers and from transistors to power supplies [1-3]. Packaging constraints and electronic considerations, as well as device or system operating modes, lead to a wide variety of complex heat dissipation profiles along the channel walls. In many cases of interest, however, a symmetric isothermal or isoflux boundary representation, or use of an isothermal/isoflux boundary together with an insulated boundary condition along the adjoining plate, can yield acceptable accuracy in the prediction of the thermal performance of such configurations.

Elenbaas [4] was the first to document a detailed study of the thermal characteristics of one such configuration, and his experimental results for isothermal plates in air were later confirmed numerically [5] and shown to apply as well to the constant heat flux conditions [6]. More recently, Aung and co-workers [7, 8] and Miyatake and co-workers [9, 10] extended the available results to include both asymmetric wall temperature and heat flux boundary conditions, including the single insulated wall.

From these and complementary studies emerges a unified picture of thermal transport in such a vertical channel. In the inlet region and in relatively short channels, individual momentum and thermal boundary layers are in evidence along each surface, and heat transfer rates approach those associated with laminar flow along isolated plates in infinite media. Alternatively, for long channels, the boundary layers merge near the entrance and fully developed flow prevails along much of the channel.

In this fully developed regime, the local heat transfer coefficient is constant (neglecting the temperature dependence of fluid properties) and equal to the well-documented forced convection values [11]. However, because the local fluid temperature is not known explicitly, it is customary to reexpress the fully developed heat transfer coefficient in terms of the ambient or inlet temperature. The Nusselt number Nu for isothermal plates appropriate to this definition can be derived from the "incompressible natural convection" form of the Navier-Stokes equations. This was done semianalytically by Elenbaas [4], confirmed by the laborious numerical calculations of Bodoia and Osterle [5], and extended to asymmetric heating by Aung [7] and Miyatake et al. [9, 10]. In a subsequent section, the limiting relations for fully developed laminar flow, in a symmetric isothermal or isoflux channel as well as in a channel with an insulated wall, will be rederived by use of a straightforward integral formulation.

The analytic relations for the isolated plate (or inlet region) limit and the fully developed (or exit region) limit can be expected to bound the Nu values over the complete range of flow development. Intermediate values of Nu can be obtained from detailed experimental and/or numerical studies or by use of the correlating expression suggested by Churchill and Usagi [12] for smoothly varying transfer processes. This correlation technique relies on the analytic expressions at the two boundaries and a limited number of data points to derive a highly accurate composite correlation; its use will be demonstrated later.

13.3 FULLY DEVELOPED LIMIT–NATURAL CONVECTION

13.3.1 Momentum Considerations

In laminar, fully developed flow between parallel plates as shown in Fig. 13.1, the pressure drop is given by [13]:

$$\frac{dP}{dx}\bigg|_{\text{loss}} = -\frac{12\mu w}{\rho b^3} \tag{13.1}$$

where the symbols are identified in the nomenclature list. For free convection flow, this flow resistance is balanced by the buoyant potential expressible as [13]

$$\frac{dP}{dx}\bigg|_{\text{buoy}} = (\bar{\rho} - \rho_0)g = -\bar{\rho}\beta g(\bar{T}_f - T_0) \tag{13.2}$$

Equating Eqs. (13.1) and (13.2), the flow rate per unit width w in the channel is found to equal

$$w = \frac{\bar{\rho}^2 g\beta b^3(\bar{T}_f - T_0)}{12\mu} \tag{13.3}$$

13.3.2 Nusselt Number–Symmetric, Isothermal Plates

An energy balance on the differential volume, shown in Fig. 13.1, equating heat transferred from two isothermal walls with that absorbed in the flow, yields

$$wc_p \, dT = 2h(T_w - T_f) \, dx \tag{13.4}$$

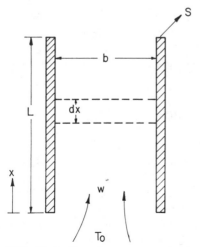

FIG. 13.1   Coordinate system for flow in a vertical channel.

From continuity considerations the flow rate w is constant, and in fully developed flow with temperature-independent properties the local heat transfer coefficient h, as well as c_p, are constant. Consequently, $wc_p/2h$ can be considered constant along the channel and Eq. (13.4) can be simply integrated to yield the local fluid temperature

$$T_f = T_w - (T_w - T_0)e^{-\Gamma x} \qquad (13.5)$$

where Γ has replaced $2h/wc_p$.

To accommodate the desire to obtain a Nusselt number based on the temperature difference between the wall and the ambient fluid, Nu_0 can be defined as

$$Nu_0 \equiv \left(\frac{q/A}{T_w - T_0}\right)\frac{b}{k} \qquad (13.6)$$

The heat transfer rate q can be determined from the flow rate and temperature rise in the channel by the use of Eqs. (13.4) and (13.5), with the latter evaluated at $x = L$ to find the exit temperature. The average fluid temperature in the channel can be found by integrating Eq. (13.5) from $x = 0$ to $x = L$ and dividing by the length of the channel, L. Following these operations,

$$q = \left[\frac{c_p\bar{\rho}^2 g\beta b^3 S}{12\mu}(T_w - T_0)\left(1 - \frac{1 - e^{-\Gamma L}}{\Gamma L}\right)\right][(T_w - T_0)(1 - e^{-\Gamma L})] \qquad (13.7)$$

Inserting Eq. (13.7) into Eq. (13.6) with the surface area A equal to $2LS$, the desired Nusselt number is found as

$$Nu_0 = \left[\frac{c_p\bar{\rho}^2 g\beta b^4 (T_w - T_0)}{24\mu kL}\right]\left[\left(1 - \frac{1 - e^{-\Gamma L}}{\Gamma L}\right)(1 - e^{-\Gamma L})\right] \qquad (13.8)$$

The combination of parameters appearing in the first term of Eq. (13.8) is recognizable as the channel Rayleigh number, Ra'. Consequently, at the fully developed limit, where $L \to \infty$ and the second term in Eq. (13.8) approaches unity, Nu_0 is seen to equal $Ra'/24$. This result agrees exactly with the previously cited analytical and numerical results [4, 5, 7].

13.3.3 Nusselt Number—Asymmetric, Isothermal Plates

For fully developed flow in a channel formed by a single isothermal plate and an insulated plate, only the isothermal surface is involved in heat transfer and Nu_0 must be based on that surface alone. Modifying Eqs. (13.4), (13.5), and (13.6) in this vein, the derivation of Nu_0 is found to parallel exactly the development of Eq. (13.8) and yield $Nu_0 = Ra'/12$ at the limit of $L \to \infty$, in agreement with [10].

A somewhat different approach must be taken when channel asymmetry is associated with different plate temperatures. Defining an asymmetry parameter r_T,

$$r_T \equiv \frac{T_{w,1} - T_0}{T_{w,2} - T_0} \tag{13.9}$$

Aung [7] found the average Nu_0, based on the area of both plates, to be given by the values indicated in Table 13.1.

It is interesting to note that the variation in $\overline{Nu}_0$ resulting from heating asymmetry is thus quite modest (less than a 7% variation over the full range of r_T). Furthermore, when Nu_0 is based on the single heated surface, there is apparently little difference between the Nu_0 associated with the second surface being at ambient temperature ($Nu_0 = 0.0889 \, Ra'$) or insulated ($Nu = 0.0833 \, Ra'$, as before).

13.3.4 Nusselt Number–Symmetric, Isoflux Plates

In a channel formed by two constant heat flux plates, the fluid temperature increases linearly along the channel and Eq. (13.5) can be replaced by

$$T_f = T_0 + \frac{2q''x}{wc_p} \tag{13.10}$$

Since, in many electronic cooling applications, it is the maximum channel wall temperature that is of critical importance, it is desirable to define the Nusselt number in the isoflux configuration according to

$$Nu_0 \equiv \left(\frac{q''}{T_{w,L} - T_0} \right) \frac{b}{k} \tag{13.11}$$

From basic heat transfer considerations and Eq. (13.10), the defining temperature difference in Nu_0 is found as

$$T_{w,L} - T_0 = (T_{w,L} - T_f) + (T_f - T_0) = q'' \left(\frac{1}{h} + \frac{2L}{wc_p} \right) \tag{13.12}$$

Using Eq. (13.10) to find the height-averaged fluid temperature in the channel and combining Eqs. (13.3), (13.11), and (13.12) yields

$$Nu_0 = \left(\frac{1}{h} + \frac{2L}{c_p \sqrt{\rho^2 g \beta b^3 q'' L / 12 \mu c_p}} \right)^{-1} \frac{b}{k} \tag{13.13}$$

TABLE 13.1 $\overline{Nu}_0$ Values for Asymmetric, Isothermal Channel [7]

r_T	$\overline{Nu}_0 / Ra'$
1.0	1/24
0.5	17/405
0.1	79/1815
0.0	2/45

and, following algebraic manipulation,

$$\text{Nu}_0 = \left(\frac{k}{bh} + \sqrt{\frac{48 \mu L k^2}{\bar{\rho}^2 g \beta b^5 q'' c_p}} \right)^{-1} \tag{13.14}$$

The combination of parameters under the square root sign in Eq. (13.14) is recognizable as the inverse of the modified channel Rayleigh number, Ra''. For the large values of L and small values of b appropriate to the fully developed limit, the first term in Eq. (13.14) is negligible relative to the square-root term and the sought-after limiting expression is thus found to equal

$$\text{Nu}_0 = \sqrt{\frac{\text{Ra}''}{48}} = 0.144 \sqrt{\text{Ra}''} \tag{13.15}$$

This result is identical to that obtained in previously cited studies and was found in [7] to apply as well to various ratios of surface heat flux, that is, q_1''/q_2'', when Ra'' is based on the average value of q''.

When, as is sometimes the case in experimental studies, the Nu_0 is defined in terms of the mid-height (or approximately average) wall temperature, the above development yields

$$\text{Nu}_0 = \frac{q''}{T_{w,L/2} - T_0} \frac{b}{k} = \sqrt{\frac{\text{Ra}''}{12}} = 0.289 \sqrt{\text{Ra}''} \tag{13.16}$$

13.3.5 Nusselt Number—Asymmetric, Isoflux Plates

When the vertical channel under consideration is formed by an insulated plate on one side, the vertical temperature gradient in the fluid is half that indicated in Eq. (13.10). Modifying the development to reflect this change and proceeding as before, the limiting channel Nusselt number based on the maximum wall temperature is found to equal

$$\text{Nu}_0 = \sqrt{\frac{\text{Ra}''}{24}} = 0.204 \sqrt{\text{Ra}''} \tag{13.17}$$

in agreement with [9].

Alternatively, the Nu_0 based on the midheight temperature is expressible as

$$\text{Nu}_0 = \sqrt{\frac{\text{Ra}''}{6}} = 0.41 \sqrt{\text{Ra}''} \tag{13.18}$$

13.4 COMPOSITE RELATIONS FOR NATURAL CONVECTION AIR COOLING

When a function is known to vary smoothly between two limiting expressions, which are themselves well defined, and when solution for intermediate values of the function is either difficult to obtain or involves other tabulated functions, an approximate composite relation can be obtained by appropriately summing the two limiting expressions. Churchill and Usagi [12] have suggested that the frequently employed linear superposition be viewed as a special case of a more general summation of the form

$$y = [(Az^p)^n + (Bz^q)^n]^{1/n} \tag{13.19}$$

where

$y \to Az^p$ as $z \to 0$
$y \to Bz^q$ as $z \to \infty$
$n > 0$ if $p < q$
$n < 0$ if $p > q$

For the natural convection problem under consideration, the Nu_0 variation takes the form of $C_1 Ra'$ or $C_2 \sqrt{Ra''}$ for small values of the gap Rayleigh number and for large Rayleigh numbers increases toward the isolated plate limit where, in laminar flow,

$$Nu_L = C_3 (Ra^*)^{1/4} \tag{13.20a}$$

for an isothermal surface and

$$Nu_L = C_4 (Ra^*)^{1/5} \tag{13.20b}$$

for an isoflux surface. Multiplying both sides of Eqs. (13.20) by b/L, this isolated plate relation can be converted to a relation between the gap Nusselt number, hb/k, and the gap Rayleigh number, Ra' and Ra'', for isothermal and isoflux plates, respectively.

Applying Eq. (13.19) to natural convection in channels, it might thus be anticipated that the Nu_0 would vary according to

$$Nu_0 = [(C_1 Ra')^{-n} + (C_3 \sqrt[4]{Ra'})^{-n}]^{-1/n} \tag{13.21a}$$

for isothermal plates and

$$Nu_0 = [(C_2 \sqrt{Ra''})^{-n} + (C_4 \sqrt[5]{Ra''})^{-n}]^{-1/n} \tag{13.21b}$$

for isoflux plates. The correct, or most nearly correct, value of the correlating exponent n can be evaluated by comparing Eq. (13.21) with experimental data or computed values of Nu_0. The authors of [12] suggest that this can best be done by rewriting Eq. (13.19) in the form

$$\frac{y}{Az^p} = \left[1 + \left(\frac{Bz^{q-p}}{A}\right)^n\right]^{1/n}$$

or $\quad Y = (1 + Z^n)^{1/n} \tag{13.22}$

and comparing data with curves of Y or Z or Y/Z versus $1/Z$ for fixed values of n. Alternatively, when the precise value of Y is known at $Z = 1$, n can be determined directly as log 2/log Y.

13.4.1 Symmetric, Isothermal Plates

As noted previously, Nu for laminar free convection on isothermal surfaces is dependent on $(Ra')^{1/4}$, and configurational variations are generally reflected in different values of the coefficient C_3. For moderately short vertical plates in air and $10^4 < Ra < 10^9$, McAdams [14] reports C_3 to equal 0.59. This expression, together with the fully developed Nu_0 relation derived earlier, can be inserted into Eq. (13.21) to yield

$$Nu_0 = [(\tfrac{1}{24} Ra')^{-n} + (0.59 \sqrt[4]{Ra'})^{-n}]^{-1/n} \tag{13.23}$$

The Elenbaas [4] correlation for this same thermal configuration takes the form

$$Nu_0 = \tfrac{1}{24} Ra'(1 - e^{-35/Ra'})^{3/4} \tag{13.24}$$

Following Churchill and Usagi [12], Eqs. (13.23) and (13.24) are compared on a Y versus Z plot in Fig. 13.2 and are seen to yield a close fit when the correlating exponent n equals approximately 2. Furthermore, a very similar value ($n = 2.09$) is obtained when n is evaluated specifically at $Z = 1$. It may therefore be expected that Eq. (13.23) with $n = 2$ would offer very good agreement with Elenbaas' experimental results.

$$Nu_0 = \left[\frac{576}{(Ra')^2} + \frac{2.873}{\sqrt{Ra'}} \right]^{-1/2} \tag{13.25}$$

This assertion and the utility of the Churchill and Usagi approach is borne out by the close proximity of the Elenbaas data points to the composite relation Eq. (13.25) and the asymptotic equations at both limits, indicated in Fig. 13.3.

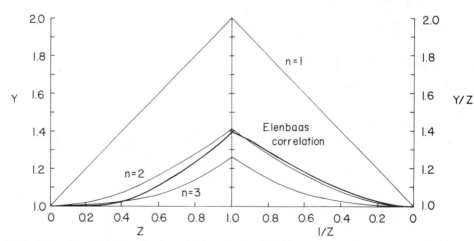

FIG. 13.2 Working plot for determining the correlating exponent for a symmetrical isothermal plate.

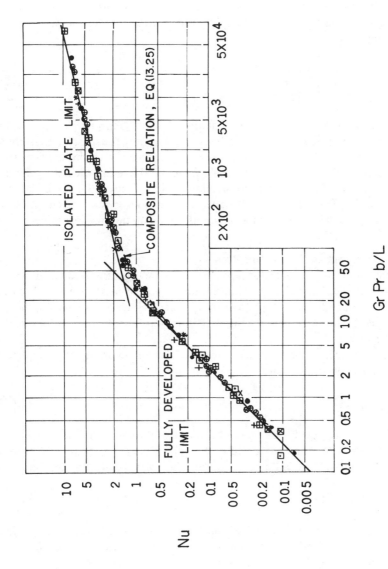

FIG. 13.3 Nusselt number variation for symmetric isothermal plates [4].

13.4.2 Asymmetric, Isothermal Plates

For vertical channels formed by an isothermal plate and an insulated plate, the asymptotic limits were previously shown to be $Nu_0 = \frac{1}{12} Ra'$ for $Ra' \to 0$ and $Nu_0 = 0.59 Ra'^{1/4}$ for $Ra' \to \infty$. Inserting these limiting expressions into Eq. (13.21), and assuming that despite channel asymmetry the symmetric correlating exponent $n = 2$ applies to this configuration as well, the composite relation for asymmetric isothermal plates is found to be

$$Nu_0 = \left[\frac{144}{(Ra')^2} + \frac{2.873}{\sqrt{Ra'}} \right]^{-1/2} \tag{13.26}$$

Comparison of Eq. (13.26) with the limited data of Nakamura [15] reported in [10] and the numerical solution of Miyatake and Fujii [10], as in Fig. 13.4, shows Eq. (13.26) to offer near-excellent agreement with the data and to improve somewhat on the predictive accuracy of the numerical solution in the region where Nu_0 displays the effects of both fully developed and developing flow. Figure 13.4 and Eq. (13.26) also reveal the Nu_0 from the thermally active surface in an asymmetric channel to be higher than a comparable surface in a symmetric configuration for low values of Ra'.

13.4.3 Symmetric, Isoflux Plates

Much of the available Nu data for heat transfer from channels formed by isoflux plates is presented in terms of the temperature difference between the wall at the channel midheight and the inlet air [e.g., 6, 8]. The analytic Nu_0 expression for the fully developed limit, appropriate to this definition, was found in the previous section to equal $\sqrt{Ra'}/12$.

Natural convection heat transfer from an isolated uniform, vertical plate is generally correlatable in the form

$$Nu_x = C_4 (Ra^*)^{1/5} \tag{13.27}$$

where, following [16], C_4 is often taken to equal 0.519 for air. Algebraically manipulating this equation to relate Nu, based on the midheight temperature difference, to the modified (isoflux) channel Rayleigh number, the isolated plate limit is found to be

$$Nu_0 = 0.60 (Ra'')^{1/5} \tag{13.28}$$

However, most reported investigations [e.g., 6, 8, 9] of channel height transfer have found this relation to underpredict the asymptotic heat transfer rate at large plate spacings. Examination of these results suggests that

$$Nu_0 = 0.73 (Ra'')^{1/5} \tag{13.29}$$

be taken as the appropriate limit.

When the two asymptotic equations, Eqs. (13.16) and (13.29), are superimposed in the manner recommended by Churchill and Usagi, using $n = 2$ as found

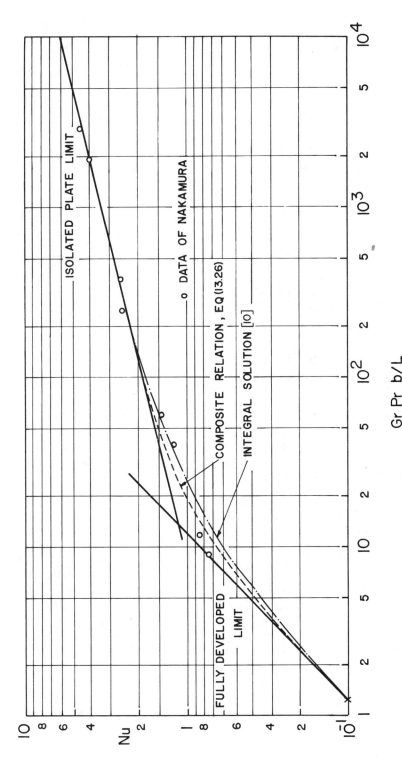

FIG. 13.4 Nusselt number variation for parallel plates, one isothermal and one insulated.

for the isothermal configuration, the composite relation for the isoflux configuration is

$$\text{Nu}_0 = \left[\frac{12}{\text{Ra}''} + \frac{1.88}{(\text{Ra}'')^{2/5}} \right]^{-1/2} \tag{13.30}$$

Comparison in Fig. 13.5 of Eq. (13.30) with typical data of Sobel et al. [6] and the results of the Engel and Mueller numerical calculations presented in [8] reveals the composite isoflux relation to have a high predictive accuracy, and no further adjustment of the correlating exponent appears to be necessary.

The larger-than-anticipated Nu values at the low Ra'' data points of Sobel et al. [6] may be explained by unaccounted for radiation and conduction losses at the channel exit, as noted by the authors.

13.4.4 Asymmetric, Isoflux Plates

When a vertical channel is formed by a single isoflux plate and an insulated plate, the desired composite relation for Nu based on the midheight temperature difference can be found by appropriately combining Eqs. (13.18) and (13.29) to yield

$$\text{Nu}_0 = \left[\frac{6}{\text{Ra}''} + \frac{1.88}{(\text{Ra}'')^{0.4}} \right]^{-1/2} \tag{13.31}$$

While Eq. (13.31) is, strictly speaking, valid only for natural convection heat transfer to air, a similar expression for other fluids can be derived by modifying Eqs. (13.27) and (13.29) to reflect the influence of Prandtl number on the heat transfer coefficient. Comparison of such a relation with the limited water data of Miyatake et al. [9] reveals similar agreement to that shown in Fig. 13.5.

13.5 OPTIMUM PLATE SPACING
FOR NATURAL CONVECTION

The composite relations derived in the previous section can be used to predict the value of the heat transfer coefficient for each of the four thermal configurations examined. No less important, however, is their potential use in optimizing the spacing between vertical, heat dissipating plates when it is desired to maximize total heat transfer from a given base area. The optimum spacing for parallel isothermal plates was found semiempirically by Elenbaas [4], and Aung [7] derived the optimum spacings for several other configurations. These results will be compared to the value to be determined in this section.

13.5.1 Symmetric, Isothermal Plates

The total heat transfer rate from an array of vertical plates, Q_T, is given by

$$Q_T = (2LS \, \Delta T_0)(m) \left(\frac{\text{Nu}_0 \, k}{b} \right) \tag{13.32}$$

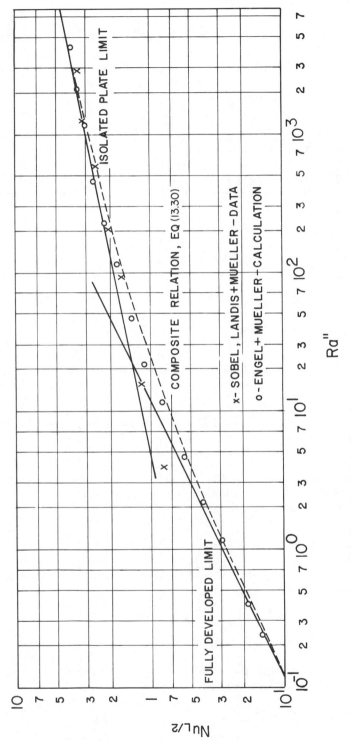

FIG. 13.5 Nusselt number variation for symmetric isoflux plates.

where m, the number of plates, equals $W/(b + d)$, b equals the spacing between adjacent plates, and d is the thickness of each plate.

Examination of Fig. 13.2 shows that the rate of heat transfer from each plate decreases as plate spacing is reduced. Since the total number of plates or total plate surface area increases with reduced spacing, Q_T may be maximized by finding the plate spacing at which the product of total plate surface area and local heat transfer coefficient is maximized. Based on his experimental results, Elenbaas determined that this optimum spacing for negligibly thick plates could be obtained by setting $Ra'_{opt} = 46$, yielding a Nu_0 of 1.2 [4].

Using Eq. (13.25) to determine Nu_0 and dividing both sides of Eq. (13.32) by the product of total fin area, temperature difference, thermal conductivity, and width of the base area yields

$$\frac{Q_t}{2LSW \, \Delta T_0 \, k} = (b + d)^{-1} b^{-1} \left(\frac{576}{P^2 b^8} + \frac{2.873}{P^{0.5} b^2} \right)^{-0.5} \tag{13.33}$$

where

$$P \equiv \frac{C_p (\bar{\rho})^2 g \beta \, \Delta T_0}{\mu k L}$$

Differentiating Eq. (13.33) with respect to b, setting the derivative to zero, and canceling common terms leads to

$$-(b + d)^{-1} - b^{-1} + \frac{1}{2} \left(\frac{576}{P^2 b^8} + \frac{2.873}{P^{0.5} b^2} \right)^{-1} \left(8 \frac{576}{P^2 b^9} + 2 \frac{2.873}{P^{0.5} b^3} \right) = 0 \tag{13.34}$$

Following additional algebraic operations, Eq. (13.34) is found to reduce to

$$(2b + 3d - 0.005 P^{1.5} b^7)_{opt} = 0 \tag{13.35}$$

Solution of Eq. (13.35) should now yield the value of b that maximizes Q_T, that is, the b_{opt} value.

In general, b_{opt} is seen to be a function of both the plate/air parameter P and the plate thickness d, but for negligibly thin plates,

$$b_{opt} = \frac{2.714}{P^{1/4}} \tag{13.36}$$

This result exceeds the Elenbaas optimum spacing by only 4% and yields optimum values of channel Rayleigh number and Nusselt number of 54.3 and 1.31, respectively.

In electronic cooling applications, it is often of interest to maximize the rate of heat transfer from individual plates or component-carrying printed circuit boards. This can be achieved by spacing the plates in such a manner that the isolated plate Nu prevails along the surface. To achieve this aim precisely requires an infinite plate spacing, but setting Nu (via Eq. (13.25)) equal to 0.99 of the isolated plate value yields $Ra' = 463$ and $b_{max} = 4.64/P^{1/4}$. This result is in general agreement with [18], where the identically defined maximum plate spacing was determined to occur at Ra' approximately greater than 600. It is of interest to note that at $Ra' = 600$, the composite Nu is found to reach 0.993 of the isolated plate value.

As might have been anticipated, the b_{max} spacing can be shown to correspond to approximately twice the boundary-layer thickness along each surface at the channel exit, that is, $x = L$. By comparison, b_{opt} corresponds to nearly 1.2 boundary-layer thicknesses at $x = L$.

13.5.2 Asymmetric, Isothermal Plates

In analyzing the asymmetric, isothermal configuration, Eq. (13.32) can again be used to calculate the total heat transfer from a given base area and to determine the optimum spacing between plates when m, the number of thermally active plates, is now set equal to $W/2(b + d)$. Proceeding as before, the governing relation for the optimum spacing is found to be

$$(2b + 3d - 0.02P^{1.5}b^7)_{opt} = 0 \qquad (13.37)$$

For negligibly thick plates, b_{opt} is then given by

$$b_{opt} = \frac{2.154}{P^{1/4}} \qquad (13.38)$$

At this optimum spacing, $Ra'_{opt} = 21.5$ and $Nu_{opt} = 1.04$.

To maximize the heat transfer rate from each individual, thermally active plate, it is again desirable to set the plate spacing such that fully developed flow does not develop in the channel and that, as a consequence, the isolated plate Nu limit is attained along the entire surface. Calculating via Eq. (13.26), Ra'_{max} at the 0.99 limit is found to equal approximately 184 and $b_{max} = 3.68/P^{1/4}$.

13.5.3 Symmetric, Isoflux Plates

When the boundary conditions along the surfaces of the parallel plates are identically or approximately equal to uniform heat flux, total heat transfer from the array can be maximized simply by allowing the number of plates to increase without limit. In most electric cooling applications, however, the plate, printed circuit board, or component surface must be maintained below a critical temperature and, as a consequence, plate spacing and Nu_0 values cannot be allowed to deteriorate to very small values.

Recalling the Nu_0 definition of Eq. (13.16) and rewriting Eq. (13.30), the relationship between the midheight temperature difference and the other parameters is found to be

$$\Delta T_{L/2} = \frac{q''b}{k}\left(\frac{12}{Ra''} + \frac{1.88}{Ra''^{2/5}}\right)^{1/2} \qquad (13.39)$$

Thus, when both the surface heat flux and the allowable temperature difference are specified, Eq. (13.39) can be used to solve for the requisite interplate spacing.

Alternatively, when only the heat flux is specified, it is of interest to determine the plate spacing yielding the lowest possible surface temperature. This condition corresponds to a spacing that is sufficiently large to avoid boundary-layer interference and, by the method described previously, is found to occur at Ra'' equal to approximately 17,000 and $b_{max} = 7.02R^{-0.2}$ with R defined after Eq. (13.40).

In distinction to the b_{max} value and the plate spacing obtained via Eq. (13.39), the optimum b value for an array of isoflux plates can be defined to yield the maximum volumetric (or prime area) heat dissipation rate per unit temperature difference. Thus, when Eq. (13.30) is used to evaluate Nu_0 in the Eq. (13.32) formulation of total array heat transfer, the optimizing equation for the symmetric, isoflux configuration takes the form

$$\frac{d}{db}\left(\frac{Q_T}{2LSW\,\Delta T_{L/2}k}\right) = \frac{d}{db}\left[(b^2 + db)^{-1}\left(\frac{12}{R}b^{-5} + \frac{1.88}{R^{0.4}}b^{-2}\right)^{-1/2}\right] = 0$$

$$(13.40)$$

where

$$R \equiv \frac{c_p \bar{\rho}^2 g\beta q''}{\bar{\mu}Lk^2}$$

Following differentiation, the governing relation for b_{opt} is found to be

$$\left(\frac{12}{R}b - \frac{3.76b^4}{R^{0.4}} + \frac{36}{R}d\right)_{opt} = 0 \tag{13.41}$$

For negligibly thick plates—that is, $d = 0$—the optimum isoflux plate spacing is

$$b_{opt} = 1.472R^{-0.2} \tag{13.42}$$

The value of Ra''_{opt} is thus 6.9, and Nu_0 at the optimum spacing is 0.62.

13.5.4 Asymmetric, Isoflux Plates

By analogy to the symmetric, isoflux configuration, the requisite plate spacing for specified values of q'' and $\Delta T_{L/2}$ on the thermally active surface can be obtained by appropriate solution of Eq. (13.31).

Similarly, Eq. (13.31) can be used to determine the lowest Ra'' at which the prevailing Nu_0 is indistinguishable from the isolated plate limit. This condition is found to occur at Ra'' equal to approximately 5400 and to yield the plate spacing required to obtain the lowest surface temperature, b_{max}, equal to $5.58R^{-0.2}$.

Finally, when the relation governing the total heat dissipation of an array of alternating isoflux and insulated plates is differentiated relative to the plate spacing and the derivative set equal to zero, the optimum value of b for this configuration can be found by solving

$$\left(\frac{6}{R}b - \frac{3.76b^4}{R^{0.4}} + \frac{18}{R}d\right)_{opt} = 0 \qquad (13.43)$$

For negligibly thin plates,

$$b_{opt} = 1.169R^{-0.2} \qquad (13.44)$$

The optimum modified channel Rayleigh number is thus 2.2, yielding a Nu_0 of 0.49.

13.6 DISCUSSION OF NATURAL CONVECTION

The preceding has established an analytical, albeit approximate, structure for determining the channel width, or spacing between surfaces forming a two-dimensional channel, appropriate to various thermal constraints for symmetric and asymmetric, isothermal and isoflux boundary conditions. With the relations presented and subject to the stated assumptions, it is thus possible to select the interplate spacing that will maximize heat transfer from the individual, thermally active surfaces or, alternatively, to choose the spacing that yields the maximum heat dissipation from the entire array. In the absence of a large body of verified experimental results, the agreement found between both the composite and optimum spacing relations for symmetric, isothermal plates and the classic Elenbaas [4] data serves to verify the credibility and engineering accuracy of the approach. Several noteworthy features of the composite and optimizing relations are discussed below.

13.6.1 Asymmetric Versus Symmetric, Fully Developed Limit

Comparison of the derived relations for the fully developed Nu_0 reveals the asymmetric value to exceed the symmetric value by a factor of 2 for isothermal surfaces and a factor of $\sqrt{2}$ for the isoflux condition. At first glance this experimentally verified result [9, 10] appears counterintuitive, since the thicker thermal boundary layer in the asymmetrically heated channel (equal to the interplate spacing) could be expected to yield *lower* heat transfer coefficients than encountered with the thinner boundary layers of the symmetrically heated configuration. Although this assertion is correct for Nusselt numbers based on the local wall-to-fluid temperature difference, it must be recalled that Nu_0 is defined in such a way as to include the temperature rise in the convecting air. As a result, Nu_0 can be expected to reflect the "helpful" influence of reduced heat addition in the asymmetric case and to yield the observed higher values.

13.6.2 Asymmetric Versus Symmetric Optimization Arrays

The higher Nu_0 to be expected in asymmetric configurations has led some thermal designers to suggest that, whenever possible, this configuration be preferred over a symmetric distribution of the heat dissipation on the array of parallel plates. Examination of the results for both maximum and optimum plate spacing reveals the error inherent in such an approach.

For isothermal plates, b_{max} was found to equal $4.64P^{-0.25}$ in the symmetric configuration and $3.68P^{-0.25}$ in the asymmetric configuration. Similarly, b_{max} equals $7.02R^{-0.2}$ for symmetric, isoflux plates and $5.58R^{-0.2}$ when the channel is formed by an isoflux plate and an adiabatic plate. Since the plate spacing required for maximum heat transfer from each surface in the asymmetric configuration is thus substantially greater than 50% of the symmetric value, the total dissipation of an asymmetric array subject to the same constraints must fall below the heat dissipation capability of a symmetric array.

Examination of the optimum plate spacing relations supports this conclusion. The optimum spacing between alternating isothermal and adiabatic plates is, indeed, found to be appropriately smaller (by approximately 26%) than the spacing between two identical isothermal plates. However, as a result of the consequent reduction in Nu_0 (from 1.31 to 1.04) and the "halving" of the thermally active surface area relative to symmetric isothermal plates, it can be shown that, for a given array base area or volume, an optimum array of negligibly thick isothermal plates alternating with insulated plates cannot dissipate more than 63% of the heat dissipated by an optimum array of isothermal plates. This finding is reinforced by the results obtained by Aung [7], which indicate that thermal asymmetry reduces total heat dissipation to approximately 65% of the comparable symmetric configuration when every second plate is at the ambient temperature.

Use of the derived optimum spacing and optimum Nu_0 values for symmetric and asymmetric isoflux channels yields a nearly identical reduction in total heat dissipation for the asymmetric configuration as encountered in isothermal plates.

13.6.3 Three-Dimensional Flow Effects

In the development of design equations for the spacing between isothermal and isoflux plates, no attempt has been made to address the influence of three-dimensional flow, that is, side inflow or lateral edge effects, on the anticipated Nu_0 values nor on the recommended optimum spacings. Clearly such effects can be expected to become progressively greater as the ratio of interplate spacings to channel height is reduced. In a recently published study by Sparrow and Bahrami [19], the lateral edge effects were found to be of no consequence for Ra' values greater than 10 but to produce deviations of up to 30% or more in the equivalent Nu_0 when Ra' was below 4. Consequently, whereas the asymptotic approach of the Elenbaas data [4] to the analytical two-dimensional fully developed flow limit—as shown in Fig. 13.2—may be fortuitous, there is little likelihood of three-dimensional flow effects in the Ra' region corresponding to the optimum and maximum interplate spacings that have been derived.

13.6.4 Summary

The complexity of heat dissipation in vertical parallel plate arrays encountered in electronic cooling applications frequently dissuade thermal analysts and designers from attempting even a first-order analysis of anticipated temperature profiles, and little theoretical effort has been devoted to thermal optimization of the relevant packaging configurations. This section has aimed at establishing an analytical structure for such analyses while presenting and verifying useful relations for heat distribution patterns identical to or approaching isothermal or isoflux boundary conditions.

13.7 FORCED CONVECTION COOLING IN CHANNELS

Whereas the heat transfer rates associated with natural convection suffice to meet the thermal requirements of a wide variety of electronic components, much of air cooling technology involves forced flow through channels formed by component-carrying cards or boards. This configuration can often be represented by a two-dimensional channel, as shown schematically in Fig. 13.6, in which steady-state heat transfer prevails. However, the geometric parameters defined in Fig. 13.6 can undergo significant variations. Thus, the channel length L may vary from approximately 15 to 50 cm; the spacing between the tops of opposing packages b from 0.5 to 2.5 cm; and often components are encountered on only one side of the channel. Package lengths as small as 0.5 cm and as large as 8 or even 10 cm are not uncommon; thicknesses t may vary from 0.2 to 1.0 cm; and adjacent packages may be almost contiguous, that is, $S = l$, or spaced so that only one package is bonded to each card. The basic thermal parameters, including the inlet air temperature T_∞, the surface temperature of the packages T_w, and the dissipated heat flux q_w'' are also generally determined by the specific application. The inlet temperature is typically found to vary from 20 to 110°C and the component surface temperature from 40 to 120°C. The surface heat fluxes may attain values as high as 2 W/cm^2, but more often equal 0.1 to 0.5 W/cm^2.

A detailed analysis of forced convection cooling of component-carrying cards via analytical expressions is beyond the state of the art, even for the two-dimensional idealization. Although a thermal map of sufficient accuracy can, in principle, be obtained by use of a numerical solution technique, such techniques often require specification of the local heat transfer coefficient. This parameter is, however, strongly influenced by the physical and thermal boundary conditions, as well as by the flow regime and state of boundary-layer development. Consequently, use of a Nusselt number obtained from fully developed pipe flow correlations and/or analysis may lead to grossly erroneous results, and care must be taken to base the analytical or numerical calculation on the value or values of the heat transfer coefficient appropriate to the thermofluid classification.

The variety in the geometry and packaging distribution of electronic components used in commercial and military equipment precludes a description of all the possible combinations of physical, thermal, and fluid parameters. It is neverthe-

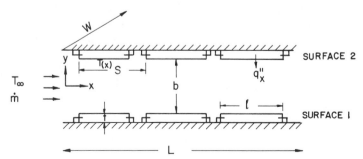

FIG. 13.6 Coordinate system for two-dimensional channel.

less, useful to classify, in somewhat idealized form, the various configurations according to:

Physical boundary condition (smooth, "sand grain" roughness, protuberance)
Thermal, circumferential boundary condition (symmetric, variable, one-side insulated)
Thermal, axial boundary condition (isothermal, isoflux, axial temperature profile, axial flux profile)
Flow regime (laminar, transition, turbulent)
Fluid temperature profile in channel (developing, fully developed)
Fluid velocity profile in channel (developing, fully developed)

This rather restricted, idealized configurational classification contains more than 430 distinct combinations, and it is clearly not possible to analyze each one in detail within the limits of this chapter. Instead, following a detailed exploration of thermal transport in a smooth-walled, symmetrically isothermal channel with fully developed laminar flow, that is, the basic channel, attention will be turned to the more common variations, including the particular contribution of turbulent flow, an insulated wall, developing flow, and delayed initiation of thermal boundary-layer growth.

13.7.1   Thermal Analysis of Basic Channel

FLUID MECHANICS

The flow of a relatively cool fluid between two heated parallel plates results in the formation and growth of momentum and temperature boundary layers along each plate surface (see Chap. 6). In laminar flow under steady-state conditions at a sufficiently downstream location to ensure that the respective boundary layers from opposite channel walls have merged, momentum conservation and flow continuity equations can be used to establish the relationship between the velocity and pressure fields in the channel [13] :

$$\frac{\partial^2 v}{\partial y^2} = (\mu)^{-1} \frac{dP}{dx} \tag{13.45}$$

For this fully developed flow, the velocity profile is symmetric about the channel centerline and attains a zero value at the channel side walls. The velocity profile for this so-called Poiseuille flow can be obtained by integrating Eq. (13.45) and using the boundary conditions

$$y = 0 \qquad \frac{\partial v}{\partial y} = 0$$
$$y = y_0 \qquad v = 0 \tag{13.46}$$

to evaluate the constants of integration, yielding

$$v = -\frac{1}{2\mu} \frac{dP}{dx} (y_0^2 - y^2) \tag{13.47}$$

or, with the average velocity V defined by

$$V \equiv \frac{1}{y_0} \int_0^{y_0} v \, dy \tag{13.48}$$

the pressure drop in the channel per unit length can be shown to be

$$\frac{\Delta P}{L} = \frac{3\mu V}{y_0^2} \tag{13.49}$$

In the analysis of flow in tubes and channels (see Chaps. 6 and 11), it is common to define a friction factor f relating the pressure drop to the velocity head according to[1]

$$f \equiv \frac{\Delta P/L}{4(\rho V^2/2)(1/D_e)} \tag{13.50}$$

Substituting the pressure drop from Eq. (13.49) and recalling the definition of the hydraulic diameter D_e,

$$D_e \equiv \frac{4A}{P} = 4y_0 \qquad \text{for a parallel-plate channel} \tag{13.51}$$

the friction factor for this configuration is found to be

$$f = \frac{24\mu}{\rho V D_e} = \frac{24}{\text{Re}} \tag{13.52}$$

It is interesting to note that in round pipes with fully developed laminar flows, $f = 16/\text{Re}$.

TEMPERATURE FIELD

In a channel where both the velocity and temperature boundary layers have merged, the temperature profile is invariant in the axial direction. Thus, following Rohsenow and Choi [13],

$$\frac{T_w - T}{T_w - T_m} = f(y) \tag{13.53}$$

$$\frac{\partial}{\partial x}\left(\frac{T_w - T}{T_w - T_m}\right) = 0 \tag{13.54}$$

---

[1] Other definitions exist. One of them omits the 4 in the denominator, yielding values of f four times greater than given by Eq. (3.50) and making $f = 64/\text{Re}$ for fully developed laminar flow.

where the mean fluid temperature T_m is defined according to

$$T_m \equiv \frac{\int_0^{y_0} vT2\pi r\, dr}{\int_0^{y_0} v2\pi r\, dr} \tag{13.55}$$

Although the temperature profile is presumed to be axially invariant, the individual temperatures (i.e., wall, local fluid, and mean fluid) may vary in the x direction subject to the constraints established by the expansion of Eq. (13.54):

$$\left(\frac{\partial T_w}{\partial x} - \frac{\partial T}{\partial x}\right) - \left(\frac{T_w - T}{T_w - T_m}\right)\left(\frac{\partial T_w}{\partial x} - \frac{\partial T_m}{\partial x}\right) = 0 \tag{13.56}$$

For fully developed, laminar flow at low or moderate velocities, the energy balance on a differential element yields [13]

$$\frac{\partial^2 T}{\partial y^2} = \frac{v}{x}\frac{\partial T}{\partial x} \tag{13.57}$$

When the channel walls are isothermal, that is, $\partial T_w/\partial x = 0$, the axial temperature gradient is found via Eq. (13.56) to be

$$\left(\frac{\partial T}{\partial x}\right)_T = \frac{T_w - T}{T_w - T_m}\frac{\partial T_m}{\partial x} \neq f(x) \tag{13.58}$$

Alternatively, when both channel walls dissipate an equal and uniform heat flux, Newton's cooling law with a uniform heat transfer coefficient leads to a uniform wall-to-fluid mean temperature difference, that is, $T_w - T_m$, along the channel and, thus, via Eqs. (13.54) and (13.56),

$$\frac{\partial T}{\partial x} = \frac{\partial T_w}{\partial x} = \frac{\partial T_m}{\partial x} \neq f(x) \tag{13.59}$$

HEAT TRANSFER COEFFICIENT

The convective heat transfer coefficient defined by Newton's cooling law was shown in Chap. 6 to be expressible in terms of the transverse fluid temperature gradient at the wall; that is,

$$h \equiv \frac{q''}{T_w - T_m} = \frac{-k(\partial T/\partial y)_w}{T_w - T_m} \tag{13.60}$$

Differentiation of the fully developed temperature profile, Eq. (13.53), shows the wall temperature gradient to be

$$\left(\frac{\partial T}{\partial y}\right)_w = -(T_w - T_m)[f'(y)]_w \tag{13.61}$$

Combining Eqs. (13.60) and (13.61), h is seen to be

$$h = k[f'(y)]_w \qquad (13.62)$$

and thus to be invariant in the axial direction.

More precisely, when Eq. (13.57) is integrated with the velocity profile derived in Eq. (13.47) and the axial gradient obtained via Eq. (13.58) to provide the transverse fluid temperature gradient at an isothermal channel surface, the Nusselt number is found to be

$$Nu_T \equiv \frac{hD_e}{k} = -D_e \frac{(\partial T/\partial y)_w}{T_w - T_m} = 7.6 \qquad (13.63)$$

Similarly, for an isoflux channel with Eq. (13.59) for the axial gradient,

$$Nu_H = 8.23 \qquad (13.64)$$

HEAT TRANSFER/FRICTION ANALOGY

In analyzing transport phenomena, it is often convenient to utilize the Stanton number, defined as

$$St \equiv \frac{Nu}{Re\,Pr} \qquad (13.65)$$

For the isothermal channel, the product of St and Pr thus equals

$$(St\,Pr)_T = \frac{Nu_T}{Re} = \frac{7.6}{Re} \qquad (13.66)$$

whereas for the isoflux channel,

$$(St\,Pr)_H = \frac{Nu_H}{Re} = \frac{8.23}{Re} \qquad (13.67)$$

The friction factor for this configuration was previously found to equal $24/Re$; consequently, in laminar, fully developed channel flow,

$$St\,Pr \approx \frac{f}{3} \qquad (13.68)$$

It is significant to note that a similar calculation for flow in a circular tube yields $St\,Pr \approx f/4$.

13.7.2 Turbulent, Fully Developed, Smooth Channel Flow

FLUID MECHANICS

In turbulent flow, eddy motion and velocity fluctuations increase the shear stress above the values associated with laminar flow. It is convenient to relate this increase to a so-called momentum eddy diffusivity ϵ_m, and thus to express the relationship between the shear stress and the velocity gradient as

$$\frac{\tau}{\rho} = (\nu + \epsilon_m) \frac{dv}{dy} \tag{13.69}$$

A simple force balance on the control volume shown in Fig. 13.7 indicates that the wall shear stress equals

$$\tau_w = \frac{2y_0 \, \Delta\rho}{2L} \tag{13.70}$$

Inserting the definition of the friction factor expressed in Eq. (13.47) with $D_e = 4y_0$ yields

$$\tau_w = \left(\frac{f}{2}\right)\rho V^2 \tag{13.71}$$

The friction factor in turbulent pipe flow was found by Nikuradse [20] to vary as

$$(f)^{-1/2} = 4 \log(\mathrm{Re}\sqrt{f}) - 0.4 \tag{13.72}$$

In the midrange of the turbulent regime, $3 \times 10^4 < \mathrm{Re} < 5 \times 10^5$, Eq. (13.72) can be well approximated by

$$f = 0.046 \, \mathrm{Re}^{-0.2} \tag{13.73}$$

Both Eq. (13.72) and Eq. (13.73) can be applied as well to turbulent flow in channels formed by parallel plates when the Reynolds number is based on the hydraulic diameter D_e [13].

HEAT TRANSFER COEFFICIENT

The heat flux in the transverse (y) direction for turbulent flow in a pipe can be expressed as

$$\frac{q''}{\rho c_p} = -(\alpha + \epsilon_h) \frac{dT}{dy} \tag{13.74}$$

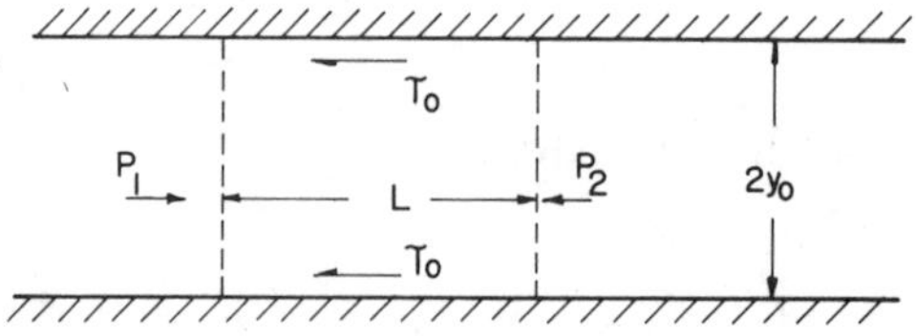

FIG. 13.7 Force balance on control volume in a channel.

where ϵ_h is the thermal eddy diffusivity and is analogous to ϵ_m introduced in Eq. (13.69). Dividing this relation by the turbulent shear stress equation, a relation between q'' and τ is found as

$$\frac{q''}{c_p \tau} = \frac{\alpha + \epsilon_h}{\nu + \epsilon_m} \frac{\partial T}{\partial \upsilon} \tag{13.75}$$

For gases, where Pr is approximately unity, it can be assumed that $(\alpha + \epsilon_h)/(\nu + \epsilon_m)$ is near unity. Furthermore, both q'' and τ are found to vary linearly with y, and their ratio is thus found to be independent of the transverse coordinate [13]. Introducing these two simplifications into Eq. (13.75), separating variables, and integrating from the wall to the point in the channel where the velocity is equal to V, the mean velocity, and the temperature equals T_m, that is

$$\frac{q''_w}{c_p T_w} \int_w^m d\upsilon = - \int_w^m dT \tag{13.76}$$

yields

$$\frac{q''_w V}{c_p T_w} = T_w - T_m \tag{13.77}$$

or, with $q''_w = h(T_w - T_m)$ and τ_w from Eq. (13.71),

$$\frac{h}{\rho c_p V} = \frac{f}{2} \tag{13.78}$$

The left side of this relation is identical to the Stanton number, and it is thus apparent that the relationship between St and the friction factor in turbulent flow differs from that encountered in laminar flow, where $St \approx f/3$.

Equation (13.78) was derived on the assumption of equality between $\alpha + \epsilon_h$ and $\nu + \epsilon_m$. Experimental results suggest that the ratio of the turbulent diffusivities, that is, $(\alpha + \epsilon_h)/(\nu + \epsilon_m)$, is, in fact, a function of the Prandtl number [13], which for conventional fluids may be approximated by $Pr^{-2/3}$. Consequently, for Pr values other than unity, the relationship between the heat transfer coefficient and the friction factor can be better stated as

$$\frac{h}{\rho c_p V} Pr^{2/3} = \frac{f}{2}$$
$$St\, Pr^{2/3} = \frac{f}{2} \tag{13.79}$$

Introducing the definition of the Stanton number and the empirical relation for the friction factor stated in Eq. (13.73) into Eq. (13.79), the Nusselt number for turbulent pipe flow is found to equal

$$Nu = 0.023\,Re^{0.8}\,Pr^{1/3} \qquad\qquad\qquad (13.80)$$

This equation is identical with the widely used Colburn correlation for pipe flow [21] and differs from the familiar McAdams correlation [14] only by a slightly lower (0.33 versus 0.4) Pr exponent.

In turbulent, as opposed to laminar flow of conventional fluids, the boundary conditions exert only a very minor influence on the surface heat transfer coefficient. This behavior can be related to the differences between the laminar and turbulent transverse temperature profile in the fluid, and makes it possible to use a single Nu correlation for both isothermal and isoflux boundary conditions for all Pr greater than 0.7, with a maximum anticipated deviation of less than a few percent. The occurrence of the primary fluid thermal resistance in the region immediately adjacent to the surface also facilitates the application of Eq. (13.80) to most conduit geometries, as well as to turbulent flow between parallel plates, when care is taken to base both the Nusselt and Reynolds numbers on the hydraulic diameter D_e.

As a further refinement and following the discussion in Sec. 6.3, Eq. (13.80) can be translated to flow in a two-dimensional channel via the average two-ray length, yielding (with $D_e = 2b = 4y_0$)

$$\frac{hb}{k} = 0.0194 \left(\frac{\rho V b}{\mu} \right)^{0.8} Pr^{1/3} \qquad\qquad (13.81)$$

13.7.3 Asymmetrically Heated Channel

In the analysis and design of air-cooled card stacks, it is not uncommon to encounter heating asymmetries in the intercard channel. The heat transfer coefficients associated with axially invariant but different heat fluxes on each of the two surfaces and similar variations for axially invariant isothermal surfaces have been studied in detail by Lundberg et al. [11] as part of a comprehensive examination of asymmetric heating in annuli. The results of this study were presented in terms of the Nusselt number on the inner surface, Nu_i, and the outer surface, Nu_o, of the annular space (see Fig. 13.8) at different radii ratios, r^*, equal to r_i/r_o. The reported Nusselt number values at $r^* = 1$ correspond to the parallel plate channel and can be applied to the card stack configuration.

Several thermal configurations involving imposed heat flux asymmetries will be

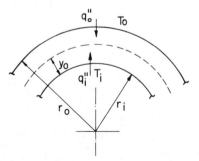

FIG. 13.8 Coordinate system for annular channel.

examined in succeeding subsections. In all of these, the results will be presented in terms of the Nusselt numbers at surfaces 1 and 2, that is,

$$\text{Nu}_1 = \frac{h_1 D_e}{k}$$

$$\text{Nu}_2 = \frac{h_2 D_e}{k}$$

(13.82)

where h_1 and h_2 are themselves based on the difference between the local surface temperature and the local mean fluid temperature.

Following [11] the theoretical equations for Nu_1 and Nu_2 will be expressed in the form

$$\text{Nu}_1 = \frac{\text{Nu}_{11}}{1 - (q_2''/q_1'')\theta_1^*}$$

(13.83)

where Nu_{11} and θ_1^* are tabulated in [11, 22]. For a parallel-plate channel, $\text{Nu}_{22} = \text{Nu}_{11}$ and $\theta_1^* = \theta_2^*$. In Eq. (13.83), Nu_{11} can be seen to correspond to the Nu on surface 1 when only that surface is heated (i.e., $q_2'' = 0$) and θ_1^* may be interpreted to represent the influence coefficient of surface 2 on the Nu at surface 1.

FULLY DEVELOPED LAMINAR FLOW

For fully developed laminar flow in a smooth parallel-plate channel, the values of the Nusselt numbers can be expressed by [22]

$$\text{Nu}_1 = \frac{5.385}{1 - 0.346(q_2''/q_1'')}$$

(13.84)

Thus, when the channel is heated symmetrically ($q_2'' = q_1''$), $\text{Nu}_1 = 8.23$ as in Eq. (13.64), but when one channel wall is insulated ($q_2'' = 0$), the nondimensional heat transfer coefficient at the active surface decreases to 5.385.

As a consequence of the heat transfer coefficient being defined in terms of the local wall-to-mean fluid temperature difference, Eq. (13.84) can yield both infinite and negative Nu values. Thus, for $q_2'' = 2.9q_1''$, the denominator in Eq. (13.84) equals zero and Nu_1 is found to be infinitely large. A more exacting analysis, however, reveals that for this ratio of heat fluxes while surface 2 is above the mean fluid temperature, the temperature of surface 1 is precisely equal to the mean fluid temperature, and the temperature difference is then equal to zero. Similarly, for even larger heat flux ratios, Nu_1 as well as $T_1 - T_m$ become negative.

FULLY DEVELOPED TURBULENT FLOW

In fully developed turbulent flow, the values of Nu_{11} and θ_1^* are dependent on both the Prandtl number and the Reynolds number. Their progression from laminar to highly turbulent flow of air (Pr = 0.7) is depicted in Table 13.2, based on values presented in [22]. Not surprisingly, Nu_{11} —representing the Nusselt number when only one surface dissipates heat and the other surface is insulated—is seen to increase with

TABLE 13.2 Nu_{11} and θ_1^*, for Turbulent Flow of Air [22]

Re	Nu_{11}	θ_1^*
Laminar	5.385	0.346
10^4	27.8	0.220
3×10^4	61.2	0.192
10^5	155	0.170
3×10^5	378	0.156
10^6	1030	0.142

increasing Re. Alternatively, the influence coefficient, θ_1^*, is seen to decrease at higher values of Re, reflecting the diminishing influence of surface 2 on surface 1 as Re increases and the boundary layers on each surface become thinner.

To illustrate this point, it may be noted that whereas in laminar flow the ratio between Nu for symmetric heating and one insulated wall was 1.53, for Re = 10^5 the ratio is 1.21 and at Re = 10^6 this ratio drops to 1.17.

13.7.4 Developing Temperature and Velocity Profiles

In contrast to most heat exchangers in chemical process equipment and thermal power stations, the channels formed by electronic component cards or heat sink structures are generally relatively short and the ratio of the channel length to the spacing between the cards is rarely larger than 100. Distinct temperature and velocity boundary layers exist over much of the length of such channels and, especially for laminar flow, the fully developed solutions cannot be expected to apply. Rather, solutions for developing temperature and velocity profiles must be sought.

LAMINAR FLOW, ISOFLUX WALLS

The local Nusselt numbers to be expected during developing laminar flow between isoflux walls were determined by Heaton et al. [23] and are summarized in Table 13.3 using the form of Eq. (13.83) for Pr = 0.7 (representing air). As can be seen in Table 13.3, the influence coefficient θ_1^* increases and the value of Nu_{11} decreases as the boundary layers grow along the channel walls.

For values of X^+ greater than 0.2, the values of Nu_{11} and θ_1^* are seen to corre-

TABLE 13.3 Nu_{11} and θ_1^* for Developing Laminar Flow of Air Through Isoflux Channel (Pr = 0.7) [23]

X^{+a}	Nu_{11}	θ_1^*
0.002	18.5	0.037
0.010	9.62	0.096
0.020	7.68	0.154
0.10	5.55	0.327
0.20	5.40	0.345
∞	5.39	0.346

$^a X^+ = 2(x/D_e)/Re\,Pr.$

spond to the fully developed values presented in Table 13.2. Using the definition of X^+, it is thus apparent that for a typical laminar Re of 1000, developing flow effects persist up to x/b ratios of approximately 140. Consequently, local Nusselt numbers for air channels encountered in electronic cooling applications can be expected to exceed significantly the fully developed values.

As an example, consider the channel shown in Fig. 13.6, formed by two multi-layer boards carrying flat packs, but in this case dissipating heat from just one side while the other side is insulated. At the center of the first flat pack in the channel at $X^+ = 0.002$, corresponding to $x/b = 1.4$ at Re $= 10^3$, the Nusselt number can be expected to reach 18.5 versus 5.385 in a location where fully developed flow has been achieved. Similarly, for both channel walls dissipating the same heat flux, $Nu_1 = Nu_2 = 19.2$, which is 2.33 times the fully developed value for an isoflux channel. Although the value of Nu decreases toward the fully developed value at increasingly more interior points, it should be noted that for Re $= 10^3$, even at $x/b = 28$, the Nu for an isoflux channel is still 10% higher than the fully developed value.

LAMINAR FLOW, ISOTHERMAL WALLS

The variation of the Nusselt number along a channel formed by isothermal walls was investigated in detail by Sparrow [24]. The resulting dependence of the mean Nu on the nondimensional channel length for Pr $= 0.72$ (air) is depicted in Fig. 13.9. As in the case of the isoflux channel, the fully developed Nu is achieved asymptotically but is approached at an x/D_e Re Pr value of approximately 0.1, corresponding to a channel length-to-spacing ratio of more than 140. Alternatively, at a more common value of x/b in air-cooled electronic assemblies of approximately 10, the mean Nu can be expected to average nearly 10 or 33% higher than the fully developed value of 7.6.

TURBULENT FLOW

Velocity and temperature profile development in turbulent flow generally occurs relatively close to the channel inlet, and at an intermediate, turbulent Re of 10^5 an

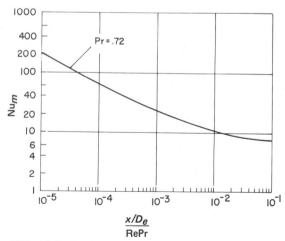

FIG. 13.9 Variation of mean Nusselt number for laminar flow in entrance region between parallel plates [24].

x/b ratio of 25 can be expected to ensure fully developed heat transfer coefficients. However, as shown in Fig. 13.10, at lower Reynolds numbers the transition to fully developed, turbulent air flow occurs at progressively larger x/b values, and Nu significantly above the values calculated via Eqs. (13.80) or (13.81) can be expected to prevail on the surfaces of heat-dissipating components at $\text{Re} < 10^4$ located at x/b values less than or equal to 30.

13.7.5 Wall Roughness and Protuberances

Although the presence of roughness elements and/or protuberances on the channel surfaces, as are almost always present on electronic component-carrying cards or boards, can be expected to have only a minor influence on the laminar flow heat transfer coefficients, in turbulent flow this influence can be most considerable. It is common to relate the differences between laminar and turbulent flow in this regard to the relative fragility of the laminar sublayer, which constitutes the major thermal resistance in turbulent flow. Unlike the laminar boundary layer itself, the sublayer in turbulent flow can be easily disturbed by surface discontinuities, resulting in substantially improved heat transfer coefficients and an accompanying increase in the channel friction factor.

The presence of protuberances on the surface results not only in local thinning of the boundary layer but may well lead to boundary-layer separation, secondary flows, and the initiation of new boundary layers at downstream locations. The increased wall friction factor associated with flow past roughened surfaces may thus reflect both increased wall shear stress (thinner and new boundary layers) and form drag (separated, secondary flows) around the elements protruding into the flow. However, although increased shear stress may well be accompanied by a parallel improvement in the heat transfer coefficient—reasoning via the Reynolds analogy and Eq. (13.79)—the increased pressure drop in the channel resulting from form drag is of little or no thermal consequence. Combining these two effects, it would appear

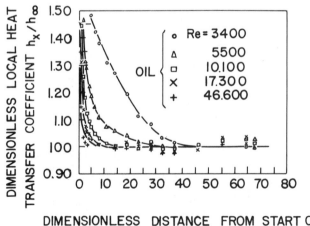

FIG. 13.10 Entrance effect on heat transfer coefficient [25].

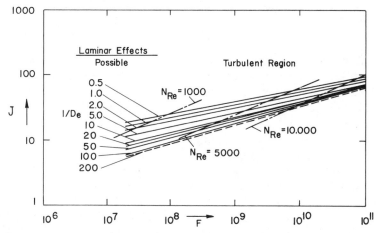

FIG. 13.11   Idealized performance map for air flow in rough channels [27].

likely that the roughness-induced enhancement of the heat transfer coefficient would generally fall below the increase in the pressure drop or friction coefficient for the same channel (see also Chap. 19).

Experimental results for small-scale roughness elements, metal rings, and wires support this assertion, and in the flow of air through pipes, Nunner [26] found the relationship

$$\frac{\text{Nu}}{\text{Nu}_{\text{smooth}}} = \left(\frac{f}{f_{\text{smooth}}}\right)^{0.5} \quad \text{for } 1 < \frac{f}{f_{\text{smooth}}} < 10$$

$$5 \times 10^3 < \text{Re} < 8 \times 10^4 \tag{13.85}$$

A more general approach to the prediction and documentation of the relationship between the heat transfer and friction coefficients is offered by LaHaye et al. [27]. In this work the flow length l between major boundary-layer disturbances was found to be the key parameter characterizing roughened surfaces. Using the ratio of this flow length to the hydraulic diameter, it was shown possible to present the results for turbulent flow past all surface geometries on a single idealized performance map shown in Fig. 13.11. These results, expressed in terms of a heat transfer factor J and a pumping power F, are based on convective heat exchange between various fin configurations and flowing gases, and Fig. 13.11 thus offers a reasonable first estimate of the value of J/F for flow of air past electronic components.

For more detailed analysis and/or correlation of the heat transfer coefficient and friction factor for air-cooled electronic boards and cards, it would appear possible to use the form of Fig. 13.11 with a more precise definition of l/D_e and consideration of the influence of laterally adjacent packages as well as the thickness of the individual packages relative to the hydraulic diameter. It is of importance to note that in a performance map of Fig. 13.11, or any equivalent consideration of heat transfer and pressure drop in channels, the smooth channel expressions, that is, Eqs. (13.52) and (13.68) for laminar flow and Eqs. (13.72) and (13.80) for turbulent flow, may be expected to constitute a lower bound on the J curves and be approached asymptotically as l/D_e increases.

13.7.6 Axial Variations

The preceding discussion of flow development, heating asymmetry, and wall roughness effects on the heat transfer rate at the channel wall were based on the assumption of axial, that is, in the flow direction, isothermality or isoflux conditions. This idealization is, of course, rarely realized in electronic cooling applications, and attention must now be turned to examining the influence of axial temperature and heat flux variations on the local Nusselt number.

It is of interest to recall the earlier results for isothermally and isoflux heated channels in laminar flow, which indicated a distinct and calculable difference in the fully developed Nu for these two cases (7.6 and 8.23, respectively). The constant heat flux condition yields a linear axial increase in fluid temperature and, for fully developed flow, where Nu is a constant, an identical linear increase in the wall temperature. The fully developed, constant heat flux channel is, thus, analogous to a channel with an axially linear temperature profile, and the consequent difference in Nu can be interpreted to reflect the sensitivity of the local heat transfer coefficient in laminar flow to axial variations. Alternatively, the near-equivalence of the isothermal and isoflux Nusselt number in turbulent flow suggests that, for all but very low Nusselt numbers, the turbulent Nu is generally unaffected by axial variations.

LAMINAR FLOW, ISOTHERMAL SEGMENT

The analysis of the heat transfer rate from an isothermal segment, on an otherwise insulated channel wall, located at a location sufficiently downstream to ensure that flow development has proceeded to the fully developed velocity profile, offers both a convenient and useful starting point for the exploration of axial effects. Following Kays [22], both the local and average Nusselt numbers for this configuration can be expressed in terms of tabulated constants and eigenvalues yielding the Nu_x and Nu_m values presented in Table 13.4. In this table, X^+ is the nondimensional heated length, equal to $2(x/D_e)/Re\,Pr$, and is taken as zero at the start of the isothermal segment, as shown in Fig. 13.12.

Examination of Table 13.4 reveals once again the importance of taking into account the development of the thermal boundary layer. The average Nusselt number on an isothermal segment or discrete heat source washed by a fully developed velocity

TABLE 13.4 Nusselt Numbers for Laminar Flow, Isothermal Segment [22]

X^+	Nu_{x^+}	Nu_m
0.001	13.5	22.9
0.005	9.9	13.8
0.01	8.5	11.5
0.05	7.6	8.5
0.10	7.6	8.0
∞	7.6	7.6

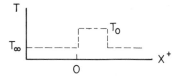

FIG. 13.12 Discrete isothermal heat source along channel wall.

profile and of length equal approximately to 0.75 times the channel spacing for air and $\mathrm{Re} \approx 10^3$, corresponding to an X^+ of 0.001, would equal more than three times the thermally fully developed value of 7.6. This latter value, encountered in the earlier discussion of fully developed flow, would be obtained only for X^+ greater than 0.10.

Interestingly, comparison of Nu_m from Table 13.4 with the values of Fig. 13.9 for the parallel development of both the velocity and temperature boundary layers shows only minor differences between the average Nusselt numbers attained in these two thermal configurations for the flow of air and other near-unity Pr fluids. It is, consequently, not essential to ascertain that flow development has progressed to the fully developed velocity profile when using the values of Table 13.4 to estimate the Nu on a discrete heat source. Furthermore, values midway between those shown in Fig. 13.9 and Table 13.4 can be used when the extent of flow development is unknown.

LAMINAR FLOW, ARBITRARY TEMPERATURE AND HEAT FLUX PROFILE

The linear, homogeneous nature of the energy equation governing heat transfer in convection makes it possible to superimpose solutions for a series of isothermal or isoflux segments or an arbitrary temperature profile in order to establish the heat transfer rate at various points along a channel in which fully developed flow is known or assumed to prevail. A detailed presentation of this approach for a circular tube appears in [22], but its extension to parallel-plate channels is beyond the scope of the present treatment.

LAMINAR FLOW, ISOFLUX SEGMENT

In the analysis of air cooling of electronic components, it is often necessary to evaluate the heat transfer coefficient for a discrete component, positioned at a downstream channel location and dissipating a uniform heat flux from its surface. This configuration can be idealized by an isoflux segment at locations where the fully developed velocity profile obtains and use made of solutions first reported in [11] and presented as well in [22]. These results for the isoflux segment, shown in Fig. 13.13, are presented in the form of Eq. (13.83) in Table 13.5 and thus allow the

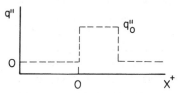

FIG. 13.13 Discrete isoflux heat source along channel wall.

TABLE 13.5 Nu_{11} and θ_1^* for Laminar Flow, Isoflux Segment [22]

X^+	Nu_{11}	θ_1^*
0.0005	23.5	0.01175
0.005	11.2	0.0560
0.02	7.49	0.1491
0.10	5.55	0.327
0.25	5.39	0.346
∞	5.38	0.346

determination of Nu for various asymmetric combinations of surface heat flux on the two channel walls, including an insulated or zero heat flux surface on one side.

Comparison of the values shown in Table 13.5 and Table 13.3 suggests that the Nu to be expected for a developing thermal boundary layer in the presence of a fully developed velocity profile is nearly identical to that encountered in the presence of a developing velocity profile. Some minor differences do exist, however, for X^+ less than 0.10, and for more precise Nusselt numbers on relatively small, discrete components the values of Table 13.5 are preferred. Recalling earlier developing flow results, the average Nu for a discrete component, sited on one channel wall, a bit smaller in size than the channel spacing and cooled by air at a Re of 1000, is found from Table 13.5 to equal 23.5 or nearly 4.5 times greater than the fully developed, one wall insulated Nu of 5.38.

TURBULENT FLOW, FLUX AND
TEMPERATURE VARIATIONS

The occurrence of the primary resistance to turbulent convective heat transfer in the fluid layer immediately adjacent to the wall results in the near-independence of the surface Nusselt number from the influence of axial temperature and heat flux variations for Pr greater than unity. This relative independence is reflected in the common use of a single Nu correlation—for example. Eq. (13.80)—for both isoflux and isothermal boundary conditions for the flow of air, other gases, and most nonmetal liquids.

This behavior is reflected, as well, in the rapid convergence of the developing nondimensional heat transfer coefficient to the fully developed turbulent Nu values, as shown in Fig. 13.10. In the interest of completeness, however, and for analyses where precision is desired, the axial Nu variation to be expected along an isoflux or isothermal segment, washed by a fully developed velocity profile, is presented next.

Solutions for thermally developing, turbulent flow between parallel plates for both asymmetric temperature and heat flux boundary conditions and a fully developed velocity profile were obtained by Hatton and Quarmby [28], and the constants and eigenvalues necessary for calculating the Nu were reported by Kays [22]. Table 13.6 summarizes the local and mean Nusselt numbers for a fluid with Pr = 1 for two values of Reynolds number for an isothermal segment along one channel wall (as in Fig. 13.12) while the opposite channel wall is insulated. Table 13.7 presents the Nu_{11} and θ_1^* values for calculating the Nu for an isoflux segment, for various ratios of surface heat flux, via Eq. (13.84).

TABLE 13.6 Local and Mean Nu for Isothermal Segment in Fully Developed, Pr = 1, Turbulent Flow [22] (other side of channel insulated)

X^+	Re $\approx 7 \times 10^3$		Re $\approx 7 \times 10^4$	
	Nu_{x^+}	Nu_m	Nu_{x^+}	Nu_m
0.0001			178.3	208.3
0.0005	34.4	40.8	152.6	170.6
0.001	31.3	37.4	148.5	159.8
0.005	26.4	29.6	147.95	150.4
0.01	26.1	27.6	147.95	149.2
0.05	26.1	26.3	147.95	148.2
0.10	26.0	26.2		

The relatively minor differences between the local Nu values of Table 13.6 and Nu_{11} of Table 13.7 serve to confirm the previously noted near independence of Nusselt number from axial variations. Similarly, comparison of Nu for symmetric isoflux segments with the values calculated via Eq. (13.80) for a Pr = 1 fluid, reveals that for Re = 7×10^4, the Nu at an x/b ratio of 20 is only 3.5% higher than the fully developed value.

13.7.7 Closure-Forced Convection

The preceding sections have attempted to provide the reader with an appreciation for the magnitude of the possible deviations from conventional channel heat transfer coefficients induced by heating asymmetry, flow and thermal development, and axial heating profiles. The tabulated Nusselt number values can serve to refine analytical or computer-based calculations requiring specification of surface heat transfer coefficients but cannot be expected to yield highly precise values.

Two recent developments in convective heat transfer research may, however, prove valuable in the future analysis of forced convection cooling of component-carrying printed circuit boards. As revealed in [29-31], it has now become possible to solve the so-called *conjugate* heat transfer problem, that is, solving the governing momentum and energy equations directly for the surface temperature without the decoupling involved in the use of heat transfer coefficients. As numerical techniques

TABLE 13.7 Nu_{11} and θ_1^* for an Isoflux Segment in Fully Developed, Pr = 1, Turbulent Flow [22]

x/b	Re $\approx 7 \times 10^3$		Re $\approx 7 \times 10^4$	
	Nu_{11}	θ_1^*	Nu_{11}	θ_1^*
2	47.3	0.013	234	0.005
6	37.9	0.033	203	0.018
20	31.5	0.089	177	0.049
60	28.0	0.173	160	0.114
200	27.0	0.200	152	0.155

are refined and computing costs diminish, this may become a viable approach for routine thermal support of electronic packaging efforts.

The similarity of packaged printed circuit boards to large-roughness flat plates has spawned an experimental and analytical study of this geometry using conventional roughness theory. In [32], heat transfer coefficients and temperature superposition functions accounting for the thermal wake were determined as a function of position downstream of a single heated block in an array for a range of velocities and channel geometries. Using the techniques proposed in [32], it may be possible in the near future to use inter-board, channel pressure drop data to calculate component surface temperatures. As data for commercial geometries become available, general correlations for the average component heat transfer coefficient, as well as for the thermal wake influence, may also become available.

13.8   NOMENCLATURE

Roman Letter Symbols

A	area or cross-sectional channel area, m
b	channel width or plate spacing, m
B	a coefficient, dimensionless
c_p	specific heat at constant pressure, J/kg °C
C	a coefficient, dimensionless
d	fin thickness, m
D	a coefficient, dimensionless
D_e	equivalent or hydraulic diameter, m
f	friction factor, dimensionless
F	pumping power, N/m s
g	gravitational acceleration, m/s^2
h	heat transfer coefficient, W/m^2 °C
k	thermal conductivity, W/m °C
L	channel or plate length, m
m	number of fins
Nu	Nusselt number, dimensionless
Nu$_0$	channel Nusselt number, dimensionless
n	an exponent, dimensionless
p	pressure, N/m^2; or an exponent, dimensionless
P	perimeter, m
Pr	Prandtl number, dimensionless
q	heat flow, W; or an exponent, dimensionless
q''	heat flux, W/m^2
Q_T	total heat flow, W
Ra	Rayleigh number, dimensionless
Ra*	modified Rayleigh number, dimensionless
Ra$'$	channel Rayleigh number, dimensionless
Ra$''$	modified channel Rayleigh number, dimensionless
Re	Reynolds number, dimensionless
S	plate width, m
St	Stanton number, dimensionless

T	temperature, °C
v	velocity in the axial direction, m/s
V	average axial velocity, m/s
w	mass flow rate per unit width, kg/m s
W	width of prime area, m
x	length or axial coordinate, m
X^+	nondimensional axial coordinate
y	transverse coordinate, m; or generalized parameter (see Eq. (13.19)
y_0	channel half-width, m
Y	a modified generalized parameter (see Eq. (13.22)
z	a generalized parameter (see Eq. (13.19)
Z	a modified general parameter (see Eq. (13.22)

Greek Letter Symbols

α	thermal diffusivity, m^2/s
β	volumetric coefficient of thermal expansion, K^{-1}
Γ	a thermal parameter, m^{-1}
Δ	indicates a difference between variables
ϵ	eddy diffusivity, m^2/s
θ^*	influence coefficient, dimensionless
μ	dynamic viscosity, kg/m s
ν	kinematic viscosity, m^2/s
ρ	density, kg/m^3
τ	shear stress, N/m^2

Subscripts

buoyant	indicates natural-convection driving force
f	indicates fluid
h	indicates thermal condition
H	indicates constant heat flux
loss	indicates pressure drop or pressure loss
m	indicates momentum or mean condition
opt	indicates optimum condition
smooth	indicates smooth condition
T	indicates constant temperature condition
w	indicates wall condition
x	indicates local value
0	indicates entrance or ambient value
1	indicates surface 1
2	indicates surface 2
11	indicates surface 1 when surface 2 is adiabatic

Superscripts

$+$	indicates a nondimensional axial coordinate
$*$	indicates an influence coefficient or modified value
$'$	indicates channel value
$''$	indicates heat flux or modified channel value

13.9 REFERENCES

1 Bar-Cohen, A., Fin Thickness for an Optimized Natural Convection Array of Rectangular Fins, *J. Heat Transfer*, vol. 101, pp. 564–566, 1979.

2 Kraus, A. D., *Cooling Electronic Equipment*, Prentice-Hall, Englewood Cliffs, N.J., 1965.

3 Aung, W., Kessler, T. J., and Beitin, K. I., Free Convection Cooling of Electronic Systems, *IEEE Trans. Parts, Hybrids and Packaging*, vol. PHP-9, no. 2, pp. 75–86, 1973.

4 Elenbaas, W., Heat Dissipation of Parallel Plates by Free Convection, *Physica*, vol. 9, no. 1, pp. 665–671, 1942.

5 Bodoia, J. R., and Osterle, J. F., The Development of Free Convection Between Heated Vertical Plates, *J. Heat Transfer*, vol. 84, pp. 40–44, 1964.

6 Sobel, N., Landis, F., and Mueller, W. K., Natural Convection Heat Transfer in Short Vertical Channels Including the Effect of Stagger, *Proc. 3d Int. Heat Transfer Conf., Chicago*, vol. 2, pp. 121–125, 1966.

7 Aung, W., Fully Developed Laminar Free Convection Between Vertical Plates Heated Asymmetrically, *Int. J. Heat Mass Transfer*, vol. 15, pp. 1577–1580, 1972.

8 Aung, W., Fletcher, L. S., and Sernas, V., Developing Laminar Free Convection Between Vertical Flat Plates with Asymmetric Heating, *Int. J. Heat Mass Transfer*, vol. 15, pp. 2293–2308, 1972.

9 Miyatake, O., Fujii, T., Fujii, M., and Tanaka, H., Natural Convective Heat Transfer Between Vertical Parallel Plates—One Plate with a Uniform Heat Flux and the Other Thermally Insulated, *Heat Transfer Jpn. Res.*, vol. 4, pp. 25–33, 1973.

10 Miyatake, O., and Fujii, T., Free Convective Heat Transfer Between Vertical Parallel Plates—One Plate Isothermally Heated and the Other Thermally Insulated, *Heat Transfer Jpn. Res.*, vol. 3, pp. 30–38, 1972.

11 Lundberg, R. E., Reynolds, W. C., and Kays, W. M., Heat Transfer with Laminar Flow in Concentric Annuli with Constant and Variable Wall Temperature with Heat Flux, NASA TN D-1972, National Technical Information Service, Springfield, Va., August 1963.

12 Churchill, S. W., and Usagi, R., A General Expression for the Correlation of Rates of Transfer and Other Phenomena, *AIChE J.*, vol. 18, no. 6, pp. 1121–1138, 1972.

13 Rohsenow, W. M., and Choi, H., *Heat, Mass and Momentum Transfer*, Prentice-Hall, Englewood Cliffs, N.J., 1961.

14 McAdams, W. H., *Heat Transmission*, McGraw-Hill, New York, 1954.

15 Nakamura, H., Preprint #120, 42nd Natl. Meeting Jpn. Soc. Mech. Eng., vol. 5, 1964.

16 Sparrow, E. M., and Gregg, J. L., Laminar Free Convection from a Vertical Plate with Uniform Surface Heat Flux, *ASME Trans. C*, pp. 435–440, 1956.

17 Aung, W., Heat Transfer in Electronic Systems with Emphasis on Asymmetric Heating, *Bell System Tech. J.*, vol. 52, pp. 907–925, 1973.

18 Levy, E. K., Optimum Plate Spacings for Laminar Natural Convection Heat Transfer from Parallel Vertical Isothermal Flat Plates, *J. Heat Transfer*, vol. 93, pp. 463–465, 1971.

19 Sparrow, E. M., and Bahrami, P. A., Experiments on Natural Convection from Vertical Parallel Plates with Either Open or Closed Edges, *J. Heat Transfer*, vol. 102, pp. 221–227, 1980.

20 Nikuradse, J., Turbulente Strömung in Nichtkreinförmigen Rohren, *Ing.-Arch.*, vol. 1, pp. 306–332, 1930.

21 Colburn, A. P., A Method of Correlating Forced Convection Heat Transfer Data and a Comparison with Fluid Friction, *Trans. AIChE*, vol. 29, pp. 174–210, 1933.

22 Kays, W. M., *Convective Heat and Mass Transfer*, McGraw-Hill, New York, 1966.

23 Heaton, H. S., Reynolds, W. C., and Kays, W. M., Heat Transfer in Annular Passages. Simultaneous Development of Velocity and Temperature Fields in Laminar Flow, *Int. J. Heat Mass Transfer*, vol. 7, pp. 763–781, 1964.

24 Sparrow, E. M., Analysis of Laminar Forced-Convection Heat Transfer in Entrance Region of Flat Rectangular Ducts, Natl. Advisory Comm. Aeronaut. Tech. Note 3331, National Technical Information Service, Springfield, Va., January 1955.

25 Hartnett, J. P., Experimental Determination of the Thermal-Entrance Length for the Flow of Water and Oil in Circular Pipes, *J. Heat Transfer*, vol. 77, no. 7, pp. 1211–1220, 1955.

26 Nunner, W., Warmeubergang und Druckabfall in Rauhen Rohren, *Z. Deutsch. Ing. Forschung.*, vol. 455, pp. 39–48, 1956.

27 LaHaye, P. G., Neugebauer, F. J., and Sakhuja, R. K., A Generalized Prediction of Heat Transfer Surfaces, *J. Heat Transfer*, vol. 96, pp. 511–517, 1974.

28 Hatton, A. P., and Quarmby, A., Heat Transfer in the Thermal Entry Length with Laminar Flow in an Annulus, *Int. J. Heat Mass Transfer*, vol. 6, pp. 903–914, 1963.

29 Sparrow, E. M., and Acharya, S., A Natural Convection Fin with a Solution-Determined Monotonically Varying Heat Transfer Coefficient, *J. Heat Transfer*, vol. 103, pp. 218–225, 1981.

30 Sparrow, E. M., and Chyu, M. K., Conjugate Forced Convection-Conduction Analysis of Heat Transfer in a Plate Fin, *J. Heat Transfer*, vol. 104, pp. 204–206, 1982.

31 Brosh, A., Degani, D., and Zalmanovich, S., Conjugated Heat Transfer in a Laminar Boundary Layer with Heat Source at the Wall, *J. Heat Transfer*, vol. 104, pp. 90–95, 1982.

32 Arvizu, D. E., and Moffat, R. J., Experimental Heat Transfer from an Array of Heated Cubical Elements on an Adiabatic Channel Wall, Report No. HMT-33, Thermosciences Div. Stanford University, Stanford, California, 1982.

14

■ extended surfaces

14.1 INTRODUCTION

Inspection of the heat transfer "rate" equation $q = hS\theta_m$ shows that for a fixed rate of heat dissipation q, the magnitude of the temperature difference θ_m is governed by the product hS. To obtain high heat dissipation at a relatively low temperature difference, one requires either a component with substantial surface area for dissipation or a coolant possessing superior heat transfer properties.

The trend in component design for airborne and space applications has been and will continue to be toward microminiaturization. Ordinarily, miniaturized electronic equipment is also quite small. Furthermore, air—which is inexpensive and often readily available—does not have outstanding attributes as a heat transfer medium. The designer of electronic equipment cooling systems is often faced with the problem of cooling miniaturized, high heat-dissipating components to a rather low temperature with a fluid having definite heat transfer limitations. This dilemma can be summarized as being one of low hS product.

The coefficient of heat transfer can be improved in two ways:

1 Use of a better fluid. This is often impossible because of weight and installation requirements. Use of a liquid coolant, for example, requires a pump, a heat exchanger, piping, valves, and possibly an expansion tank or other appurtenances required for handling the ultimate heat sink fluid.
2 Use of the available coolant fluid at a higher velocity. This is often impractical because of the increased power required to force the fluid through the system. Inspection of the several correlations in Chap. 6 will show that a twofold increase in heat transfer coefficient requires a more than twofold increase in fluid velocity. At the same time, the twofold increase in fluid velocity results in almost a fourfold increase in pressure loss and possibly as much as an eightfold increase in power required. Power is weight, and because the fluid is circulated by a pump, fan, or blower, large penalties in weight must be expected under these circumstances.

Increase of the hS product is not easily achieved by increasing the heat transfer coefficient. A considerably more effective method of accomplishing this increase is by augmenting the surface by employing fins. Such augmentation can result in as much as an order-of-magnitude increase in the hS product without appreciably altering the heat transfer coefficient.

As shown in Fig. 14.1, several types of extended surfaces (fins) can be used. However, in electronic cooling applications the designer strives to use a type of fin that lends itself readily to analysis and to easy manufacture. In this category are the longitudinal fin of rectangular profile, the radial fin of rectangular profile, and the cylindrical spine (pin fin). The longitudinal fin of triangular profile is more difficult

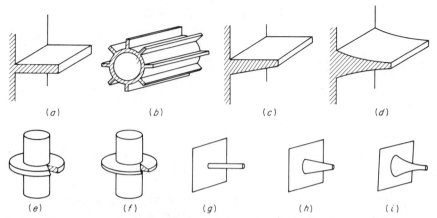

FIG. 14.1 Some typical examples of extended surfaces. (*a*) Longitudinal fin of rectangular profile. (*b*) Cylindrical tube equipped with longitudinal fins. (*c*) Longitudinal fin of trapezoidal profile. (*d*) Longitudinal fin of truncated concave parabolic profile. (*e*) Cylindrical tube equipped with radial fin of rectangular profile. (*f*) Cylindrical tube equipped with radial fin of truncated triangular profile. (*g*) Cylindrical spine. (*h*) Truncated conical spine. (*i*) Truncated concave parabolic spine.

to analyze and to manufacture, but is often used because its installation weight is about one-half that of the fin of rectangular profile.

In the ensuing analysis the following simplifying assumptions are made [1, 2] .

1 The heat flow is steady; that is, the temperature at any point in the fin does not vary with time.
2 The fin material is homogeneous, and the thermal conductivity is constant and uniform.
3 The coefficient of heat transfer is constant and uniform over the entire face surface of the fin.
4 The temperature of the surrounding fluid is constant and uniform. Because one is dealing with cooling, this temperature is always assumed to be lower than that at any point on the fin.
5 There is no temperature gradients within the fin other than along its height. This requires that the fin length and height be great when compared with the width.
6 There is no bond resistance to the flow of heat at the base of the fin.
7 The temperature at the base of the fin is uniform and constant.
8 There are no heat sources within the fin itself.
9 Unless otherwise noted, there is a negligible amount of heat transferred by convection from the end and sides of the fin. Note that in this terminology, the faces of the fin are the surfaces that dissipate heat.

14.2 THE LONGITUDINAL FIN OF RECTANGULAR PROFILE

In the longitudinal fin of rectangular profile shown with its terminology in Fig. 14.2, the length coordinate x is taken as having its origin at the tip (edge) and positive in a direction toward the base of the fin.

The differential equation for the temperature excess θ is formulated by considering a differential element dx having a cross-sectional area normal to the path of heat flow, with $\delta = \delta_0$

$$A = \delta L$$

and a surface area

$$dS = 2(L + \delta)\, dx$$

which reduces to

$$dS = 2L\, dx$$

if assumptions 5 and 9 in Sec. 14.1 are honored.

The temperature excess is defined as the temperature difference between the fin and the constant-temperature surrounding fluid. Hence at any point on the fin,

$$\theta \equiv T - T_s$$

and $d\theta = dT$

and at the base of the fin,

$$\theta_0 = T_0 - T_s$$

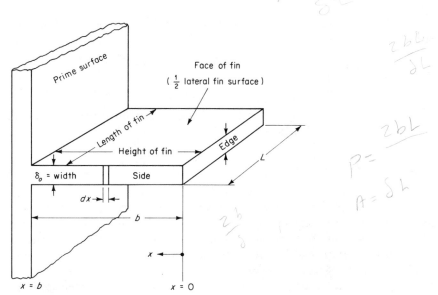

FIG. 14.2 Terminology and coordinate system for longitudinal fin of rectangular profile.

The principle of conservation of energy dictates that the heat entering the element by conduction must equal the sum of that leaving by conduction and convection. Hence

$$-kA \frac{dT}{dx} = -kA \frac{dT}{dx} - \frac{d}{dx}\left(kA \frac{dT}{dx}\right) dx + 2hL(T-T_s)\, dx$$

and the differential equation for temperature excess is

$$\frac{d^2\theta}{dx^2} - m^2\theta = 0 \tag{14.1}$$

where $T - T_s$ has been replaced by θ, dT by $d\theta$, A by δL, and where

$$m = \left(\frac{2h}{k\delta}\right)^{1/2}$$

The general solution to Eq. (14.1) is

$$\theta = C_1 e^{mx} + C_2 e^{-mx}$$

where the arbitrary constants C_1 and C_2 are evaluated from the boundary conditions

$$\theta = \theta_0 \qquad \text{at } x = b$$

$$\frac{d\theta}{dx} = 0 \qquad \text{at } x = 0$$

Use of these boundary conditions yields the particular solution for the temperature excess:

$$\theta = \frac{\theta_0 \cosh mx}{\cosh mb} \tag{14.2}$$

The heat flow through the base is obtained from the derivative of Eq. (14.2) multiplied by $-k\delta L$ and evaluated at $x = b$:

$$q_0 = k\delta L m \theta_0 \tanh mb \tag{14.3}$$

Equation (14.2) shows that the temperature excess decreases with distance from the fin base. This decreasing temperature excess as a function of fin height leads to a problem in using the rate equation, $q = hS\theta_m$, where θ_m is difficult to calculate. It is evident that designs based on the temperature excess existing at the base of the fin will be optimistic. Similarly, one can see that a design based on any other temperature excess will be quite uncertain, because the selected temperature difference will occur at only one point along the height of the fin.

It is customary to use the fin efficiency in applying extended surfaces in heat transfer analysis and heat exchanger design. The fin efficiency is defined as the ratio of the actual heat dissipated by the fin to the ideal heat dissipated if the entire fin is operating at the temperature difference existing at the base. The actual heat dissipated is the heat passing through the base of the fin, q_0, and the ideal heat dissipation is equal to $q_i = hS\theta_0$. Hence, the efficiency is defined as

$$\eta \equiv \frac{q_0}{q_i} \tag{14.4}$$

For the longitudinal fin of rectangular profile,

$$\eta \equiv \frac{q_0}{q_i} = \frac{k\delta Lm\theta_0 \tanh mb}{hS\theta_0}$$

or, since $S = 2Lb$,

$$\eta = \frac{k\delta Lm\theta_0 \tanh mb}{2hLb\theta_0} = \frac{k\delta m \tanh mb}{2hb}$$

The final expression for the efficiency is obtained by noting that $1/m^2 = k\delta/2h$, so that

$$\eta = \frac{\tanh mb}{mb} \tag{14.5}$$

The ideal limit for efficiency is, of course, unity (100%). If a configuration has an efficiency of 65%, for example, one can consider 65% of the total extended surface as operating at the base temperature excess. A tabulation of Eq. (14.5) is available [3].

14.3   THE RADIAL FIN OF RECTANGULAR PROFILE

The analysis of the radial fin of rectangular profile involves a rather wild journey through the realm of Bessel functions. The goal of such an analysis is the development of a mathematical expression for the fin efficiency. Tables of fin efficiency for this fin have been developed [3]. To use the tables, one must carefully consider the terminology and coordinate system for this fin, which is presented in Fig. 14.3. The tables utilize two parameters, ρ and ϕ, which are defined as a radius ratio,

$$\rho \equiv \frac{r_0}{r_e} \tag{14.6}$$

and a fin performance factor,

$$\phi \equiv (r_e - r_0)^{3/2} \left(\frac{2h}{kA_p}\right)^{1/2} \tag{14.7}$$

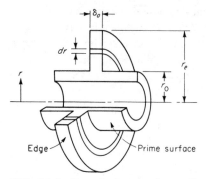

FIG. 14.3   Terminology and coordinate system for radial fin of rectangular profile.

where A_p is the fin profile area,

$$A_p = (r_e - r_0)\delta_0 \tag{14.8}$$

14.4   FINS OF OTHER PROFILES

The fin efficiency tables [3] list the efficiencies of several fins and spines as a function of the parameter mb for the fins shown in Fig. 14.4, where it may be observed that all have height b. To formulate m, the following applies:

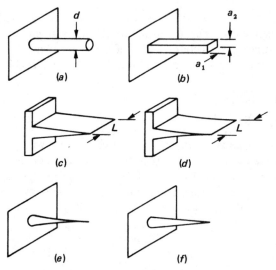

FIG. 14.4   Several spines and longitudinal fins whose efficiencies are tabulated in [3]. All have length L, fin height b, and base width δ_0, unless otherwise noted. (*a*) Cylindrical spine with diameter d. (*b*) Rectangular spine with sides a_1 and a_2 (if $a_1 = a_2$, the spine becomes square). (*c*) Longitudinal fin of triangular profile. (*d*) Longitudinal fin of concave parabolic profile. (*e*) Spine of concave parabolic profile. (*f*) Conical spine.

1 Cylindrical spine:

$$m = \left(\frac{4h}{kd} \right)^{1/2} \tag{14.9}$$

2 Rectangular spine:

$$m = \left(\frac{hP}{kA} \right)^{1/2} \tag{14.10}$$

where

$$P = 2(a_1 + a_2) \tag{14.11a}$$

and $A = a_1 a_2$ $\tag{14.11b}$

3 Longitudinal fin of triangular profile:[1]

$$m = \left(\frac{2h}{k\delta_0} \right)^{1/2} \tag{14.12}$$

4 Longitudinal fin of concave parabolic profile:

$$m = \left(\frac{2h}{k\delta_0} \right)^{1/2} \tag{14.13}$$

5 Concave parabolic spine:

$$m = \left(\frac{2h}{k\delta_0} \right)^{1/2} \tag{14.14}$$

6 Conical spine:

$$m = \left(\frac{2h}{k\delta_0} \right)^{1/2} \tag{14.15}$$

14.5 OPTIMUM DIMENSIONS

It is possible to develop the optimum dimensions for longitudinal fins of rectangular and triangular profile. For a given profile area A_p, there will be a particular value of fin height and fin width, b or δ, that will yield a maximum heat dissipation.

First consider the fin of rectangular profile. The heat flow through the base is given by Eq. (14.3), which if written in terms of the profile area A_p, yields ($\delta = \delta_0$)

$$q_0 = L\theta_0 (2hk)^{1/2} \delta^{1/2} \tanh \left[A_p \left(\frac{2h}{k} \right)^{1/2} \delta^{-3/2} \right] \tag{14.16}$$

---

[1] A close approximation that neglects the taper angle.

The value of fin width that yields a maximum heat dissipation is obtained at the point where the derivative of Eq. (14.16) vanishes. Performance of this laborious differentiation [3] gives

$$3\beta_R \, \mathrm{sech}^2 \, \beta_R = \tanh \beta_R$$

where

$$\beta_R = A_p \left(\frac{2h}{k} \right)^{1/2} \delta^{-3/2}$$

and because

$$\frac{2 \tanh \beta_R}{\mathrm{sech}^2 \, \beta_R} = \frac{2(\sinh \beta_R / \cosh \beta_R)}{1/(\cosh^2 \beta_R)} = 2 \sinh \beta_R \, \cosh \beta_R = \sinh 2\beta_R$$

one obtains

$$\beta_R = \tfrac{1}{6} \sinh 2\beta_R$$

Solution of this equation yields

$$\beta_R = A_p \left(\frac{2h}{k} \right)^{1/2} \delta^{-3/2} = 1.419$$

from which the optimum fin width is determined as

$$\delta = 0.791 \left(\frac{2hA_p^2}{k} \right)^{1/3} \tag{14.17}$$

Under these circumstances, the optimum fin height will be

$$b = \frac{A_p}{\delta} = 1.262 \left(\frac{kA_p}{2h} \right)^{1/3} \tag{14.18}$$

For the longitudinal fin of triangular profile, analysis [3] shows that the heat dissipated by the fin is given by

$$q_0 = \frac{2h\theta_0 I_1(2mb)}{m I_0(2mb)} \tag{14.19}$$

where m is given by Eq. (14.12) and I_0 and I_1 are modified Bessel functions. To find the optimum dimensions, one may define

$$\beta_T = 2mb = 2 \left(\frac{2h}{k\delta_0} \right)^{1/2} \left(\frac{2A_p}{\delta_0} \right) = 4A_p \left(\frac{2h}{k} \right)^{1/2} \delta_0^{-3/2}$$

so that Eq. (14.19), which is the expression for the heat flow, can be represented as

$$q_0 = [4A_p(2h^2)k]^{1/3} \theta_0 L\beta_T^{-1/3} \left[\frac{I_1(\beta_T)}{I_0(\beta_T)} \right]$$

Differentiation with respect to β_T [3] yields the optimization equation when the result is set equal to zero and then rearranged:

$$I_0(\beta_T)I_2(\beta_T) + \frac{2}{3} \frac{I_0(\beta_T)I_1(\beta_T)}{\beta_T} = I_1^2(\beta_T)$$

This has a real root at $\beta_T = 2.6188$, and gives the optimum base width

$$\delta_0 = 1.328 \left[A_p^2 \left(\frac{2h}{k} \right) \right]^{1/3} \tag{14.20}$$

and optimum fin height

$$b = \frac{2A_p}{\delta_0} = 1.506 \left(\frac{A_p k}{2h} \right)^{1/3} \tag{14.21}$$

14.6   COMPARISON OF LONGITUDINAL FINS AND KEYS TO GOOD DESIGN

Longitudinal fins of rectangular and triangular profile can be compared to determine which profile requires the least profile area for the dissipation of a given amount of heat. The comparison is accomplished by a substitution of the optimum width into the expression for heat flow through the base in each case.

For the rectangular profile, one substitutes Eq. (14.17) into the expression for heat flow through the base, Eq. (14.3), and obtains

$$q_0 = 1.26(h^2 A_p k)^{1/3} \theta_0$$

with the recognition that, for optimum conditions, $mb = 1.419$. From this, a simple rearrangement yields the profile area for given base heat flow and temperature excess:

$$A_p = \frac{0.500}{h^2 k} \left(\frac{q_0}{\theta_0} \right)^3 \tag{14.22}$$

The same procedure may be followed for the triangular profile. Equation (14.20), which represents the optimum base width, is substituted into the equation for heat flow through the base, Eq. (14.19), noting that $2mb = \beta_T = 2.6188$:

$$q_0 = 1.422(h^2 A_p k)^{1/3} \theta_0$$

from which

$$A_p = \frac{0.347}{h^2 k} \left(\frac{q_0}{\theta_0} \right)^3 \qquad (14.23)$$

Inspection of Eqs. (14.22) and (14.23) shows that the profile area required for a given heat dissipation is a function of the cube of the base heat flow to base temperature excess ratio, q_0/θ_0. These equations also show that the required profile area will be inversely proportional to the thermal conductivity of the fin material and to the square of the heat transfer coefficient between the fin and the surrounding fluid.

Three significant conclusions can be drawn from further consideration of Eqs. (14.23) and (14.24). The first is that, for the same material, environmental conditions (heat transfer coefficient), and ratio of base heat flow to base temperature excess, the triangular profile requires only about 69% as much material as the longitudinal profile. This shows that, if the triangular profile can be manufactured without undue difficulty, its use is to be encouraged.

The second conclusion concerns the choice of material for any particular profile. The equations show that the profile area is inversely proportional to the thermal conductivity of the fin material. The total weight of the fin is proportional to the profile area and to the specific weight of the material used. Hence, the fin weight is directly proportional to the specific weight γ, and inversely proportional to the thermal conductivity k.

Finally, it is seen that the profile area, and hence the fin volume, increases as the cube of the heat flow. If it is desired to double the heat flow, one may make the choice of using two identical fins or making one fin eight times as large. It is obvious that the designer will tend to use more stubby fins rather than fewer higher fins.

There is, however, a limit as to the number of fins that can be placed on a surface. This limit is posed by the characteristics of the convection process, the thickness of adjacent boundary layers, and the permissible fluid pressure loss.

14.7 ANALYSIS OF ARRAYS OF EXTENDED SURFACE

An array of extended surface may be defined as a combination of at least two fins. The fins are connected at what may be called node points or merely nodes. Examples of arrays of extended surface may be found in stacked cold plates (see Chap. 15) and in transistor applications (see Chap. 21), where the array may also be called a heat sink.

Consider Fig. 14.5, which shows an array of four fins. They are numbered consecutively from 1 to 4, and it may be noted that the highest number (4) designates the fin that is closest to the base of the array, that is, the point where the heat is injected. Of interest to the analyst is the ratio of the temperature excess at the base of the array to the heat flowing into the array. This ratio is designated as the array input impedance, and for Fig. 14.5 this ratio is

$$Z_I = \frac{\theta_{b4}}{q_{b4}} \qquad (14.24)$$

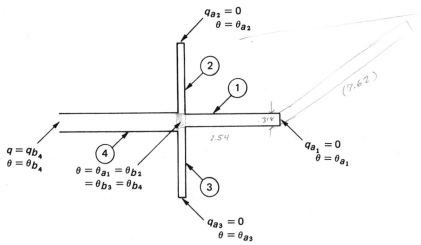

FIG. 14.5 Array of extended surface composed of four fins. All fins have their own particular width δ, height b, and origins located at the fin tip with positive direction toward the fin base. The length of the array is L.

Each fin in the array, which is composed exclusively of longitudinal fins of rectangular profile, has its own differential equation for temperature excess given by Eq. (14.1):

$$\frac{d^2\theta}{dx^2} - m^2\theta = 0 \tag{14.1}$$

with a general solution

$$\theta = C_1 e^{mx} + C_2 e^{-mx}$$

For the array of Fig. 14.5, four temperature excess equations may be written and the two arbitrary constants in each general solution can be evaluated by considering a pair of boundary conditions. In addition, the analyst is aided by the fact that the equations may be uncoupled with the observation that the heat flowing out of the tip of fin 4 must divide and flow into fins 1, 2, and 3.

Analyses such as that described in the foregoing paragraph can be and have been made [3]. Here, however, deviation is made from this procedure and a newer algorithm is described [4]. Keeping in mind that the base of each fin in the array is at $x = b$, the theory of differential equations guarantees that Eq. (14.1) has two independent solutions $\lambda_1(x)$ and $\lambda_2(x)$ that satisfy the *initial* (not boundary) conditions for the fins shown in Fig. 14.5.

$$\lambda_1(b) = 1 \qquad \frac{d\lambda_1}{dx}\bigg|_{x=b} = 0 \tag{14.25a}$$

$$\lambda_2(b) = 0 \qquad \frac{d\lambda_2}{dx}\bigg|_{x=b} = \frac{1}{k\delta L} \tag{14.25b}$$

The temperature excess and heat flow at any point on the fin will then be given in terms of these values at the base of the fin by the equations

$$\theta(x) = \theta_b \lambda_1(x) + q_b \lambda_2(x) \tag{14.26}$$

and $$q(x) = k\delta L [\theta_b \lambda_1'(x) + q_b \lambda_2'(x)] \tag{14.27}$$

where the primes designate derivatives.

Equations (14.26) and (14.27) produce a *thermal transmission matrix* Γ, which gives tip conditions (where $x = 0$), designated by θ_a and q_a, in terms of the base conditions

$$\begin{bmatrix} \theta_a \\ q_a \end{bmatrix} = [\Gamma] \begin{bmatrix} \theta_b \\ q_b \end{bmatrix} = \begin{bmatrix} \gamma_{11} & \gamma_{12} \\ \gamma_{21} & \gamma_{22} \end{bmatrix} \begin{bmatrix} \theta_b \\ q_b \end{bmatrix} \tag{14.28}$$

where

$$\begin{aligned}
\gamma_{11} &= \lambda_1(a) \\
\gamma_{12} &= \lambda_2(a) \\
\gamma_{21} &= k\delta L \lambda_1'(a) \\
\gamma_{22} &= k\delta L \lambda_2'(a)
\end{aligned} \tag{14.29}$$

Solution of Eq. (14.1) as in initial-value problem with the initial conditions given by Eqs. (14.25a) and (14.25b) yields the thermal transmission matrix

$$[\Gamma] = \begin{bmatrix} \cosh mb & -Z_0 \sinh mb \\ \dfrac{-\sinh mb}{Z_0} & \cosh mb \end{bmatrix}$$

where, as before, $m = (2h/k\delta)^{1/2}$ and where Z_0 is defined as the characteristic impedance of the fin:

$$Z_0 \equiv \frac{1}{(2hk\delta)^{1/2} L} \tag{14.30}$$

The basic tool for analyzing an array of fins is provided by the following mathematical observation, which is easily verified: If tip and base conditions of temperature excess and heat flow are related by the thermal transmission matrix of Eq. (14.28), then the q/θ ratio of the tip and the base are related by the bilinear transformation

$$\frac{q_b}{\theta_b} = f\left(\frac{q_a}{\theta_a}\right) = \frac{\gamma_{21} - \gamma_{11}(q_a/\theta_a)}{-\gamma_{22} + \gamma_{12}(q_a/\theta_a)} \tag{14.31}$$

To see how this is used, consider the array of Fig. 14.5. It has the important feature that the ratio q/θ is immediately computable for every remote tip (fins 1, 2, and 3). Either the tips dissipate heat or they do not. Hence

$$\frac{q_a}{\theta_a}\bigg|_1 = h\delta_1 L \quad \text{or} \quad \frac{q_a}{\theta_a}\bigg|_1 = 0$$

$$\frac{q_a}{\theta_a}\bigg|_2 = h\delta_2 L \quad \text{or} \quad \frac{q_a}{\theta_a}\bigg|_2 = 0$$

$$\text{and} \quad \frac{q_a}{\theta_a}\bigg|_3 = h\delta_3 L \quad \text{or} \quad \frac{q_a}{\theta_a}\bigg|_3 = 0$$

Now, by continuity,

$$q_{a4} = q_{b1} + q_{b2} + q_{b3}$$

$$\text{and} \quad \theta_{a4} = \theta_{b1} = \theta_{b2} = \theta_{b3}$$

Hence

$$\frac{q_a}{\theta_a}\bigg|_4 = \frac{q_b}{\theta_b}\bigg|_1 + \frac{q_b}{\theta_b}\bigg|_2 + \frac{q_b}{\theta_b}\bigg|_3$$

and the final result can be obtained by employing Eq. (14.31).

For example, consider the array of Fig. 14.5 fabricated of an aluminum with $k = 173$ W/m °C operating in an environment with $h = 45$ W/m² °C. Suppose that fin 1 has a height of 2.540 cm and a width of 0.318 cm. Further suppose that fins 2 and 3 each have a height of 1.905 cm and a width of 0.159 cm. Let fin 4 be 3.175 cm high and 0.476 cm wide. For an array length of 7.62 cm, it is desired to compute the array input impedance if the tips of fins 1, 2, and 3 are considered as adiabatic (no tip heat dissipation).

For fin 1,

$$m = \left(\frac{2h}{k\delta}\right)^{1/2} = \left[\frac{2(45)}{(173)(0.318/100)}\right]^{1/2} = 12.7904 \text{ m}^{-1}$$

$$Z_0 = \frac{1}{(2hk\delta)^{1/2}L} = \frac{1}{[2(45)(173)(0.318/100)]^{1/2}(7.62/100)} = 1.8650 \text{ °C/W}$$

$$mb = (12.7904)\left(\frac{2.54}{100}\right) = 0.3249$$

Because $q_a/\theta_a|_1 = 0$, calculation of γ_{11} and γ_{12} is not necessary.

$$\gamma_{21} = -\frac{\sinh mb}{Z_0} = -\frac{\sinh (0.3249)}{1.8650}$$

$$= -\frac{0.3306}{1.2650} = -0.1773 \text{ W/°C}$$

and $\gamma_{22} = \cosh mb = \cosh(0.3249) = 1.0532$

Then by Eq. (14.31) with $q_a/\theta_a|_1 = 0$,

$$\left.\frac{q_b}{\theta_b}\right|_1 = \frac{\gamma_{21}}{-\gamma_{22}} = \frac{-0.1773}{-1.0532} = 0.1683 \text{ W/m}^\circ\text{C}$$

For fins 2 and 3,

$$m = \left(\frac{2h}{k\delta}\right)^{1/2} = \left[\frac{2(45)}{(173)(0.159/100)}\right]^{1/2} = 18.0884 \text{ m}^{-1}$$

$$Z_0 = \frac{1}{(2hk\delta)^{1/2}L} = \frac{1}{[2(45)(173)(0.159/100)]^{1/2}(7.62/100)} = 2.6376 \text{ }^\circ\text{C/W}$$

$$mb = (18.0884)\left(\frac{1.905}{100}\right) = 0.3446$$

Again because $q_a/\theta_a|_2 = q_a/\theta_a|_3 = 0$, all that is required is γ_{21} and γ_{22}.

$$\gamma_{21} = -\frac{\sinh mb}{Z_0} = -\frac{\sinh(0.3446)}{2.6376}$$

$$= -\frac{0.3514}{2.6376} = -0.1332 \text{ W/}^\circ\text{C}$$

and $\gamma_{22} = \cosh mb = \cosh(0.3446) = 1.0600$

With the tip q/θ ratios equal to zero, these make

$$\left.\frac{q_b}{\theta_b}\right|_2 = \left.\frac{q_b}{\theta_b}\right|_3 = \frac{\gamma_{21}}{-\gamma_{22}} = \frac{-0.1332}{-1.0600} = 0.1257 \text{ W/}^\circ\text{C}$$

For fin 4, by continuity,

$$\left.\frac{q_a}{\theta_a}\right|_4 = \left.\frac{q_b}{\theta_b}\right|_1 + \left.\frac{q_b}{\theta_b}\right|_2 + \left.\frac{q_b}{\theta_b}\right|_3$$

$$= 0.1683 + 0.1257 + 0.1257 = 0.4197 \text{ W/}^\circ\text{C}$$

In addition, for fin 4,

$$m = \left(\frac{2h}{k\delta}\right)^{1/2} = \left[\frac{2(45)}{173(0.476/100)}\right]^{1/2} = 10.4543 \text{ m}^{-1}$$

$$Z_0 = \frac{1}{(2hk\delta)^{1/2}L} = \frac{1}{[2(45)(173)(0.476/100)]^{1/2}(7.62/100)} = 1.5244 \text{ }^\circ\text{C/W}$$

$$mb = (10.4543)\left(\frac{3.175}{100}\right) = 0.3319$$

Then

$$\gamma_{11} = \gamma_{22} = \cosh mb = \cosh(0.3319) = 1.0556$$

$$\gamma_{12} = -Z_0 \sinh mb = -(1.5244)\sinh(0.3319)$$

$$= -(1.5244)(0.3381) = -0.5153 \ ^{\circ}\text{C/W}$$

and $\gamma_{21} = -\dfrac{\sinh mb}{Z_0} = -\dfrac{0.3381}{1.5244} = -0.2218 \ \text{W/}^{\circ}\text{C}$

Finally, Eq. (14.31) is used with $q_a/\theta_a|_4 = 0.4197 \ \text{W/}^{\circ}\text{C}$.

$$\frac{q_b}{\theta_b}\bigg|_4 = \frac{\gamma_{21} - \gamma_{11}(q_a/\theta_a|_4)}{-\gamma_{22} + \gamma_{12}(q_a/\theta_a|_4)}$$

$$= \frac{-0.2218 - 1.0556(0.4197)}{-1.0556 + (-0.5153)(0.4197)}$$

$$= \frac{-0.6648}{-1.2719} = 0.5227 \ \text{W/}^{\circ}\text{C}$$

This is the required result. Its impact can be established by considering that if the array is required to dissipate 30 W, a temperature excess of

$$\theta_{b4} = \frac{q_{b4}}{q_b/\theta_b|_4} = \frac{30}{0.5227} = 57.4\,^{\circ}\text{C}$$

will be required. In an environment at, say, 71°C, for a 30 W dissipation, the base of the array will operate at a temperature of $71 + 57.4 = 128.4\,^{\circ}\text{C}$.

14.8 NOMENCLATURE

Roman Letter Symbols

a	side of rectangular spine, m
A	cross-sectional area, m^2; or, with subscript p, profile area, m^2
b	fin height, m
C	arbitrary constant
d	spine diameter, m
f	designates a function
h	heat transfer coefficient, W/m^2 °C
I	modified Bessel function of first kind
k	thermal conductivity, W/m °C
L	fin length, m
m	fin performance factor, m^{-1}
P	perimeter, m
q	heat flow, W
S	surface area, m^2

T	temperature, °C
x	length coordinate, m
z	plate spacing in Elenbaas correlation, m
Z	thermal impedance, °C/W

Greek Letter Symbols

β	fin optimizing parameter, dimensionless
γ	specific weight, kg/m³ ; or element of thermal transmission matrix
Γ	thermal transmission matrix
δ	fin width, m
Δ	indicates change in variable
η	fin efficiency, dimensionless
θ	temperature excess or temperature difference, °C
λ	a solution to a second-order differential equation
ρ	radial fin radius ratio, dimensionless
ϕ	a radial fin performance factor defined by Eq. (14.7), dimensionless

Subscripts

e	indicates edge
i	indicates ideal
I	indicates input
p	indicates profile area
R	indicates rectangular profile
s	indicates surroundings
T	indicates triangular profile
0	indicates base condition or characteristic value

14.9 REFERENCES

1 Murray, W. M., Heat Dissipation Through an Annular Disk or Fin of Uniform Thickness, *J. Appl. Mech.*, vol. 5, pp. A78–A80, 1938.
2 Gardner, K. A., Efficiency of Extended Surfaces, *Trans. ASME*, vol. 67, pp. 621–631, 1945.
3 Kern, D. Q., and Kraus, A. D., *Extended Surface Heat Transfer*, McGraw-Hill, New York, 1972.
4 Kraus, A. D., Snider, A. D., and Doty, L. F., An Efficient Algorithm for Evaluating Arrays of Extended Surface, *J. Heat Transfer*, vol. 100, pp. 288–293, 1978.

15

■ cold plates

15.0 A NOTE ON UNITS

Because the material in this chapter is based heavily on dimensional data for heat exchanger surfaces investigated by Kays and London [1], use is made of the units in the English engineering system.

15.1 INTRODUCTION

In spacecraft, aircraft, and missiles, volume and weight are used sparingly. It is essential in these vehicles that onboard heat interchange duties be accomplished in equipment that is as compact and light as possible. In cryogenic systems operating with warmer and cooler fluids but both at low temperatures, as in the liquefaction of permanent gases, there is also a stringent need for the heat exchangers to be compact. Only in this way is it possible to minimize the containment surfaces through which heat might leak into the system. Where very low temperatures are involved, the removal of unwanted heat is both difficult and expensive. There are also many other applications for compact heat exchangers between these two extreme examples.

This chapter describes a procedure for evolving the design of a forced convection heat exchanger for use in aerospace electronics cooling. This type of exchanger is often called a cold plate or forced cooled electronic chassis, and it should not be confused with a heat exchanger used to transfer heat between two working fluids. It is characterized by operation with a single working fluid, which usually acts as a receiver for the dissipated heat; it is further characterized, when components are judiciously placed, by its approach to an isothermal surface. It can also be used to remove heat from a circulating coolant, as in the case of heat transmittal to the shell or skin of a space vehicle.

Experience has shown that the performance of actual hardware designed from the equations presented in this chapter comes surprisingly close to the predicted values calculated from these equations—provided that two heat dissipating requirements are met:

1 The heat is evenly applied so that temperature gradients within (along and across) the forced cooled chassis itself are minimized. This is referred to as *even loading*.
2 The heat is applied on both sides of the forced cooled chassis. Application of heat to only one side will lead to slightly erroneous results, which—fortunately—for an air-cooled application lead to negligible error. The problem of single-side loading should be discussed, however, and this is done in Sec. 15.8.

15.2 COMPACT HEAT EXCHANGER SURFACES

The designer or analyst always has a choice as to whether a cold plate is to be fabricated in his own shop or purchased from an outside source. In-house fabrication is not only costly for a "single design shot" but likely to be risky because the heat transfer coefficient and friction factor correlations may not be applicable without extrapolation to the proposed design.

For example, in a cold plate containing small rectangular passages where the flow is almost certain to be in the laminar regime, can the Sieder-Tate correlation [2]

$$\frac{hd_e}{k_b} = 1.86 \left[\left(\frac{d_e G}{\mu_b} \right) \left(\frac{c_p \mu}{k} \right)_b \left(\frac{d_e}{L} \right) \right]^{1/3} \left(\frac{\mu}{\mu_w} \right)^{0.14}$$

be applied? Perhaps yes, but a broad-brush application to smaller and smaller surfaces may be an invitation to disappointment.

In this chapter, only compact surfaces will be considered. Several organizations manufacture a variety of compact surfaces on a short delivery schedule. Compact heat exchanger surfaces are described in the literature by geometric factors that have been standardized largely through the extensive work of Kays and London [1]. These factors and the relationships between them are essential for the application of the basic heat transfer and flow friction data to a particular design problem.

A description of the Kays-London surfaces has been presented in Chap. 10.

15.3   DESIGN EQUATIONS

The design of a cold plate depends on a balance of the *rate equation*,

$$q = h\eta_w S(\text{LMTD}) \tag{15.1}$$

where LMTD is the logarithmic mean temperature difference and η_w is the overall passage efficiency, and the energy equation for the coolant fluid,

$$q = Wc_p(t_2 - t_1) \tag{15.2}$$

15.4   THE LOGARITHMIC MEAN TEMPERATURE DIFFERENCE AND THE EFFECTIVENESS ϵ AS A FUNCTION OF N_{tu}

Figure 15.1a displays a temperature-surface profile for a heat exchanger with a constant temperature source and a rising temperature temperature receiver fluid. It is one of the two extremes to be considered. The other is shown in Fig. 15.1b, which shows the temperature-surface profile for a constant heat flux at the dissipating surface. Because of conduction in the base surface, which is the wall of the cold plate, the base surface tends to be isothermal and actual cold plate operation tends to resemble the situation shown in Fig. 15.1a.

The logarithmic mean temperature difference, LMTD in Eq. (15.1), is given by

$$\text{LMTD} = \frac{(T_s - t_1) - (T_s - t_2)}{\ln (T_s - t_1)/(T_s - t_2)} = \frac{t_2 - t_1}{\ln (T_s - t_1)/(T_s - t_2)} \tag{15.3}$$

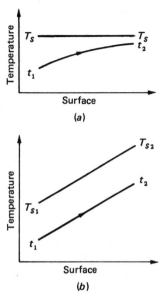

FIG. 15.1 Two temperature-surface profiles for the cold plate heat exchanger. (*a*) Constant temperature source. (*b*) Constant heat flux source.

When this is put into Eq. (15.1), Eqs. (15.1) and (15.2) can be combined to yield

$$q = h\eta_w S \left[\frac{t_2 - t_1}{\ln (T_s - t_1)/(T_s - t_2)} \right] = Wc_p(t_2 - t_1)$$

Now a little algebra shows that

$$\ln \frac{T_s - t_1}{T_s - t_2} = \frac{h\eta_w S}{Wc_p}$$

and with $N_{tu} \equiv h\eta_w S/Wc_p$,

$$\frac{T_s - t_1}{T_s - t_2} = e^{N_{tu}} = F \tag{15.4}$$

A solution for the surface temperature T_s is easy to obtain:

$$T_s - t_1 = F(T_s - t_2)$$
$$Ft_2 - t_1 = (F - 1)T_s$$

and

$$T_s = \frac{Ft_2 - t_1}{F - 1} \tag{15.5}$$

The exchanger effectiveness is defined as

$$\epsilon \equiv \frac{t_2 - t_1}{T_s - t_1}$$

and an expression for the effectiveness of the cold plate can be obtained by starting with Eq. (15.4),

$$\frac{T_s - t_1}{T_s - t_2} = e^{Ntu} \tag{15.4}$$

and performing a little algebra. Add and subtract t_1 to the denominator on the left side of Eq. (15.4).

$$\frac{T_s - t_1}{T_s - t_1 - t_2 + t_1} = e^{Ntu}$$

Then

$$\frac{(T_s - t_1)}{(T_s - t_1) - (t_2 - t_1)} = \frac{(T_s - t_1)}{(T_s - t_1)[1 - (t_2 - t_1)/(T_s - t_1)]} = e^{Ntu}$$

and $$\frac{1}{1 - \epsilon} = e^{Ntu}$$

$$1 - \epsilon = e^{-Ntu}$$

and $$\epsilon = 1 - e^{-Ntu} \tag{15.6}$$

15.5 THERMAL TRANSPORT PROPERTIES FOR AIR

Thermal transport properties for air at atmospheric pressure are plotted in Fig. 6.10. These properties should be taken at the average film temperature in the fluid passage. The average fluid temperature is, of course,

$$t_a = \frac{t_1 + t_2}{2}$$

and the average film temperature is the arithmetic average of the surface temperature and the average fluid temperature,

$$T_f = \frac{T_s + t_a}{2} = \frac{1}{2}\left[T_s + \frac{1}{2}(t_1 + t_2)\right]$$

or $$T_f = \frac{1}{4}(2T_s + t_1 + t_2)$$

When dealing with air, it makes little difference whether the average (bulk) or the average film temperature is used. In the sample problems that follow, use is made of the average film temperature.

15.6 DESIGN PROCEDURE FOR THE COLD PLATE HEAT EXCHANGER

The design procedure applies to either the rectangular passage or the compact core; it is a logical progression involving the following steps:

1 Assume a configuration.
2 Compute the physical data, free areas, surfaces, and equivalent diameter where required.
3 Determine the average film or the average (bulk) fluid temperature and the required fluid properties at that temperature.
4 Evaluate the heat transfer coefficient.
5 Formulate N_{tu} and compute the surface temperature T_s, which is the temperature attainable for the imposed conditions.
6 Compare the surface temperature computed in step 5 with that required. If the computed temperature is higher than that required, the cold plate exchanger is too small and the process must be repeated from step 1. If the computed temperature is much lower than the required temperature, the cold plate exchanger is over-surfaced and, in the interest of weight economy, the process should be repeated from step 1.
7 Compute the pressure drop and see if the value obtained falls below the specified limit.

15.7 SAMPLE PROBLEM: EVEN LOADING, BOTH SIDES

It is desired to use one of the Kays and London compact plain plate fin heat exchanger cores as a cold plate to handle the heat dissipation of a package of electronic components. The design specifications are as follows:

1 Dissipation: 400 W
2 Allowable chassis temperature: 125°C (257°F) max
3 Cold plate width: 4 in max
4 Cold plate depth: 6 in min, 9 in max
5 Cold plate height: 0.30 in max
6 Coolant: air
7 Flow rate: 1.35 lb/min max
8 Inlet temperature: 160°F
9 Allowable pressure loss: 2.25 in H_2O

The solution to the problem is twofold:

(a) Design the cold plate.
(b) Estimate the air flow rate when the inlet temperature is 100°F. Chassis temperature is to be 257°F or less.

The worksheet shown as Table 15.1 has been constructed to show how this problem may be solved in a step-by-step manner. Six columns of figures may be noted. The first column represents a design for a 19.86 plate fin core, which must be discarded

TABLE 15.1 Cold Plate Heat Exchanger Design and Performance Worksheet—Even Loading on Both Sides[a]

Dissipation: 400 W = 1,365 Btu/h
Req'd. chassis temperature = 257°F
Head loading: One side ☐ Two sides ☑

Restrictions on dimensions:
Width–4 in max
Depth–6 in min, 9 in max
Coolant: air; Flow rate: 1.35 lb/min

Item	Dimensions	How computed	Trial					
			I	II	III	IV	V	VI
Physical data								
Type		Given or assumed	19.86	11.1				
Plate separation b	in	Given or assumed	0.25	0.25				
Plate thickness a	in	Table 10.3	0.01	0.01				
Fin thickness δ	in	Assume	0.006	0.006				
β	ft²/ft³	Table 10.3	561	367				
Hydraulic radius r_h	ft	$r_h = (1/4)d_e$ (Table 10.3)	0.001538	0.00253				
Equivalent diameter d_e	ft	$4r_h$	0.00615	0.01012				
S_f/S	ft²/ft²	Table 10.3	0.849	0.756				
Cold plate width W	in	Assume	4	4				
Cold plate depth D	in	Assume	6	8				
Cold plate height H	in	Assume	0.27	0.27				
Cold plate volume V	ft³	V (1/1,728)WDH	0.00375	0.00500				
Cold plate material		Assume	Aluminum	Aluminum				
Thermal conductivity	Btu/ft h °F	Standard handbook	117	117				
α	ft²/ft³	$\alpha = [b/(b + 2a)](\beta)$	520	340				
σ	ft²/ft²	$\sigma = \alpha r_h$	0.80	0.86				
Frontal area A_f	ft²	$A_f = (1/144)WH$	0.0075	0.0075				
Flow area A	ft²	$A = \sigma A_f$	0.0060	0.00645				
Surface S	ft²	$S = \alpha V$	1.95	1.70				
Heat balance fluid properties								
Coolant flow w	lb/min	Given or assumed	1.35	1.35	0.65	0.70	0.75	0.80
Coolant flow W	lb/h	$W = 60w$	81	81	39	42	45	48

	Units	Formula						
Δt	°F	$\Delta t = q/Wc_p$	70.4	70.4	146	135.8	126.5	118.8
t_1	°F	Given or assumed	160.0	160.0	100	100	100	100
t_2	°F	$t_2 = t_1 + \Delta t$	230.4	230.4	246	235.8	226.5	218.8
T_f	°F	$T_f = (1/4)[2T_s + (t_1 + t_2)]$	226	226	215	212.5	210	208
μ at T_f	lb/h ft	Fig. 6.10	0.0516	0.0516	0.0509	0.0507	0.0505	0.0503
$(\mathrm{Pr})^{2/3}$ at T_f		Fig. 6.10	0.792	0.792	0.792	0.792	0.792	0.792
Coefficients								
G	lb/ft² h	$G = W/A$	13,500	12,500	6,050	6,510	6,980	7,450
Re		$\mathrm{Re} = d_e G/\mu$	1,608	2,462	1,205	1,300	1,400	1,498
j		Fig. 10.4	0.00451	0.00446	0.00495	0.00484	0.00472	0.00466
f		Fig. 10.4	0.0139	0.0119				
h	Btu/ft² h °F	$h = Gc_p j/(\mathrm{Pr})^{2/3}$	18.52	16.98	9.12	9.59	10.02	10.56
Surface efficiency								
m	1/ft	$m = \sqrt{24h/k\delta}$	25.2	24.1	17.68	18.08	18.51	19.01
b^*	ft	$b^* = (1/24)b$	0.01041	0.01041	0.01041	0.01041	0.01041	0.01041
mb^*		Compute	0.263	0.251	0.184	0.188	0.193	0.199
η_f		$\eta_f = (\tanh mb^*)/mb^*$	0.977	0.979	0.989	0.988	0.987	0.986
$1 - \eta_f$		Compute	0.023	0.021	0.011	0.012	0.013	0.014
$S_f/S(1 - \eta_f)$		Compute	0.020	0.016	0.008	0.009	0.010	0.011
η_w		$\eta_w = (S_f/S)(1 - \eta_f)$	0.980	0.984	0.992	0.991	0.990	0.989
Surface temperature								
N_{tu}		$N_{tu} = h\eta_w S/Wc_p$	1.814	1.452	1.636	1.595	1.556	1.536
F		$F = e^{N_{tu}}$	6.12	4.28	5.13	4.93	4.74	4.65
Ft_2	°F	Compute	1,410	988	1,262	1,162	1,074	1,018
t_1	°F	Given	160	160	100	100	100	100
$Ft_2 - t_1$	°F	Compute	1,250	828	1,162	1,062	974	918
$F - 1$		Compute	5.12	3.28	4.13	3.93	3.74	3.65
T_s	°F	$T_s = (Ft_2 - t_1)/(F - 1)$	244	253	281	270	260	251.5
Pressure loss								
G'	lb/ft² s	$G' = G/3{,}600$	3.75	3.49				
$(G')^2$	lb²/ft⁴ s²	Compute	14.08	12.18				
ρ_1	lb/ft³	$\rho_1 = 39.55/(460 + t_1)$	0.0638	0.0638				

(See footnote on page 366.)

TABLE 15.1 Cold Plate Heat Exchanger Design and Performance Worksheet—Even Loading on Both Sides (Continued)

Dissipation: 400 W = 1,365 Btu/h
Req'd. chassis temperature = 257°F
Head loading: One side □ Two sides ☑

Restrictions on dimensions:
Width—4 in max
Depth—6 in min, 9 in max
Coolant: Air; Flow rate: 1.35 lb/min

Item	Dimensions	How computed	Trial					
			I	II	III	IV	V	VI
Pressure loss								
ρ_2	lb/ft³	$\rho_2 = 39.55/(460 + t_2)$	0.0573	0.0573				
v_1	ft³/lb	$v_1 = 1/\rho_1$	15.69	15.69				
v_2	ft³/lb	$v_2 = 1/\rho_2$	17.43	17.43				
v_m	ft³/lb	$v_m = (1/2)(v_1 + v_2)$	16.56	16.56				
v_m/v_1		Compute	1.058	1.058				
v_2/v_1		Compute	1.114	1.114				
K_c		Fig. 10.8	1.01	0.28				
K_e		Fig. 10.8	−0.645	−0.13				
$1 + K_c - \sigma^2$		Compute	1.37	054				
$2[(v_2/v_1) - 1]$		Compute	0.23	0.23				
$f(S/A)(v_m/v_1)$		Compute	4.77	3.32				
$-(1 - \sigma^2 - K_e)(v_2/v_1)$		Compute	−1.12	−0.43				
Σ		Sum of 4 previous items	5.25	3.66				
$(G')^2 v_1/2g_c$		Compute	3.43	2.97				
Δp	in H₂O	$0.1925\,(G')^2(v_1/2g_c)[\Sigma]$	3.47	2.09				
Remark as to adequacy			Heat transfer too good, Δp too high. Use 11.1 and make longer	OK				

a Refer to Chap. 10 for heat exchanger nomenclature.

because of excessive pressure loss. The second column is the actual design to be proposed. This uses an 11.1 plate fin core, and although the performance obtained is not as good, the pressure loss is now within the required limit.

The last four columns of computation in Table 15.1 represent performance runs for assumed air flow values. It may be noted that none of these runs yields a chassis temperature of exactly 257°F. Based on these data, however, a graphical solution can be evolved as shown in Fig. 15.2. Use of four runs minimizes the chance of error for the curve, which must be plotted from at least three valid points. If one point is in error, the error will be noticed immediately and the point may then be discarded. Note that because this example considers finned exchanger cores, Table 15.1 utilizes the relationship

$$N_{tu} \equiv \frac{h\eta_w S}{Wc_p}$$

where η_w is the weighted overall surface efficiency.

15.8   MODIFICATION FOR SINGLE–SIDE HEAT LOADING

If heat is dissipated by only one side of the cold plate, the nondissipating side must be treated as a pair of fins as shown in Fig. 15.3. The efficiency of such an arrangement may be computed by using the procedure outlined in Sec. 14.7.

For example, consider the 11.1 plain plate fin with cover plates $\frac{1}{8}$ in thick. For the repeating section shown in Fig. 15.3, fins 1 and 2 have adiabatic tips, and pertinent dimensions are as follows:

$$b_c = \frac{1}{2(11.1)(12)} = 0.00375 \text{ ft}$$

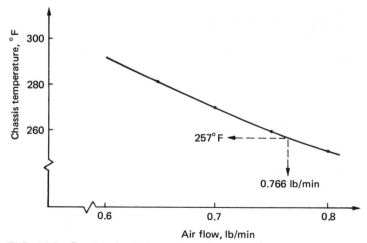

FIG. 15.2   Graphical solution of sample problem based on points computed in Table 15.1.

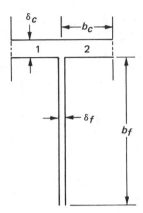

FIG. 15.3   Repeating section of 11.1 plain plate fin heat exchanger surface. Fins 1 and 2 form the cover plate where no heat is injected.

$$\delta_c = \frac{0.125}{12} = 0.01042 \text{ ft}$$

$$b_f = \frac{0.25}{12} = 0.02083 \text{ ft}$$

$$\delta_f = \frac{0.006}{12} = 0.00050 \text{ ft}$$

For an aluminum exchanger ($k = 117$ Btu/ft h °F), one may plot the weighted overall efficiency by assuming several values of h. For example, if $h = 10$ Btu/ft² h °F:

For fins 1 and 2, for a 1-ft length,[1]

$$m_c = \left(\frac{h}{k\delta_c}\right)^{1/2} = \left[\frac{10}{117(0.01042)}\right]^{1/2} = 2.8640 \text{ ft}^{-1}$$

$$m_c b_c = (2.8647)(0.00375) = 0.0108$$

$$Y_{oc} = (hk\delta_c)^{1/2} L = [10(117)(0.01042)]^{1/2}(1) = 3.9416 \text{ Btu/h °F}$$

and $$Z_{oc} = \frac{1}{Y_{oc}} = 0.2864 \text{ °F h/Btu}$$

Equation (14.31) is now employed,

$$\left.\frac{q_b}{\theta_b}\right|_1 = \left.\frac{q_b}{\theta_b}\right|_2 = \frac{\gamma_{21} - \gamma_{11}(q_a/\theta_a)}{-\gamma_{22} + \gamma_{12}(q_a/\theta_a)}$$

and because $q_a/\theta_a|_1 = q_a/\theta_a|_2 = 0$, only γ_{21} and γ_{22} need to be computed.

$$\gamma_{21} = -\frac{\sinh mb}{Z_0} = -\frac{\sinh(0.0108)}{0.2864} = -\frac{0.0108}{0.2864} = -0.0375 \text{ Btu/h °F}$$

[1] No factor of 2 appears in m_c and Z_{oc}. This is because only one side of fins 1 and 2 dissipate to the coolant stream.

$$\gamma_{22} = \cosh mb = \cosh (0.0108) = 1.0001$$

and $\left.\dfrac{q_b}{\theta_b}\right|_1 = \left.\dfrac{q_b}{\theta_b}\right|_2 = \dfrac{\gamma_{21}}{-\gamma_{22}} = \dfrac{-0.0375}{-1.0001} = 0.0375$ Btu/h °F

This makes

$$\left.\frac{q_a}{\theta_a}\right|_2 = 2(0.0375) = 0.0751 \text{ Btu/h °F}$$

For fin 3,

$$m_f = \left(\frac{2h}{h\delta_f}\right)^{1/2} = \left[\frac{2(10)}{117(0.0005)}\right]^{1/2} = 18.4900 \text{ ft}^{-1}$$

$$m_f b_f = (18.4900)(0.02083) = 0.3852$$

$$Y_{of} = (2hk\delta_f)^{1/2} = 1.0817 \text{ Btu/h °F}$$

and $\quad Z_{of} = 0.9245$ °F h/Btu

In this case,

$$\gamma_{11} = \gamma_{22} = \cosh mb = \cosh (0.3852) = 1.0751$$

$$\gamma_{12} = -Z_0 \sinh mb = -0.9245 \sinh (0.3852)$$

$$= -(0.9245)(0.3948) = -0.3650 \text{ °F h/Btu}$$

$$\gamma_{21} = -\frac{\sinh mb}{Z_0} = -\frac{\sinh (0.3852)}{0.9245}$$

$$= -\frac{0.3948}{0.9245} = -0.4270 \text{ Btu/h °F}$$

Hence, from Eq. (14.31),

$$\left.\frac{q_b}{\theta_b}\right|_1 = \frac{\gamma_{21} - \gamma_{11}(q_a/\theta_a)}{-\gamma_{22} + \gamma_{12}(q_a/\theta_a)}$$

$$= \frac{-0.4270 - (1.0751)(0.0751)}{-1.0751 + (-0.3650)(0.0751)}$$

$$= \frac{-0.5078}{-1.1025}$$

or, finally,

$$\left.\frac{q_b}{\theta_b}\right|_1 = 0.4606 \text{ Btu/h °F}$$

The surface contained in one repeating section of fins is computed to be

$$S_f = [2(0.02083) + 2(0.00375)](1) = 0.04917 \text{ ft}^2$$

and if the entire finned surface were to operate at the base temperature excess, one would have the ideal heat dissipation

$$q_i = hS_f\theta_b$$

or $\left.\dfrac{q_b}{\theta_b}\right|_i = hS_f = 10(0.04917) = 0.4917 \text{ Btu/h }^\circ\text{F}$

This makes the fin efficiency

$$\eta_f = \frac{q_b/\theta_b}{q_b/\theta_b|_i} = \frac{q}{q_i} = \frac{0.4606}{0.4917} = 0.937$$

Table 10.3 shows that for the 11.1 plain plate fin surface, $S_f/S = 0.756$. However, this figure applies to a case of equal dissipation through each cover plate. To obtain a similar figure for single-sided dissipation, observe that the prime surface to total surface ratio for the two-sided dissipation may be obtained from

$$S_b = S - S_f$$

or $\dfrac{S_b}{S} = 1 - \dfrac{S_f}{S} = 1 - 0.756 = 0.244$

For the case of single-sided dissipation, half of this becomes finned surface, or

$$\frac{S_f}{S} = 0.756 + \frac{1}{2}(0.244) = 0.878$$

Now Eq. (14.15) may be applied to determine the weighted overall efficiency:

$$\eta_0 = 1 - (1 - \eta_f)\frac{S_f}{S}$$
$$= 1 - (1 - 0.937)(0.878)$$
$$= 1 - 0.055 = 0.945$$

A plot of the weighted overall efficiency for single 11.1 and 19.86 plain plate fin compact heat exchanger cores with dissipation on only one side is shown in Fig. 15.4. These plots can be used only for an outer surface that is $\frac{1}{8}$ in wide.

15.9 SAMPLE PROBLEM: HEAT LOADING, ONE SIDE

An 11.1 plain plate fin aluminum heat exchanger core that is 8 in wide and 15 in deep is to carry 1.56 lb/min of air at an inlet temperature of 55°F. Estimate the surface temperature and the air pressure loss for an even dissipation on one side of 1 kW if the outer surface is $\frac{1}{8}$ in wide.

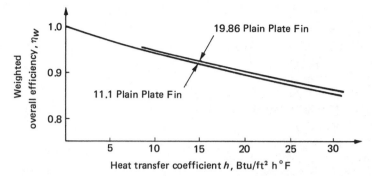

FIG. 15.4 Weighted overall efficiency for 11.1 and 19.86 plain plate fin surfaces with single-sided heat dissipation. No heat is injected at the cover plate, which is $\frac{1}{8}$ in wide. These curves may not be used for cover plates of a different width.

Although the worksheet of Table 15.1 is useful and can be employed, the procedure here will be merely to display the calculations.

1 Physical data:
   Type: 11.1 plain plate fin
   Plate separation $b = \frac{1}{4}$ in (Table 10.3)
   Plate thickness $a = \frac{1}{8}$ in (given)
   Fin thickness $\delta = 0.006$ in (Table 10.3)
   $\beta = 367$ ft^2/ft^3 (Table 10.3)
   Hydraulic radius $r_h = 0.00253$ ft (Table 10.3)
   Equivalent diameter $d_e = 4r_h = 0.01012$ ft
   S_f/S: not needed
   Width $W = 8$ in (given)
   Depth $D = 15$ in (given)
   Height $H = b + a = 0.25 + 0.125 = 0.375$ in
   Volume $V = WDH/1728 = 0.0260$ ft^3
   Material: aluminum (given)
   Thermal conductivity $k = 117$ Btu/ft h °F
   $\alpha = [b/(b + a)]\beta = (0.25/0.375)367 = 244.67$ ft^2/ft^3
   $\sigma = \alpha r_h = 0.619$
   Frontal area $A_f = WH/144 = 0.0208$ ft^2
   Flow area $A = \sigma A_f = 0.0129$ ft^2
   Surface $S = \alpha V = 6.372$ ft^2
2 Heat balance and fluid properties:
   Flow $w = 1.56$ lb/min (given)
   Flow $W = 60w = 93.6$ lb/h
   $\Delta t = \text{watts}/4.21W = 1000/4.21w = 152.3^*$
   $t_1 = 55°$F (given)

$^*q = Wc_p \Delta t$, where W is in lb/h, $c_p = 0.241$ Btu/lb °F, and q is in Btu/h, $q = 4.21w \Delta t$, where w is in lb/min and q is in watts.

$t_2 = t_1 + \Delta t = 207.3°F$

$T_f = t_1 + \Delta t/2 = 131.1°F$

$\mu @ T_f = 0.0482 \text{ lb/h ft (Fig. 6.10)}$

$(\text{Pr})^{2/3} @ T_f = 0.795 \text{ (Fig. 6.10)}$

3 Heat transfer coefficient:

$G = W/A = 93.6/0.0129 = 7255.8 \text{ lb/ft}^2 \text{ h}$

$\text{Re} = d_e G/\mu = (0.0101)(7255.8)/(0.0482) = 1523.4$

$j = 0.00457 \text{ (Fig. 10.4)}$

$f = 0.0146 \text{ (Fig. 10.4)}$

$h = jGc_p/(\text{Pr})^{2/3} = 10.06 \text{ Btu/ft}^2 \text{ h °F}$

4 Weighted overall efficiency:

$\eta_w = 0.943 \text{ (Fig. 15.4)}$

5 Surface temperature:

$N_{tu} = hS\eta_w/Wc_p = 2.680$

$F = e^{N_{tu}} = 14.581$

$T_s = (Ft_2 - t_1)/(F - 1) = 218.5°F$

6 Pressure loss:

$G' = G/3600 = 2.016$

$(G')^2 = 4.062$

$v_1 = (55 + 460)/39.55 = 13.02$

$v_2 = (207.3 + 460)/39.55 = 16.87$

$v_m = (v_1 + v_2)/2 = 14.95$

$v_m/v_1 = 1.148$

$v_2/v_1 = 1.296$

$K_c = 1.11 \text{ (Fig. 10.8)}$

$K_e = -0.42 \text{ (Fig. 10.8)}$

$\psi_1 = 1 + K_c - \sigma^2 = 1.727$

$\psi_2 = 2(v_2/v_1 - 1) = 0.591$

$\psi_3 = (fS/A)(v_m/v_1) = 8.278$

$\psi_4 = (1 - K_e - \sigma^2)(v_2/v_1) = 1.343$

$\psi_5 = \psi_1 + \psi_2 + \psi_3 - \psi_4 = 9.253$

$\psi_6 = (G')^2 v_1/2g_c = 0.822$

$\Delta p = 0.1925\psi_5\psi_6 = 1.464 \text{ in H}_2\text{O}$

The results may be summarized:

$T_s = 218.5°F$

$\Delta p = 1.464 \text{ in H}_2\text{O}$

15.10 CORE STACKING

The compact heat exchanger surfaces may be stacked to form what is sometimes called a *sandwich*. Typical stacks are the double and the triple sandwiches as shown in Fig. 15.5.

The procedure for cold plate analysis and design is the same regardless of the number of stacks employed. However, the overall passage efficiency will be different

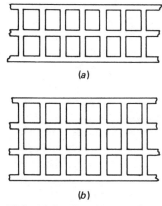

(a)

(b)

FIG. 15.5 Two heat exchanger stacks. (*a*) Double stack. (*b*) Triple stack.

for each stacking arrangement. It can be evaluated by employing the procedure outlined in Sec. 14.7.

In the event that the stacking consists of a double stack with equal heat distribution on opposite sides, Fig. 15.3 represents the situation and the calculation procedure follows the procedure outlined in Sec. 15.9. Here, however, the widths of fins 1 and 2 will be half the separation plate width.

For a triple stack with equal heat dissipation on opposite sides, the configuration is shown in Fig. 14.5 and the calculation procedure of Sec. 14.6 is used. Here, however, the width of the separation plate is used with heat dissipation on both sides (fins 2 and 3), and the fin height used for fin 1 is half the plate spacing.

15.11   A NOTE ON THE ISOTHERMAL SURFACE

It has been pointed out that the cold plate performance comes surprisingly close to that of an isothermal surface when the dissipated heat is distributed evenly over the surface of the cold plate. Indeed, this fact has served as the basis for the discussions and illustrative examples in this chapter. It is well, at this point, to consider the effects of a departure from this rather ideal situation.

In Fig. 15.6, curve *aa* is the temperature-length profile of the coolant fluid and

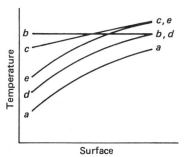

FIG. 15.6 Some temperature-surface profiles for a cold plate heat exchanger.

curve *bb* is the isothermal surface. Suppose, however, that the heat loading or film coefficients, individually or in combination, have an effect such that the surface temperature profile is as shown by curve *cc*. Then the methods of computation as outlined in this chapter will yield an optimistic design because the logarithmic mean temperature difference between curve *cc* and curve *aa* may be lower than that between *bb* and *aa*.

If it is suspected that the temperature of the surface will depart significantly from the isothermal, a safe way of designing the cold plate is to assume that the surface will have the temperature profile shown by curve *dd* of Fig. 15.6. Then when the cold plate is actually installed and the surface temperature profile takes the shape of curve *ee*, all conditions are satisfied and the cold plate is more than meeting its service requirements.

If, however, the cold plate is or is close to an isothermal surface, then operation along curves *dd* and/or *ee* represents an overdesign of the cold plate. For this reason, the designer must take a long, hard look at the service requirements, the temperatures of both coolant and cold plate surface, and the heat distribution in the cold plate before beginning a design.

15.12   NOMENCLATURE

Roman Letter Symbols

a	cover plate width, ft
A	area, ft^2
b	fin height, ft
c	specific heat, Btu/lb $^\circ$F
d	diameter, ft
D	exchanger depth, ft
f	friction factor, dimensionless
F	a factor defined by Eq. (15.4)
G	mass flow velocity, lb/ft^2 h
h	heat transfer coefficient, Btu/ft^2 h $^\circ$F
H	exchanger height, ft
j	heat transfer factor, dimensionless
k	thermal conductivity, Btu/ft h $^\circ$F
K	loss coefficient, dimensionless
L	length, ft
LMTD	logarithmic mean temperature difference, $^\circ$F
m	fin performance factor, ft^{-1}
N_{tu}	number of transfer units, dimensionless
Nu	Nusselt number, dimensionless
Pr	Prandtl number, dimensionless
q	heat flow, Btu/h
Re	Reynolds number, dimensionless
r	radius, ft
S	surface area, ft^2
t	temperature, $^\circ$F

T	temperature, °F
v	specific volume, ft³/lb
V	volume, ft³
w	flow of fluid, lb/min
W	width of exchanger, ft; or flow of fluid, lb/h
Y	fin admittance, Btu/h °F
Z	fin impedance, °F h/Btu

Greek Letter Symbols

α	ratio of total surface area on *one* side of exchanger to total volume on *both* sides of exchanger, dimensionless
β	ratio of total surface area on *one* side of exchanger to total volume on *one* side of exchanger, dimensionless
γ	an element of thermal transmission matrix, dimensions vary
δ	fin or cover plate width, ft
Δ	indicates change in varaible
ϵ	exchanger effectiveness, dimensionless
η	fin or passage efficiency, dimensionless
θ	temperature excess, °F
μ	dynamic viscosity, lb/h ft
ρ	density, lb/ft³
σ	ratio of free flow area to frontal area, dimensionless
ψ	indicates a grouping of terms in pressure loss calculations, dimensions vary

Subscripts

a	indicates tip of fin
b	indicates base of fin
c	indicates cover or entrance (contraction) condition
e	indicates equivalent or exit condition
f	indicates film or fin or frontal condition
h	indicates hydraulic
i	indicates ideal condition
m	indicates mean condition
p	indicates constant pressure condition
s	indicates surroundings
w	indicates wall or weighted condition

Superscript

$'$	indicates mass flow velocity in lb/ft² s

15.13 REFERENCES

1 Kays, W. M., and London, A. L., *Compact Heat Exchangers*, McGraw-Hill, New York, 1964.
2 Sieder, E. N., and Tate, G. E., Heat Transfer and Pressure Drop of Liquids in Tubes, *Ind. Eng. Chem.*, vol. 28, pp. 1429–1436, 1936.

16

■ immersion cooling

16.1 INTRODUCTION

Thermal control of operational electronic components by direct immersion in low-boiling-point, dielectric fluids dates back to the late 1940s [1]. During the past 35 years, however, this technique has been used extensively only for stabilizing the temperature of high-power radar components typified by klystron tubes and high-voltage power supplies dissipating 1 to 5 kW in a volume of approximately 0.03 m³ (1 ft³) [2–6]. More recently, the growing use of large-scale and very large-scale integration (LSI and VLSI) microelectronic technologies, with chip dissipation approaching 2000 mW and chip surface heat fluxes exceeding 10 W/cm², has focused renewed attention on immersion cooling. In both these applications, the use of this technique can provide a benign, local ambient and accommodate substantial spatial and temporal power variations while minimizing temperature excursions and component failure rates.

16.2 CLASSIFICATION

In examining immersion cooling systems, it is convenient to classify existing and/or potential configurations by the thermal processing of the generated vapor. In the simplest "evaporator" configuration, shown schematically in Fig. 16.1, no attempt is made to recycle the gaseous phase and evaporation is allowed to proceed until the liquid supply is depleted. The boiling point of the dielectric fluid and hence the range of the operating temperature of the components can be regulated by the use of a vapor line control valve set to open at the desired pressure. However, the weight and volume penalties associated with the liquid reservoir limit the utility of such simple evaporators to relatively low heat release systems—briefly dissipating a very high heat flux and/or requiring only a modest heat removal rate during extended operation. When both high dissipation rates and long operating periods must be accommodated, the vapor generated in the immersion cooler must be condensed and the liquid returned to the reservoir.

   Many systems that rely on fluid recycle employ "remote condensers"—connected by appropriate piping to the immersion module—or condensing surfaces placed directly in the vapor space above the evaporating/boiling liquid, as shown in Figs. 16.2 and 16.3. Regrettably, there are two major drawbacks associated with both remote and vapor space condensers: Packaging height is limited to the liquid height at minimum temperature and the least favorable orientation, and the presence of even small quantities of air in the vapor space can cause a dramatic degradation in condenser performance. These liabilities are exacerbated by the relatively large volumetric expansion coefficient of many candidate fluids (approximately 1 to $1\frac{1}{2}$% per 10°C rise in fluid

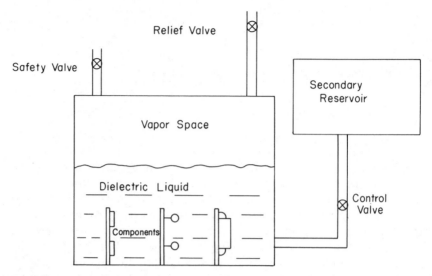

FIG. 16.1 Simple evaporator module for electronic components.

temperature) and the high solubility of air in these fluids, approaching 0.5 cm³ air at STP/1 cm³ liquid at room temperature. To cope with this latter factor, it is necessary to specify elaborate filling and degassing procedures for such immersion cooling units, thus decreasing overall reliability and raising maintenance and field repair costs.

The limitations inherent in the design and operation of remote and vapor space

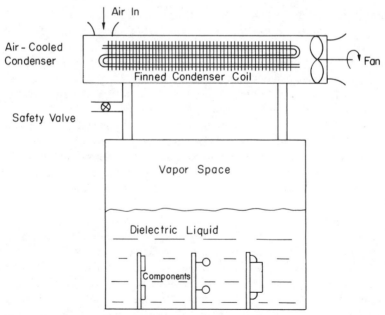

FIG. 16.2 Remote condenser/immersion cooling module for electronic components.

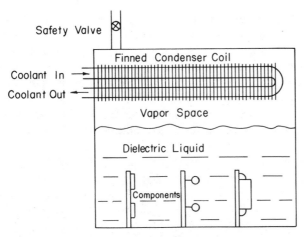

FIG. 16.3 Vapor space condenser/immersion cooling module for electronic components.

condensers can be overcome by submerging the condenser in the liquid, as in Fig. 16.4. In this configuration, the condenser surface serves primarily to subcool the enclosed fluid and vapor bubbles generated by boiling at the surfaces of the dissipative electronic components rise and begin to condense in the fluid. The presence of non-condensables dissolved in the liquid reduces the collapse rate of the vapor bubbles, but a slight inclination of the surface can be used to direct small, low-vapor-content bubbles impinging on the submerged condenser toward a gas accumulator or expansion chamber.

In the absence of such gas control measures, much of the noncondensable gas remaining after moderate degassing of the liquid is concentrated in small, migrating gas bubbles, with the remainder dissolved in the subcooled liquid.

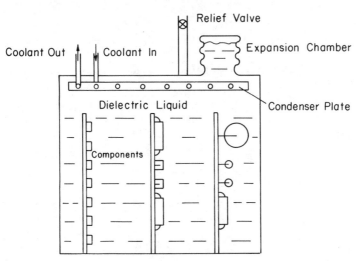

FIG. 16.4 Horizontal submerged condenser/immersion cooling module for electronic components.

When operating in this largely convective or liquid heat exchanger mode, the submerged condenser is thus not affected substantially by the presence of noncondensables in the liquid. This is in sharp distinction to the remote and vapor space condensers, in which the noncondensables accumulate in the vapor space, impede the flow of vapor toward the condenser surface, and lead to deterioration of the condensive heat transfer coefficient. The elimination of the vapor space can therefore be expected to reduce significantly the effect of noncondensables on the thermal performance of immersion modules and result in considerable economies in volume and weight [2, 3].

Similar advantages can often be realized by utilizing the side walls of the container as the primary cooling surfaces, as shown in Fig. 16.5, or as secondary cooling surfaces in conjunction with the horizontal submerged condenser. The substantial submerged condenser area made available in this configuration reduces the average surface heat flux and can make possible the use of air rather than liquid cooling to remove the heat dissipated within the liquid-filled enclosure [5]. That is, although extremely high heat fluxes and nucleate boiling may prevail on the limited area of the operating components, the heat flux at the multiple submerged condenser surfaces can be sufficiently low to allow forced or even free convection air cooling of the entire module.

The volumetric constraints applied to the electronic cooling system impose a further distinction in the classification of vapor condensing, immersion cooling modules. Much of the research and development effort, as well as the equipment marketed to date, has focused on constant or near-constant pressure modules with expansion chambers and/or pressure relief valves used to meet this requirement. Alternatively, in some applications and a notable experimental investigation [7], the module volume was held nearly constant and the pressure allowed to rise as the mass fraction of the gas and vapor in both the liquid and the vapor space increased.

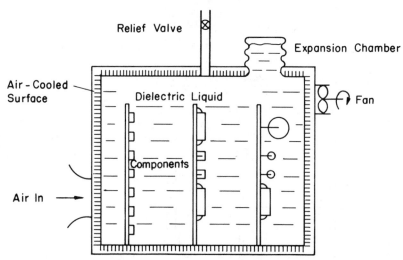

FIG. 16.5 Multiple-surface submerged condenser/immersion cooling module for electronic components.

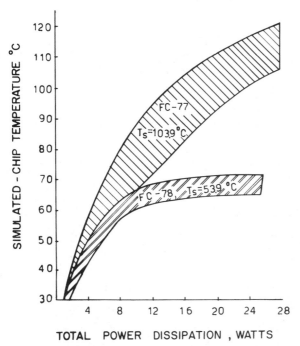

FIG. 16.6 Simulated chip surface temperature in immersion cooling (nine chips on 8.5 cm^2 substrate) [9].

16.3 PUBLISHED LITERATURE

In investigating analytically or experimentally the thermal performance of immersion cooling modules, attention must be directed toward the heat dissipating surfaces, thermal transport at the submerged or exposed condenser surface(s), and the interaction between the vapor bubbles and the enclosed liquid when ebullient heat transfer prevails on the component surfaces.

Although the open literature contains only limited data on the boiling characteristics of immersed electronic components, the available results for ebullient cooling of klystron microwave tubes [6], ferrite phase shifters [8], thick-film microelectronic resistors [9], and power transistors [10], as well as scores of proprietory company reports, confirm the viability of this thermal control technique. Empirical temperature curves, such as those shown in Fig. 16.6, for a nine-test chip array on an 8.5-cm^2 alumina substrate immersed alternately in FC-77 and FC-78 [9], indicate that pool boiling of Freon and Fluorocarbon dielectric fluids corresponds, in general, to the sequence of physical mechanisms outlined in Sec. 8.3 and represented in the boiling curve of Fig. 8.4. Furthermore, as shown in Table 16.1, these fluids cover a wide range of boiling points at atmospheric pressure, and an appropriate fluid can generally be found for every application of interest.

As previously noted, however, the wetting characteristics and extraordinarily high air solubilities of many room-temperature dielectric fluids lead to somewhat anomalous behavior. The initial surface temperature excursion associated with delayed

TABLE 16.1 Properties of Candidate Fluids for Immersion Cooling[a],[b]

	FC-88	Freon E1	Freon C-51-12	Freon TF	FC-78	FCX 326	FC-77	FCX 327	FC-75	Freon E2	FCX 328	Freon E3	FC-43	N-43	Freon E4	Freon E5
Boiling point, 1 atm (°C)	31	40.8	45	47.6	50	76	97.2	102	102.2	104.4	129	152.3	173.9	177	193.8	224.2
Heat of vaporization (J/g)	87.9	96.3[c]	89.6[c]	146.8	95.3		83.7		88.3	72.81[c]		60.7[c]	69.7	69.0	52.3[c]	46.0[c]
Density (g/cm³)	1.64	1.54	1.67	1.56	1.70	1.80	1.78	1.85	1.76	1.66	1.87	1.73	1.87	1.87	1.76	1.79
Thermal conductivity (W/m K)	0.056	0.063	0.059	0.074	0.062		0.064		0.064	0.064[c]		0.065	0.067		0.066[c]	0.067
Specific heat (J/g K)	1.05	1.03[c]	1.13	0.89	1.00		1.05		1.05	1.01[c]		1.00	1.13	1.13	0.99[c]	0.99[c]
Viscosity (cp)	0.4	0.5	0.98	0.69	0.44		0.80		0.82	1.1		2.2	2.6	2.7	4.1	7.0
Viscosity (cs)		0.3	0.59							0.6		1.3			2.3	3.9
Surface tension (dynes/cm)	13[c]	10.4		17.3	13		15		15	12.9		14.2	16	16.1	15.2	15.9
Dielectric constant		3.02[c]	1.85[c]		1.81	1.70	1.86	1.75	1.86	2.75[c]	2.03	2.58[c]	1.90	1.86	2.5[c]	2.45[c]
Volumetric thermal expansion (K⁻¹)	0.0016	0.0019	0.0026	0.0016	0.0016	0.0016	0.0016		0.0016	0.0014		0.0013	0.0014	0.0012	0.0012	0.0011[c]
Manufacturer	3M	Du Pont	Du Pont	Du Pont	3M	Du Pont	3M	Du Pont	3M	Du Pont	Du Pont	Du Pont	3M	3M	Du Pont	Du Pont

[a] Liquid properties at 21–25°C.
[b] Based on information supplied by the manufacturer indicated.
[c] Estimated value.

bubble nucleation and often referred to as a hysteresis effect in the most prominent of these anomalies and is shown in Fig. 8.5. Some research effort has been devoted to defining the nature of this phenomenon [8, 11], but more recent interest has focused on various surface treatments that appear to prevent hysteresis while reducing the wall superheat required to achieve a specified ebullient heat transfer rate [10, 12, 13]. The modified R-113 nucleate boiling curves shown in Fig. 16.7, for a variety of micro-porous surfaces developed by the Mitsubishi Electric Corporation of Japan [10], are typical of published results. When applied to the base of a 20-W power transistor immersed in R-113, the thermal resistance of the Mitsubishi microporous surface, formed at room temperature directly on the transistor, was found to be only one-fifth of the value for the original smooth surface.

Natural convection heat transfer from arrays of electronic components immersed in dielectric fluids has not been studied in detail and there is, as yet, no mechanistic model or broadly based correlation that addresses the influence of such factors as component geometry and distribution on the heat transfer coefficient to be encountered along the surface of each component. The large body of multifluid data available for more common configurations suggests that use of the appropriate fluid Prandtl number in existing correlations should provide acceptable predictive accuracy for individual components that can be modeled as isolated flat plates, cylinders, spheres, or cubes (see, e.g., Table 6.9). The natural convection thermal transport rates at submerged condenser surfaces or at the immersion module walls can be similarly predicted to a first approximation by available flat plate correlations (see, e.g., Table 6.9).

In the pursuit of a detailed thermal analysis, it is the overall heat transfer coefficient between the immersed component and the enclosure surface that is sought. Often, a conservative estimate of this parameter can be obtained by summing the individual thermal resistances—that is, component to fluid and fluid to wall—or, when

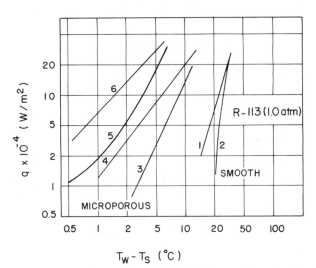

FIG. 16.7 Boiling curves for R-113 on smooth and microporous enhanced surfaces [10].

appropriate, using available correlations for the overall heat transfer coefficient across a fluid gap. Thus, for example, a study of natural convection from a simulated, nine-chip, integrated circuit package across a narrow (0.35 to 0.7 cm) liquid-filled enclosure found conventional correlations to underpredict measured coefficients by as much as 50% [14].

After years of neglect, thermal transport rates from objects confined within fluid-filled enclosures have begun to attract research interest, and a successful correlation of the overall heat transfer coefficient between an individual sphere, cylinder, or cube and a spherical enclosure is now available [15]. The correlation

$$\mathrm{Nu}_b = 0.744 \left(\mathrm{Ra}_b \, \frac{L}{R_i} \right)^{0.222} \tag{16.1}$$

for $\qquad 0.07 < \dfrac{L}{R_i} < 2.67 \qquad 4.6 \times 10^5 < \mathrm{Ra}_b < 4 \times 10^{10}$

is based on an average gap width L between the immersed heat source and the enclosure, and an effective boundary layer length b along the immersed body, and was found to offer a 13% average deviation for air, water, and glycerine data in a 25-cm-diameter spherical enclosure [15]. Although the thermal designer is unlikely to encounter an immersion cooling module of spherical geometry, the form of Eq. (16.1) can be used to correlate natural convection data for more relevant configurations and, even with the coefficient and exponent shown, may offer a convenient initial value for design and development studies.

In the absence of noncondensable gas, the vapor space or remote condensation of dielectric fluid vapors can be expected to follow well-known formulations, and both the ASHRAE and ASME literature abound with data and correlations for the geometries and pressures of interest. When, however, fluid and system degassing is insufficient to provide less than approximately 0.5% mass fraction of air in the vapor space, the condensation heat transfer coefficient can be expected to deteriorate to $\frac{1}{5}$ less of the values when air is not present [26]. The heat transfer rate at a submerged surface is claimed not to suffer this noncondensable gas limitation, but the lack of explicit information on significant design parameters and predictive relations for the prevailing heat transfer coefficients has been a major obstacle to the use of submerged condenser systems.

Fairbanks et al. were the first to examine in detail the nature and parametric variation of heat transfer at a horizontal "submerged" condenser [4]. Their investigation of a finned condenser submerged in turn in water, FC-75, and a mixture of water/ethylene glycol established the influence of several fluid properties and a container geometric factor on the heat transfer coefficient [4]. Simons and Seely [16] examined the relative performance of several immersion cooling systems for a particular packaging configuration. Their results supported the viability of horizontal submerged condenser use, but found the peak submerged condenser operating point to lie somewhat below the vapor space condensation rate [8, 16].

A comprehensive analytical and experimental investigation, published by Markowitz and Bergles in 1972 [17], defined more clearly the operational limits of horizontal submerged condensers and related their thermal performance to a nondimensional vapor bubble collapse distance. In a more recent study, the Markowitz and Bergles approach was extended to a multiple-surface submerged condenser configuration and was used successfully to interpret and correlate heat transfer rates at vertical submerged condenser surfaces [18, 19]. The methodology and results of these investigations can thus be used to establish a structure for the analysis and design of submerged condenser modules and are discussed in detail in succeeding sections.

16.4 THERMAL CHARACTERISTICS AND OPERATIONAL LIMITS

16.4.1 Performance Maps

The thermal performance of submerged condenser systems can be best understood by reference to families of constant condenser temperature operating curves—relating the temperature difference, $T_h - T_c$, between the heat-dissipating elements and the condenser surface to the overall heat dissipation or the condenser heat flux q_c''. Such operating curves for a horizontal, water-cooled condenser, submerged alternately in degassed water and R-113 contained in a 15-cm cubical enclosure with several different distributions of 0.65-cm diameter, cylindrical heaters that were immersed horizontally in the working fluid, were obtained by Markowitz and Bergles and are reproduced in Figs. 16.8 through 16.11 [17, 20]. Despite differences in working fluid and heater distribution, these four figures reveal a similar relationship between the overall temperature difference and the condenser heat flux. Data obtained in a

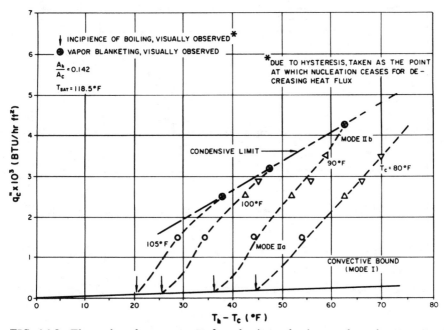

FIG. 16.8 Thermal performance map for a horizontal submerged condenser system, R-113, one heater [20].

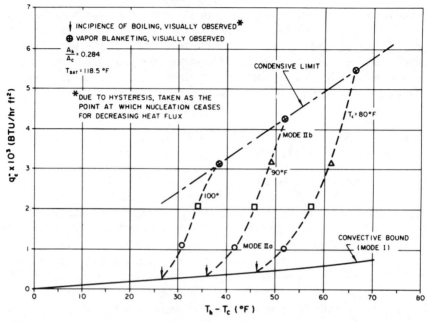

FIG. 16.9 Thermal performance map for a horizontal submerged condenser system, R-113, two heaters [20].

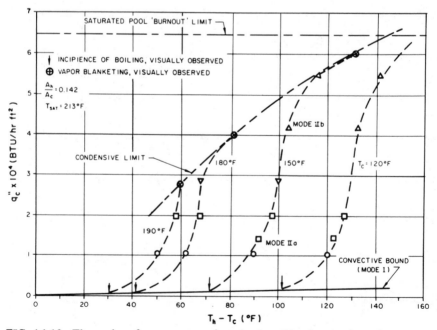

FIG. 16.10 Thermal performance map for a horizontal submerged condenser system, water, one heater [20].

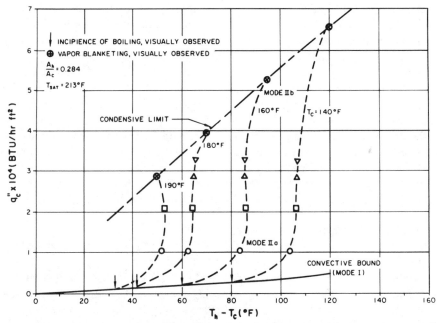

FIG. 16.11 Thermal performance map for a horizontal submerged condenser system, water, two heaters [20].

multiple-surface submerged condenser study can be shown to display similar trends to those shown in these figures [21].

Examining any one of these performance maps in detail reveals that the temperature difference prevailing in a submerged condenser module is well bounded from below and from above and in the bounded region is only moderately sensitive to the total heat dissipation in the module. The lower bound of operation corresponds to heat transfer by natural convection at both the heated and condenser surfaces and is associated with relatively low values of submerged condenser heat flux. As q_c'' increases, while condenser surface temperature T_c is held constant, the heater or heat-generating component surface temperature T_h increases past the incipience temperature and subcooled boiling is initiated on the surface. For further increase in q_c'', fully developed boiling is attained, resulting in bubble pumping of the enclosed fluid and enhanced natural convection heat transfer at the submerged condenser surface. As q_c'' is increased still further, the generated vapor bubbles begin to impinge and condense on the condenser surface; and for large values of q_c'', heat transfer at that surface is primarily by condensation. The locus of $T_h - T_c$ at a constant condenser surface temperature essentially follows a boiling curve, but its exact shape is influenced somewhat by the liquid temperature or, more precisely, the degree of bulk liquid subcooling.

The overall heat dissipation or surface heat flux in a submerged condenser system is subject to one of two possible upper bounds, depending on the specific system parameters. For relatively low values of condenser surface temperature and/or low heater-to-condenser surface area ratios, the upper bound is established by the critical or "burn-out" boiling heat flux at the component surface. Alternately, high values of T_c (i.e., approaching the fluid saturation temperature), and/or area ratios approaching unity, result in an experimentally observed upper bound that is apparently due to a condensation limit associated with vapor blanketing of the condenser surface. This sequence of heat transfer mechanisms is illustrated in the series of photographs and accompanying sketches in Fig. 16.12.

It is significant to note that, in addition to identifying clearly the upper and lower bounds on submerged condenser operation, the performance maps highlight as well the unique behavior of the overall heat transfer coefficient in the region bounded by the convective and condensive limits. Examination of the performance maps reveals that, at fixed condenser temperature, the overall heat transfer coefficient increases dramatically with increased heat dissipation or condenser heat flux, typically yielding less than a 25% increase in $T_h - T_c$ for a factor of 3 increase in q_c''.

16.4.2 Bulk Liquid Temperature

The temperature of the bulk fluid essentially determines the temperature of the low-dissipation and thermally passive elements in the immersion cooling module. Whereas the operating curves relating $T_h - T_c$ to q_c'' contain the basic system information, the bulk temperature or subcooling is only implicitly presented through its effect on the locations of the constant T_c curves. More precise information on the liquid temperature variation in horizontal submerged condenser modules is contained in Figs. 16.13 through 16.16, where the liquid temperature variation for one and two heaters immersed in water and R-113 [20] is displayed.

Examination of Fig. 16.13, for one heater in water, in greater detail, reveals that the bulk temperature increases asymptotically toward the saturation temperature and attains this value at the condensation limit. At the left of the graph is the locus of the liquid temperature that could be expected to result from pure natural convection heat transfer at the condenser surface. The experimentally determined bulk temperature profile thus gives clear evidence of augmented convective heat transfer even at heat fluxes considerably below the condensive limit. This same behavior is apparent in the appropriate figures for two heaters in water (Fig. 16.14) as well as one and two heaters in R-113 (Figs. 16.15 and 16.16), though the magnitude of the convective augmentation effect appears to vary with these parameters.

Clearly, some care must be taken in defining the bulk liquid temperature in an immersion module. Detailed measurements in a horizontal submerged condenser configuration found the region between the plane of the heaters and the condenser (but external to the respective boundary layers) to be nearly isothermal, and it was this temperature that was defined as the appropriate bulk value [17, 20]. Detailed measurements in a multiple-surface submerged condenser module point to a far more complex temperature field and the existence, in this configuration, of two distinct temperature zones [18, 21].

In the experimental apparatus of the Bar-Cohen/Distel study—a 30 cm on each side, externally finned and air-cooled steel module—horizontal and vertical temperature variations *above* the plane of the uppermost heater were generally quite modest, reaching peak values of 3.5, 8, and 3°C for water, FC-77, and R-113, respectively [18, 21]. These peak temperature differences were typically encountered at one-half

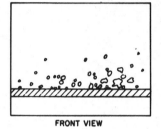
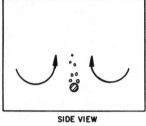

FRONT VIEW   **SIDE VIEW**

(a)

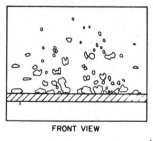
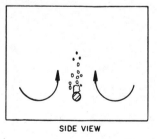

FRONT VIEW   **SIDE VIEW**

(b)

FIG. 16.12 Horizontal submerged condenser operating modes [20]. (a) Submerged condenser operating mode IIa—water, one heater ($q_c'' = 1.4 \times 10^4$ Btu/ft^2 h, $T_b = 206°$F. (b) Submerged condenser operating mode IIa—water, one heater ($q_c'' = 1.4 \times 10^4$ Btu/ft^2 h, $T_b = 208°$F).

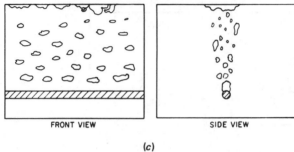

FRONT VIEW   SIDE VIEW

(c)

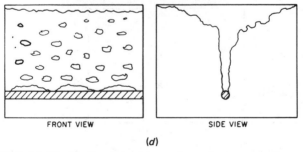

FRONT VIEW   SIDE VIEW

(d)

FIG. 16.12 Horizontal submerged condenser operating modes [20] (*Cont.*) (c) Submerged condenser operating mode IIb—water, one heater ($q_c = 1.4 \times 10^4$ Btu/ft^2 h, $T_b = 210.5°$F). (d) Submerged condenser condensation limit— water, one heater ($q_c = 1.4 \times 10^4$ Btu/ft^2 h, $T_b = 213°$F).

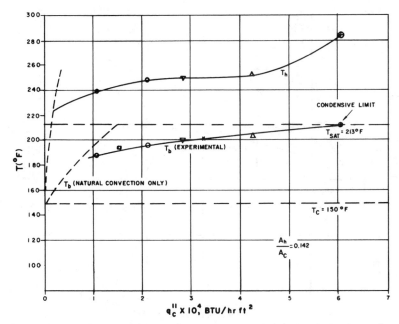

FIG 16.13 Bulk and heater surface temperature variation in horizontal submerged condenser system, water, one heater [20].

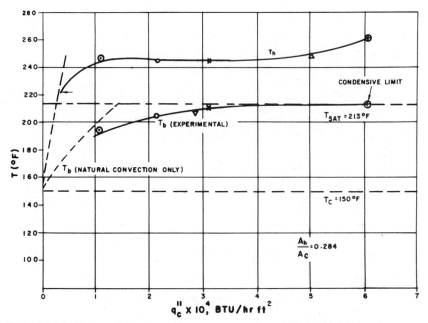

FIG. 16.14 Bulk and heater surface temperature variation in horizontal submerged condenser system, water, two heaters [20].

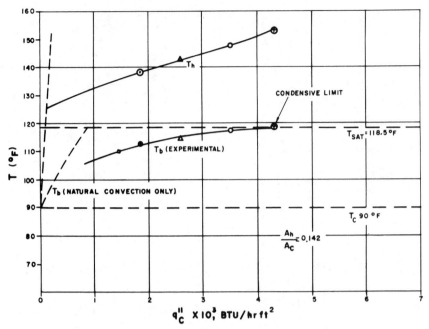

FIG. 16.15 Bulk and heater surface temperature variation in horizontal submerged condenser system, R-113, one heater [20].

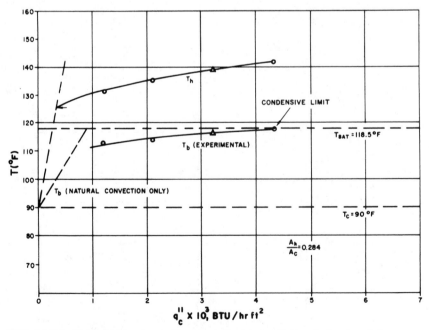

FIG. 16.16 Bulk and heater surface temperature variation in horizontal submerged condenser system, R-113, two heaters [20].

the maximum module dissipation rate and were found to decrease with further increases in power dissipation. The fluid below the plane of the lowermost heater was found to display considerable thermal stratification. Although horizontal temperature gradients were again modest, temperature differences between the cold fluid along the very bottom of the enclosure and the warm fluid adjacent to the nearest active heating element reached values of 25, 37, and 10.5°C for water, FC-77, and R-113, respectively, at intermediate heat dissipations.

As a more representative measure of spatial temperature variations, consider the temperature differences between the fluid adjacent to the top surface of the enclosure and the liquid slightly below the lowest immersed heater shown in Fig. 16.17. The dependence of this predominantly vertical temperature variation on working fluid is again in evidence, and the anisothermality is seen to be most severe at intermediate heat rejection rates, decreasing substantially at both low and high heat dissipations.

This thermal stratification is reflected as well in the dependence of the average or bulk liquid temperature on the heat dissipation rate of a multiple-surface submerged condenser. Typical results, shown in Fig. 16.18 for an air-cooled module, display once again the augmentation of natural convection heat transfer but, unlike the horizontal submerged condenser, the bulk temperature in this configuration does not reach the saturation value.

Regardless of the precise thermal mechanisms active in the enclosed fluid, the liquid temperature cannot exceed the saturation temperature nor fall below the inlet air temperature. Consequently, as borne out by the data of Fig. 16.17, the maximum temperature variation in an enclosed, low-boiling-point fluid can be expected to lie significantly below the comparable ΔT for liquids of higher saturation temperature.

The dependence of the vertical temperature variation on heat dissipation and the

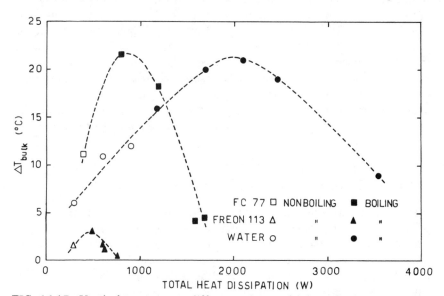

FIG. 16.17 Vertical temperature differences in a multiple-surface submerged condenser module [18, 21].

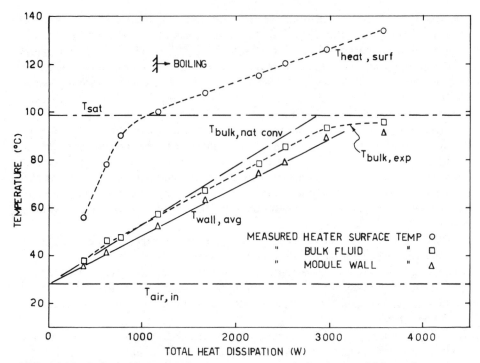

FIG. 16.18 Bulk and heater surface temperature variation in a multiple-surface submerged condenser module [18, 21].

characteristic curve traced out by the data points for each fluid in Fig. 16.17 can be related to the interplay between convective thermal stratification and ebullient liquid circulation induced by vapor bubble motion and collapse. Although the buoyant rise of heated fluid to the top of the immersion module acts to establish a progressively stronger vertical temperature gradient with increasing heat dissipation, bubble pumping does not commence until boiling incipience, and requires a substantial decrease in bulk subcooling to become an effective liquid stirring mechanism. The peak vertical temperature difference can thus be expected to occur at heat dissipations somewhat greater than are needed to initiate boiling, as is clearly shown in Fig. 16.17.

16.5 PREDICTING SUBMERGED CONDENSER PERFORMANCE

The preceding has identified the thermal characteristics and operating limits of immersion cooling modules and bounded the range of liquid anisothermality in multiple-surface submerged condenser configurations. Although this information is sufficient to establish the viability and approximate envelope dimensions of a thermal control module employing immersion cooling, detailed design and prediction of component temperatures requires specific design equations for the relevant parameters.

16.5.1 Component Surface Temperature

Natural convection heat transfer rates from a single component immersed in a large volume of liquid can be estimated by use of conventional Nusselt number correlations for simple geometric shapes immersed in infinite fluids as, for example, tabulated in Table 6.9, or by use of available enclosure correlations [see Sec. 6.5.3 and Eq. (16.1)], but precise values must be obtained empirically for each geometry of interest.

Similarly, although boiling curves such as in Figs. 8.4, 16.6, and 16.7 can guide the thermal designer and the boiling correlation of Eq. (8.14) can be used to investigate parametric trends, a detailed prediction of ebullient heat transfer requires at least several empirical data points for the fluid-surface combination of interest.

16.5.2 Submerged Condenser Surface

PHENOMENOLOGICAL MODEL

In contrast to the more common transport phenomena at the surface of the immersed components, bubble pumped heat transfer at the submerged condenser surface is largely unique to immersion cooling modules. Although the precise mechanism by which the bubble/liquid interaction augments natural convection has yet to be determined, the dependence of the phenomenon on the vapor fraction in the enclosure can hardly be doubted. It can, furthermore, be argued that (at least for a horizontal submerged condenser) the bubble collapse length and, more specifically, the ratio of collapse length to the distance separating the components from the condenser surface, plays a crucial role in determining the rate and mechanism of condenser surface heat transfer.

For values of the collapse distance L_c [observed visually or determined approximately via Eq. (8.48)] much smaller than the separation distance W, all the generated vapor bubbles collapse near the heated surface. Consequently, the bubbles can—in this case—be expected to have only a negligible effect on heat transfer at the condenser surface, and the observed transfer rates should correspond approximately to those associated with natural convection. For $L_c/W \leqslant 1$, the vapor bubbles approach or impinge on the condenser surface and the substantial vapor fraction in the fluid may enhance the convective exchange rate; the augmentation of natural convection improving with increased values of L_c/W. As this ratio increases past unity, the bubbles begin to form vapor pockets at the horizontal submerged condenser surface, and the heat transfer rate can be expected to reflect a progressively greater influence of vapor space condensation, asymptotically attaining precisely the Nusselt numbers associated with condensation on the underside of a horizontal plate as L_c/W becomes very large.

In Fig. 16.19 the experimentally determined variation of the heat transfer coefficient at a horizontal submerged condenser, h_c, is related to the calculated value of L_c/W for one heater immersed in water and separated by a distance of 11.2 cm from the condenser [17]. In this apparatus, natural convection in water was predicted to yield coefficients of approximately 1200 W/cm^2 °C and vapor space condensation h values of nearly 8500 W/m^2 °C. It would thus appear that for $L_c/W < 0.15$, heat transfer at the condenser was achieved almost exclusively by natural convection. The region $0.15 < L_c/W < 6$ appears to correspond to the region of bubble pumped augmented convection, whereas values of $L_c/W > 6$ seem to lie in the condensation zone.

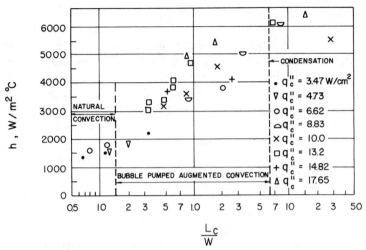

FIG. 16.19 Dependence of heat transfer coefficient at a horizontal submerged condenser surface on the nondimensional bubble collapse length [17, 20].

As suggested by the proposed phenomenological model, a definite progression from natural convection to augmented convection and then to vapor space condensation is apparent in Fig. 16.19. Prediction of pure natural convection at the submerged condenser can be achieved by use of available correlations, subject to the previously discussed limitations. Bubble pumped convection and vapor space condensation are discussed in the following subsections.

BUBBLE PUMPED CONVECTION

The rate of heat transfer in free convection at the condenser surface can be related to the thickness of the thermal boundary layer along that surface. The boundary-layer thickness is itself dependent on the rate of fluid circulation, induced by an unstable density gradient and its equivalent buoyant force, normally resulting from a temperature gradient in the enclosed liquid. However, the presence of a vapor fraction in the fluid introduces an additional buoyant force, which can increase dramatically the rate of fluid circulation and hence the rate of heat transfer at the submerged condenser surface.

As discussed in Chap. 6, the rate of heat transfer in natural convection can be correlated in an expression of the form

$$\text{Nu}_{nc} = C(\text{Gr}\,\text{Pr})^n = C(\text{Ra})^n \tag{16.2}$$

where $n = \frac{1}{4}$ for laminar flow, $n = \frac{1}{3}$ for turbulent flow, and C is a configurational constant but different for turbulent and laminar flow. It may thus be appropriate to expect that the augmented heat transfer rate can be expressed as

$$\text{Nu}_{\text{aug}} = C(\text{Ra}_{\text{aug}})^n \tag{16.3}$$

Or, defining the augmentation ratio $\text{Nu}^* \equiv \text{Nu}_{\text{aug}}/\text{Nu}_{nc}$ and dividing Eq. (16.3) by Eq. (16.2),

$$Nu^* = \left(\frac{Ra_{aug}}{Ra_{nc}}\right)^n \tag{16.4}$$

For thermally induced natural convection, the circulation driving force (contained within the Ra) can be defined as a buoyant force per unit volume, F_t', equal to

$$F_t' = g\,\Delta\rho = g\rho\beta(T_b - T_c) \tag{16.5}$$

Alternately, for vapor fraction-induced natural convection, the volumetric buoyant force F_v' equals

$$F_v' = g(\rho_l - \rho_v)\frac{V_v}{V_e} \tag{16.6}$$

Accounting for the presence of both driving forces and defining the vapor fraction $\alpha \equiv V_v/V_e$, the buoyant driving force for augmented natural convection, F_{aug}', can now be expressed as

$$F_{aug}' = g\rho\beta(T_b - T_c) + \theta g(\rho_l - \rho_v)\alpha \tag{16.7}$$

where θ is a configuration factor whose value depends on the distribution of the vapor fraction.

Substituting F_{aug}' from Eq. (16.7) into Eq. (16.4), the augmentation ratio is found to be

$$Nu^* = \left[1 + \frac{\theta\alpha(\rho_l - \rho_v)}{\rho\beta(T_b - T_c)}\right]^n \tag{16.8}$$

The vapor fraction appearing in this formulation is governed by the rate of vapor generation at the boiling surfaces and the rate of vapor condensation in the subcooled liquid, or

$$\alpha = \frac{1}{V_e}\; [(\text{bubble generation rate})(\text{bubble residence time})$$

$$\times\;(\text{average bubble volume})]$$

$$= \frac{1}{V_e}\left[\left(\frac{\text{vapor generation rate}}{\text{bubble departure volume}}\right)(\text{bubble residence time})\right.$$

$$\left.\times\left(\frac{\text{average radius}}{\text{departure radius}}\right)^3 (\text{bubble departure volume})\right] \tag{16.9}$$

This equation can be cast in approximate analytic form by utilizing expressions for the vapor generation rate and bubble collapse characteristics (residence time and timewise variation of the radius) presented and discussed in Chap. 8, notably Eqs.

(8.13), (8.43), and (8.44). Following this procedure and inserting the resulting relation for α into Eq. (16.8), the derived expression for the augmentation ratio can be shown to take the form [20]

$$\text{Nu}^* = \left\{ 1 + 0.0840\theta\eta \left[\frac{(\rho_l - \rho_v)qR_0^2}{V_e\rho_v\rho_l h_{fg}\kappa\beta \text{Ja}^2(T_b - T_c)} \right] \right\}^n \tag{16.10}$$

where η is the fraction of the heat flux involved in vapor generation.

Because of the nature of the configuration factor θ and the vapor generation parameter η, the $\theta\eta$ product appearing in Eq. (16.10) is not analytically determinate and is best left to empirical evaluation. Thus, the data for augmented natural convection at submerged condenser surfaces should be correlatable in the form of Eq. (16.10) with a single semiempirical factor λ, (equal to $0.0840\theta\eta$) for each working fluid and geometry combination of interest. For submerged condenser configurations in which heater heat fluxes span the entire nucleate boiling range (from incipience to critical heat flux), the consequent variation in η may necessitate several λ values or a continuously varying λ function to achieve acceptable correlation accuracy.

A successful correlation of the horizontal submerged condenser, for water data obtained by Markowitz and Bergles [17, 20] in the form of Eq. (16.10), is shown in Fig. 16.20. As anticipated, each heater configuration yields a distinct value of λ. The augmentation ratio data for a vertical condenser submerged alternately in water and R-113 is shown in Fig. 16.21 and is seen to reflect the influence of both working fluid and geometry on the correlating coefficient λ. Despite the complexity of the model and the use of simplifying assumptions, nearly all of the Nu* data points are within 10% of the correlation, and the maximum disparity is less than 25%. Although this agreement is most rewarding, the model's present inability to determine a priori

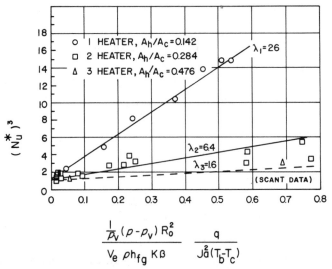

FIG. 16.20 Correlation of bubble pumped natural convection of water at a horizontal submerged condenser surface [17, 20].

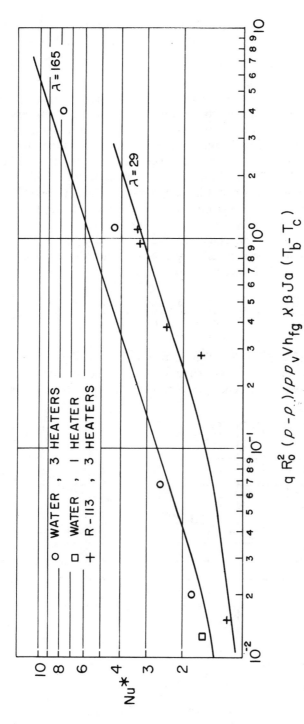

FIG. 16.21 Correlation of bubble pumped natural convection at a vertical submerged condenser surface [19, 20].

the relevant value of λ and the significant range of λ values apparent in Figs. 16.20 and 16.21 necessitates experimental determination of λ for each fluid/geometric combination of interest.

CONDENSATION LIMIT

As the heat dissipation of a submerged condenser module increases toward its operating limit, the vapor region adjacent to the horizontal submerged condenser—in both horizontal surface and multiple-surface modules—grows and finally blankets the surface. Any further increases in the total heat dissipation beyond the vapor blanketing point result in a rapidly thickening vapor layer and explosive ejection of liquid into the expansion chamber. This is the condensive limit apparent in the previously discussed performance maps and bulk temperature plots and can be related to the thermal transport rate for film condensation on a downward-facing (with respect to gravity) horizontal surface.

Film condensation on the underside of a horizontal surface is governed by thermal conduction through the thin condensate layer. Following Gerstmann and Griffith [22], the heat transfer coefficient can be expressed as

$$\frac{h}{k}\left[\frac{g_0\,\sigma}{g(\rho-\rho_v)}\right]^{1/2} = 0.26\left\{\frac{g\rho(\rho-\rho_v)h'_{fg}}{k\mu(T_s-T_c)}\left[\frac{g_0\,\sigma}{g(\rho-\rho_v)}\right]^{3/2}\right\}^{1/4} \qquad (16.11)$$

where $h'_{fg} = h_{fg} + (3/8c)(T_s - T_c)$. This relation is plotted in Fig. 16.22 and compared with the horizontal surface condensive limit data. The agreement with both R-113 and water is excellent, leaving little doubt as to the nature of the condensive limit.

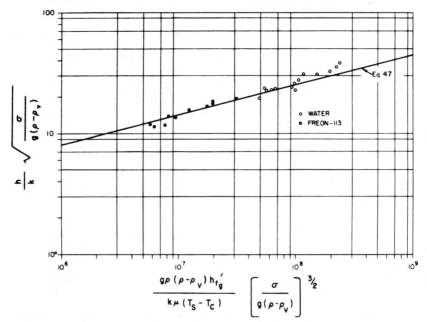

FIG. 16.22 Condensation limit on a horizontal submerged condenser [17, 20].

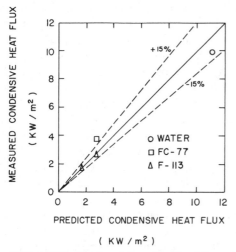

FIG. 16.23 Horizontal surface heat flux at the condensation limit for a multiple-surface submerged condenser module [18, 21].

The limiting heat fluxes at the upper horizontal surface of the Bar-Cohen and Distel multiple-surface submerged condenser module [19] are compared in Fig. 16.23 with Eq. (16.11). Although the measurement error introduced by the conductive interaction between the horizontal and continuous vertical surfaces makes precise comparison difficult, the relative proximity of the measured and theoretical values for R-113, FC-77, and water confirms the generality of the vapor space condensation limit on submerged condenser operation.

16.6 SUBMERGED CONDENSER APPLICATIONS

16.6.1 High-Power Microwave Components
Immersion cooling and submerged condenser techniques were first applied to the thermal stabilization of high-power-dissipation microwave components [1, 2], and this application category, typified by the dissipation of one to several kilowatts in a relatively large volume of 0.01 to 0.04 m^3, continues to be of primary interest.

The thermal performance of a multiple-surface submerged condenser module 0.3 m on each side, filled with FC-77 and cooled externally by natural convection, was reported in [23]. This test apparatus using thermally simulated electronic components and operated at atmospheric pressure was found to attain the condensation limit at 800 W for a room ambient temperature of 29°C. At this operating point, the liquid temperature averaged 101°C and the finned module side walls nearly 96°C, or 5°C below the liquid temperature. These results compare favorably with those discussed in [5], where a module of approximately one-half the volume and externally finned on the four side walls was found to dissipate a maximum of 400 W from a power supply operating in a 50°C ambient at a 20 psia system pressure while maintaining a 5.5°C temperature difference between the fluid and side walls.

A considerable improvement in heat dissipation capability can be achieved by using forced circulation of air to cool the external fin array. Reference [5] details the performance of a multiple-surface submerged condenser module that maintained the surface temperature of a Klystron tube at approximately 120°C while dissipating 1.8 kW in a nominal 0.03 m³ module filled with FC-77. These results were obtained at an internal operating pressure slightly above atmospheric and the flow of 0.083 m³/s of 60°C air past densely stacked fin arrays on three of the four vertical walls. A similarly sized test module filled with FC-77 with less effective external heat exchange was found to attain the condensive limit at 1.43 kW for 0.095 m³/s of 30°C air and to be limited by critical heat flux at the three cylindrical heaters at 1.7 kW for an air flow of 0.155 m³/s of 28°C air [18]. When the module was filled with R-113, this same air flow rate at an inlet temperature of 25°C yielded a peak dissipation of 790 W with a bulk average temperature of 49°C and a heater surface temperature of 60°C [18].

Water or ethylene glycol cooling of the condenser surfaces can provide further improvement in heat transfer capability. Reference [6] presents some details of liquid-cooled horizontal submerged condensers used to thermally stabilize high-voltage power supplies. The high-dissipation module, at 930 W, required 0.065 m² of condenser coil, through which flowed 158 cm³/s of ethylene glycol at an average temperature of 61°C, to maintain FC-77 at 105°C [6]. In a later study, a 0.021-m² water-cooled horizontal submerged condenser maintained at 27°C provided a peak dissipation capability of 403 W for components immersed in R-113.

16.6.2   Logic and Memory Devices

The thermal control needs of a growing number of electronic logic and memory devices can best be met by direct immersion of the components in a dielectric coolant chosen to provide boiling heat transfer at the heated surfaces. Whereas the power dissipation of typical individual integrated circuit packages is today in the range of 1 to 10 W and may increase to 10 to 30 W with the advent of VLSI technology, the total power dissipation of a "computer" is typically of the order of 0.5 to 1.5 kW and can in some cases approach 5 kW. The operational limits of submerged condenser units, cited in Sec. 16.6.1, suggest that such power dissipations could be accommodated in modules of modest volume (typically 0.03 m³; 1 ft³), but the need to package many individual circuit packages or circuit cards densely may make it difficult to realize the full benefit of submerged condenser heat exchange.

Although the optimum spacing between packages or immersed circuit boards in the preboiling natural convection mode can perhaps be determined in a manner analogous to that presented in the Chap. 13 discussion of natural convection cooling in air, the open literature contains insufficient information to determine the optimum spacing in ebullient heat transfer. For modest dissipations and relatively short inter-card channels, bubble departure diameter [as determined by Eq. (8.10)] can serve as a convenient reference value [24], and the minimum card spacing can be taken to equal twice or perhaps three times the departure diameter.

This same approach can be applied to the spacing between individual components and adjacent circuit boards or nearby components. However, with increasing heat flux from individual components and/or increasing vapor fraction (or quality) in the fluid flowing past the dissipative surfaces, as may occur in long, high-flux intercard channels, greater attention must be directed toward the influence of the two-phase flow pattern on the surface temperature. The flow boiling of water, including the effects of flow pattern and channel geometry on boiling incipience and critical heat flux, has been extensively studied and documented. The present dearth of comparable data for the flow boiling of dielectric fluids in channels formed by electronic components and circuit card boards makes it most difficult to determine theoretically the thermally optimum spacings for immersion-cooled logic and memory components.

Nevertheless, it would appear that for many applications of interest, where the component dissipation rate is significantly below the critical heat flux, component-to-component or component-to-board spacings of 0.5 to 1 cm should be adequate. At these values, packaging density would not be seriously constrained by boiling considerations and condensation and/or component interconnection limitations could be expected to dominate. As an example, it is interesting to note that, based on boiling considerations *alone*, a total dissipation of 2.25 kW—from ten, 13 × 13 × 0.5 cm thick cards, each carrying nine, 25-W integrated circuit packages with an intercard and card-to-module wall spacing of 1 cm—could be accommodated in a 15 × 15 × 15 cm module filled with FC-77. However, to attain this packaging density of nearly 670 kW/m^3 (22 kW/ft^3), it would be necessary to provide external liquid cooling of the module walls so as to provide a highly efficient multiple-surface submerged condenser.

16.6.3 Microprocessors

The rapid maturation of microprocessor technology and the expanding use of microprocessors as integral parts of mechanical systems demands that attention be devoted to the thermal control requirements of an electronic system consisting of a single module or integrated circuit package. The absence of an engineered cooling system for the microprocessor within its "mechanical" host (e.g., car engine, sewing machine) necessitates reliable electronic operation in a frequently hostile environment. Thus, thermal limitations can be expected to constrain the future growth of this technology.

A miniaturized submerged condenser system consisting of a substrate, the appropriate discrete and monolithic components, a metallic lid, and a liquid (filling the gap between the lid and substrate) can be envisioned to provide substantial heat dissipation capability for a microprocessor when the lid (possibly finned) is air cooled by natural or forced convection. Such a design, utilizing a flexible lid cover or internal compressible volume and carefully degassed fluid, can be anticipated to operate at near-constant pressure. The absence of such measures will result in a constant-volume submerged condenser, as discussed in [7].

Use of the relations and results discussed in previous sections suggests that a vertical package, 3 cm on each side, filled with FC-78 at atmospheric pressure, and possessing an externally roughened or finned metallic lid, could dissipate approximately 10 W to ambient air at 21°C while maintaining the substrate or junction temperature at 65°C. Under these conditions, the lid temperature could be expected to approach 38°C and microprocessor dissipation would be constrained by the condensation limit. Use of FC-77 as the working fluid would facilitate the dissipation of 20 W at a 124°C junction temperature, but removal of 20 W from the lid at the stated ambient and lid temperatures would require extraordinary thermal control measures, which may not be compatible with microprocessor applications. It is most significant to note that these theoretical submerged condenser heat-removal rates represent a dramatic improvement over the 2 to 3 W dissipation at 125°C junction temperature of "air-filled" integrated circuit packages of similar dimensions in the stated ambient.

Realization of these most promising submerged condenser dissipation rates may, however, be tempered by the omnipresent dissolved air in the dielectric fluids and the influence of the narrow substrate-to-lid gap on the boiling process. The release of dissolved gas with increasing bulk fluid temperature can be expected to reduce significantly the condensation limit and raise the bulk and chip temperature at sublimit dissipation rates. For the desirably small separation distance between substrate and lid, the critical ebullient heat flux may be significantly below that associated with boiling in an unconstrained fluid [25] and thus place a lower-than-anticipated limit on the maximum dissipation rate.

16.7   CONCLUDING REMARKS

Discussion of immersion cooling of electronic components in dielectric fluids has served to elucidate the thermal mechanisms active in the transport of heat from the immersed components to the enclosed fluid and from the fluid to the container walls. The relatively high heat transfer coefficients associated with boiling and condensation processes, and to a lesser extent liquid natural convection, allow immersion cooling to accommodate extremely high surface and volumetric heat release rates and/or packaging densities.

The results cited and the relations presented provide the thermal analyst and designer with the ability to perform preliminary analysis or conceptual design of immersion cooling systems. Furthermore, the operational limits and mechanistic models for submerged condenser behavior in Secs. 16.4 and 16.5 offer a convenient structure for evaluating and correlating data from new submerged condenser configurations.

The relatively recent appearance of immersion cooling and the unique two-phase flow and heat exchange configurations encountered in submerged condenser modules distinguish these thermal control techniques from the more conventional air- and liquid-cooling technologies. The attainment of a detailed understanding and precise modeling of thermal stabilization by component immersion in dielectric fluids and submerged condenser operation continue to challenge the heat transfer community. Broadening application of these techniques can, however, be expected to offer diverse opportunities for the refinement and validation of existing models and serve as well to identify new areas of uncertainty or inadequate modeling.

16.8   NOMENCLATURE

Roman Letter Symbols

a	acceleration, m/s^2
A	heat transfer area, m^2
b	effective boundary layer length, m
c	specific heat, J/kg °C

C	configurational constant in Eq. (16.2)
F_t'	thermal buoyant force per unit volume, N/m^3
F_v'	vapor fraction-induced buoyant force per unit volume, N/m^3
g	gravitational acceleration, m/s^2
g_0	gravitational constant, $1 \ m \, kg/N \, s^2$
Gr	Grashof number, dimensionless, $\equiv \rho^2 g\beta L^3 \, \Delta T/\mu^2$
h	heat transfer coefficient, $W/m^2 \, °C$
h_{fg}	latent heat of vaporization, J/kg
Ja	Jakob number, $c_p \rho(T_s - T_b)/h_{gh}\rho_v$
k	thermal conductivity, $W/m \, °C$
L	characteristic length, m
L_c	bubble collapse length, m
n	exponent
Nu	Nusselt number, dimensionless
Nu_b	average Nu based on temperature difference between heat source and enclosure in Eq. (16.1), dimensionless
Nu^*	ratio, Nu_{aug}/Nu_{nc}, dimensionless
Pr	Prandtl number, dimensionless, $\equiv c_p \mu/k$
q	heat flow, W
R	bubble radius, m
R_i	volumetrically equivalent radius of immersed heat source, m
Ra	Rayleigh number, dimensionless, $\equiv Pr \, Gr$
Ra_b	specific Ra based on length b, dimensionless
W	separation distance between immersed heat source and enclosure, m

Greek Letter Symbols

α	volumetric vapor fraction
β	thermal coefficient of volumetric expansion, $°C^{-1}$
η	vapor generation coefficient defined in Eq. (8.13)
θ	configuration factor introduced in Eq. (16.7)
κ	thermal diffusivity, m^2/s
λ	combined configuration factor defined in Eq. (16.10)
μ	dynamic viscosity, $N \, s/m^2$
ρ	density, kg/m^3
σ	surface tension, kg/m

Subscripts

aug	indicates augmented condition
b	indicates bulk fluid
c	indicates condenser surface condition
crit	indicates critical heat flux condition
e	indicates enclosure
h	indicates heated surface condition
l	indicates liquid
nc	indicates natural convection
o	indicates departure condition

s   indicates saturated condition
v   indicates vapor condition

16.9 REFERENCES

1 Greene, A. O., and Wightman, J. C., Cooling Electronic Equipment by Direct Evaporation of Liquid Refrigerant, Air Material Command Rept. PB 136065, Wright-Patterson Air Force Base, Fairborn, Oh. National Technical Information Service, Springfield, Va., 1948.

2 Mark, M. M., Stephenson, M., and Goltsos, C. E., An Evaporative-Gravity Technique for Airborne Equipment Cooling, *IRE Trans.*, vol. ANE-5, pp. 47–52, 1953.

3 Goltsos, C. E., and Mark, M., Packaging with a Flexible Container for Oil-Filled or Evaporative-Cooled Electronic Equipment, *IRE Trans.*, vol. PEP-6, pp. 44–48, 1962.

4 Fairbanks, D. R., Goltsos, C. E., and Mark, M., The Submerged Condenser, ASME Publication 67-HT-15, 1967.

5 Paradis, L. R., Simplified Transmitter Cooling System, *Proc. 8th Int. Electronic Packaging Symp. (IEEE-WESCON), San Francisco*, vol. 8, pp. 5/5–5/12, August 21–22, 1967.

6 Cochran, D. L., Boiling Heat Transfer in Electronics, *Electronic Packaging and Production*, vol. 8, no. 7, pp. CL-3–CL-7, 1968.

7 Bravo, H. V., and Bergles, A. E., Limits of Boiling Heat Transfer in a Liquid-Filled Enclosure, *Proc. 1976 Heat Transfer and Fluid Mechanics Inst.*, eds. J. W. Baughn and H. A. Dwyer, pp. 114–127, Stanford University Press, Palo Alto, Calif., 1976.

8 Seely, J., and Chu, R., *Heat Transfer in Microelectronic Equipment*, chap. VI, Marcel Dekker, New York, 1972.

9 Megerlin, F., and Markowitz, A., Advanced Cooling and Thermal Control Techniques, Paper P744, Raytheon Company, Bedford, Mass., 1971.

10 Fujii, M., Nishiyama, E., and Yamanaka, G., Nucleate Pool Boiling Heat Transfer from Micro-Porous Heating Surface, 18th Natl. Heat Transfer Conf., San Diego, Calif., August 1978, published in *Advances in Enhanced Heat Transfer*, ed. J. M. Chenoweth, J. Kaellis, J. W. Michaels, and S. S. Shenkman, pp. 45–51, ASME, New York, 1979.

11 Bergles, A. E., Bakhru, N., and Shires, J. W., Jr., Cooling High-Power Density Computer Components, Rept. no. DSR-70712-60, Department of Mechanical Engineering, Massachusetts Institute of Technology, Cambridge, Mass., 1968.

12 Czikk, A. M., and O'Neill, P. S., Correlation of Nucleate Boiling from Porous Metal Films, 18th Natl. Heat Transfer Conf., San Diego, Calif., August 1978, published in *Advances in Enhanced Heat Transfer*, ed. J. M. Chenoweth, J. Kaellis, J. W. Michaels, and S. S. Shenkman, pp. 53–60, ASME, New York, 1979.

13 Oktay, S., and Schmeckenbecher, A. F., Preparation and Performance of Dendritic Heat Sinks, *J. Electrochem. Soc.*, vol. 21, pp. 912–918, 1974.

14 Megerlin, F. E., and Vingerhoet, P., Thermal Control of Densely Packaged Microelectronics in Dielectric Fluids, Proc. Natl. Aeronautical Electronics Conf. Sponsored by the IEEE Professional Group on Aeronautical and Navigational Electronics, Dayton, Oh., pp. 254–259, 1971.

15 Powe, R. E., Warrington, R. O., and Scanlan, J. A., Natural Convection Heat Transfer Between Bodies and Their Spherical Enclosures, in *Internal Natural Convection Flows*, ed. I. Catton and K. E. Torrance, ASME, New York, 1980.

16 Simons, R. E., and Seely, J. H., A Survey of Vapor Phase Cooling Systems, *Electronic Design*, vol. 17, pp. 53–56, 1969.

17 Markowitz, A., and Bergles, A. E., Operational Limits of a Submerged Condenser, in *Progress in Heat and Mass Transfer*, vol. 6, Pergamon Press, Oxford, pp. 701–716, 1972.

18 Bar-Cohen, A., and Distel, H., Thermal Characteristics of Multiple-Surface Submerged Condenser Modules, *Proc. 1978 Heat Transfer and Fluid Mechanics Inst.*, ed. C. T. Crowe and W. L. Grosshandler, Stanford University Press, Stanford, Calif., 1978.

19 Bar-Cohen, A., and Distel, H., Bubble-Pumped Augmented Natural Convection in Submerged Condenser Systems, *Proc. 6th Int. Heat Transfer Conf.*, Toronto, vol. 3, pp. 197–202, 1978.

20 Markowitz, A., Boiling and Condensation in a Liquid-Filled Enclosure, Ph.D. dissertation, Massachusetts Institute of Technology, Cambridge, Mass., 1971.

21 Distel, H., Thermal Control Module for Electronic Components, M.S. thesis, Ben-Gurion University of the Negev, Beer Sheva, Israel, 1977 (in Hebrew).

22 Gerstmann, J., and Griffith, P., Laminar Film Condensation on the Underside of Horizontal and Inclined Surfaces, *Int. J. Heat Mass Transfer*, vol. 10, pp. 561–580, 1967.

23 Bar-Cohen, A., Immersion and Heat-of-Fusion Thermal Control Techniques for Electronic Components, Ben-Gurion University of the Negev, Department of Mechanical Engineering Rept. MED-14/76, Beer Sheva, Israel, 1976.

24 Armstrong, R. J., Evaporative Cooling with Freon Dielectric Liquids, *Dupont Technical Bull. EL-11*, Wilmington, Del., 1966.

25 Katto, Y., and Kosko, Y., Critical Heat Flux of Saturated Natural Convection Boiling in a Space Bounded by Two-Horizontal Co-Axial Disks and Heated from Below, *Int. J. Multi-Phase Flow*, vol. 5, pp. 219–224, 1979.

26 Minkowycz, W. J., and Sparrow, E. M., Condensation Heat Transfer in the Presence of Non-condensables, Interfacial Resistance, Superheating, Variable Properties, and Diffusion, *Int. J. Heat Mass Transfer*, vol. 19, pp. 1125–1144, 1966.

17

■ heat pipes

17.1 INTRODUCTION

A heat pipe is a self contained heat pump that has the capability of transporting heat at a high rate over fairly substantial distances with no external pumping power. It has been used with considerable success in many applications ranging from the temperature control of the permafrost layer under the Alaska pipeline to the cooling of surgical probes. Its use in electronic cooling applications has met with some success and it is the intent of this chapter to describe the types of heat pipes, to discuss the selection of heat pipe working fluids, and to consider in some detail, with the assistance of a comprehensive example, the theory of operation and the limitations on heat pipe performance.

The heat pipe is a relatively "new" device. Its principle of operation was conceived by Gaugler [1] and Trefethen [2] but was not very widely publicized until 1964 when Grover [3] independently reinvented the concept, demonstrated the effectiveness of the heat pipe, gave the device its name and developed some of its first applications.

17.2 OPERATING PRINCIPLES

As shown in Fig. 17.1, the heat pipe is composed of three sections; an evaporator, an adiabatic section, and a condenser. A porous capillary wick covers the inside surface and extends over the length of the pipe. The heat pipe chamber may take on almost any shape (round, square, or rectangular) and the wick is designed to be saturated with the liquid phase of the working fluid. The remaining volume of the chamber is termed the vapor space and it contains the vapor phase of the working fluid.

Heat is applied at the evaporator end and is removed at the condenser end. At the evaporator end, heat applied by the external source vaporizes that portion of the working fluid located at the evaporator. This creates a difference in pressure between the evaporator and condenser ends, which drives the vaporized fluid through the adiabatic section. At the condenser end, heat is removed by condensation and is ultimately dissipated through an external heat sink such as a finned array.

The depletion of the liquid in the wick at the evaporator end with the associated increase in vapor pressure in the vapor space at the evaporator end causes the liquid-vapor interface at the evaporator end to retreat into the wick as shown in Fig. 17.2. This causes a capillary pressure to develop in the wick at the evaporator end and it is this capillary pressure that causes the liquid to be pumped through the wick from the condenser end to the evaporator end.

Reference to Chap. 8 will show that heat transfer coefficients in evaporation and condensation are among the highest attainable and heat transport in a heat pipe that

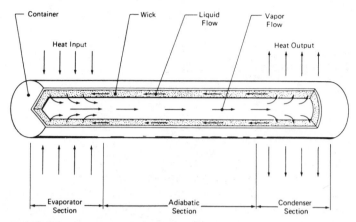

FIG. 17.1 Components and principle of operation of a conventional heat pipe. (From Chi [6], by permission.)

exploits these modes of heat transfer can be an order of magnitude higher than that obtainable in a conventional convective device. It should be recognized that the capillary pressure created at the evaporator end will cause a continuous flow of liquid in the wick and replenishment of the liquid at the evaporator end. Moreover, because heat flow by evaporation and condensation requires small temperature differences, and because the vapor flowing through the adiabatic section experiences a negligible temperature drop, certain heat pipes exhibit thermal characteristics that are even better than solid conductors of the same size.

However, heat pipe performance depends strongly on size, shape, material, method of construction, wick structure, and working fluid. Heat pipes possess heat transfer limitations and sometimes show troublesome start-up problems. However, when the heat pipe is properly designed, its use can be advantageous in certain applications.

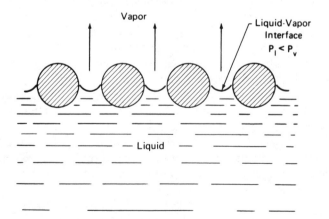

FIG. 17.2 Development of capillary pressure at a liquid-vapor interface. (From Chi [6], by permission.)

17.3 WORKING FLUIDS

Figure 17.3 displays the normal boiling points of several substances and a designation of the three heat pipe categories. Observe that the use of a heat pipe in an electronic cooling application most often will require a heat pipe in the moderate temperature range employing ammonia, Freon-21, methanol, water or some other fluid whose normal boiling point is in the 250 to 375 K range. The boiling point, however, is only a preliminary consideration. Working fluids must also possess other characteristics such as a high surface tension before they become useful in heat pipe applications.

17.4 WICK STRUCTURES

The wick provides a means for the flow of liquid from the condenser to the evaporator sections of the heat pipe. It also provides surface pores that are required at the liquid-

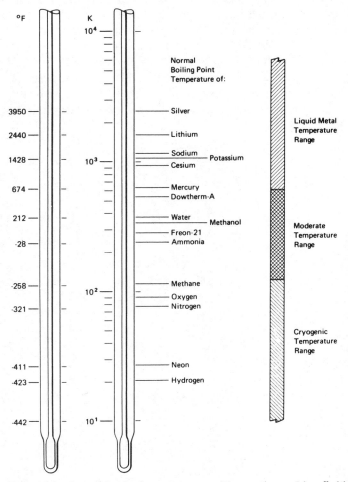

FIG. 17.3 Logarithmic thermometers with sample working fluids for cryogenic, moderate-temperature, and liquid metal heat pipes indicated. (From Chi [6], by permission.)

vapor interface for the development of the required capillary pressure. The wick structure also has an impact on the radial temperature drop at the evaporator end between the inner heat pipe surface and the liquid-vapor surface.

Thus, an effective wick requires large internal pores in a direction normal to the liquid flow path. This will minimize liquid flow resistance. In addition, small surface pores are required for the development of high capillary pressure and a highly conductive heat flow path for the minimization of the radial surface to liquid-vapor interface temperature drop. To satisfy these requirements, two types of wick structures have been developed. These are the homogeneous wicks made of a single material, examples of which are shown in Fig. 17.4, and the composite wicks containing two or more materials, with some typical examples displayed in Fig. 17.5.

The most common wick structure is probably the wrapped screen wick shown in Fig. 17.4a. This type of wick is designated by its mesh number, which is an indication of the number of pores per unit length or unit surface area. The surface pore size is inversely proportional to the mesh number and the liquid flow resistance can be controlled by the tightness of the wrapping. This is attractive but, because of the interruptions of the wick metal by a liquid of low thermal conductivity in the moderate range heat pipe, the radial temperature drop from the inner pipe surface to the liquid-vapor surface at the evaporator end can be quite high. This problem can be alleviated through the use of the sintered metal wick structure (Fig. 17.4b). Notice here that the pore size is small but the small pores will make it more difficult for the liquid to flow from the condenser to the evaporator.

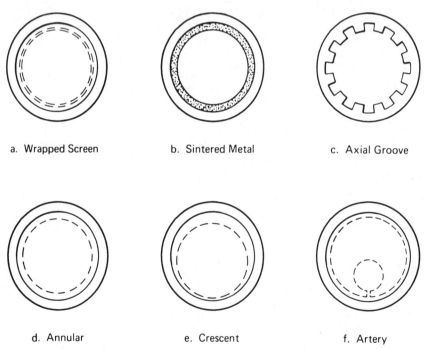

a. Wrapped Screen   b. Sintered Metal   c. Axial Groove

d. Annular   e. Crescent   f. Artery

FIG. 17.4 Cross sections of homogeneous wick structures. (From Chi [6], by permission.)

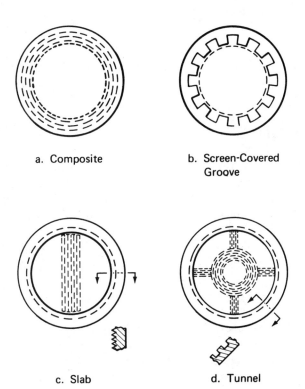

a. Composite   b. Screen-Covered
                                   Groove

c. Slab   d. Tunnel

FIG. 17.5 Cross sections of composite wick structures. (From Chi [6], by permission.)

The axially grooved wick (Fig. 17.4c) possesses highly conductive metal paths for the minimization of the radial temperature drop. Here, however, present manufacturing techniques make pore sizes difficult to control. Annular and crescent wicks (Fig. 17.4d and Fig. 17.4e) have small resistance to liquid flow but are vulnerable to liquids of low conductivity. The artery wick, shown in Fig. 17.4f, was developed to reduce the thickness of the radial heat flow path through the structure and to provide a low resistance path for the liquid flow from condenser to evaporator. However, these wicks may lead to operational problems if they are not self priming; the arteries must fill automatically with liquid at start up or after a dry out.

All composite wicks have a separate structure for the development of the capillary pressure and liquid flow. Notice that in some of the structures shown in Fig. 17.5, a separation of the heat flow path from the liquid flow path can be provided. For example, the screen covered groove wick shown in Fig. 17.5b has a fine mesh screen for high capillary pressure, axial grooves to reduce liquid flow resistance and a metal structure to reduce the radial temperature drop. The slab wick displayed in Fig. 17.5c is inserted into an internally threaded container. High capillary pressure is derived from a layer of fine mesh screen at the surface. Liquid flow is assisted by the coarse screen inside the slab and the threaded grooves tend to provide uniform circumferential distribution of liquid and enhanced radial heat transfer.

17.5   CLASSIFICATION BY TYPE OF CONTROL

Figure 17.3 indicates that heat pipes are categorized or classified by the temperature range of the working fluid into three types: liquid metal, moderate temperature, and cryogenic. Heat pipes may also be classified by the type of control employed. Control, of course, is often necessary because a heat pipe without control will self-adjust its operating temperature in accordance with the heat source (at the evaporator) and heat sink (at the condenser) conditions.

Applications will vary. For example, it may be desirable to control the temperature to a prescribed range in the presence of wide variations in heat source and heat sink temperatures. On the other hand, it may be required to permit the passage of heat under one set of conditions and completely block the heat flow under another set of conditions. This leads to a consideration of the performance of heat pipes known as thermal switches or thermal diodes.

There are four major control approaches and these are described in the following subsections.

17.5.1   Gas-Loaded Heat Pipe

As discussed on pages 190 and 191, the presence of a noncondensable gas has a marked effect on the performance of a condenser. This effect can be exploited for heat pipe control. Any noncondensable gas present in the vapor space is swept to the condenser section during operation and gas will block a portion of the condenser surface. The heat flow at the condenser can be controlled by controlling the volume of the noncondensable gas. Examples of self-controlled devices, those that are controlled by the vapor pressure of the working fluid are shown in Fig. 17.6a, b, c. Examples of feedback-controlled devices are shown in Fig. 17.6d, e.

17.5.2   Excess-Liquid Heat Pipe

The foregoing subsection has pointed out that control can be provided by condenser blocking with a noncondensable gas. Control can also be attained by condenser flooding with excess working fluid. In the excess-liquid heat pipe, excess working fluid in the liquid phase is swept into the condenser and blocks a portion of the condenser. Observe in Fig. 17.7a that a decrease in vapor temperature will expand the control fluid in the bellows, which forces excess liquid to block a portion of the condenser. An example of a thermal diode is displayed in Fig. 17.7b.

17.5.3   Vapor Flow-Modulated Heat Pipe

The performance of the heat pipe can be controlled by controlling the vapor flow through the adiabatic section. Observe that, in Fig. 17.8a, an increase in heat input or an increase in heat source temperature felt at the surface of the evaporator causes a rise in the temperature and pressure of the vapor in the evaporator section. The flow of this vapor through the throttling valve creates a temperature and pressure drop that results in a reduction of the magnitudes of these quantities in the condenser section. Thus, the condensing temperature and pressure can be held at values that

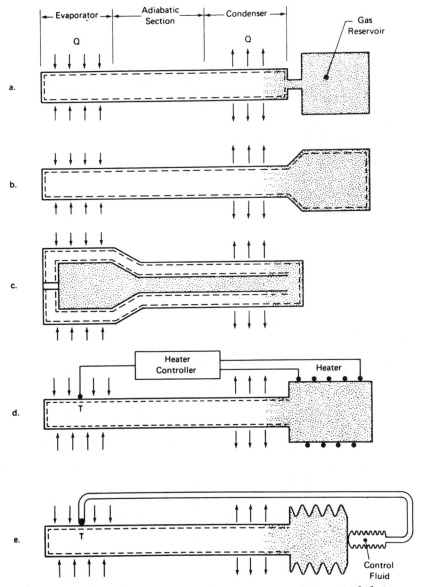

FIG. 17.6 Some representative gas-loaded heat pipes. (From Chi [6], by permission.)

yield the required condenser performance even though the temperature at the heat source has increased. In the event that the heat input increases, the condenser can keep pace with this increase and adjust its performance by means of the throttling valve.

Figure 17.8*b* shows a thermal switch where the flow of vapor through the throttling valve can be cut off entirely.

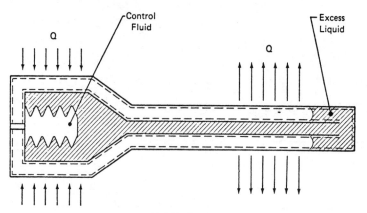

a. Variable Conductance

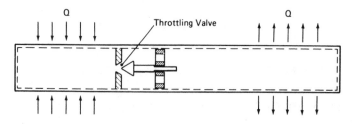

b. Thermal Diode

FIG. 17.7 Some representative excess liquid heat pipes. (From Chi [6], by permission.)

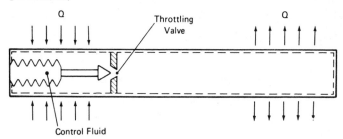

Q Q

a. Vapor-Modulated Variable Conductance

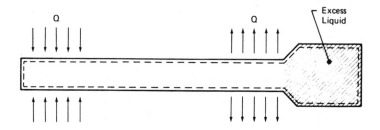

Q Q

b. Vapor-Modulated Thermal Diode

FIG. 17.8 Some representative vapor flow-modulated heat pipes. (From Chi [6], by permission.)

17.5.4 Liquid Flow-Modulated Heat Pipe

Liquid flow control is also an effective way of maintaining control over heat pipe performance. One way of controlling the liquid flow is through the use of a liquid trap as shown in Fig. 17.9*a*. This trap is a wick lined reservoir located at the evaporator end. The wick in the trap is called the trap wick and this wick is not connected to the operating wick in the rest of the heat pipe. In the normal mode of operation with the heat pipe operating in standard fashion, the trap wick is dry. If the heat input increases or the attitude of the heat pipe changes, condensation may occur in the trap and the liquid trap may become an alternate condensing end of the heat pipe. As liquid accumulates in the trap, the main wick begins to starve and this, of course, causes a reduction in heat pumping capability.

An example of a heat pipe with the evaporator section below the condenser section is shown in Fig. 17.9*b*. This, in itself, is a type of control because the heat pipe can function as a thermal diode providing the wick is appropriately designed. Observe that the condensed liquid is returned to the evaporator section with the assistance of the gravitational force.

17.5.5 An Application

Figure 17.10 is an artist's conception of three dielectric heat pipes used to cool high voltage components in an electronic package. The heat pipes can withstand high voltage differences and provide electrically insulated cooling of the components. The finned heat sinks at the condenser end (which is above the evaporator end) operate at ground potential.

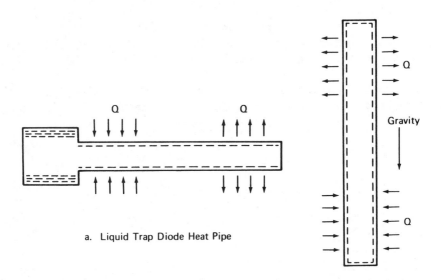

a. Liquid Trap Diode Heat Pipe

b. Gravity-Operated Diode Heat Pipe

FIG. 17.9 Some representative liquid flow-modulated heat pipes. (From Chi [6], by permission.)

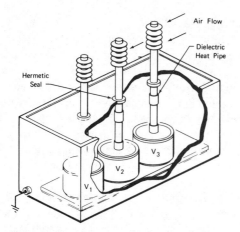

FIG. 17.10   Sketch of dielectric heat pipes used for cooling high-voltage components. (From Chi [6], by permission.)

17.6   LIMITATION ON HEAT TRANSPORT

The heat transport capability of the heat pipe is a function of the ease of circulation of the working fluid. Four types of limitations on the circulation of the working fluid can occur, and these must be investigated in turn:

1   The capillary limitation q_c, which is linked to the capillary pumping capability
2   The sonic limitation q_s, which is due to the choking of the vapor flow
3   The entrainment limitation q_e, which is due to the tearing of the liquid off the liquid-vapor interface by the vapor flowing at a high velocity
4   The boiling limitation q_b, which deals with the disruption of the liquid flow at the evaporator by nucleate boiling in the wick

The following section provides a step-by-step procedure for the determination of these limitations. The discussion is based on the dimensional terminology shown in Fig. 17.11.

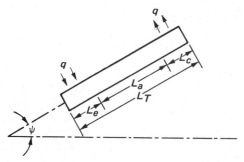

FIG. 17.11   Length and inclination terminology for heat pipes.

17.7   CALCULATION OF HEAT TRANSPORT LIMITATIONS

17.7.1   Capillary Limitation

1. Obtain the effective capillary radius r_c in meters from Table 17.1 and the maximum capillary pressure p_{cm}.

$$p_{cm} = \frac{2\sigma}{r_c} \quad (\text{N/m}^2) \tag{17.1}$$

2. Determine the normal and axial hydrostatic pressures, $\Delta p_\perp$ and $\Delta p_\parallel$ in newtons per square meter.

$$\Delta p_\perp = \rho_l g d_v \cos \psi \quad (\text{N/m}^2) \tag{17.2a}$$

$$\Delta p_\parallel = \rho_l g L_v \sin \psi \quad (\text{N/m}^2) \tag{17.2b}$$

where $g = 9.807$ m/s^2. Observe that in heat pipes employing wick structures that permit no circumferential communication of liquid (as in the rectangular axial groove structure), $\Delta p_\perp = 0$.

3. The maximum effective pumping pressure is

$$p_{pm} = p_{cm} - \Delta p_\perp - \Delta p_\parallel \tag{17.3}$$

4. Calculate the frictional coefficients for liquid and vapor flow:

$$F_l = \frac{\mu_l}{KA_w \lambda \rho_l} \quad (\text{N/W m}^3) \tag{17.4}$$

where K is the wick permeability determined in accordance with Table 17.2, and

$$F_v = \frac{C(f_v \, \text{Re}_v)\mu_v}{2(r_{h,v})^2 A_v \rho_v \lambda} \quad (\text{N/W m}^3) \tag{17.5}$$

TABLE 17.1   Expressions for the Effective Capillary Radius r_c for Several Wick Structures

Structure	r_c	Data
Circular cylinder (artery or tunnel wicks)	$r_c = r$	r = radius of liquid flow passage
Rectangular groove	$r_c = w$	w = groove width
Triangular groove	$r_c = \dfrac{w}{\cos \beta}$	w = groove width β = half-included angle
Parallel wires	$r_c = w$	w = wire spacing
Wire screens	$r_c = \dfrac{w + d_w}{2} = \dfrac{1}{2N}$	N = screen mesh number w = wire spacing d_w = wire diameter
Packed spheres	$r_c = 0.41 r_s$	r_s = sphere radius

TABLE 17.2 Wick Permeability K for Several Wick Structures

Structure	K	Data
Circular cylinder (artery or tunnel wick)	$K = \dfrac{r^2}{8}$	r = radius of liquid flow passage
Open rectangular grooves	$K = \dfrac{2er_{h,l}^2}{(f_l\,Re_l)}$	e = wick porosity = w/s w = groove width s = groove pitch δ = groove depth $r_{h,l} = 2w\delta/(w + 2\delta)$ $(f_l\,Re_l)$ from Fig. 17.12
Circular annular wick	$K = \dfrac{2r_{h,l}^2}{(f_l\,Re_l)}$	$r_{h,l} = r_1 - r_2$ (see Fig. 17.13) $(f_l\,Re_l)$ from Fig. 17.13
Wrapped screen wick	$K = \dfrac{d_w^2 e^3}{122(1 - e)^2}$	d_w = wire diameter $e = 1 - (\pi SNd_w/4)$ S = wick crimping factor N = mesh number
Packed sphere	$K = \dfrac{r_s^2 e^3}{37.5(1 - e)^2}$	r_s = sphere radius e = porosity (dependent on packing mode)

where the hydraulic radius $r_{h,v}$ is half of d_v and the value of $(f_v\,Re_v)$ depends on the flow regime (laminar or turbulent) and whether the vapor flow is compressible or incompressible (Mach number Ma < 0.2). In order to determine these criteria, the Reynolds number

$$\text{Re} = \frac{2r_{h,v}q_c}{A_v\mu_v\lambda} \tag{17.6}$$

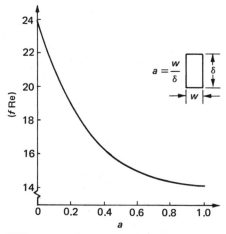

FIG. 17.12 Frictional coefficients as a function of aspect ratio for laminar flow in rectangular tubes. (From Chi [6], by permission.)

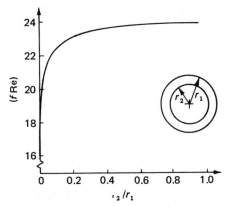

FIG. 17.13 Frictional coefficients as a function of radius ratio for laminar flow in circular annuli. (From Chi [6], by permission.)

and the Mach number

$$\text{Ma} = \frac{q_c}{A_v \rho_v \lambda (\gamma_v R_v T_v g)^{1/2}} \tag{17.7}$$

must be calculated. In Eq. (17.7), R_v must bear the units J/kg K; and in Eqs. (17.6) and (17.7), any assumed q_c must be verified by checking the eventual calculated q_c.

For circular vapor core cross sections, the following apply:

$Re_v \leqslant 2300, Ma_v \leqslant 0.2$:

$$(f_v Re_v) = 16$$

$$C = 1.00$$

$Re_v \leqslant 2300, Ma_v > 0.2$:

$$(f_v Re_v) = 16$$

$$C = C_1 = \left[1 + \left(\frac{\gamma_v - 1}{2} \right) M_v^2 \right]^{-1/2}$$

$Re_v > 2300, Ma_v \leqslant 0.2$:

$$(f_v Re_v) = 0.038$$

$$C = C_2 = \left(\frac{2r_{h,v} q}{A_v \lambda \mu_v} \right)^{3/4}$$

$Re_v > 2300, Ma_v > 0.2$:

$$(f_v Re_v) = 0.038$$

$$C = C_1 C_2$$

5. Obtain the capillary heat transport factor.

$$(qL)_c = \frac{P_{pm}}{F_l + F_v} \quad \text{(W m)} \tag{17.8}$$

6. Determine the capillary heat transfer limit.

$$q_c = \frac{(qL)_c}{(1/2)L_c + L_a + (1/2)L_e} \quad \text{(W)} \tag{17.9}$$

where L_c, L_a, and L_e are, respectively, the lengths of condenser, adiabatic section, and evaporator (see Fig. 17.1).

17.7.2 Sonic Limitation

$$q_s = A_v \rho_v \lambda \left[\frac{\gamma_v R_v T_v}{2(\gamma_v + 1)} \right]^{1/2} \quad \text{(W)} \tag{17.10}$$

17.7.3 Entrainment Limitation

$$q_e = A_v \lambda \left(\frac{\sigma \rho_v}{2r_{h,s}} \right)^{1/2} \quad \text{(W)} \tag{17.11}$$

where $r_{h,s}$ is the hydraulic radius of the wick surface pores, equal to half the wire spacing for screen wicks, the width of the groove for grooved wicks, and $0.41r_s$ for packed-sphere wicks.

17.7.4 Boiling Limitation

$$q_b = \frac{2\pi L_e k_e T_v}{\lambda \rho_v \ln (r_i/r_v)} \left(\frac{2\sigma}{r_n} - P_{cm} \right) \quad \text{(W)} \tag{17.12}$$

where k_e is the effective thermal conductivity for liquid-saturated wicks determined in accordance with Table 17.3 and r_n is the nucleation radius, which may be taken conservatively between 2.54×10^{-5} and 2.54×10^{-7} m for conventional heat pipes.

17.8 EXAMPLE

Specification: Design a heat pipe with water as the working fluid at 200°F (93.33°C, 366.33 K) to operate at sea level. Available for use is a copper tube $\frac{7}{8}$ in OD ($d_o = 0.02223$ m) $\times$ 18 BWG (wall thickness = 0.00124 m), 1 ft ($L_t = 0.3048$ m) long. The desired heat transport is 60 Btu/h (17.58 W). The application requires installation at an inclination of 5° ($\psi = 5°$).

Objective: Determine if 17.58 W exceeds any of the four limitations.
Selection (tube):

$$d_o = 0.02223 \text{ m} \quad \text{(specified)}$$

$d_i = d_o - 2(\text{wall}) = 0.01974$ m

$L_t = 0.3048$ m (maximum specified)

$L_c = 0.0508$ m

$L_a = 0.1778$ m

$L_e = 0.0627$ m

$\psi = 5°$ (specified)

Selection (wick):

Wire screen wick, copper ($k_w = 389.25$ W/m K)
Mesh number $N = 240$/in (9.448×10^3 m^{-1})
Wire diameter $d_w = 0.0025$ in (6.350×10^{-5} m)
Crimping factor $S = 1.05$
Stack 8 layers and 8 spaces (spacing = 6.350×10^{-5} m)
Wick thickness $t_w = 16 (6.350 \times 10^{-5}) = 1.016 \times 10^{-3}$ m
Vapor core diameter $d_v = d_i - 2t_w = 0.01770$ m

Water properties at 366.33 K (liquid):

Density $\rho_l = 963.12$ kg/m^3
Viscosity $\mu_l = 3.051 \times 10^{-4}$ kg/m s
Thermal conductivity $k_l = 0.678$ W/m K
Surface tension $\sigma = 0.0614$ N/m
Latent heat of vaporization $\lambda = 2.273 \times 10^6$ J/kg

TABLE 17.3 Effective Thermal Conductivity k_e for Liquid-Saturated Wick Structures[a]

Wick structure	k_e
Wick and liquid in series	$k_e = \dfrac{k_l k_w}{e k_w + k_l(1 - e)}$
Wick and liquid in parallel	$k_e = e k_l + k_w(1 - e)$
Wrapped screen	$k_e = \dfrac{k_l[(k_l + k_w) - (1 - e)(k_l - k_w)]}{(k_l + k_w) + (1 - e)(k_l - k_w)}$
Packed spheres	$k_e = \dfrac{k_l[(2k_l + k_w) - 2(1 - e)(k_l - k_w)]}{(2k_l + k_w) + (1 - e)(k_l - k_w)}$
Rectangular grooves	$k_e = \dfrac{(w_f k_l k_w \delta) + w k_l(0.185 w_f k_w + \delta k_l)}{(w + w_f)(0.185 w_f k_f + \delta k_l)}$

[a] k_e = effective thermal conductivity
  k_l = liquid thermal conductivity
  k_w = wick material thermal conductivity
  e = wick porosity
  w_f = groove fin thickness
  w = wick thickness
  δ = groove depth

Water properties at 366.33 K (vapor):

Density $\rho_v = 0.4762$ kg/m^3
Viscosity $\mu_v = 1.293 \times 10^{-5}$ kg/m s
Thermal conductivity $k_v = 0.0234$ W/m K
Vapor is triatomic, $\gamma = 1.33$
Gas constant $R_v = 8314/18 = 461.89$ J/kg K

Capillary limitation:

Capillary radius (Table 17.1)

$$r_c = \frac{1}{2N} = \frac{1}{2(9.448 \times 10^3)} = 5.292 \times 10^{-5} \text{ m}$$

Maximum capillary pressure [Eq. (17.1)]

$$p_{cm} = \frac{2\sigma}{r_c} = \frac{2(0.0614)}{5.292 \times 10^{-5}} = 2.320 \times 10^3 \text{ N/m}^2$$

Normal and axial hydrostatic pressures [Eq. (17.2)]

$$\Delta p_\perp = \rho_l g d_v \cos \psi$$
$$= (963.12)(9.807)(0.01770)(\cos 5°) = 1.666 \times 10^2 \text{ N/m}^2$$

$$\Delta p_\| = \rho_l g L_t \sin \psi$$
$$= (963.12)(9.807)(0.3048)(\sin 5°) = 2.509 \times 10^2 \text{ N/m}^2$$

Maximum effective pumping pressure [Eq. (17.3)]

$$p_{pm} = p_{cm} - \Delta p_\perp - \Delta p_\|$$
$$= 2.320 \times 10^3 - 1.666 \times 10^2 - 2.509 \times 10^2$$
$$= 1903 \text{ N/m}^2$$

Wick area

$$A_w = \frac{\pi}{4}(d_i^2 - d_v^2)$$

$$= \frac{\pi}{4}[(0.01974)^2 - (0.01770)^2] = 5.975 \times 10^{-5} \text{ m}^2$$

Wick porosity (Table 17.2)

$$\epsilon = 1 - \frac{\pi S N d_w}{4}$$

$$= 1 - \frac{\pi(1.05)(9.448 \times 10^3)(6.350 \times 10^{-5})}{4} = 1 - 0.495 = 0.505$$

Wick permeability (Table 17.2)

$$K = \frac{d_w^2 \epsilon^3}{122(1-\epsilon)^2}$$

$$= \frac{(6.350 \times 10^{-5})^2 (0.505)^3}{122(1-0.505)^2}$$

$$= 1.741 \times 10^{-11} \text{ m}^2$$

Liquid frictional coefficient [Eq. (17.4)]

$$F_l = \frac{\mu_l}{KA_w\lambda\rho_l}$$

$$= \frac{3.051 \times 10^{-4}}{(1.741 \times 10^{-11})(5.975 \times 10^{-5})(5.975 \times 10^6)(963.12)}$$

$$= 1.340 \times 10^2 \text{ N/W m}^3$$

Vapor hydraulic radius

$$r_{h,v} = \tfrac{1}{2} d_v = \tfrac{1}{2}(0.01770) = 0.00885 \text{ m}$$

Vapor space area

$$A_v = \frac{\pi}{4} d_v^2 = \frac{\pi}{4}(0.01770)^2 = 2.462 \times 10^{-4} \text{ m}^2$$

Reynolds number ($q = 17.58$ W) [Eq. (17.6)]

$$\text{Re} = \frac{2r_{h,v}q_c}{A_v\mu_v\lambda}$$

$$= \frac{2(0.00885)(17.58)}{(2.462 \times 10^{-4})(1.293 \times 10^{-5})(2.273 \times 10^6)}$$

$$= 43.0 \ll 2300$$

Mach number [Eq. (17.7)]

$$\text{Ma} = \frac{q_c}{A_v\rho_v\lambda(\gamma_v R_v T_v g)^{1/2}}$$

$$= \frac{17.58}{(2.462 \times 10^{-4})(0.4762)(2.273 \times 10^6)[1.33(461.89)(366.33)(9.807)]^{1/2}}$$

$$= 4.441 \times 10^{-5} \ll 0.2$$

With Re $<$ 2300 and Ma $<$ 0.2

$$(f_v \text{Re}_v) = 16$$

$$C = 1.00$$

Vapor frictional coefficient [Eq. (17.5)]

$$F_v = \frac{C(f_v \, \text{Re}_v)\mu_v}{2(r_{h,v})^2 A_v \rho_v \lambda}$$

$$= \frac{1.00(16)(1.293 \times 10^{-5})}{2(0.00885)^2 (2.462 \times 10^{-4})(0.4762)(2.273 \times 10^6)}$$

$$= 4.386 \times 10^{-5}$$

Capillary heat transport factor [Eq. (17.8)]

$$(qL)_c = \frac{P_{pm}}{F_l + F_v}$$

$$= \frac{1903}{1.340 \times 10^2 + 4.386 \times 10^{-5}}$$

$$= \frac{1903}{134.0} = 14.203$$

Capillary heat transfer limit [Eq. (17.9)]

$$q_c = \frac{(qL)_c}{(1/2)L_c + L_a + (1/2)L_e}$$

$$= \frac{14.203}{(1/2)(0.0508) + (0.1778) + (1/2)(0.0627)}$$

$$= \frac{14.203}{0.2346} = 60.55 \, \text{W} > 17.58 \, \text{W}$$

The selected heat pipe will not be limited by capillary transport and, because the Reynolds number (Re = 4.7 ≪ 2300) and Mach number (Ma = 3.973 × 10⁻⁵ ≪ 0.2) calculated for 17.58 W are so low, there is no need to recalculate these for 60.55 W.

Sonic limitation [Eq. (17.10)]:

$$q_s = A_v \rho_v \lambda \left[\frac{\gamma_v R_v T_v}{2(\gamma_v + 1)} \right]^{1/2}$$

$$= (2.462 \times 10^{-4})(0.4762)(2.273 \times 10^6) \left[\frac{1.33(461.89)(366.33)}{2(1.33 + 1)} \right]^{1/2}$$

$$= 58,553 \, \text{W} > 17.58 \, \text{W}$$

Entrainment limitation:

Hydraulic radius of wick surface pores

$$r_{h,s} = \tfrac{1}{2}(6.350 \times 10^{-5}) = 3.175 \times 10^{-5} \, \text{m}$$

Entrainment limit [Eq. (17.11)]

$$q_c = A_v \lambda \left(\frac{\sigma \rho_v}{2r_{h,s}} \right)^{1/2}$$

$$= (2.462 \times 10^{-4})(2.273 \times 10^6) \left[\frac{0.0614(0.4762)}{2(3.175 \times 10^{-5})} \right]^{1/2}$$

$$= 12{,}006 \text{ W} > 17.58 \text{ W}$$

Boiling limitation:

Take $r_n = 2.54 \times 10^{-6}$ m
Inside radius $r_i = \frac{1}{2} d_i = \frac{1}{2}(0.01974) = 0.00987$ m
Vapor space radius $r_v = \frac{1}{2} d_v = \frac{1}{2}(0.01770) = 0.00885$ m
Effective thermal conductivity of wick material (Table 17.3)

$$k_e = \frac{k_l[(k_l + k_w) - (1 - \epsilon)(k_l - k_w)]}{(k_l + k_w) + (1 - \epsilon)(k_l - k_w)}$$

$$= \frac{0.678[(0.678 + 389.25) - (1 - 0.505)(0.678 - 389.25)]}{[(0.678 + 389.25) + (1 - 0.505)(0.678 - 389.25)]}$$

$$= 1.998 \text{ W/m K}$$

Boiling limitation [Eq. (17.12)]

$$q_b = \frac{2\pi L_e k_e T_v}{\lambda \rho_v \ln (r_i/r_v)} \left(\frac{2\sigma}{r_n} - p_{cm} \right)$$

$$= \frac{2\pi (0.0627)(1.998)(366.33)}{(2.273 \times 10^6)(0.4762) \ln (0.00987/0.00885)} \left[\frac{2(0.0614)}{2.54 \times 10^{-6}} - 2.32 \times 10^3 \right]$$

$$= 112.8 \text{ W} > 17.58 \text{ W}$$

Of the four possible limits, the capillary limit is the smallest. In any event, the heat pipe will handle 17.58 W.

17.9   HEAT PIPE PERFORMANCE

If the duty (heat load) of the heat pipe is not restricted by any of the foregoing four limitations, the designer may proceed to a determination of the two significant temperatures of interest. These are the temperature at the surface of the evaporator, T_e, and the temperature at the surface of the condenser, T_c.

An electrothermal analog of the heat pipe is shown in Fig. 17.14. Observe that there are six resistances to the flow of heat from evaporator to environment.

1 R_{pe}, the resistance of the pipe wall at the evaporator
2 R_{we}, the resistance of the wick at the evaporator
3 R_a, the resistance of the adiabatic section of the pipe
4 R_{wc}, the resistance of the wick at the condenser

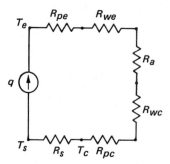

FIG. 17.14   Electrothermal analog for heat pipe.

5 R_{pc}, the resistance of the pipe at the condenser
6 R_s, the resistance between the condenser section and the environment

Because the analyst must know what the component that is mounted to the evaporator "sees," a precise evaluation of these resistances is necessary so that T_e can be determined. The balance of this section provides some guidance in this regard, and the next section presents a fairly detailed example.

1 R_{pe}, the resistance of the pipe wall at the evaporator:

$$R_{pe} = \frac{\ln (d_o/d_i)}{2\pi L_e k_m} \quad (°C/W) \tag{17.13}$$

for circular pipes, where k_m is the thermal conductivity of the pipe wall.
2 R_{we}, the resistance of the wick at the evaporator:

$$R_{we} = \frac{\ln (d_i/d_o)}{2\pi L_e k_e} \quad (°C/W) \tag{17.14}$$

for circular pipes, where k_e is the effective conductivity of the wick material.
3 R_a, the resistance of the adiabatic section of the pipe:

$$R_a = \frac{T_v(p_{ve} - p_{vc})}{\rho_v \lambda J q} \quad (°C/W) \tag{17.15}$$

where p_{ve} and p_{vc} are the vapor pressures at the evaporator and condenser, respectively, and J is the mechanical equivalent of heat (Joule's constant). Because there is little vapor pressure loss in the adiabatic section, R_a is usually negligible.
4 R_{wc}, the resistance of the wick at the condenser:

$$R_{wc} = \frac{\ln (d_i/d_v)}{2\pi L_c k_e} \quad (°C/W) \tag{17.16}$$

for circular pipes.

5 R_{pc}, the resistance of the pipe wall at the condenser:

$$R_{pc} = \frac{\ln (d_o/d_i)}{2\pi L_c k_m} \quad (°C/W) \tag{17.17}$$

for circular pipes.

6 R_s, the resistance between the condenser section and the environment:

$$R_s = \frac{1}{h S_t} \quad (°C/W) \tag{17.18}$$

where h is the heat transfer coefficient between the pipe outer wall and the environment and S_t is the total surface at the condenser end. If the condenser end is finned, $S_t = S_b + \eta S_f$, where S_b is based on prime surface, S_f is fin surface, and η is fin efficiency.

17.10 EXAMPLE

Consider the heat pipe of the previous example. The condenser end is to be equipped with six copper fins with outer diameter 0.0381 m ($1\frac{1}{2}$ in) and width 0.00254 m ($\frac{1}{10}$ in) on a spacing of 0.00508 m ($\frac{1}{5}$ in). The condenser will be exposed to air at 104°F (40°C, 313 K) at a flow velocity of 3 m/s.

1 R_{pe} [Eq. (17.13)]

From the previous example (Sec. 17.8),

$d_i = 0.01974$ m

$d_o = 0.02223$ m

$L_e = 0.0627$ m

For copper, $k_m = 379$ W/m K.

$$\begin{aligned}
R_{pe} &= \frac{\ln (d_o/d_i)}{2\pi L_e k_m} \\
&= \frac{\ln (0.02223/0.01974)}{2\pi(0.0627)(379)} \\
&= 7.956 \times 10^{-4} \ °C/W
\end{aligned}$$

2 R_{we} [Eq. (17.14)]

From the previous example,

$d_v = 0.01770$ m

$k_e = 1.998$ W/m K

$$R_{we} = \frac{\ln (d_i/d_v)}{2\pi L_e k_e} = \frac{\ln (0.01974/0.01872)}{2\pi(0.0627)(1.998)} = 1.381 \times 10^{-1} \ °C/W$$

3 R_a is negligible.
4 R_{wc} [Eq. (17.16)]
From the previous example,

$$L_c = 0.0508 \text{ m}$$

$$\begin{aligned}
R_{wc} &= \frac{\ln (d_i/d_v)}{2\pi L_c k_e} \\
&= \frac{\ln (0.01974/0.01770)}{2\pi(0.0508)(1.998)} \\
&= 1.704 \times 10^{-1} \text{ °C/W}
\end{aligned}$$

5 R_{pc} [Eq. (17.17)]

$$\begin{aligned}
R_{pc} &= \frac{\ln (d_o/d_i)}{2\pi L_c k_m} \\
&= \frac{\ln (0.02223/0.01974)}{2\pi(0.0508)(379)} \\
&= 9.820 \times 10^{-4} \text{ °C/W}
\end{aligned}$$

6 R_s [Eq. (17.18)]
The Briggs and Young correlation [5] is

$$\frac{h}{d_o k_a} = 0.134 \left(\frac{\rho_a V_a d_o}{\mu_a}\right)^{0.681} \left(\frac{c_{pa}\mu_a}{k_a}\right)^{1/3} \left(\frac{s}{l}\right)^{0.200} \left(\frac{s}{\delta}\right)^{0.1134}$$

where k_a, c_{pa}, ρ_a, and μ_a are taken at the air bulk stream temperature, l is the fin height, s is the fin spacing, and δ is the fin width.
Air properties at 313 K:

Thermal conductivity $k_a = 0.0277 \text{ W/m K}$
Specific heat $c_{pa} = 1.0090 \times 10^3 \text{ W s/kg K}$
Viscosity $\mu_a = 1.9305 \times 10^{-5} \text{ kg/m s}$
Density $\rho_a = 1.1056 \text{ kg/m}^3$
Prandtl number $= 0.7032$

Air velocity: 3 m/s
Fin data:

$$D_o = 0.0381 \text{ m}$$

$$l = \tfrac{1}{2}(D_o - d_o) = \tfrac{1}{2}(0.0381 - 0.02223) = 0.00794 \text{ m}$$

$$\delta = 0.00254 \text{ m}$$

$$s = 0.00508 - 2\left(\frac{\delta}{2}\right) = 0.00508 - 0.00254 = 0.00254 \text{ m}$$

$$d_o = 0.02223 \text{ m}$$

Reynolds number:

$$\text{Re} = \frac{\rho_a V_a d_o}{\mu_a} = \frac{(1.1056)(3)(0.02223)}{1.9305 \times 10^{-5}} = 3819.5$$

Nusselt number:

$$\frac{hd_o}{k_a} = 0.134(\text{Re})^{0.681}(P_r)^{1/3}\left(\frac{s}{l}\right)^{0.200}\left(\frac{s}{\delta}\right)^{0.1134}$$

$$= 0.134(3819.5)^{0.681}(0.7032)^{1/3}\left(\frac{0.00254}{0.00794}\right)^{0.200}\left(\frac{0.00254}{0.00254}\right)^{0.1134}$$

$$= 26.09$$

Heat transfer coefficient:

$$h = \text{Nu}\,\frac{k}{d_o} = 26.09\left(\frac{0.0277}{0.02223}\right)$$

$$= 32.51\ \text{W/m}^2\ \text{K}$$

Surface (number of fins $n = 6$):

$$S_b = \pi d_o(L_c - n\delta)$$

$$= \pi(0.02223)[0.0508 - 6(0.00254)]$$

$$= 2.483 \times 10^{-3}\ \text{m}^2$$

$$S_f = 2\,\frac{\pi}{4}\,n[D_o^2 - d_o^2]$$

$$= 2\left(\frac{\pi}{4}\right)(6)[(0.0381)^2 - (0.02223)^2]$$

$$= 9.024 \times 10^{-3}\ \text{m}^2$$

Fin efficiency:

Diameter ratio

$$\rho = \frac{d_o}{D_o} = \frac{0.02223}{0.0381} = 0.583$$

Profile area

$$A_p = l\delta = 0.00794(0.00254) = 2.017 \times 10^{-5}\ \text{m}^2$$

Fin performance factor:

$$\phi = (l)^{3/2}\left(\frac{2h}{k_m A_p}\right)^{1/2} = (0.00794)^{3/2}\left[\frac{2(32.51)}{379(2.017 \times 10^{-5})}\right]^{1/2} = 0.0653$$

Fin efficiency [5] :

$$\eta = 0.996$$

R_s [Eq. (17.18)] :

$$R_s = \frac{1}{hS_t} = \frac{1}{h(S_b + \eta S_f)}$$

$$= \frac{1}{32.51[2.483 \times 10^{-3} + 0.996(9.024 \times 10^{-3})]}$$

$$= 2.6815\ ^\circ C/W$$

Total resistance:

$$R_t = R_{pe} + R_{we} + R_{wc} + R_{pc} + R_s$$
$$= 7.956 \times 10^{-4} + 1.381 \times 10^{-1} + 1.704 \times 10^{-1} + 9.82 \times 10^{-4} + 2.6815$$
$$= 2.9918\ ^\circ C/W$$

Total temperature rise:

$$\Delta T = qR_t = (17.58)(2.9918) = 52.59^\circ C$$

Temperature at surface of evaporator:

$$T_e = 40 + 52.59 = 92.59^\circ C \approx 93.33^\circ C$$

It may be remarked that the resistance of the heat pipe is quite small when compared to the total resistance in the heat flow path. The resistance between T_e and T_c (see Fig. 17.14) is $2.9918 - 2.6815 = 0.3103\ ^\circ C/W$, and the temperature drop through the heat pipe is $T_e - T_c = 0.3103(17.58) = 5.45^\circ C$. This shows that the heat pipe is close to being an isothermal device.

Indeed, one may define an overall coefficient of heat transfer for the heat pipe itself in accordance with

$$q = UA\ \Delta T$$

where A is the cross-sectional area of the pipe structure. This makes

$$U \equiv \frac{q}{A\ \Delta T} \quad (W/m^2\ K)$$

and in this case, with

$$A = \frac{\pi}{4}\ d_o^2 = \frac{\pi}{4}\ (0.02223)^2 = 3.8812 \times 10^{-4}\ m^2$$

U is seen to be

$$U = \frac{q}{A \, \Delta T} = \frac{17.58}{(3.8812 \times 10^{-4})(5.45)} = 8311 \text{ W/m}^2 \text{ K}$$

a rather remarkable number.

As a final point, it should be observed that the heat pipe will operate at a point very close to the temperature for which it was designed.

17.11 NOMENCLATURE

Roman Letter Symbols

a	aspect ratio, dimensionless
A	area, m^2
c	specific heat, J/kg K
C	a constant, defined where used
d	diameter, m
D	fin diameter, m
f	drag coefficient defined in Eq. (17.5), dimensionless
F	frictional coefficient, N/W m^3
g	gravitational acceleration, 9.807 m/s^2
h	heat transfer coefficient, W/m^2 K
J	mechanical equivalent of heat, 4.184 kg/cal
k	thermal conductivity, W/m K
K	wick permeability, m^2
l	fin height (radial fins), m
L	length, m
Ma	Mach number, dimensionless
N	screen mesh number, dimensionless
p	pressure, N/m^2
q	heat flow, W
qL	with subscript c, the capillary heat transport factor defined by Eq. (17.8), W m
R	thermal resistance, K/W; or universal gas constant, J/kg K
Re	Reynolds number, dimensionless
r	radius, m
s	groove pitch, m; or fin spacing, m
S	crimping factor, dimensionless
t	thickness, m
T	temperature, K
U	overall heat transfer coefficient, W/m^2 K
V	velocity, m/s
w	groove width, m; or wire spacing, m

Greek Letter Symbols

β	half of included angle, degrees or radians
γ	specific heat ratio, dimensionless

δ	groove depth, m; or fin width, m
ϵ	wick porosity, dimensionless
η	fin efficiency, dimensionless
λ	latent heat of vaporization, J/kg
μ	dynamic viscosity, kg/m s
ρ	density, kg/m^3
σ	surface tension coefficient, N/m
ϕ	fin performance factor, dimensionless
ψ	angle of inclination, degrees or radians

Subscripts

a	indicates adiabatic section or air
b	indicates boiling limitation or base surface
c	indicates capillary, capillary limitation, condenser, or condenser section
cm	indicates maximum capillary pressure
e	indicates entrainment limitation, effective or equivalent value, or evaporation section
f	indicates fin
i	indicates inner quantity
l	indicates liquid
m	indicates mean quantity or metal
n	indicates nucleation
o	indicates outer quantity
p	indicates profile
pc	indicates pipe at condenser section
pe	indicates pipe at evaporation section
pm	indicates maximum pumping pressure
s	indicates sonic limitation, sphere, or surroundings
t	indicates total
v	indicates vapor
w	indicates wire
wc	indicates wick at condenser
we	indicates wick at evaporator
$\parallel$	indicates axial hydrostatic pressure
$\perp$	indicates normal hydrostatic pressure

17.12 REFERENCES

1 Gaugler, R. S., Heat Transfer Devices, U.S. Patent 2350348, 1944.
2 Trefethen, L., *On the Surface Tension Pumping of Liquids or a Possible Role of the Candlewick in Space Exploration*, GE Tech. Inf. Ser. No. 615-D114, General Electric Company, Schenectady, N.Y., February 1962.
3 Grover, G. M., Cotter, T. P., and Erickson, G. F., Structures of Very High Thermal Conductivity, *J. Appl. Phys.*, vol. 35, pp. 1190–1191, 1964.
4 Briggs, D. E., and Young, E. H., Convection Heat Transfer and Pressure Drop of Air Flowing Across Triangular Pitch Banks of Finned Tubes, *Chem. Eng. Prog. Symp. Ser.*, vol. 41, no. 59, pp. 1–10, 1963.
5 Kern, D. Q., and Kraus, A. D., *Extended Surface Heat Transfer*, McGraw-Hill, New York, 1972.
6 Chi, S. W., *Heat Pipe Theory and Practice*, Hemisphere, Washington, D.C., 1976.

18

■ thermoelectric coolers

18.1 INTRODUCTION

Thermoelectric coolers have considerable utility in several specific applications where precise temperature control is required. One example of a thermoelectric solution of a temperature control problem is that of cooling infrared detectors, where a cascaded thermoelectric cooler is used to attain a temperature of the order of $-100°C$ in a room-temperature environment.

Other examples include the use of an array of coolers integral with the enclosing walls of a package of electronic equipment to control the interior to a precise temperature, removal of hot spots from components such as gyros and accelerometers, and cooling an entire avionics system to a temperature *below* that of the ambient. In addition, thermoelectric air conditioning installations exist in many shipboard applications.

Because thermoelectric coolers are becoming more and more useful and because their design and performance are not well understood, this chapter addresses thermoelectric effects and the thermal design and performance of thermoelectric coolers. Examples of cooler design are given using the worksheet approach.

18.2 THERMOELECTRIC EFFECTS

The phenomenon of thermoelectricity, discovered by Peltier, has been known to physicists since 1834. It commanded little sustained attention among engineers and scientists until the middle 1950s because the poor thermoelectric qualities of known materials made them unsuitable for use in a practical refrigerating device. However, the advent of the semiconductor removed the thermoelectric cooler from the area of the laboratory curiosity, and in some applications it now has distinct advantages over other cooling systems.

18.2.1 Definitions

Consider Fig. 18.1, which shows a pair of materials, A and B. The junction between materials A and B is held at some temperature that differs from the temperature at the voltmeter. The voltmeter exhibits a reading in microvolts. If the temperature T_1 is varied, the voltmeter reading will vary. The *Seebeck effect* concerns this net conversion of thermal energy into electrical energy under zero current conditions. The voltage read on the voltmeter is called the Seebeck voltage, or Seebeck electromotive force, emf, and is proportional to the temperature. Hence

$$dE_S \sim dT$$

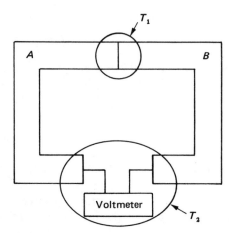

FIG. 18.1 Seebeck effect. The difference in temperature between T_1 and T_2 at the junctions of the two dissimilar materials A and B causes the voltmeter to read the Seebeck voltage.

Insertion of a proportionality constant α, called the *Seebeck coefficient*, yields

$$dE_S = \pm \alpha \, dT$$

where the $\pm$ indicates a reliance on materials A and B to yield a positive or negative voltage. Obviously,

$$E_S = \pm \int_{T_1}^{T_2} \alpha \, dT$$

and $\alpha = \lim\limits_{\Delta T \to 0} \dfrac{\Delta E_S}{\Delta T} = \dfrac{dE_S}{dT}$ (V/°C)

The *Peltier effect* concerns the reversible evolution or absorption of heat that takes place when an electric current traverses a nonhomogeneous conductor or crosses the junction between two dissimilar materials. Consider Fig. 18.2, which shows two dissimilar materials A and B, with their junctions held at temperatures T_1 and T_2. Note a flow of heat into junction 1 and an efflux of heat from junction 2. This suggests that $T_1 < T_2$. The *Peltier heat* is proportional to the current flow:

$$dQ_P \sim I \, dt$$

The proportionality constant, called the *Peltier coefficient* and having the dimension of volts (for this reason, the coefficient is often called the *Peltier voltage*), is inserted to yield

$$dQ_P = \pm \pi I \, dt \quad \text{or} \quad q_p = \pm \pi I \quad \text{(W)}$$

A thermodynamic analysis of these effects prompted William Thomson to predict that an electromotive force must exist between different parts of the same material if they are held at different temperatures. Thomson made this prediction because he was unable to show in the laboratory that the Peltier and Seebeck voltages were equal. He demonstrated that if a uniform metal bar is heated at the middle and a current is sent through it from end to end from an external source, the heat will be conducted unequally along the two halves.

The *Thomson effect*, therefore, concerns the reversible evolution or absorption of heat occurring whenever an electric current traverses a single, homogeneous conductor across which a temperature gradient is maintained. This Thomson effect will be apparent regardless of how the current is introduced; the *Thomson heat* is therefore represented by the proportionality

$$dQ_T \sim I \, dT \, dt$$

Insertion of the proportionality constant σ as the Thomson coefficient gives

$$dQ_T = \pm \sigma I \, dT \, dt \quad \text{or} \quad dq_T = \pm \sigma I \, dT \quad \text{(W)}$$

The quantity $\sigma \, dT$ is obviously a voltage, so the Thomson voltage E_T can be defined by

$$E_T = \pm \int_{T_1}^{T_2} \sigma \, dT$$

In summary, the Peltier and Thomson effects concern reversible effects. The evolution and absorption of heat at junctions 1 and 2 of Fig. 18.2 demonstrate that there must be a difference of potential at places where two dissimilar materials are in

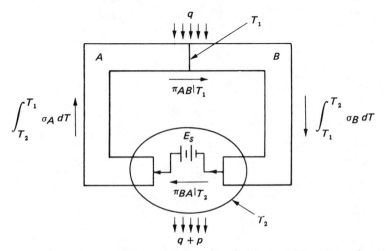

FIG. 18.2 Thermoelectric device showing Peltier and Thomson voltages.

contact. If both junctions are at the same temperature, the thermal emf's at the two junctions are in opposite directions and annul each other. However, if there is a temperature difference, these two emf's do not balance, and their resultant, together with the Thomson emf's, is due to, or establishes, a current in the circuit.

Two other effects are present. Indeed, these effects are more apparent than the thermoelectric effects cited above. They concern the irreversible effects of heat evolution known as the *Joule effect* (I^2R loss) and heat conduction, the *Fourier effect*. Hence

$$q_J = I^2R \quad \text{(W)}$$

$$q_F = \frac{kA}{L}\,\Delta T \quad \text{(W)}$$

18.2.2 Balance Among Seebeck, Peltier, and Thomson Voltages

Now consider Fig. 18.3, which again shows two materials A and B. Their junctions are at two different temperatures T_1 and T_2. The battery supplies a current I at some external voltage E_x. Kirchhoff's voltage law can be applied, taking the sum of the voltage drops around the circuit:

$$E_x = \pi_{AB}|_{T_2} + \pi_{BA}|_{T_2} + \int_{T_2}^{T_1} \sigma_A\,dT + \int_{T_1}^{T_2} \sigma_B\,dT + IR_A + IR_B$$

This may be adjusted to yield

$$E_x = \pi_{AB}|_{T_1} - \pi_{AB}|_{T_2} + \int_{T_2}^{T_1} \sigma_A\,dT - \int_{T_2}^{T_1} \sigma_B\,dT + IR_A + IR_B$$

$$= \pi_{AB}|_{T_1} - \pi_{AB}|_{T_2} + \int_{T_2}^{T_1} (\sigma_A - \sigma_B)\,dT + I(R_A + R_B)$$

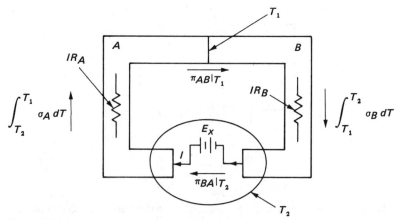

FIG. 18.3 Thermoelectric device with applied voltage.

Inspection of this equation indicates that there is a back emf in this circuit. This back emf requires that the battery provide more voltage than is just necessary to overcome the IR losses in the system. The back emf is equal to the Seebeck voltage. This fact can be derived experimentally. Hence

$$E_S = \int_{T_2}^{T_1} \alpha \, dT = \pi|_{T_1} - \pi|_{T_2} + \int_{T_2}^{T_1} (\sigma_A - \sigma_B) \, dT \tag{18.1}$$

18.2.3 Relationship Among α, π, and σ

Equation (18.1) provides the lead for the determination of the relationships among α, π, and σ. In considering a completely reversible process and one where thermal equilibrium is attained, Kirchhoff's law is again applied to give

$$E_S = \int_{T_2}^{T_1} \alpha \, dT = \pi|_{T_1} - \pi|_{T_2} + \int_{T_2}^{T_1} (\sigma_A - \sigma_B) \, dT$$

$$= \Delta\pi_{12} + \int_{T_2}^{T_1} (\sigma_A - \sigma_B) \, dT$$

and, upon differentiation with respect to T, one obtains

$$\frac{dE_S}{dT} = \frac{d\pi}{dT} + \sigma_A - \sigma_B$$

from which

$$\sigma_A - \sigma_B = \frac{dE_S}{dT} - \frac{d\pi}{dT}$$

Therefore,

$$\sigma_A - \sigma_B = \alpha - \frac{d\pi}{dT} \tag{18.2}$$

The second law of thermodynamics concerning the net change of entropy in a completely reversible process ($\Delta S = 0$) may also be applied with T, the temperature expressed in kelvins:

$$\Delta S = \int \frac{dq}{T} = \sum \frac{\Delta q}{T} = 0$$

$$= \frac{q_P}{T} + \frac{q_T}{T} = \frac{\pi I}{T_1} - \frac{\pi I}{T_2} + \int_{T_2}^{T_1} \frac{(\sigma_A - \sigma_B) I \, dT}{T} = 0$$

and again by differentiating with respect to T and eliminating I, one obtains

$$\frac{d}{dT}\left(\frac{\pi}{T}\right) + \frac{\sigma_A - \sigma_B}{T} = 0$$

from which

$$\sigma_A - \sigma_B = -T\frac{d}{dT}\left(\frac{\pi}{T}\right) \tag{18.3}$$

Equations (18.2) and (18.3) are now equated:

$$\alpha = -T\frac{d}{dT}\left(\frac{\pi}{T}\right) + \frac{d\pi}{dT}$$

$$= -T\left(-\frac{\pi}{T^2} + \frac{1}{T}\frac{d\pi}{dT}\right) + \frac{d\pi}{dT} = \frac{\pi}{T}$$

and hence

$$\pi = \alpha T \tag{18.4}$$

With Eq. (18.4) substituted into Eq. (18.3), one obtains

$$\sigma_A - \sigma_B = \Delta\sigma = -T\frac{d}{dT}\left(\frac{\pi}{T}\right)$$

$$\Delta\sigma = -T\frac{d\alpha}{dT}$$

$$= -T\frac{d}{dT}\left(\frac{dE_S}{dT}\right)$$

$$= -T\frac{d^2E_S}{dT^2} \tag{18.5}$$

The Seebeck coefficient α is usually represented as a polynomial

$$\alpha = a + bT + \cdots$$

where a and b are constants. Then

$$E_S = \int_{T_2}^{T_1} \alpha\, dT = \int_{T_2}^{T_1} (a + bT + \cdots)\, dT = aT + \frac{b}{2}T^2 + \cdots \Big|_{T_2}^{T_1} \tag{18.6}$$

$$\pi = \alpha T = (a + bT + \cdots)T = aT + bT^2 + \cdots \tag{18.7}$$

$$\Delta\sigma = -T\frac{d^2E_S}{dT^2} = -T\frac{d^2}{dT^2}\left(aT + \frac{bT^2}{2} + \cdots\right) = -bT + \cdots \tag{18.8}$$

18.3 THE BASIC EQUATION FOR NET HEAT ABSORBED AT THE COLD JUNCTION OF THE COOLER

In designing thermoelectric coolers, it is customary to neglect the effects of the Thomson coefficient and Thomson heat. The error that occurs from such neglect is not serious and permits a simple, rapid design procedure, which would be tedious and complicated if the Thomson effect were included.

Consider Fig. 18.4, which represents a single thermoelectric element of material A or B. Let A_x be the cross-sectional area of the element (normal to the flow of current and heat) and consider the incremental element of length Δx. The origin of the coordinate system is taken at the end of the element where the current enters, and the length coordinate is positive in the direction of current flow.

There are two boundary conditions,

$$T = T_0 \quad \text{at } x = 0$$

$$T = T_e \quad \text{at } x = L$$

and in the steady state, three heat quantities with respect to the element of length Δx may be noted:

1 The heat entering at point x,

$$q_x = -kA_x \frac{dT}{dx}\bigg|_x$$

2 The heat generated within the increment Δx,

$$q_J = \frac{I^2 \rho \, \Delta x}{A_x}$$

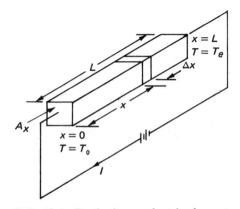

FIG. 18.4 Single thermoelectric element.

3   The heat leaving at point $x + \Delta x$,

$$q_{x+\Delta x} = -kA_x \left.\frac{dT}{dx}\right|_{x+\Delta x}$$

A steady-state heat balance requires that the heat leaving must be equal to the heat entering plus the heat generated:

$$q_x + q_J = q_{x+\Delta x}$$

$$-kA_x \left.\frac{dT}{dx}\right|_x + \frac{I^2 \rho\, \Delta x}{A_x} = -kA_x \left.\frac{dT}{dx}\right|_{x+\Delta x}$$

or $$kA_x \left(\left.\frac{dT}{dx}\right|_{x+\Delta x} - \left.\frac{dT}{dx}\right|_x\right) + \frac{I^2 \rho\, \Delta x}{A_x} = 0$$

Dividing through by Δx gives

$$kA_x \left[\frac{(dT/dx)|_{x+\Delta x} - (dT/dx)|_x}{\Delta x}\right] + \frac{I^2 \rho}{A_x} = 0$$

and in the limit as Δx approaches zero,

$$kA_x \frac{d^2 T}{dx^2} + \frac{I^2 \rho}{A_x} = 0$$

or $$\frac{d^2 T}{dx^2} + \frac{I^2 \rho}{kA_x^2} = 0 \tag{18.9}$$

The general solution to Eq. (18.9) is obtained through the use of a double integration:

$$\frac{dT}{dx} + \frac{I^2 \rho x}{kA_x^2} + C_1 = 0$$

and $$T + \frac{I^2 \rho x^2}{2kA_x^2} + C_1 x + C_2 = 0$$

When $x = 0$, $T = T_0$, and hence

$$C_2 = -T_0$$

and $$T - T_0 + \frac{I^2 \rho x^2}{2kA_x^2} + C_1 x = 0$$

When $x = L$, $T = T_e$, and

$$T_e - T_0 + \frac{I^2 \rho L^2}{2kA_x^2} + C_1 L = 0$$

is obtained by substitution. This then yields C_1,

$$C_1 = \frac{T_0 - T_e}{L} - \frac{I^2 \rho L}{2kA_x^2}$$

The particular solution for the temperature in the element is therefore given by

$$T - T_0 + \frac{I^2 \rho x^2}{2kA_x^2} + \left(\frac{T_0 - T_e}{L}\right) x - \left(\frac{I^2 \rho L}{2kA_x^2}\right) x = 0$$

or $\quad T(x) = -\left(\frac{I^2 \rho}{2kA_x^2}\right) x^2 - \left(\frac{T_0 - T_e}{L} - \frac{I^2 \rho L}{2kA_x^2}\right) x + T_0 \qquad (18.10)$

which indicates a parabolic temperature distribution.

The point of maximum temperature designated by X will be closer to the hot junction and will be at the point where dT/dx vanishes. Hence

$$\frac{dT}{dx} = -\frac{I^2 \rho X}{kA_x^2} - \frac{T_0 - T_e}{L} + \frac{I^2 \rho L}{2kA_x^2} = 0$$

from which

$$X = \frac{(I^2 \rho L / 2kA_x^2) - (T_0 - T_e)/L}{I^2 \rho / kA_x^2}$$

or, since $T_0 - T_e = \Delta T$,

$$x = \frac{L}{2} - \frac{k \Delta T A_x^2}{I^2 \rho L} \qquad (18.11)$$

Now the temperature at X will be greater than either T_0 or T_e. The Joulean heat will therefore be transferred to both junctions. That part which is transferred to the cold junction is the heat generated between the point X and the cold junction.

Assume that the cold junction is at $x = L$ and designate the fraction transferred to the cold junction as f,

$$f = \frac{L - X}{L} = \frac{1}{2} + \frac{k \Delta T A_x^2}{I^2 \rho L^2}$$

or $\quad f = \frac{1}{2} + \frac{K \Delta T}{I^2 R}$

where $K = kA_x/L$ and $R = \rho L/A_x$.

For a thermoelectric cooler with the cold junction at $x = L$, $T_0 > T_e$ and ΔT is positive. The net heat absorbed is therefore equal to the Peltier heat minus the $I^2 R$ loss chargeable to the cold junction. This $I^2 R$ loss is equal to the fraction transferred times the total $I^2 R$ loss. Hence, with T_c now designated as the cold temperature,

$$q = \alpha I T_c - fI^2 R$$

or $\quad q = \alpha I T_c - \left(\dfrac{1}{2} + \dfrac{K\,\Delta T}{I^2 R}\right) I^2 R$

$$= \alpha I T_c - \frac{1}{2} I^2 R - K\,\Delta T \tag{18.12}$$

Equation (18.12) applied to all elements in the cooler is the expression for the net heat absorbed at the cold junction.

18.4 DESIGN EQUATIONS FOR MAXIMUM HEAT PUMPING

Consider Fig. 18.5, which shows a pair of thermoelectric elements operating at a given cold junction temperature T_c K, and a given hot junction temperature, T_h K. This pair of junctions will hereafter be referred to as a *couple*, which is the standard term of reference. The materials themselves are designated as materials A and B (subscript). Each has a finite electrical resistivity ρ (Ω cm), thermal conductivity k (W/cm $^\circ$C), and Seebeck coefficient α (V/$^\circ$C).

The temperatures T_h and T_c establish a temperature difference across the couple:

$$\Delta T = T_h - T_c \tag{18.13}$$

The net heat absorbed at the cold junction is equal to the summation of the Peltier heat,

$$q_P = \pi T_c = \alpha I T_c$$

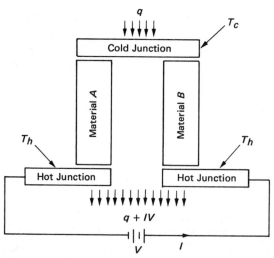

FIG. 18.5 Thermoelectric cooler.

minus one-half of the Joule heat (I^2R loss),

$$q_J = \tfrac{1}{2}I^2R$$

and minus the heat "regained" at the cold junction due to leakage from the hot junction to the cold junction through the elements due to the temperature difference ΔT.

$$q_F = K\,\Delta T$$

Hence

$$q = \alpha I T_c - \tfrac{1}{2}I^2R - K\,\Delta T \tag{18.12}$$

where the total resistance of the couple (when no junction resistance is present) is

$$R = \frac{\rho_A L_A}{A_A} + \frac{\rho_B L_B}{A_B} \tag{18.14}$$

and where the overall heat conductance is given by

$$K = \frac{k_A A_A}{L_A} + \frac{k_B A_B}{L_B} \tag{18.15}$$

and where

$$\alpha = |\alpha_A| + |\alpha_B| \tag{18.16}$$

In order to power the device, one must apply a voltage equal to the sum of the Seebeck voltages in the couple and the resistive voltage drop, namely,

$$V = \alpha T_h - \alpha T_c + IR = \alpha\,\Delta T + IR \tag{18.17}$$

which may be used to evaluate the input power as

$$P = IV = I(\alpha\,\Delta T + IR) \tag{18.18}$$

The heat rejected at the hot junction is

$$q_R = q + P \quad \text{(W)} \tag{18.19}$$

The coefficient of performance of the couple is the ratio of the net cooling effect to the power required:

$$\eta = \frac{q}{P} = \frac{\alpha T_c I - (1/2)I^2R - K\,\Delta T}{\alpha\,\Delta T I + I^2R} \tag{18.20}$$

Now return to Eq. (18.12), which defines the net heat pumping capacity. The current that yields the maximum heat pumping capacity is the current that satisfies the condition $dq/dI = 0$:

$$\frac{dq}{dI} = \alpha T_c - IR$$

from which

$$I = I_m = \frac{\alpha T_{c,MIN}}{R} \tag{18.21}$$

where the subscript m is (and will be) used to designate a value that yields maximum heat pumping capacity.

Substitution of Eq. (18.21) into Eq. (18.12) results in

$$q_m = \frac{\alpha^2 T_c^2}{2R} - K\,\Delta T = 0 \tag{18.22}$$

When I_m is used in the couple with the cold junction thermally insulated, one has the case where the Peltier effect just balances the Joule loss and the "back" heat leakage. This yields the maximum temperature differential obtainable. Hence, when Eq. (18.22) is rearranged,

$$0 = \frac{\alpha^2 T_c^2}{2R} - K\,\Delta T_m$$

or $$\Delta T_m = \frac{\alpha^2 T_c^2}{2KR} \tag{18.23}$$

Observe that the maximum temperature differential can be enhanced by minimizing the product KR, which may be written as

$$KR = \left(\frac{\rho_A L_A}{A_A} + \frac{\rho_B L_B}{A_B}\right)\left(\frac{k_A A_A}{L_A} + \frac{k_B A_B}{L_B}\right)$$

which may be expanded to give, for $L_A = L_B$,

$$KR = k_A \rho_A + k_B \rho_A \frac{A_B}{A_A} + k_A \rho_B \frac{A_A}{A_B} + k_B \rho_B$$

and then optimized by finding the point where the derivative with respect to the area ratio, A_A/A_B, vanishes:

$$\frac{dKR}{d(A_A/A_B)} = k_A \rho_B - k_B \rho_A \left(\frac{A_B}{A_A}\right)^2 = 0$$

This equation may be solved to yield the area ratio for optimized ΔT_m:

$$\frac{A_A}{A_B} = \sqrt{\frac{\rho_A k_B}{\rho_B k_A}} \tag{18.24}$$

The area ratio given by Eq. (18.24) can be used to obtain the optimum value of KR, which is designated as ϕ:

$$\phi = k_A \rho_A + 2\sqrt{k_A \rho_A k_B \rho_B} + k_B \rho_B$$

or $\quad \phi \equiv (\sqrt{k_A \rho_A} + \sqrt{k_B \rho_B})^2 \tag{18.25}$

the only restrictions being that $L_A = L_B$ and that the area ratio is defined as shown by Eq. (18.24).

Equation (18.23) for ΔT_m may now be written

$$\Delta T_m = \tfrac{1}{2} z T_c^2 \tag{18.26}$$

which serves to define the figure of merit of the materials:

$$z \equiv \frac{\alpha^2}{\phi} \tag{18.27}$$

The figure of merit depends on selecting the proper area ratio, as given by Eq. (18.24), and making the element lengths equal. If this is done, the figure of merit is a function of the properties of the elements, k_A, k_B, ρ_A, ρ_B, α_A, and α_B.

Before proceeding to the coefficient of performance, q_m must be expressed in terms of ΔT and ΔT_m, using Eqs. (18.22) and (18.23):

$$q_m = \frac{\alpha^2 T_c^2}{2R} - K\Delta T = \frac{K\alpha^2 T_c^2}{2KR} - K\Delta T$$

$$= K\Delta T_m - K\Delta T = K(\Delta T_m - \Delta T)$$

$$= K\Delta T_m \left(1 - \frac{\Delta T}{\Delta T_m}\right) \tag{18.28}$$

Equation (18.28) also shows that attaining $\Delta T = \Delta T_m$ is impossible unless $q_m = 0$.

The power expended may also be written in terms of the temperatures. Begin with Eq. (18.18) and substitute Eqs. (18.21) and (18.23):

$$P_m = \alpha I_m \Delta T + I_m^2 R = \frac{\alpha^2 T_c \Delta T}{R} + \frac{\alpha^2 T_c^2}{R} = \frac{\alpha^2 T_c^2}{R}\left(\frac{\Delta T}{T_c} + 1\right)$$

$$= \frac{2K\alpha^2 T_c^2}{2KR}\left(1 + \frac{\Delta T}{T_c}\right) = 2K\Delta T_m \left(1 + \frac{\Delta T}{T_c}\right) \tag{18.29}$$

Now obtain the coefficient of performance when the current is adjusted for maximum cooling by dividing Eq. (18.28) by Eq. (18.29):

$$P_m = I_m V_m$$

$$\eta_m = \frac{q_m}{P_m} = \frac{K \, \Delta T_m (1 - \Delta T/\Delta T_m)}{2K \, \Delta T_m (1 + \Delta T/\Delta T_c)}$$

$$= \frac{1 - \Delta T/\Delta T_m}{2(1 + \Delta T/T_c)} \tag{18.30}$$

A plot of the maximum temperature differential as given in Eq. (18.26) is shown in Fig. 18.6. It may be noted that for a reasonable cold-side temperature of, say, 300 K (27°C) the maximum temperature difference attainable with present materials (bismuth telluride, 1981, $z \approx 0.003/°C$) is 135°C. It may also be noted that as the state of the art in materials advances, higher values of maximum temperature differential will be attainable. For example, when $z = 0.004/°C$, $\Delta T_m = 180°C$ for $T_c = 300$ K. These facts, when considered in conjunction with Eq. (18.30), lead to the conclusion that the coefficient of performance of a particular cooler will increase as the materials improve.

Figure 18.7 presents a graphical picture of Eq. (18.12). It shows the variation of the heat pumped, the current, and the temperature differential between the hot and cold junction for a couple having the following characteristics [1]:

Materials	Bismuth Telluride, N and P
Figure of merit	$z = 0.00308/°C$
Cold-side temperature	$T_c = 308.6$ K (96°F)
Couple resistance	$R = 0.00173 \ \Omega$
Couple conductivity	$K = 0.0364$ W/°C
Couple Seebeck coefficient	$\alpha = 0.00044$ V/°C

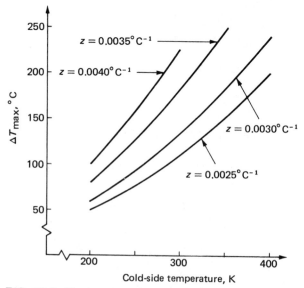

FIG. 18.6 Maximum cooler temperature differential as a function of cooler cold-side temperature.

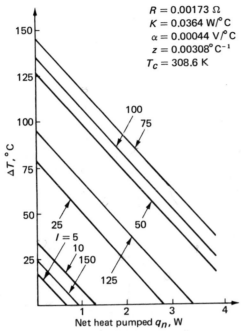

$$R = 0.00173 \ \Omega$$
$$K = 0.0364 \ W/^{\circ}C$$
$$\alpha = 0.00044 \ V/^{\circ}C$$
$$z = 0.00308^{\circ} \ C^{-1}$$
$$T_c = 308.6 \ K$$

FIG. 18.7   Cooler temperature differential plotted against heat pumped for several values of current.

It can be seen that for a particular value of current, the maximum temperature differential is obtained when no heat is pumped, that is, when $q = 0$. Indeed, the maximum temperature differential for 25 A falls nowhere near the maximum possible temperature differential obtained from Fig. 18.6 ($\Delta T_m \approx 145^{\circ}C$). The current for maximum temperature differential can be computed from Eq. (18.21):

$$I_m = \frac{\alpha T_c}{R} = \frac{4.4 \times 10^{-4}(308.6)}{1.73 \times 10^{-3}} = 78.5 \ A$$

and the maximum temperature differential will occur when the net heat pumped is zero:

$$\Delta T_m = \frac{\alpha T_c I_m}{K} - \frac{1}{2} I_m^2 \frac{R}{K}$$

$$= \frac{(4.4 \times 10^{-4})(308.6)(78.5)}{3.64 \times 10^{-2}} - \frac{1}{2}(78.5)^2 \left(\frac{1.73 \times 10^{-3}}{3.64 \times 10^{-2}} \right)$$

$$= 292.8 - 146.4 = 146.4^{\circ}C$$

which checks with Eq. (18.26):

$$\Delta T_m = \frac{1}{2} z T_c^2 = \frac{1}{2}(3.08 \times 10^{-3})(308.6)^2 = 146.7^{\circ}C$$

In addition, this current gives the maximum heat pumped when the temperature differential is zero:

$$q_m = \alpha T_c I_m - \tfrac{1}{2} I_m^2 R$$
$$= (4.4 \times 10^{-4})(308.6)(78.5) - \tfrac{1}{2}(78.5)^2 (1.73 \times 10^{-3})$$
$$= 10.66 - 5.33 = 5.33 \text{ W}$$

If the current were increased to, say, 90 A, a value higher than that for maximum temperature differential, the performance of the cooler would degrade. In this case the maximum temperature differential would be (for the case of no heat pumped)

$$\Delta T_m = \frac{\alpha T_c I}{K} - \frac{1}{2} I^2 \frac{R}{K}$$
$$= \frac{(4.4 \times 10^{-4})(308.6)(90)}{3.64 \times 10^{-2}} - \frac{1}{2}(90)^2 \left(\frac{1.73 \times 10^{-3}}{3.64 \times 10^{-2}} \right)$$
$$= 335.7 - 192.5 = 143.2°\text{C}$$

and the heat pumped would be (for the case of no temperature differential)

$$q_m = \alpha T_c I - \tfrac{1}{2} I^2 R$$
$$= (4.4 \times 10^{-4})(308.6)(90) - \tfrac{1}{2}(90)^2 (1.73 \times 10^{-3})$$
$$= 12.22 - 7.01 = 5.21 \text{ W}$$

The foregoing shows that for a given temperature differential, a given heat pumping capacity can be obtained using two values of current. For the case considered, maximum heat pumping occurs when the current is 78.5 A (no temperature differential). Less heat pumping occurs when the current is set at a value less than 78.5 A. Indeed, for a fixed temperature differential, one can see this by taking Eq. (18.12),

$$q = \alpha I T_c - \tfrac{1}{2} I^2 R - K \, \Delta T \tag{18.12}$$

and solving it for I:

$$\tfrac{1}{2} R I^2 - \alpha T_c I + q + K \, \Delta T = 0$$
$$I^2 - \frac{2\alpha T_c}{R} I + \frac{2}{R}(q + K \, \Delta T) = 0$$
$$I = \frac{\alpha T_c}{R} \pm \frac{1}{2} \sqrt{ \left(\frac{2\alpha T_c}{R} \right)^2 - \frac{8}{R}(q + K \, \Delta T) }$$

This equation is plotted in Fig. 18.8 for a cooler having the physical properties given previously and for a $\Delta T = 50°\text{C}$.

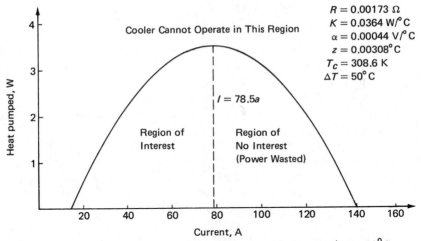

FIG. 18.8   Heat pumped as a function of current for a cooler $\Delta T = 50°C$.

It is obvious that the designer will wish to keep the design point and, indeed, the operation of the cooler to values of current less than, in this case, $I_m = 78.5$ A.

The coefficient of performance for the cooler having these physical properties is plotted in Fig. 18.9. In this figure the coefficient of performance is plotted as a function of applied current and operating temperature differential. It is to be noted that the current for maximum heat pumping, 78.5 A, does not yield the maximum coefficient of performance.

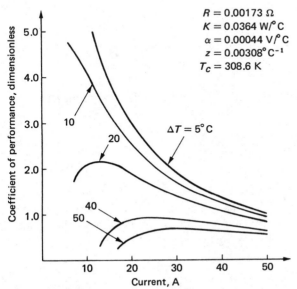

FIG. 18.9   Cooler coefficient of performance as a function of applied current for several values of cooler temperature differential.

18.5   PROCEDURE FOR COOLER DESIGN–MAXIMUM HEAT PUMPING

A worksheet giving the procedure to be followed for designing a cooler for maximum heat pumping is shown in Table 18.1. Use of this worksheet enables the designer to design a cooler quickly and accurately for a variety of conditions.

An example of the use of the worksheet is shown in column 4 of Table 18.1. This illustrative example, which serves as a guide to the worksheet, is a design for the following problem.

Problem: Design a cooler for a refrigeration load of 300 W at 35.6°C for ground cooling operation, where the amount of power expended to operate the cooler is not of paramount importance. A heat sink is available at 73.3°C that can dissipate up to 1500 W.

18.6   DESIGN EQUATIONS FOR OPTIMUM COEFFICIENT OF PERFORMANCE

Again consider Fig. 18.5, which shows a couple composed of elements A and B operating at a cold junction temperature T_c K and a hot junction temperature T_h K. The temperature difference between the hot and cold junctions is

$$\Delta T = T_h - T_c \tag{18.13}$$

and, as given previously, a net heat transfer occurs at the cold junction,

$$q = \alpha I T_c - \tfrac{1}{2} I^2 R - K \Delta T \tag{18.12}$$

upon application of electrical power

$$P = I(\alpha \Delta T + IR) \tag{18.18}$$

The coefficient of performance of the couple is the ratio of the heat withdrawn to the power expended:

$$\eta = \frac{\alpha I T_c - (1/2) I^2 R - K \Delta T}{I(\alpha \Delta T + IR)} \tag{18.20}$$

The coefficient of performance, given by Eq. (18.20), can be maximized by applying the optimum current I_0. This current may be found by taking the derivative $d\eta/dI$ and determining where this derivative vanishes. Because of the mathematical tedium involved, the result is given as Eq. (18.31):

$$I_0 = \frac{\alpha \Delta T}{R\{\sqrt{1 + z[(T_h + T_c)/2]} - 1\}} \tag{18.31}$$

and the derivation is presented in Sec. 18.7.

TABLE 18.1   Design of a Thermoelectric Cooler for Maximum Heat Pumping

Item	Dimensions	Obtained from	Computation
		Material Properties	
Material A		Assumed N type	$Bi_2 Te_3$
Material B		Assumed P type	$Bi_2 Te_3$
α_A	V/°C	Property	2.3×10^{-4}
α_B	V/°C	Property	-2.1×10^{-4}
ρ_A	Ω cm	Property	10^{-3}
ρ_B	Ω cm	Property	10^{-3}
k_A	W/cm °C	Property	1.7×10^{-2}
k_B	W/cm °C	Property	1.45×10^{-2}
		Element Dimensions	
A_A/A_B		$A_A/A_B = \sqrt{k_B\rho_A/k_A\rho_B}$	0.924
A_B	cm²	Assume	0.385 *.0196 6.2 mm × 6.2 mm*
A_A	cm²	$A_A = \sqrt{k_B\rho_A/k_A\rho_B}\, A_B$	0.356 *.0196 5.96 mm × 5.96 mm*
d_B	cm	$d_B = 2\sqrt{A_B/\pi}$	0.70
d_A	cm	$d_A = 2\sqrt{A_A/\pi}$	0.67
$L_A = L_B$	cm	Assume	0.3176 *(.1143)*
L_A/A_A	1/cm	Compute	0.896
L_B/A_B	1/cm	Compute	0.829
A_A/L_A	cm	Compute	1.116
A_B/L_B	cm	Compute	1.208
		Couple Figure of Merit	
α	V/°C	$\alpha = |\alpha_A| + |\alpha_B|$	4.4×10^{-4}
$\sqrt{k_A\rho_A}$		Compute	4.13×10^{-3}
$\sqrt{k_B\rho_B}$		Compute	3.805×10^{-3}
$(\sqrt{k_A\rho_A} + \sqrt{k_B\rho_B})$		Compute	7.94×10^{-3}
$(\sqrt{k_A\rho_A} + \sqrt{k_B\rho_B})2$		Compute	6.3×10^{-5}
z	1/°C	$z = \alpha^2/(\sqrt{k_A\rho_A} + \sqrt{k_B\rho_B})^2$	0.00308
		Physical Data	
R	Ω	$R = (\rho_A L_A/A_A) + (\rho_B L_B/A_B)$	1.73×10^{-3}
K	W/°C	$K = (k_A A_A/L_A) + (k_B A_B/L_B)$	3.64×10^{-2}
		Performance Data	
T_h	K	Given	346.7 *323 (50 C)*
T_c	K	Given	308.6 *268 (-5 C)*
ΔT	K	$\Delta T = T_h - T_c$	38.1 *55*
		Electrical Data	
I_m	A/couple	$I_m = \alpha T_c/R$	78.5 *68.16*
V_m	V/couple	$V_m = \alpha T_h$	0.1526 *0.14212*
P	W/couple	$P = I_m V_m$	11.98 *9.68689*

TABLE 18.1   Design of a Thermoelectric Cooler for Maximum Heat Pumping (*Continued*)

Item	Dimensions	Obtained from	Computation
		Thermal Data	
q_T	W	Given	*20* 300
$\frac{1}{2}\alpha^2 T_c^2$	V²	Compute	*0.006954* 0.00925
$\frac{1}{2}(\alpha^2 T_c^2/R)$	W/couple	Compute	*4.0188* 5.35
$K\,\Delta T$	W/couple	Compute	*2.002* 1.39
q_m	W/couple	$q_m = \frac{1}{2}(\alpha^2 T_c^2/R) - K\,\Delta T$	*2.0169* 3.96
No. of couples		$n = q_T/q_m$	*9.9* 76
Total heat rejected	W	$q_R = q_T + nP$	*116* 1206
η		$\eta = q_m/P$	*0.208* 0.331

The current for optimum coefficient of performance as given by Eq. (18.31) can be inserted into the expression for coefficient of performance as given by Eq. (18.20) to give the optimum coefficient of performance, η_0, in terms of the temperatures and the figure of merit:

$$\eta_0 = \frac{\alpha T_c I_0 - (1/2)I_0^2 R - K\,\Delta T}{I_0(\alpha\,\Delta T + I_0 R)}$$

Using $\gamma = \sqrt{1 + (z/2)(T_h + T_c)}$ [and hence $z = 2(\gamma^2 - 1)/(T_h + T_c)$],

$$\eta_0 = \frac{\alpha T_c[\alpha\,\Delta T/R(\gamma-1)] - (1/2)[\alpha\,\Delta T/R(\gamma-1)]^2 R - K\,\Delta T}{[\alpha\,\Delta T/R(\gamma-1)]\{\alpha\,\Delta T + [\alpha\,\Delta T/R(\gamma-1)]R\}}$$

and, upon expansion,

$$\eta_0 = \frac{[\alpha^2\,\Delta T^2/R(\gamma-1)^2]\,[(\gamma-1)(T_c/\Delta T) - (1/2) - KR(\gamma-1)^2/\alpha^2\,\Delta T]}{\alpha^2\,\Delta T^2\gamma/R(\gamma-1)^2}$$

With $KR/\gamma^2 = 1/z = (T_h - T_c)/2(\gamma^2 - 1)$ substituted in the last term within the second set of brackets in the numerator,

$$\eta_0 = \frac{1}{\gamma}\left[(\gamma-1)\frac{T_c}{\Delta T} - \frac{1}{2} - \frac{1}{2}\frac{(\gamma-1)^2}{\gamma^2-1}\frac{T_h + T_c}{\Delta T}\right]$$

Because $(\gamma-1)^2/(\gamma^2-1) = (\gamma-1)/(\gamma+1)$, one obtains, after factoring $T_c/\Delta T$,

$$\eta_0 = \frac{T_c}{\gamma\,\Delta T}\left[\gamma - 1 - \frac{1}{2}\frac{\Delta T}{T_c} - \frac{1}{2}\left(\frac{\gamma-1}{\gamma+1}\right)\left(\frac{T_h + T_c}{T_c}\right)\right]$$

or $$\eta_0 = \frac{T_c}{\gamma\,\Delta T}\left[\gamma - 1 - \frac{1}{2}\left(\frac{T_h}{T_c} - 1\right) - \frac{1}{2}\left(\frac{\gamma-1}{\gamma+1}\right)\left(\frac{T_h}{T_c} + 1\right)\right]$$

Finding a least common denominator yields

$$\eta_0 = \frac{T_c}{\gamma \, \Delta T} \left\{ \frac{2(\gamma - 1)(\gamma + 1) - (\gamma + 1)[T_h/T_c) - 1] - (\gamma - 1)[(T_h/T_c) + 1]}{2(\gamma + 1)} \right\}$$

or, after simplification,

$$\eta_0 = \frac{T_c}{\gamma \, \Delta T} \left[\frac{2\gamma - 2\gamma(T_h/T_c)}{2(\gamma + 1)} \right]$$

Here then is the final expression for optimum coefficient of performance when the optimum current is applied:

$$\eta_0 = \frac{T_c}{\Delta T} \left[\frac{\gamma - (T_h/T_c)}{\gamma + 1} \right] \tag{18.32}$$

18.7 DERIVATION FOR CURRENT THAT MAXIMIZES COEFFICIENT OF PERFORMANCE

The coefficient of performance is given by Eq. (18.20):

$$\eta = \frac{\alpha I T_c - (1/2) I^2 R - K \, \Delta T}{I(IR + \alpha \, \Delta T)} \tag{18.20}$$

and the value of I that causes $d\eta/dI$ to vanish is sought. Hence,

$$\frac{I(IR + \alpha \, \Delta T)(\alpha T_c - IR) - [\alpha I T_c - (1/2) I^2 R - K \, \Delta T](2IR + \alpha \, \Delta T)}{I^2 (IR + \alpha \, \Delta T)^2} = 0$$

or $\quad (I^2 R + \alpha I \, \Delta T)(\alpha T_c - IR) = (\alpha I T_c - \frac{1}{2} I^2 R - K \, \Delta T)(2IR + \alpha \, \Delta T)$

Expansion of terms gives

$$I^2 R \alpha T_c - I^3 R^2 + \alpha^2 I T_c \, \Delta T - \alpha I^2 R \, \Delta T = 2I^2 R \alpha T_c + \alpha^2 I T_c \, \Delta T - I^3 R^2$$
$$- \tfrac{1}{2} \alpha I^2 R \, \Delta T - 2IRK \, \Delta T - \alpha K \, \Delta T^2$$

or $\quad \tfrac{1}{2} \alpha I^2 R \, \Delta T + \alpha T_c I^2 R - 2KRI \, \Delta T - K\alpha \, \Delta T^2 = 0$

which is really a quadratic in I:

$$R\alpha \left(\frac{\Delta T}{2} + T_c \right) I^2 - (2KR \, \Delta T)I - (K\alpha \, \Delta T^2) = 0$$

Solution for I gives

$$I = \frac{2KR \, \Delta T \pm \sqrt{(2KR \, \Delta T)^2 + 4KR\alpha^2 \, \Delta T^2 [(\Delta T/2) + T_c]}}{2R\alpha[(\Delta T/2) + T_c]}$$

Observe that the value of the radical must always exceed the value of $2KR\,\Delta T$ because it contains $(2KR\,\Delta T)^2$ plus some other positive number. In order to assure cooling there can be no negative current, and hence for cooling

$$I = \frac{2KR\,\Delta T + \sqrt{(2KR\,\Delta T)^2 + 4KR\alpha^2\,\Delta T^2\,[(\Delta T/2) + T_c]}}{2R\alpha[(\Delta T/2) + T_c]}$$

and upon factoring $4(KR)^2\,\Delta T^2$ from under the radical, one obtains

$$I = \frac{2KR\,\Delta T + 2KR\,\Delta T\sqrt{1 + (\alpha^2/KR)[(\Delta T/2) + T_c]}}{2R\alpha[(\Delta T/2) + T_c]}$$

or $$I = KR\,\Delta T\,\frac{1 + \sqrt{1 + z[(\Delta T/2) + T_c]}}{R\alpha[(\Delta T/2) + T_c]}$$

Now

$$\frac{\Delta T}{2} + T_c = \frac{T_h - T_c}{2} + T_c = \frac{T_h + T_c}{2} = \bar{T}$$

so that

$$I = \frac{K\,\Delta T}{\alpha}\left\{\frac{1 + \sqrt{1 + z[(T_h + T_c)/2]}}{(1/2)(T_h + T_c)}\right\}$$

This expression may be adjusted by multiplying by

$$\frac{R\alpha}{R\alpha}\left(\frac{\sqrt{1 + z\bar{T}} - 1}{\sqrt{1 + z\bar{T}} - 1}\right) = 1$$

to obtain

$$I = \frac{KR\alpha\,\Delta T}{R\alpha^2}\left(\frac{1 + \sqrt{1 + z\bar{T}}}{\bar{T}}\right)\left(\frac{\sqrt{1 + z\bar{T}} - 1}{\sqrt{1 + z\bar{T}} - 1}\right)$$

and with $1/z = KR/\alpha^2$, expansion gives

$$I = \frac{\alpha\,\Delta T}{R}\left[\frac{z\bar{T}}{z\bar{T}(\sqrt{1 + z\bar{T}} - 1)}\right]$$

and the final expression for the current that optimizes the coefficient of performance:

$$I = \frac{\alpha\,\Delta T}{R(\sqrt{1 + z\bar{T}} - 1)}$$

18.8   PROCEDURE FOR COOLER DESIGN— OPTIMUM COEFFICIENT OF PERFORMANCE

A worksheet for the design of a cooler for optimum coefficient of performance is given in Table 18.2. Column 4 of this worksheet presents the solution for the following design problem (stated on p. 458).

TABLE 18.2 Design of a Thermoelectric Cooler for Maximum Coefficient of Performance

Item	Dimensions	Obtained from	Computation	
		Material Properties		
α_A	V/°C	Property	2.3×10^{-4}	
α_B	V/°C	Property	2.1×10^{-4}	
ρ_A	Ω cm	Property	1.0×10^{-3}	
ρ_B	Ω cm	Property	1.0×10^{-3}	
k_A	W/cm °C	Property	0.0170	
k_B	W/cm °C	Property	0.0145	
		Element Dimensions		
A_A/A_B		$A_A/A_B = \sqrt{k_B\rho_A/k_A\rho_B}$	0.9235	
A_B	cm²	Assume	0.3848	
A_A	cm²	$A_A = (k_B\rho_A/k_A\rho_B)A_B$	0.3554	
d_B	cm	$d_B = 2\sqrt{A_B/\pi}$	0.700	
d_A	cm	$d_A = 2\sqrt{A_A/\pi}$	0.6727	
$L_A = L_B$	cm	Assume	0.318	
L_A/A_A	1/cm	Compute	0.8948	
L_B/A_B	1/cm	Compute	0.8264	
A_A/L_A	cm	Compute	1.118	
A_B/L_B	cm	Compute	1.210	
		Couple Figure of Merit		
α	V/°C	$\alpha = |\alpha_A| + |\alpha_B|$	4.4×10^{-4}	
$\sqrt{k_A\rho_A}$		Compute	0.00412	
$\sqrt{k_B\rho_B}$		Compute	0.0038	
$(\sqrt{k_A\rho_A} + \sqrt{k_B\rho_B})$		Compute	0.00792	
$(\sqrt{k_A\rho_A} + \sqrt{k_B\rho_B})^2$		Compute	6.27×10^{-5}	
z	1/°C	$z = \alpha^2/(\sqrt{k_A\rho_A} + \sqrt{k_B\rho_B})^2$	3.078×10^{-3}	
		Physical Data		
R	Ω	$R = (\rho_A L_A/A_A) + (\rho_B L_B/A_B)$	1.721×10^{-3}	
K	W/°C	$K = (k_A A_A/L_A) + (k_A A_B/L_B)$	0.03655	
		Performance Data		
T_h	K	Given	346.7	*323*
T_c	K	Given	308.6	*268*
ΔT	K	$\Delta T = T_h - T_c$	38.1	*55*
$T_h + T_c$	K	Compute	655.3	*591*
$(z/2)(T_h + T_c)$		Compute	1.006	*0.9095*
$1 + (z/2)(T_h + T_c)$		Compute	2.006	*1.9095*
γ		$\gamma = \sqrt{1 + (z/2)(T_h + T_c)}$	1.4170	*1.3818*
		Electrical Data		
I_0	A/couple	$I_0 = \alpha \Delta T/[R(\gamma - 1)]$	23.16	*36.83*
V_0	V/couple	$V_0 = \alpha \Delta T + I_0 R$	0.05648	*.08758*
P	W/couple	$P = I_0 V_0$	1.308	*3.225*

TABLE 18.2   Design of a Thermoelectric Cooler for Maximum Coefficient of Performance (*Continued*)

Item	Dimensions	Obtained from	Computation
		Thermal Data	
q_T	W	Given	*20* 300.0
q_P	W/couple	$q_P = \alpha I_0 T_c$	*4.343* 3.144
q_J	W/couple	$q_J = (1/2)I_0^2 R$	*1.1672* 0.4616
q_F	W/couple	$q_F = K\,\Delta T$	*2.0102* 1.3810
$q_J + q_F$	W/couple	Compute	*3.1774* 1.8426
q_0	W/couple	$q_0 = q_P - (q_J + q_F)$	*1.1655* 1.301
No. of couples		$n = q_T/q_0$	*17.16* 231
Total heat rejected	W	$q_R = q_{RT} + nP$	*75* 601.6
COP (η)		$COP = q_0/P$	*0.3614* 0.9946

45.3 C

Problem: Design a cooler for a refrigeration load of 300 W at 96°F for an airborne application where electrical power is at a premium. A heat sink is available at *71.4* 164°F that can dissipate up to 900 W. Bismuth telluride elements having the same properties as those given in Sec. 18.5 are to be used.

18.9   CASCADED COOLERS

It is often desirable to stage or cascade coolers. Figure 18.10 shows a two-stage cooler where the refrigeration load q_1 enters the first stage at its cold junction, which is

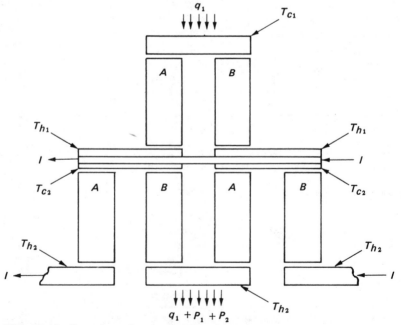

FIG. 18.10   Two-stage thermoelectric cooler.

held at temperature T_{c1}. The heat rejected from the first stage is the sum of the refrigeration load q_1 and the power expended in the first stage P_1, and this heat rejection occurs at temperature T_{h1}. For the first stage, the heat pumped is

$$q_1 = \alpha T_{c1} I_1 - \tfrac{1}{2} I_1^2 R_1 - K_1 (T_{h1} - T_{c1})$$

the power expended is

$$P_1 = I_1 \left[\alpha(T_{h1} - T_{c1}) + I_1 R_1 \right]$$

and the coefficient of performance is

$$\eta_1 = \frac{q_1}{P_1}$$

The heat rejected at the hot junction of stage 1 is equal to the refrigeration load for stage 2,

$$q_2 = q_1 + P_1$$

and in the absence of losses, the stage 1 hot junction temperature is equal to the stage 2 cold junction temperature $(T_{h1} = T_{c2})$.

For the second stage with hot junction T_{h2}, the heat pumped is

$$q_2 = q_1 + P_1 = \alpha T_{c2} I_2 - \tfrac{1}{2} I_2^2 R_2 - K_2 (T_{h2} - T_{c2})$$

the power expended is

$$P_2 = I_2 \left[\alpha(T_{h2} - T_{c2}) + I_2 R_2 \right]$$

and the coefficient of performance (COP or η) is

$$\eta_2 = \frac{q_2}{P_2} = \frac{q_1 + P_1}{P_2}$$

The overall coefficient of performance for the two-stage cooler may be represented by

$$\eta = \frac{q_1}{P_1 + P_2}$$

Now proceed to a consideration of n stages, where the ith stage has a hot junction temperature T_{hi}, a cold junction temperature T_{ci}, a heat pumping capacity q_i, and a power expenditure P_i. Recurrence relationships may be written

$$T_{c,i+1} = T_{h,i} \qquad i = 1, 2, 3, \ldots \tag{18.33}$$

$$q_{i+1} = q_i + P_i \qquad i = 1, 2, 3, \ldots \tag{18.34}$$

$$\eta_i = \frac{q_i}{P_i} \qquad i = 1, 2, 3, \ldots \tag{18.35}$$

and $\eta = \dfrac{q_1}{\displaystyle\sum_{i=1} P_i}$

Algebraic manipulation of Eq. (18.34) shows that

$$q_{i+1} = q_i + P_i = q_i \left(1 + \frac{P_i}{q_i}\right) = q_i \left(1 + \frac{1}{\eta_i}\right) \tag{18.36}$$

and this leads to an evaluation of q_{i+2},

$$q_{i+2} = q_{i+1} + P_{i+1} = q_{i+1} \left(1 + \frac{P_{i+1}}{q_{i+1}}\right)$$

$$= q_{i+1} \left(1 + \frac{1}{\eta_{i+1}}\right)$$

$$= q_i \left(1 + \frac{1}{\eta_i}\right) \left(1 + \frac{1}{\eta_{i+1}}\right)$$

and finally, for the nth stage,

$$q_n = q_{n-1} + P_{n-1} = q_{n-1} \left(1 + \frac{P_{n-1}}{q_{n-1}}\right) = q_{n-1} \left(1 + \frac{1}{\eta_{n-1}}\right)$$

where q_n can be represented in terms of the coefficient of performance of all previous stages and the refrigeration load q_1.

Observe that for the total cascade,

$$q_n = q_1 \left(1 + \frac{1}{\eta_1}\right) \left(1 + \frac{1}{\eta_2}\right) \cdots \left(1 + \frac{1}{\eta_{n-1}}\right) \tag{18.37}$$

But the heat pumping capacity of the nth stage can be represented by the sum of the refrigeration load q_1 and the total power consumed by all stages, $\Sigma P = P_1 + P_2 + P_3 + \cdots + P_n$:

$$q_n = q_1 + \Sigma P = q_1 \left(1 + \frac{\Sigma P}{q_1}\right) = q_1 \left(1 + \frac{1}{\eta}\right) \tag{18.38}$$

where η is defined as the coefficient of performance of the entire cascade.

Equations (18.37) and (18.38) can then be equated:

$$q_1 \left(1 + \frac{1}{\eta}\right) = q_1 \left(1 + \frac{1}{\eta_1}\right) \left(1 + \frac{1}{\eta_2}\right) \cdots \left(1 + \frac{1}{\eta_{n-1}}\right)$$

and hence

$$1 + \frac{1}{\eta} = \prod_{i=1}^{i=n} \left(1 + \frac{1}{\eta_i}\right)$$

or $$\eta = \frac{1}{\displaystyle\prod_{i=1}^{i=n} (1 + 1/\eta_i) - 1}$$ (18.39)

For the case of the two-stage cooler, $n = 2$, and Eq. (18.39) becomes

$$\eta = \frac{1}{(1 + 1/\eta_1)(1 + 1/\eta_2) - 1}$$

which can be adjusted to

$$\eta = \frac{\eta_1 \eta_2}{(\eta_2 + 1)(\eta_1 + 1) - \eta_1 \eta_2}$$

or $$\eta = \frac{\eta_1 \eta_2}{\eta_1 + \eta_2 + 1}$$ (18.40)

Rittner [2] has shown that the optimum coefficient of performance for the entire stage will occur when $\eta_1 = \eta_2 = \eta_3 = \cdots = \eta_n$. LeBlanc [3] has pointed out that in a two-stage cooler where there are no temperature losses between stages, the midstage temperature $(T_{h1} = T_{c2} = T_m)$ is given for bismuth telluride elements by

$$T_m = \sqrt{(T_{h2})(T_{c1})}$$ (18.41)

to an accuracy of about 95%. Indeed, inspection of Eq. (18.40) shows that the overall coefficient of performance of a two-stage cooler is greater than the coefficient of performance of a single-stage cooler operating over the same temperature extremes.

18.10   EFFECT OF SIMPLIFYING ASSUMPTIONS

In the development of the foregoing material, several simplifying assumptions have been made in order to allow the design of a thermoelectric cooler with a minimum of mathematical tedium. Three of these assumptions will be discussed here, and an assessment will be made of how these assumptions have altered the accuracy of the design equations given in this chapter.

18.10.1   Neglect of Thomson Voltage

Return to the differential equation for the heat pumped as derived in Sec. 18.3 and consider the Thomson heat in the incremental element Δx in Fig. 18.4. A heat balance that equates the heat entering with the heat leaving *plus* the heat absorbed due to the Thomson effect is obtained, namely,

$$kA_x \frac{d^2T}{dx^2} - \sigma I \frac{dT}{dx} + \frac{I^2\rho}{A_x} = 0$$

or $$\frac{d^2T}{dx^2} - \frac{\sigma I}{kA_x} \frac{dT}{dx} + \frac{I^2\rho}{kA_x^2} = 0 \qquad (18.42)$$

Note the difference between Eq. (18.42) and Eq. (18.9):

$$\frac{d^2T}{dx^2} + \frac{I^2\rho}{kA_x^2} = 0 \qquad (18.9)$$

and recognize the increased difficulty in obtaining its solution—particularly since k, ρ, and α are functions of temperature. Indeed, only a few special cases of Eq. (18.42) can be solved explicitly. Burshtein [4], for example, has solved this equation for k and σ constant (independent of temperature). Sherman et al. [5] have developed a numerical solution utilizing a digital computer program.

Sherman et al.'s program [5] has given exact solutions that depart from the average parameter solutions for an optimum coefficient of performance design by -14% to $+20\%$. In some cases the coefficient of performance by the exact solution is lower than that determined from the average parameter solutions; in other cases, the reverse is true. This range of comparison is due to the materials assumed. Only in the case where the exact approach yielded a coefficient of performance 14% higher than the approach using the "standard," average equations is the material similar to those presently available. Heikes and Ure [6] point out that this is due partially to the fact that the sign of the Thomson coefficient allows the Thomson heat to assist the Peltier heat at the cold junction. Indeed, this can be considered as the Thomson heat absorbing some of the I^2R loss before this loss gets to the cold junction to be overcome by the Peltier heat.

Although these results and this discussion are rather cursory, it appears that there is evidence that, for a material similar to those used in cooling devices at the present state of the art, neglect of the Thomson coefficient and Thomson heat permits a conservative cooler design.

18.10.2 Temperature Losses

The effect of temperature losses between the thermoelectric junctions and the cold and hot reservoirs are now considered by using some familiar numbers.

It has been shown that in an optimum coefficient of performance design, the coefficient of performance is a function of only the hot and cold junction temperatures and the material figure of merit. An illustrative example using $T_c = 308.6\,\text{K}$, $T_h = 346.7\,\text{K}$, and $z = 0.00308/°\text{C}$ gave a coefficient of performance of unity.

The temperature pattern in a typical thermoelectric cooler is shown in Fig. 18.11. Note the drop in temperature between the hot junction and the base of the hot-side heat exchanger. A similar difference in temperature exists on the cold side.

If the temperature difference between the junctions and the exchangers (or the chassis for the case of avionics cooling) is, say, $2°\text{C}$, then the cooler is apparently working between temperature extremes of $310.6\,\text{K}$ and $344.7\,\text{K}$. For a coefficient of

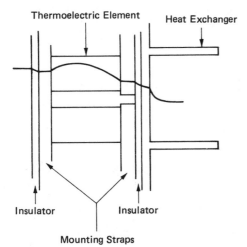

FIG. 18.11 Temperature profile in a typical thermoelectric cooler.

performance of unity, we may use Eq. (18.32) to show that the apparent figure of merit is 0.00269/°C. This means that if there were no temperature losses between junctions and reservoir, a coefficient of performance of unity working between the end point temperatures attained could be obtained with a material having a figure of merit of 0.00269°C.

This apparent figure of merit, which considers the device as a whole (outside to outside rather than junction to junction), is a very realistic quantity. It shows that the designer has two choices: (1) He can use the material figure of merit in the design equations developed in this chapter, but he must specify junction temperatures that consider the losses from junction to reservoirs. (2) He can use the apparent figure of merit and design to the actual temperatures.

As an example, if the designer wants 100°F and 160°F as the extreme temperatures, the first because of avionics equipment reliability considerations and the second because of heat exchanger capability, he can design to 96°F and 164°F (308.6 K and 346.7 K) using the actual material figure of merit. In this case he has accounted for the temperature losses.

On the other hand, he could design to 100°F and 160°F using an apparent figure of merit and obtain the same result. However, because the apparent figure of merit will never be known exactly, the former method is recommended.

18.10.3 Effect of Junction Resistance and Parallel Thermal Paths

Section 18.10.2 discussed temperature losses between the cooler junctions and the heat reservoirs. These are "series" losses, because they consider the temperature losses between the outside surfaces of the cooler and the junctions. There will also be "parallel" heat losses, because the space between the cooler elements must have finite thermal conductivity. In addition, a finite electrical resistance must exist at each junction with a resultant dumping of I^2R loss right on top of the Peltier effect and an associated degradation of heat pumping capacity.

If one considers the figure of merit of the cooler,

$$z = \frac{\alpha^2}{\phi}$$

one may propose a figure of merit that includes losses:

$$z' = \frac{\alpha^2}{\psi} \tag{18.43}$$

The value of ψ can be derived in a manner similar to that for ϕ (Sec. 18.4):

$$\psi = (1 + \Lambda_R)(1 + \Lambda_K)\phi \tag{18.44}$$

where Λ_R is the ratio of junction resistance to element resistance and Λ_K is the ratio of parallel heat path conductance to element conductance.

Using Eqs. (18.43) and (18.44), one sees that for a junction resistance of 1% of the element resistance, and a parallel heat flow path conductance of 1% of the element conductance,

$$z' = \frac{\alpha^2}{\psi} = \frac{\alpha^2}{(1 \pm \Lambda_R)(1 + \Lambda_K)\phi} = \frac{z}{(1 + \Lambda_R)(1 + \Lambda_K)}$$

and $z' = \left[\dfrac{1}{(1.01)(1.01)} \right] z$

or the figure of merit of the cooler material with losses taken into account will be

$$z' = \frac{1}{1.0201}(0.00308) = 0.00302/°C$$

This shows that improper manufacturing technique and little regard to parallel thermal paths can severely degrade cooler performance. The illustrative figures show that when the losses that are always present are kept to a few percent, no serious degradation occurs.

18.11 NOMENCLATURE

Roman Letter Symbols

A	cross-sectional area, m^2
E	electromotive force, V
f	fraction of heat transferred to cold junction, dimensionless
I	current, A
k	thermal conductivity, W/m °C
K	thermal conductance, W/°C
L	length, m

q	heat flow, W
Q	heat, J
R	electrical resistance, Ω
S	entropy, J/K
T	temperature, K
V	voltage, V
X	point of maximum temperature, m
x	length coordinate, m
z	figure of merit, K^{-1} or $°C^{-1}$

Greek Letter Symbols

α	Seebeck coefficient, V/°C
γ	a combination of terms defined where used, dimensionless
Δ	indicates a change in quantity
η	coefficient of performance, dimensionless
Λ	a ratio defined by Eq. (18.44), dimensionless
π	Peltier coefficient, V
ρ	electrical resistivity, Ω cm
σ	Thomson coefficient, V/°C
ϕ	a combination of property values, defined by Eq. (18.25), $V^2/°C$
ψ	a combination of terms defined in the same manner as ϕ, $V^2/°C$

Subscripts

A	indicates a material
B	indicates a material
c	indicates cold junction
e	indicates end of configuration or equivalent condition
F	indicates Fourier heat
h	indicates hot junction
J	indicates Joule heat (I^2R loss)
m	indicates maximum condition
P	indicates Peltier heat
R	indicates rejected quantity
S	indicates Seebeck voltage
T	indicates Thomson heat
x	indicates external quantity or cross-sectional quantity
0	indicates origin or optimum condition

Superscripts

$'$	indicates apparent quantity
$-$	indicates mean or average quantity

18.12   REFERENCES

1 Chemical Rubber Co., *Handbook of Chemistry and Physics*, CRC, Cleveland, Oh., 1954.
2 Rittner, E. S., On the Theory of the Peltier Heat Pump, *J. Appl. Phys.*, vol. 30, pp. 702–707, 1959.

3 LeBlanc, R., Personal communication, February 1963.
4 Burshtein, A. J., An Investigation of the Steady-State Heat Flow Through a Current Carrying Conductor, *J. Sov. Phys. Tech. Phys.*, vol. 2, pp. 1397–1406, 1957.
5 Sherman, B., Heikes, R., and Ure, R., Calculation of Efficiency of Thermoelectric Devices, *J. Appl. Phys.*, vol. 31, pp. 1–16, 1960.
6 Heikes, R., and Ure, R., *Thermoelectricity: Science and Engineering*, Interscience, New York, 1961.

19

■ augmentation techniques

19.1 INTRODUCTION

The development of high-performance thermal systems such as those required by modern electronic equipment has stimulated interest in methods to augment, intensify, or enhance heat transfer. Consideration of thermal configurations in which heat flow is governed by (or can be approximated by) the relationship

$$q = hS\,\Delta T \tag{19.1a}$$

or $$\frac{q}{\Delta T} = hS \tag{19.1b}$$

where ΔT is appropriately defined as the temperature difference between the fluid and the surface that is being heated or cooled, shows that the heat flow per unit temperature difference depends on the product of the heat transfer coefficient and the surface area.

If it is desired to increase the heat flow or lower the permissible temperature difference, or accomplish both of these simultaneously, an increase in the hS product is required. This can be accomplished by increasing h, increasing S, or increasing both simultaneously.

In recent years, a great deal of research effort has been devoted to developing apparatus and conducting the research required to define the techniques and conditions under which heat and mass transfer may be augmented. The more effective and feasible techniques have graduated from the laboratory to full-scale industrial equipment. Under ideal conditions, these techniques may be capable of providing a 5- to 10-fold improvement in the hS product but often lead to a parallel increase in the friction factor or in the product of shear stress and surface area. As a consequence, when these techniques are compared to the natural state on the basis of equal pumping power, the augmentation ratio drops to between 2 and 1 and in many cases becomes less than unity [1]. Unfortunately, it does not appear to be feasible to establish a generally applicable selection criterion for augmentation techniques because of the large number of factors that enter into the ultimate selection decision. Most of the considerations revolve around economic factors such as development cost, initial cost, operating cost, and maintenance cost. Indeed, in the area of electronic equipment cooling, criteria such as high reliability and safety may appear at the forefront of selection considerations.

A great deal of the material presented in this chapter is taken from the presentation by A. E. Bergles at the National Science Foundation Research Workshop on Directions of Heat Transfer in Electronic Equipment, Atlanta, Georgia, October 17–19, 1977. The authors are indebted to Professor Bergles for permission to use this material.

Augmentation techniques may be classified as passive, those that require no external power; active, those that do require external power; and compound, those that encompass techniques from *both* the passive and active categories. A summary of the techniques is displayed in Table 19.1. Each technique listed in Table 19.1 can be applied to six "modes" that may occur in the cooling of electronic equipment. These are single-phase convection, both free and forced; boiling heat transfer; both pool and flow boiling; and condensation of both confined and forced circulated vapor. This leads to at least 84 possibilities (each of the six modes with the 14 techniques[1] listed in Table 19.1), and a survey of the entire literature is clearly impossible. Bergles et al. [2] provide a survey that lists and classifies nearly 2000 separate references; these works are discussed in papers by Bergles [3, 4].

It is the intent of the balance of this chapter to survey briefly those techniques listed in Table 19.1 that have been applied or are potentially applicable to electronic cooling considerations.

Before turning to the discussion of the actual techniques, it is important to note that some enhancement factors are naturally present in the electronic cooling environment. Thus, protuberances and the surface roughness of electronic circuit boards, the inadvertent vibrations present in electronic equipment, and electric fields within power equipment such as transformers can be expected to yield hS products higher than would be expected under thermally ideal (smooth-surface, vibration-free) conditions.

19.2 SINGLE–PHASE FREE CONVECTION

19.2.1 Passive Techniques

The literature on the effects of surface roughness (such as that encountered with components on printed circuit boards) on free or natural convection is very limited and

[1] Any compound technique counts as one.

TABLE 19.1 Classification and Summary of Augmentation Techniques

---

Passive techniques
  Treated surfaces
  Rough surfaces
  Extended surfaces
  Displaced enhancement devices
  Swirl flow devices
  Surface tension devices
  Additives for liquids
  Additives for gases
Active techniques
  Mechanical aids
  Surface vibration
  Fluid vibration
  Electrostatic or magnetic fields
  Injection or suction
Compound techniques
  Two or more of the above

---

somewhat contradictory. Jofre and Barron [5] report a 200% improvement in average coefficients with a vertical roughened plate; however, Fujii et al. [6] found only a 5% improvement in similar tests.

The use of extended surface can be considered as augmentation technique. This subject is treated in detail in Chap. 14.

19.2.2 Active Techniques

Those techniques that require external power are not normally considered for electronic cooling systems; however, several types of potential augmentation occur naturally.

Surface vibration can increase heat transfer coefficients quite substantially in laboratory tests; however, the intensities are so high that mechanical failure of the equipment would be observed in practical applications.

The data of Smith and Forbes [7] and Pak et al. [8] indicate that substantial increases in heat transfer coefficients can be expected when a liquid-filled enclosure containing heat sources is vibrated.

Coolant pulsation is unlikely to be of sufficient intensity to affect heat transfer significantly. For example, with gases, intensities above 140 db are required.

19.3 SINGLE–PHASE FORCED CONVECTION

19.3.1 Rough Surfaces

Surface roughness has a large effect on the heat transfer coefficients for irregular geometries such as integrated circuit packages. Few published data are available; however, it appears that heat transfer coefficients for printed wire boards can be as much as 10 times the smooth, flat plate predictions [9]. Transverse protuberances are frequently added at the leading edge of boards or cards to ensure a high turbulence level. Although surface roughness has been studied extensively, no unified treatment is available to predict the performance of the rough heat transfer surfaces encountered in electronics.

On the other hand, correlations are available for roughness configurations for the interior of tubes (usually externally finned) used to cool air as it passes through a computer or used in shell-and-tube heat exchangers that cool water used for air-cooled heat exchangers and cold plates. Specifically, friction factors and heat transfer coefficients for repeated-rib roughnesses can be estimated from the correlations presented by Webb et al. [10]. Many variations of this tubing are available commercially.

19.3.2 Extended Surfaces

Extended surfaces can be considered "old technology" as far as most applications are concerned. Tube-and-plate fins, plate fins, and finned tube banks are used to augment air-side heat transfer in various electronic cooling systems. Various fin configurations have been developed to increase heat transfer coefficients as well as increase surface area. One example is the perforated-plate surfaces reviewed by Shah [11].

Heat transfer inside tubes can be improved by internal fins, either straight or spiraled. Watkinson presents the most recent data for water [12].

19.3.3   Swirl Flow Devices

Vortex generators, particularly the full-length, twisted-tape tube insert, can be used to improve the heat transfer coefficient to turbulent liquid flow by as much as 100% [13].

19.3.4   Additives for Liquids

Baker [14] has suggested use of two-phase, two-component flow for heat transfer augmentation. In an experimental setup, air was injected into a narrow channel upstream of the point where the liquid coolant flowed by a chip. Substantial improvements over single-phase forced convection for Freon-113 and Dow Corning-200 silicon liquid were observed. As a practical matter, it is evident that the air supply and removal represent major design complications.

19.3.5   Additives for Gases

Evaporative spray cooling techniques were suggested many years ago for cooling of components with high dissipation rates [15]. Since the coolant must be expendable, water is preferred; but when the use of water is incompatible with the electrical isolation requirements, dielectric fluids can be used. Spray cooling results obtained with R-113 have been reported by Ruch and Holman [16] and by Goldstein and Griffith [17].

19.3.6   Active Techniques

Mechanically aided heat transfer occurs frequently in rotating electric machinery. A discussion of experimental and analytical work is given in [18].

Surface vibration and fluid vibration are not effective enough to warrant the cumbersome equipment required to bring about the effect.

Some very impressive enhancements have been recorded with electric fields, particularly in the laminar flow region. The recent studies of Savkar [19] and of Newton and Allen [20] demonstrated improvements of as much as 100% when voltages in the 10-kV range were applied to transformer oil. The fields encountered in large power transformers are sufficient to produce substantial local increases in heat transfer coefficients. In general, the enhancement is not considered in design; the electric field merely contributes to a more conservative safety margin.

19.3.7   Performance Evaluation Criteria for Single-Phase Forced Convection

Numerous factors enter into the ultimate decision to use an enhancement technique: heat duty increase, surface area reduction, pumping power requirements, initial cost, maintenance cost, safety, and reliability. Because these factors are numerous and frequently difficult to quantify, the designer must make a decision based on the particular system and requirements. Two common geometries lend themselves to simplified thermal-hydraulic analysis: arrays cooled by forced circulation of air, and heat exchangers for cooling primary or secondary liquid coolants.

Eight criteria have been suggested to define the merits of surfaces and inserts (passive techniques only) that enhance single-phase flow in channels [21]. The first four criteria noted in Table 19.2 have also been used to evaluate compact heat ex-

TABLE 19.2 Summary of Performance Evaluation Criteria for Enhanced Heat Transfer [21]

	Criterion number							
	1	2	3	4	5	6	7	8
Fixed								
Basic geometry	X	X	X	X				
Flow rate	X						X	X
Pressure drop		X				X		X
Pumping power			X		X			
Heat duty				X	X	X	X	X
Objective								
Increase heat transfer	X	X	X					
Reduce pumping power				X				
Reduce exchanger size					X	X	X	X

changer cores used in air-cooled heat exchangers [22]. The criteria are expressed quantitatively as a performance ratio representing increased heat transfer, reduced pumping power, or reduced heat exchanger size. Details are given in the papers cited.

19.4 POOL BOILING

19.4.1 Treated Surfaces

Treated surfaces involve surface condition adjustments other than those normally encountered. This excludes the well-known (but unpredictable) effects of surface material, finish, fouling, and oxidation on nucleate boiling and critical heat flux. Some techniques that are very effective with water, such as small spots of Teflon, are excluded, as water is not acceptable for direct immersion cooling of microelectronic components. The commonly used fluorocarbons wet all materials; hence, the spots cannot be nonwetting, as in the case of water.

The most effective technique for augmentation of boiling heat transfer to fluorocarbons and the elimination of the often-encountered temperature overshoot at boiling incipience is to provide a high density of doubly reentrant cavities, which serve as nucleation sites even after periods of high subcooling. One example is the brazed coating marketed commercially as High Flux [23]. Another effective technique, the "dendritic" surface developed by Oktay and Schmeckenbecher [24], involves a deposited surface. Although it is unlikely that chips can be altered to accommodate such surfaces, chip carrying substrates or transistor surfaces may be modified to include a "treated" section [25].

19.4.2 Rough Surfaces and Extended Surfaces

A number of effective augmentation techniques involve large-scale surface modifications, such as those produced by machining: grooving and deformation [26], rolled-over low-fin tubing [27], and Thermoexcel E [28].

Extended surfaces in the form of low-fin tubing generally have higher nominal heat transfer coefficients than smooth tubes. The application of these techniques to immersion-cooled electronic components is limited by packaging and material constraints.

19.4.3 Surface Tension Devices

Wicking may be utilized when there is an unreliable supply of liquid to a heated surface, for example, electronic cooling systems in aircraft undergoing violent maneuvers or spacecraft operating in near zero gravity. The hermetically sealed transformer [29], for example, uses wicking to supply coolant to the inner portions of the winding. The vaporized fluid is condensed at the cold container wall and returned to the bottom pool. A recent review of the behavior of wick-covered surfaces is given by Corman and McLaughlin [30].

Heat pipes can be included in this classification, as surface tension effects are vital to the evaporation, condensation, and liquid transport processes. Three types of applications to electronics have been considered: heat removal from large discrete components, heat removal from totally enclosed electronic packages, and thermal control of systems subject to a wide range of operational environments. Because heat pipes are treated in more detail in Chap. 17, they will not be discussed further here except to note that, to be effective, they must be thermally well connected to the heat source (evaporator end) and to the coolant (condenser end). Enhanced surfaces are frequently used to reduce the thermal resistances of these connections.

19.4.4 Liquid Additives

Binary aqueous mixtures (with small percentages of a volatile additive) often have nucleate boiling characteristics, particularly critical heat fluxes, that are far superior to those for pure water alone [31]. Unfortunately, no similar phenomenon has been observed for fluorocarbons. Mixtures are used simply to obtain a range of boiling points.

19.4.5 Active Techniques

None of the active augmentation techniques has been seriously considered for electronic cooling applications. Vaporization does not normally occur in rotating electrical or electronic equipment. Surface vibration and fluid vibration have little effect on established nucleate boiling. In the case of vibration of an enclosure containing heat sources, Fuls and Geiger [32] observed only a slight increase in nucleate-boiling heat transfer coefficients.

Electric fields produce impressive improvements in low-flux boiling (boiling curve hysteresis is eliminated) and critical heat flux for fluorocarbons [33]. However, the kilovolt-strength fields are not naturally present, and it would be cumbersome and hazardous to add them to electronic packages.

Finally, suction may be used to stimulate nucleate boiling to occur at low wall superheat [34]; however, the technique is very involved. A porous heated surface, flow control element, and a vapor removal or recirculation system are required.

19.5 FLOW BOILING

There are two situations where forced convection nucleate boiling is utilized in electronics cooling: cooling of large, high-power-density components (klystrons, accelerator targets, electromagnets) and direct immersion of components (chips) in a flowing liquid. Water is generally used in the former application and a fluorocarbon in the latter application. In the interest of maximizing the transferrable heat flux, the coolant is almost always subcooled and flows at high velocity, as the intent is to accommodate high heat fluxes rather than to generate vapor.

19.5.1 Treated Surfaces

Porous boiling surfaces do not improve subcooled flow boiling of fluorocarbons; however, the annoying boiling curve hysteresis is eliminated [35].

19.5.2 Rough Surfaces and Extended Surfaces

Surface roughness and surface finish do not have a large effect on subcooled boiling [36]. There is some indication that the critical heat flux for water is increased with moderate machined roughness [37]; however, the critical heat flux is decreased when large transverse ribs are used [38].

Cylindrical studs have been proposed as heat sinks for individual chips or other devices [39]. The dissipated power is conducted along the studs, which are immersed in a flowing liquid. Boiling can occur at high device power.

19.5.3 Displaced Enhancement Devices

Megerlin et al. [40] reported subcooled nucleate boiling and critical heat flux data for water flowing in short, packed tubes. Mesh and spiral brush inserts were selected as "displaced" enhancement devices to increase the wall heat fluxes that could be accommodated. Critical heat fluxes were increased by about 100% over the smooth tube values; however, the wall temperatures were very high.

19.5.4 Swirl Flow Devices

Inlet vortex generators of the spiral ramp or tangental slot type have been used to accommodate very high heat fluxes for subcooled flow boiling of water. A record flux of 1.73×10^8 W/m^2 was achieved by Gambill and Greene [41].

Twisted-tape inserts are quite popular because of their simplicity and adaptability to existing equipment. They are ideal for hot-spot applications, since a short tape can cure the thermal problem while having little effect on the overall pressure drop. The nucleate boiling characteristics are similar to those for empty tubes [42]; however, critical heat fluxes can be increased by 100% [43]. As a result of modification of the pressure drop, the critical heat fluxes for swirl flow are approximately twice those for straight flow, even at the same test-section pumping power.

19.5.5 Active Techniques

Active augmentation techniques are generally not very effective when applied to a flow boiling configuration, perhaps because of the highly energetic, highly agitated nature of this heat transfer mechanism. Futhermore, the equipment available for active augmentation is relatively expensive and complex, and its use may act to *reduce* the overall reliability of the electronic equipment to be cooled.

19.6   FREE CONVECTION CONDENSATION

In cooling systems involving phase change, the components may be immersed in a pool of saturated liquid. The vapor generated is condensed on upper surfaces (finned tubes or cold plates) maintained at a lower temperature by chilled water or air.

19.6.1   Treated Surfaces

No promoters for drop condensation of fluorocarbons have been found, presumably because of the extreme wettability of these fluids [44]. Attention is thus directed toward means of enhancing film condensation.

19.6.2   Rough Surfaces

It appears that no data are available for film condensation of fluorocarbons on rough surfaces. However, the favorable experience of Nicol and Medwell [45] with steam seems to suggest that the techniques might be useful for liquids used in electronics cooling. Their best-performing roughness doubled the heat transfer coefficient for steam condensing on the outside of a vertical tube.

19.6.3   Extended Surfaces

Low-fin tubing is a standard item in horizontal in-tube condensers. One of the major considerations is to shape and space the fins so that the condensate does not fill the grooves completely. A recent development is the Thermoexcel C tube described by Arai et al. [28]. An interesting application of the Thermoexcel C and Thermoexcel E surfaces to thyristor cooling is described by Nakayama et al. [46].

Thin-film condensation is promoted by a variety of surfaces developed originally for enhancement of vapor space condensation of water in multistage flash desalination systems. Designs attempt to exploit the Gregorig effect, whereby condensation occurs primarily at the tops of the ridges. The condensed film there is kept thin as surface tension forces pull the condensate into the grooves, where it runs off. The principle has been applied to both vertical and horizontal tubes.

A unique double-grooved surface was developed by Markowitz et al. [47] for horizontal plate (facing down condensers). Condensing coefficients (based on nominal surface area) for Freon-113 were improved by almost 100%. This scheme was actually developed for a submerged condenser where direct-contact condensation occurred in the bulk of the subcooled liquid. Further details may be found in Chap. 16.

19.6.4   Active Techniques

Condensing coefficients for fluorocarbons can be enhanced several hundred percent by a number of active techniques: rotating surfaces [48], electrostatic fields [49], and suction [50]. It is, however, improbable that these techniques would be considered because of the complexity, power expenditure, and potential unreliability.

19.7   FORCED FLOW CONDENSATION

Condensing of water and fluorocarbons in forced flow in tubes can be improved by many techniques: rough surface tubes, inner-fin tubes, static mixer inserts, twisted-tape inserts, acoustic vibration, and electrical fields [4]. It appears, though, that the process does not occur in cooling systems for electronics.

19.8 SUMMARY AND PROSPECTS FOR THE FUTURE

Natural augmentation of heat transfer occurs in a variety of cooling systems for electronic and electrical equipment. Techniques to augment heat transfer are used in many specialized situations, and a general increase in the number of applications has been observed.

There is a large body of experience, on a laboratory scale and in full-size systems, that will facilitate the applications of augmentation as the need arises.

19.9 NOMENCLATURE

Roman Letter Symbols

h heat transfer coefficient, W/m^2 °C
q heat flow, W
S surface area, m^2
T temperature, °C

Greek Letter Symbols

Δ indicates a change in a variable

19.10 REFERENCES

1 Bergles, A. E., Techniques to Augment Heat Transfer, in *Handbook of Heat Transfer*, ed. W. M. Rohsenow and J. P. Hartnett, McGraw-Hill, New York, 1973.

2 Bergles, A. E. Webb, R. L., Junkhan, G. H., and Jensen, M. K., Bibliography on Augmentation of Convective Heat and Mass Transfer, Iowa State University Heat Transfer Laboratory Rept. HTL-19, ISU-ERI-Ames-79206, May 1979.

3 Bergles, A. E., *Heat Transfer 1978: Proc. 6th Int. Heat Transfer Conf., Toronto*, vol. 6, pp. 89–108, Hemisphere, Washington, D.C., 1978.

4 Bergles, A. E., Survey of Augmentation of Two Phase Heat Transfer, *ASHRAE Trans.*, vol. 82, part 1, pp. 891–905, 1976.

5 Jofre, R. J., and Barron, R. F., Free Convection Heat Transfer to Rough Plate, ASME Paper 67-WA/HT-38, 1967.

6 Fujii, J., Fujii, M., and Takenchi, M., Influence of Various Surface Roughnesses on the Natural Convection, *Int. J. Heat Mass Transfer*, vol. 16, pp. 629–640, 1973.

7 Smith, G. V., and Forbes, R. E., Effect of Random Vibration on Natural Convection Heat Transfer in Rectangular Enclosures, in *Augmentation of Convective Heat and Mass Transfer*, pp. 158–162, ASME, New York, 1970.

8 Pak, H. Y., Winter, E. R. F., and Schoenhals, R. J., Convection Heat Transfer in a Contained Fluid Subjected to Vibration, in *Augmentation of Convective Heat and Mass Transfer*, pp. 148–157, ASME, New York, 1970.

9 Wenthen, F. T., Experimental Verification, Presented at NSF Research Workshop on Directions of Heat Transfer in Electronic Equipment, Atlanta, Georgia, October 1977.

10 Webb, R. L., Eckert, E. R. G., and Goldstein, R. J., Heat Transfer and Friction in Tubes with Repeated-Rib Roughness, *Int. J. Heat Mass Transfer*, vol. 14, pp. 601–618, 1971.

11 Shah, R. K., Perforated Heat Exchanger Surfaces–2: Heat Transfer and Flow Friction Characteristics, ASME Paper 75-WA/HT-9, 1975.

12 Watkinson, A. P., Miletti, D. L., and Tarassof, P., Turbulent Heat Transfer and Pressure Drop in Internally Finned Tubes, *AIChE Symp. Ser.*, vol. 69, pp. 94–103, 1973.

13 Lopina, R. F., and Bergles, A. E., Heat Transfer and Pressure Drop in Tape-Generated Swirl Flow of Single Phase Water, *J. Heat Transfer*, vol. 91, pp. 434–442, 1969.

14 Baker, E., Liquid Immersion Cooling of Small Electronic Devices, *Microelectronics and Reliability*, vol. 12, pp. 163–173, 1973.

15 Kaye, J., and Choi, H. Y., General Aspects of Cooling Airborne Electronic Equipment, *IRE Trans. Aeronaut. Navigat. Electron.*, vol. ANE-S, no. 1, pp. 1–9, 1958.

16 Ruch, M. A., and Holman, J. P., Boiling Heat Transfer to a Freon-113 Jet Impinging Upward onto a Flat, Heated Surface, *Int. J. Heat Mass Transfer*, vol. 18, pp. 51–60, 1975.

17 Goldstein, S. D., and Griffith, P., *Spray Cooling of Laser Mirrors*, ASME Paper 74-HT-17, Boston, 1974.

18 Macken, N. A., and Paul, F. W., Augmentation of Steady State and Transient Two-Phase Heat Transfer Using Mechanical Acceleration, *ASHRAE Trans.*, vol. 82, part 1, pp. 932–940, 1976.

19 Savkar, S. D., Dielectrophoric Effects in Laminar Forced Convection Between Two Parallel Plates, *Phys. Fluids*, vol. 14, pp. 2670–2679, 1971.

20 Newton, D. C., and Allen, P. H. G., Senftleben Effect in Insulating Oil under Uniform Electric Stress, *Lett. Heat Mass Transfer*, vol. 4, pp. 9–16, 1977.

21 Bergles, A. E., Blumenkrantz, A. R., and Taborek, J., Performance Evaluation Criteria for Enhanced Heat Transfer Surfaces, *Jpn. Soc. Mech. Eng.*, vol. 2, pp. 239–243, 1974.

22 Bergles, A. E., Junkhan, G. H., and Bunn, R. L., Performance Criteria for Cooling Devices on Agricultural and Industrial Machines, *SAE Trans.*, vol. 85, pp. 38–48, 1976.

23 Gottzmann, C. F., O'Neill, P. S., and Minton, P. E., High Efficiency Heat Exchangers, *Chem. Eng. Prog.*, vol. 69, no. 7, pp. 69–75, 1973.

24 Oktay, S., and Schmeckenbecher, A. F., Preparation and Performance of Dendritic Heat Sinks, *J. Electrochem. Soc.*, vol. 21, pp. 912–918, 1974.

25 Fujii, M., Nishayama, E., and Yamanaka, G., Nucleate Pool Boiling Heat Transfer from Microporous Heating Surface, *Advances in Enhanced Heat Transfer*, ASME, New York, 1979.

26 Ragi, E. G., U.S. Patent No. 3,684,007, 1970.

27 Webb, R. L., U.S. Patent No. 3,696,861, 1972.

28 Arai, N., Fukushima, T., and Arai, A., Heat Transfer Tubes Enhancing Boiling and Condensation in the Heat Exchanger of a Refrigerating Machine, *ASHRAE Trans.*, vol. 83, part 2, pp. 58–70, 1977.

29 Kaye, J., Review of Industrial Applications of Heat Transfer to Electronics, *Proc. IRE*, vol. 44, pp. 997–991, 1956.

30 Corman, J. C., and McLaughlin, M. H., Boiling Augmentation with Structural Surfaces, *ASHRAE Trans.*, vol. 82, part 1, pp. 906–918, 1976.

31 van Stralen, S. J. D., Heat Transfer to Boiling Binary Liquid Mixtures, *Brit. Chem. Eng.*, vol. 4, pp. 8–17, 78–82, 1959.

32 Fuls, G. M., and Geiger, G. E., Effect of Bubble Stabilization on Pool Boiling Heat Transfer, *J. Heat Transfer*, vol. 92, pp. 635–640, 1970.

33 Choi, H. Y., Electrodynamic Boiling Heat Transfer, Ph.D. thesis, Massachusetts Institute of Technology, Cambridge, Mass., 1962.

34 Raiff, R. J., and Wayner, P. C., Jr., Evaporation from a Porous Heat Source, *Int. J. Heat Mass Transfer*, vol. 17, pp. 1919–1930, 1973.

35 Murphy, R. W., and Bergles, A. E., Augmentation of Heat Transfer in Tubes by Use of Mesh and Brush Inserts, *Proc. 1972 Heat Transfer and Fluid Mechanics Inst.*, pp. 400–416, Stanford University Press, Palo Alto, Calif., 1972.

36 Brown, W. T., Jr., A Study of Flow Surface Boiling, Ph.D. thesis, Massachusetts Institute of Technology, Cambridge, Mass., 1967.

37 Durant, W. S., Towell, R. H., and Mirshak, S., Heat Transfer to Water Flowing in an Annulus by Roughening the Heated Wall, *Chem. Eng. Prog. Symp. Ser.*, vol. 61, no. 60, pp. 106–113, 1965.

38 Murphy, R. W., and Truesdale, K. L., The Mechanism of Flow Boiling Augmentation in Tubes with Discrete Surface Roughness Elements, Raytheon Company Rept. BR-7294, Bedford, Mass., 1972.

39 Seely, J. H., and Chu, R. C., *Heat Transfer in Microelectronic Equipment*, Marcel Dekker, New York, 1972.

40 Megerlin, F. E., Murphy, R. W., and Bergles, A. E., Augmentation of Heat Transfer in Tubes by Use of Mesh and Brush Inserts, *J. Heat Transfer*, vol. 96, pp. 145-151, 1974.

41 Gambill, W. R., and Greene, N. D., Boiling Burnout with H_2O in Vortex Flow, *Chem. Eng. Prog.*, vol. 54, pp. 68-76, 1958.

42 Lopina, R. F., and Bergles, A. E., Subcooled Boiling of Water in Tape-Generated Swirl Flow, *J. Heat Transfer*, vol. 95, pp. 281-283, 1973.

43 Gambill, W. R., Bundy, R. D., and Wansbrough, R. W., Heat Transfer, Burnout and Pressure Drop for Water in Swirl Flow Through Tubes with Internal Twisted Topes, *Chem. Eng. Prog. Symp. Ser.*, vol. 57, no. 32, pp. 127-137, 1961.

44 Iltscheff, S., Concerning Tests for Achieving Drop Condensation with Fluorinated Refrigerants, *Kalttechnik-Klimatisierung*, vol. 23, pp. 237-241, 1971.

45 Nicol, A. A., and Medwell, J. O., Effect of Surface Roughness on Condensing Steam, *Can. J. Chem. Eng.*, vol. 44, pp. 170-173, 1966.

46 Nakayama, W., Kaikoku, T., Kuwahara, H., and Kakizaki, K., High-Flux Heat Transfer Surface "Thermoexcel," *Hitachi Rev.*, vol. 24, pp. 329-334, 1975.

47 Markowitz, A., Mikic, B. B., and Bergles, A. E., Condensation on a Downward Facing Ripple Surface, *J. Heat Transfer*, vol. 94, pp. 315-320, 1972.

48 Chandra, S., Houghton, A. V., and Castonguay, T. T., The Condensation of Vapor on a Horizontal Rotating Square Tube, *Proc. 1974 Heat Transfer and Fluid Mechanics Inst.*, pp. 21-37, Stanford University Press, Palo Alto, Calif., 1974.

49 Seth, A. K., and Lee, L., The Effect of an Electric Field in the Presence of Noncondensable Gas on Film Condensation Heat Transfer, *J. Heat Transfer*, vol. 96, pp. 257-258, 1974.

50 Lienhard, J., and Dhir, V., A Simple Analysis of Laminar Film Condensation with Suction, *J. Heat Transfer*, vol. 94, pp. 334-336, 1972.

IV

■ electronic system applications

20

■ cooling inertial equipment

20.1 INTRODUCTION

In this chapter the problem of cooling and precise temperature control of inertial guidance equipment is considered. After a discussion of the types of inertial guidance platforms (often referred to as the inertial measuring unit), attention focuses on the floated gyroscope because of its utilization as an essential component in long-range submarines, supersonic aircraft, and ballistic missile and space vehicle inertial guidance systems.

20.2 THE INERTIAL PLATFORM

The inertial platform or inertial measuring unit (IMU) is the heart of an inertial guidance or inertial navigation system. With its control and logic electronics combined with a computer, it forms the inertial guidance system, and after the system has been initialized and the starting position entered into the computer, it performs all navigation or guidance functions internally, requiring no external input. Inertial platforms are of two types: gimbaled and strapdown.

20.3 THE GIMBALED INERTIAL PLATFORM

Before proceeding to a discussion of thermal considerations peculiar to a gimbaled platform, it is necessary to note that the inertial platform is an electromechanical device whose primary function is to provide electrical outputs that are proportional to the linear components of motion: acceleration, velocity, and displacement. These components of motion are further resolved into x, y, and z components (if a rectangular coordinate system is to be employed) and stabilized with respect to a specified reference frame. The accelerometer is the instrument that produces an electrical signal that is proportional to linear acceleration. Velocity and displacement may then be obtained from integrations of the acceleration signal. The gyroscope or gyro is an instrument that is space-stable and produces an electrical signal that is proportional to angular movement. A gimbal is a structural member that allows rotational freedom in one axis, and the gimbal on which the inertial components are mounted (usually the innermost gimbal) is termed the stable element or inertially stabilized platform.

There are two basic types of inertial navigation system mechanizations that employ gimbaled inertial measuring units. These are the earth-referenced system and the space-referenced (space-stable) system. The earth-referenced system may be of two types: the earth-fixed type, in which the coordinate frame remains fixed with respect to the rotating earth; and the local vertical type, in which the reference frame moves with the vehicle but where one axis always remains aligned with the local gravity vector. The space-stable systems fix the coordinate reference with respect to selected fixed stars. These systems are also of two types: the torqued gyro type, in which the gyros are torqued to compensate for the precessed gyro errors; and the free-running gyro type, in which appropriate software compensates for these errors.

In a four-gimbal IMU, the innermost gimbal is the stable element and is considered as the first gimbal. It usually contains three gyros, three accelerometers, gyro and accelerometer electronics, thermal controls, the gimbal drive motor, a resolver to resolve gyro signals into drive components for the second and third gimbal attitudes, and a slip-ring assembly to conduct electrical signals between the second and third gimbals. The second and third gimbals also contain a drive motor, a resolver, and a slip-ring assembly. Here the resolvers provide drive signals for the fourth gimbal, which, in addition to its gimbal drive motor, a slip-ring assembly to conduct signals to the housing, and a resolver for coordinate transformation, contains provision for angular or rate feedback to the second gimbal. The housing contains electrical and mechanical interfaces, purge and fill connections, and thermal controls.

Satisfactory thermal design of the gimbaled inertial platform requires removal of heat at acceptable temperature levels and control of inertial component temperatures and gradients to facilitate navigational accuracy. As a rule, limitation of temperature is accomplished with less difficulty than precise temperature control, which often requires an elaborate array of temperature sensors and heaters at the component level and less frequently at the block or case level.

20.3.1   Heat Transfer Mechanisms

Within the pressurized inertial measuring unit, heat is transferred by element conduction, radiation, convection, and gas or air conduction. These phenomena act in combination, and the geometry is sufficiently complex that individual effects cannot be isolated with precision.

It is generally true that the conduction path through the gimbals and across the rotary components is tortuous, such that the primary mechanisms of heat transfer from inner element to case are convection and radiation. In earlier natural convection-cooled platforms without fourth-gimbal shock isolation, element conduction could account for as much as 33% of the heat transfer. However, in typical forced convection-cooled platforms with shock isolation, element conduction is reduced to 10% or 15% of the total rejection.

Radiation can be a significant mode of heat rejection, particularly in the absence of forced convection, and it is subject to some measure of control by means of surface finish. Radiation is maximized with flat black surfaces and virtually eliminated with polished surfaces; varying degrees of intermediate performance are attainable.

Convection and gas conduction are primary mechanisms of platform heat transfer. Natural convection is utilized in most IMUs with conventional gyros; the higher power dissipation of the more exotic gyros requires the use of forced convection in systems of this type.

There are two distinct disadvantages to natural convection cooling. The first is that local heat transfer coefficients and gas temperature profiles are sensitive to gimbal position. Convection currents are therefore modified in the course of vehicle motion, and this alters the temperature profile, with a deleterious effect on navigational accuracy. The second disadvantage to natural convection cooling is that the gas circulation is dependent on gravity; thus thermal characteristics typical of the terrestrial environment may change significantly in an orbital environment or transfer trajectory where radiation, element conduction, and gas conduction control the heat transfer.

Forced convection cooling is less sensitive to gravitational effects, but local heat transfer coefficient variations due to gimbal shielding may be more severe than are encountered with natural convection. Moreover, the coefficient also varies across the heat transfer surface. If a stream of gas impinges against a sphere, the local heat flux is a maximum at the stagnation point and the downstream point directly behind it. It varies for all other points, reaching a minimum at the point at which impingement ceases.

From the foregoing discussion, it is apparent that the major detriments to constant, uniform temperature are lack of symmetry in the assembly, change in the relative position of the assembly elements, and variation in the convection heat transfer coefficient due to local flow anomalies. Is is not implied that the listed order represents the relative importance of these factors; they cannot be isolated and must be considered simultaneously. In a practical sense, the interaction defies precise analysis, and the designer should assume that gradient control will require some laboratory experimentation with baffling, control temperature, and, in the case of forced convection cooling, flow rate. It is axiomatic that such experimentation is imperative to the achievement of optimum design.

20.3.2 Temperature Control

Accurate navigation requires precise control of inertial component temperatures and gradients. Several levels of control may be employed within the pressurized housing of the inertial measuring unit:

1 Temperature control amplifier (TCA) heating of inertial components
2 Temperature control amplifier heating of stable element
3 Temperature control amplifier heating of platform case
4 Control of cooling gas inlet temperature (forced convection only)
5 Control of cooling gas flow rate (forced convection only)
6 Control of cooling gas flow distribution (forced or natural convection)

Temperature control amplifiers control the power supplied to strategically located electrical heaters. Power input to the heaters is modulated in response to signals from sensors located in the critical control areas to maintain a constant uniform temperature, which is unaffected by external temperature excursions. Obviously, there are limits to the external temperature range over which control can be maintained, and the control criterion is that the heater must be energized at some level of power between zero and maximum. If the temperature of a hot environment is increased, TCA heater power is decreased accordingly and control is lost at the point at which the sensor calls for the heater to be shut off. If the temperature of a cold environment is decreased, TCA heater power is increased and control is lost when the heater dissipates full power.

Physically, TCA heaters can be fabricated in any of several configurations. The heaters can be of film, wire, or cartridge construction. Inertial component heaters frequently consist of alternate helically wound sensor and heater wires.

Inertial measuring units are also provided with high-powered "fast" heaters to minimize warm-up time under cold starting conditions, and overtemperature protection devices to prevent damage in case of control malfunction.

20.3.3   Air or Gas Circulation

The primary function of air or gas circulation within the platform is heat removal, but there are unavoidable simultaneous control and gradient effects incurred as part of the heat transfer process. Consequently, good thermal design requires careful implementation of the convection process.

To some extent, hardware temperature gradients are proportional to the coolant temperature increase across the platform. Therefore the flow rate should be sufficient to limit the increase to 10°C or less. Since flow rate also affects flow distribution, local heat transfer coefficient variation, and pumping power requirements, there are trade-offs involved in flow rate optimization.

In either natural or forced convection cooling systems, it is desirable to distribute the flow as uniformly as practical to all areas of the stable element. This entails strategic location and configuration of supply and exhaust ports, baffles, and louvers. In forced flow systems, care must be taken to avoid short circuiting the flow, with consequent element bypass. Furthermore, the pressure loss characteristic of the system must be compatible with the pressure-generating characteristic of the fan to prevent excessive flow restriction.

20.3.4   Thermal Analysis

The preceding sections have stressed the difficulty of conducting a precise analysis of a gimbaled inertial measuring unit. Nevertheless, early analysis effort is required if a good design is to be achieved. Although optimization may require significant testing in the laboratory, analysis is required to assure heat rejection at satisfactory temperature levels. In forced convection systems, analysis also assures compatability between the fan static pressure flow characteristic curve and the system resistance characteristic curve.

Often, the typical analysis requires finite-difference modeling. The analyst, however, should be aware that when dealing with gimbaled platforms, the most meticulous analysis effort can never eliminate the uncertainties that result from convection coefficient variation due to changes in gimbal position. Unfortunately, there is no universal set of guidelines that are generally applicable.

20.4   THE STRAPDOWN PLATFORM

A strapdown system typically includes an inertial sensor assembly, sensor loop electronics, a power supply, and a computer. The strapdown terminology is derived from the fact that the inertial sensors are physically attached to the vehicle frame rather than to movable gimbals. The output of the system is in terms of changes in velocity and angle, which are used by the computer to calculate vehicle position, velocity, and attitude. The strapdown concept eliminates the need for rotary components such as resolvers, synchros, and slip rings, with a consequent economy in total system size and mechanical complexity.

20.4.1 Operating Temperature

The first consideration in selection of the inertial component block temperature is to list those components attached to the block assembly that require thermal control, their power dissipations, and their operating temperature. These components are the gyros, accelerometers, and the sensor electronics.

Experience has shown that the accelerometers and electronics, although thermally sensitive, are not as sensitive to absolute temperature levels as they are to thermal stability. On the other hand, most gyros exhibit optimum performance at a particular temperature. Therefore, gyro performance is usually the primary factor to be considered in establishing the block operating temperature.

Several power dissipations are associated with the gyro. Spin motor and signal generator power are relatively constant, whereas torque power and TCA power are variable. In applications where close thermal control is required, the TCA power is adjusted to account for the variation in torque power.

It has been observed that gyro-to-block thermal conductances vary significantly with altitude. Here, too, the TCA is used to compensate for this effect.

20.4.2 Block-to-Chassis Thermal Conductance Requirements

Once the inertial component block temperature has been established, it becomes necessary to determine the block to chassis thermal conductance (W/°C) required to maintain positive thermal control over the entire ambient range. To do this, it is necessary first to determine the expected chassis or structure temperature range (the block is attached to a mounting structure). This is done by adding all of the power dissipations and dividing this total into the prescribed temperature drop between block and chassis (the required block temperature minus the expected chassis temperature). Here, again, the TCA can be used to adjust the block temperature over the expected range of chassis temperatures.

20.5 THE FLOATED GYROSCOPE

The floated gyroscope, a particularly sensitive instrument, is used in conjunction with servos to stabilize a platform to which precise accelerometers are affixed. This stable platform may then be maintained fixed in space regardless of the motion of the vehicle. The accelerometers are able to measure true accelerations precisely in their respective coordinate directions, and the system is able to integrate once to establish a velocity and then a second time to determine the distance covered.

The essential mechanical elements of a floated gyroscope are a high-speed spin rotor supported by bearings inside a cylindrical enclosure, a buoyant fluid, and a housing. The cylindrical enclosure containing the spin rotor is designated the *float*, and the buoyant fluid is called the *flotation fluid*. Figures 20.1 and 20.2 present cutaway views of floated gyroscopes. Note the annular space between the float and the housing, which is filled with flotation fluid.

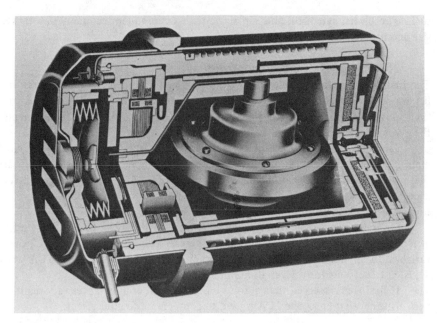

FIG. 20.1 Cutaway view of Honeywell gas-bearing gyro. (Courtesy of Honeywell, Inc.)

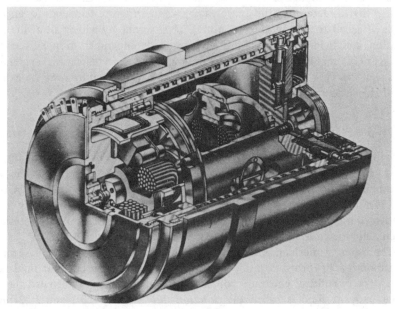

FIG. 20.2 Cutaway view of Sperry SYG-1000 gyro. (Courtesy of Sperry Rand Corp.)

The flotation fluid plays a dual role. First, it supports the float, and when its density is matched to that of the float, it gives neutral buoyancy. Under this condition, frictional drag in the float alignment bearings is reduced to a minute level. Second, the flotational fluid provides viscous damping about the rotational axis of the float. This damping is the means for obtaining the integrating effect.

20.6   THE GYRO THERMAL PROBLEM

Cooling gyros is only part of the problem confronting the engineer. The gyro dissipates heat, but a consideration of the withdrawal of this heat without regard to precise temperature control does not solve the thermal problem. Hence, one sees a difference between cooling and temperature or environmental control.

Because the density of the flotation fluid is a function of temperature, neutral buoyancy is approached only at a precise temperature level. However, this problem is not particularly severe and can be solved by use of conventional temperature control systems.

The viscosity of the flotation fluid is also a function of temperature, and a high viscosity is required to achieve the necessary damping. This damping must be predictable, and the proper value is achieved only at a precise temperature level.

Finally, the presence of thermal gradients within the flotation fluid can cause density and viscosity variations and, as a result, distortion of structural components. Such distortions produce imbalance torques that are indistinguishable from true gyroscopic torques. The control of thermal gradients is the core of the gyro thermal problem and requires very sophisticated engineering. This is because the component itself is particularly vulnerable to changes of temperature and draft in the external environment. Figure 20.3 shows how, in an uncontrolled environment, the extra- and intraplatform temperature and draft conditions ultimately result in prohibitive performance inaccuracy.

Because viscosity and density of the flotation fluid are both strong functions of fluid temperature, temperature control of the fluid to $0.02°C$ is not uncommon in gyro applications.

20.7   SOLVING THE COMPONENT THERMAL GRADIENT PROBLEM

The gyro internal thermal gradient problem exists because the environmental temperature is nonuniform and the coefficient of heat transfer on the component case varies around the case circumference as well as along the case axis. Furthermore, there is a nonuniformity of radiant heat exchange with the surrounding structure, largely because of variation in shape or arrangement factor.

Solutions to the problem do not include the mere placement of an insulating barrier on the component case, because the internal heat must be dissipated. But some form of thermal barrier must be provided, of high enough thermal conductivity to permit rejection of dissipated heat (at all environmental temperatures) but of low enough thermal conductivity to allow operation of component heaters to

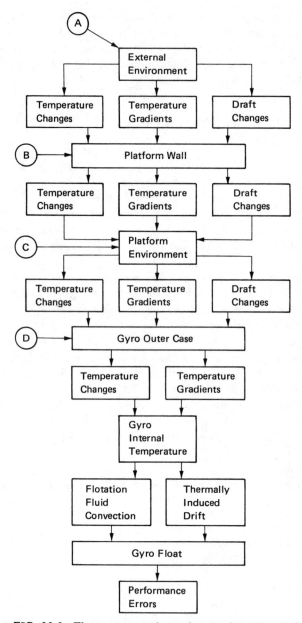

FIG. 20.3   The gyroscope thermal control problem [7].

maintain the flotation fluid temperature at a precise level without excessive power expenditure.

Barnaby et al. [1] have proposed the use of the term *attenuation* to assist in the consideration of the gyro temperature control problem. The attenuation is defined as the ratio of the housing to maximum flotation fluid temperature difference and is a measure of freedom from thermal effects.

20.8   METHODS OF ACHIEVING COMPONENT TEMPERATURE CONTROL

Barnaby et al. [1] have proposed three methods of enhancing the attenuation in a gyroscope. These are as follows:

1 Using a housing wraparound containing alternate layers of insulating and conducting material
2 Using a rotating cylinder within the housing
3 Using a housing of pyrolytic graphite

The use of the alternate layer wraparound has three advantageous effects. It allows internally generated heat to be dissipated to an outside high-temperature environment. It provides adequate thermal insulation in a cold environment; the internal heat is not dissipated so rapidly as to require a great deal of heater power. These are solutions to the cooling problem. Of equal if not more importance is the attenuation of externally induced thermal perturbations, with the result that the flotation fluid within the annular gap is essentially isothermal. This is a solution to the temperature or environmental control problem.

The use of a rotating cylinder located concentrically between the inner and outer walls of the housing has been found to eliminate temperature gradients by smoothing. Barnaby et al. [1] claim that the rotating cylinder can provide more than 100 times the thermal attenuation possible with the alternate wrap.

The pyrolytic graphite housing can be an effective thermal attenuator because of a pronounced thermal conductivity anisotrophy. This type of housing can provide a low thermal conductivity in the radial direction. This tends to resist heat flow from the environment. The thermal conductivity in the circumferential direction, being about 100 times that in the radial direction, assists circumferential smoothing and the reduction of temperature gradients.

20.9   INERTIAL EQUIPMENT TEMPERATURE CONTROL SYSTEMS

An environmental control system for an inertial guidance application must meet four basic conditions:

1 The internally generated heat, often of the order of a few hundred watts, must be removed efficiently.
2 The contents of the system within the platform must have maximum isolation from external draft or environmental changes.
3 The temperature distribution within the platform must be as uniform as possible and constant with respect to time. The thermal transient response must be as fast as possible.
4 The air flow distribution within the platform must be constant with respect to time and with respect to the attitude of the platform.

From the foregoing one can see that the effective overall thermal resistance of the platform must be such that for maximum heat load and steady-state conditions, the temperature distribution around the inner gimbal is conducive to the proper operation of the inertial components.

Reference to Fig. 20.3 shows that some means of isolation must be introduced at points B and D if system errors are to be minimized. Indeed, the task facing the thermal engineer can be summarized as follows: Given the temperature and draft conditions at A in Fig. 20.3, and assuming a suitably designed platform and outer case (points B and D), specify the temperature and draft conditions surrounding the gyro outer case at D. Then determine whether the conditions specified lead to operation within the accuracy limits of the system.

Many methods have been considered for the achievement of temperature control of inertial components. These are summarized below.

1 *Centralized system.* Each sensitive component is enclosed in a fluid coil housing and then insulated. Heat is removed by the circulating fluid and is transferred to air in a liquid-to-air heat exchanger. This system allows maximum isolation of the inertial components with effective removal of internally generated heat. Furthermore, the inertial components are insensitive to draft (air flow) changes and relatively insensitive to temperature changes.

2 *Wraparound heat exchanger.* The heat is removed from the platform casing by two wraparound compact heat exchangers. The internal air is circulated in a closed loop by a fixed-speed fan. The external air, which is the ultimate heat sink, is circulated by a variable-speed fan whose speed is controlled by a sensed temperature within the inertial components. This system promotes symmetrical temperature distribution, and, by virtue of the fixed-speed fan within the inner loop, the system is relatively impervious to internal draft changes.

3 *Thermoelectric system.* The ultimate practicability of this system looks quite encouraging. Thermoelectric elements would be mounted directly to the sensitive components, and the thermoelectric device would act alternately as a heater or cooler depending on the temperature level and the temperature tolerance required. Such a system would be at the mercy of the power converter performance as well as that of the temperature sensing device. In addition, some means of rejecting the heat to the cooling air would be necessary.

4 *On-off heaters.* The on-off heater was the first type of temperature control device used in inertial equipment. Heaters are still used with success but are limited, even with proportional control, to continuous "on" or continuous "off" operation.

5 *Other techniques.* Other techniques such as a rotating shield around the platform assembly, thermal shrouds, and a wick-vapor pressure device [2] have been proposed and have been used with success.

20.10   IRON CORE TRANSFORMERS

Because the miniaturized iron core transformers used in electronic equipment today are less efficient than their large predecessors, a significant cooling problem can occur. Not only will the transformer fail, but there may be consequent failures elsewhere in the system.

Figure 20.4 shows an electrothermal analog circuit of the heat flow in an iron core transformer. Note that conduction across the imbedment (impregnant or potting

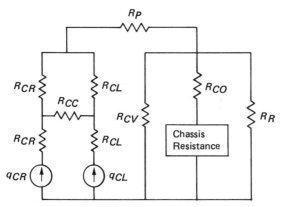

FIG. 20.4 Electrothermal analog of a transformer.

compound) in the form of the resistance R_P is shown. In the event that an imbedment is not used, the resistance R_P is short-circuited and the calculations may proceed with this modification incorporated.

The heat flow within a transformer is quite complicated. For example, it is difficult to evaluate R_{CC}, the thermal resistance between the core and the winding. Usually one considers the coil hot spot temperature, because the largest temperature drop most often occurs between the coil hot spot and the core.

Welsh [3] points out that conduction cooling can be enhanced by securely bonding the core to the chassis and then designing the chassis to remove the dissipated heat. The utmost care should be exercised to assure a good thermal bond (see Chap. 9). In the event that a potting material is used, the case must be thermally bonded to the chassis. Higher operating temperature, however, must be anticipated.

Kilham et al. [4] have studied the use of fluorochemical boiling within transformers. Use of a boiling liquid reduces the value of the resistance R_{CC} in Fig. 20.4 to a point where the dissipated heat is rejected under the potential of a small temperature difference between the case and the surrounding fluid.

20.11 FOIL–WOUND TRANSFORMERS

Foil-wound transformers are also used in electronic equipment. Indeed, solenoids to provide the focusing for traveling wave tubes' electron beams have been foil-wound, and heat dissipations in this application can be as high as a kilowatt. The foil-wound transformer is composed of alternate layers of a conducting foil such as aluminum and an insulating film such as mylar. The foil and insulating film are wound in a radial stack called a wafer.

In high heat-dissipating applications of foil-wound transformers, cooling must be provided. Air, for example, is passed through the space between the wafers. The heat transfer surface is a percentage of the wafer face surface, because the insulating film will conduct only a negligible amount of heat. If, for example, a foil of 0.00762 cm (0.003 in) is used in conjunction with a mylar film of 0.00254 cm (0.001 in) thickness, then the surface area for heat transfer will be $0.00762/0.01016 = 0.75$, and 75% of the wafer face surface area will be used in the rate equation.

Heat transfer coefficients for the case of ribbonlike laminar flow in the space between the wafers can be obtained from the work of Norris and Streid [5] and employed as shown by Cicero [6].

20.12 REFERENCES

1 Barnaby, R. E., Fenster, S. D., and R. J. Ross, Control of Thermal Drift in Floated Gyroscopes, *Sperry Eng. Rev.*, vol. 14, pp. 36–41, September 1961.
2 Hatsopoulos, G., and Kaye, J., A Novel Method of Obtaining an Isothermal Surface for Steady State and Transient Conditions, MIT Instrumentation Lab Rept. 121, May 1955.
3 Welsh, J. P., Techniques of Cooling Electronic Equipment, *Electrical Manufacturing*, vol. 62, pp. 80–87, November 1958.
4 Kilham, L. F., Ursch, R. R., and Ahearn, J. F., Fluorochemical Vapor Techniques in Electronic Equipment, *Electrical Manufacturing*, vol. 64, pp. 88–92, August 1950.
5 Norris, R. H., and D. D. Streid, Laminar-Flow Heat Transfer Coefficients in Ducts, *Trans. ASME*, vol. 62, pp. 525–533, 1940.
6 Cicero, A. D., Optimum Design of Airborne Tape-Wound Transformers Cooled by Forced Air Convection, Sylvania Electric Products, Inc., Boston Engineering Laboratory, Rept. TR-14-29, app. F, 1958.
7 LeBlanc, R., Personal communication, 1961.

21

■ cooling transistors and vacuum tubes

21.1 INTRODUCTION

It was seen in Chap. 2 that reliability considerations demand that electrical and electronic components be maintained at fairly low temperature levels in order to obtain required life expectancy. This is certainly true of semiconductor devices such as the transistor. Since its appearance on the electronic component scene in 1948, the transistor has been found to have immense utility in many applications such as communications equipment, audio and high-frequency amplifiers, switching circuits, electronic computers, and control systems. Since 1953, by virtue of its ability to operate with less power and because of its smaller size, the transistor has supplemented the range of application of vacuum tubes.

The key to the use of transistors and other semiconductor devices is temperature control, not only because reliability is an inverse function of temperature, but also because of the phenomenon known as *thermal runaway*. An increase in junction temperature beyond a certain level causes an increase in cut-off current. This in turn increases the collector current, which raises the power dissipation at the transistor junction. The increase in power dissipation further increases the junction temperature, and this cycle continues until the transistor is rendered useless as an electronic component.

This chapter addresses methods of analyzing transistor cooling problems, a cursory description of some commercially available transistor heat sinks and coolers, and several illustrative problems involving transistor heat flow analysis and heat sink design.

Cooling vacuum tubes does not represent as difficult a problem as cooling transistors, because the vacuum tube, although it dissipates more heat because of its filament power, is larger than the transistor and is usually permitted to operate at a higher temperature.

This chapter also discusses several of the types of tubes used in electronic equipment and the ways of predicting the heat transfer from them. Heat transfer from vacuum tubes under natural convection conditions is considered first. Then the various types of tube shields that are often used and the effects of these tube shields on heat transfer performance is investigated. The various heat transfer coefficient correlations are cited, and some of them are used in illustrative examples.

21.2 TRANSISTOR CLASSIFICATION

Transistors are classified in many ways. One classification is that of power level: low-, medium-, and high-power transistors. This classification is somewhat difficult to accept from the heat transfer point of view. Frequently, low- and medium-power transistors are more difficult to keep cool than high-power transistors, merely because heat transfer is such a strong function of size. Other classifications are put forward on the basis of junction construction and the actual utilization of the device.

The heat transfer engineer will probably want to classify the transistor on the basis of how simple it is to accomplish the requisite heat dissipation within the allowable temperature extremes. One needs only refer to the rate equation,

$$q = hS\theta_m = hS\,\Delta T \tag{21.1}$$

which states that the heat q must be removed within a prescribed temperature difference θ_m. This heat removal can be enhanced by increasing either the heat transfer coefficient h, or the surface S, or both.

The surface may be increased by using a plate or heat sink to which the transistor is mounted. The term "heat sink" is a misnomer, because the heat sink is always the surrounding environment or the coolant fluid if, indeed, a coolant fluid is used. In any case, as will be seen, a fairly large plate will aid in the transfer of heat to the ultimate sink, and the plate is termed a *heat sink.*

The surface may also be increased by attaching a finned configuration to the transistor case. This finned surface is usually called a *cooler.* The difference between a heat sink and a cooler is that the transistor is added to the heat sink and the cooler is added to the transistor.

Transistors may also be mounted on cold plate heat exchangers, or they may be cooled directly by placing them in a duct containing a coolant stream. In the latter case, if the heat transfer coefficient is low, it can be increased by placing turbulators upstream of the transistors.

21.3 THE ELECTROTHERMAL ANALOG CIRCUIT

The heat flow from the junction of a semiconductor device is usually analyzed by means of an electrothermal analog circuit. Many investigators have proposed versions of the type of "thermal circuit" to be used [1–3]. These differ in complexity and vary with the semiconductor under consideration. One such thermal circuit, shown in Fig. 21.1, has been found to have great utility and is essentially very simple.

The thermal resistance from junction to case, R_J, is usually specified by the transistor manufacturer. Most manufacturers use the units °C/W. The value of R_J depends on the internal construction of the junction, the length of the thermal path from junction to case, and the size and length of the lead wires.

The resistance R_m is the mounting resistance between the transistor and the metal heat sink.

Some transistors are constructed with their case electrically connected to one of the transistor junctions. This type of transistor requires the use of an insulating washer for electrical isolation. The washer must have very high electrical resistivity to provide the isolation but, unfortunately, high electrical resistivity means low thermal conductivity. The resistance R_m must therefore include the washer resistance as well as the contact resistance. Liberal application of a lubricant such as silicone grease has been found to reduce the overall value of R_m. Some applications have used

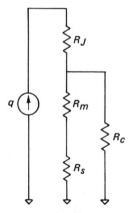

FIG. 21.1 Electrothermal analog of heat flow from a transistor.

an anodized layer on the heat sink. The anodized layer provides adequate isolation without a severe temperature drop from transistor to heat sink.

The resistance R_c is the resistance to the flow of heat from the case to the surrounding environment. This resistance may be very large because the transistor case is very small. It can be ignored entirely, because only a small fraction of the dissipated heat can be expected to flow in this path, and ignoring this heat flow path will result in a conservative design of the heat sink plate. If one does not wish to ignore the effect of the resistance R_c, its effect can be handled by the method developed in Sec. 21.5.

The heat sink resistance is the resistance R_s. This is the resistance that can be adjusted to yield the required overall resistance. It is composed of two resistances in parallel—R_{cv}, the resistance to heat flow from the heat sink by convection, and R_r, the resistance to the flow of heat from the heat sink by radiation. Although some analysts tend to ignore R_r, such a decision may result in an overdesign of the heat sink, as Fig. 21.2 clearly shows. Here a radiative heat flow fraction

$$F = \frac{q_r}{q_r + q_c}$$

is plotted for a 1-cm-diameter cylinder, 2 cm high, with surface treated to have an emissivity of 0.90 in an environment of 25°C as a function of the convective heat transfer coefficient in W/m² °C. Observe that at low values of h, where natural convection dominates, the transfer of heat by radiation is not negligible. This picture, of course, will change to some extent with component size and orientation, surface emissivity, and environmental temperature.

Based on these considerations, a thermal analog can be constructed as shown in Fig. 21.3. The resistance R_J is a parallel combination,

$$R_J = \frac{(R_{L1} + R_1)R_2(R_{L3} + R_3)}{R_2(R_{L1} + R_1) + (R_{L1} + R_1)(R_{L3} + R_3) + R_2(R_{L3} + R_3)} \tag{21.2}$$

where R_{L1} and R_{L3} are the resistances of the lead wires from emitter to case and collector to case, respectively, and R_1, R_2, and R_3 are the resistances from the junction to the case through the emitter, base, and collector, respectively.

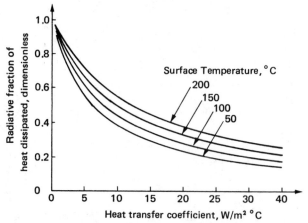

FIG. 21.2 Radiative heat flow fraction for small electronic component as a function of convective heat transfer coefficient. The component is cylindrical, 2 cm high and 1 cm in diameter, with surface treated to have an emissivity of 0.90.

Equation (21.2) is based on the particular transistor element construction shown in Fig. 21.4. Other types of construction will require a different relationship for R_J. For example, Eq. (21.2) will not apply to the element construction shown in Fig. 21.5. Fortunately, the evaluation of Eq. (21.2) can be avoided, because R_J can usually be obtained from the transistor manufacturer.

The mounting resistance R_m is the simple summation

$$R_m = R_{co} + R_w \tag{21.3}$$

where R_{co} is the contact resistance and R_w is the resistance of the isolating washer.

The case of the transistor is a distributed heat flow configuration. Its resistance, R_c, can be evaluated using a transmission-line analogy or by treating the entire case as a capped cylinder [4]. The heat sink resistance R_s is composed of the parallel combination of a convection effect, R_{cv}, and a radiation effect, R_r. The convection effect R_{cv} can be determined from manufacturer's data or by the determination of a fin array q_b/θ_b ratio as discussed in Chap. 14 or in [5]. If the heat sink is a single metal plate, then

$$R_s = R_{cv} = \frac{1}{hS\eta} \tag{21.4}$$

The efficiency term in Eq. (21.4) is not obtained in a straightforward manner if the heat sink is square or rectangular. If it is circular, the heat sink is essentially a

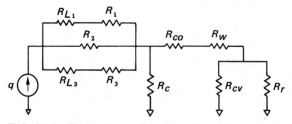

FIG. 21.3 Electrothermal analog of heat flow from a transistor.

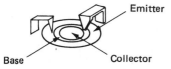

FIG. 21.4 Sketch showing one type of transistor construction.

radial fin of rectangular profile with efficiency as given in the efficiency tables provided in Kern and Kraus [5]. In using these tables it should be recalled that

$$\rho = \frac{r_0}{r_e} \qquad \phi = (r_e - r_0)^{3/2} \sqrt{\frac{2h}{kA_p}}$$

21.4 COMMERCIALLY AVAILABLE HEAT SINKS

An extensive array of transistor heat sinks and coolers can be obtained commercially, inexpensively, and on short delivery schedule. These heat sinks come in a variety of forms. Some can be clipped or clamped onto the case of the transistor. Others are used as the transistor mounting.

The Featherweight coolers shown in Fig. 21.6 are made of a beryllium copper alloy and can be used for all JEDEC-style milliwatt transistors and diodes. Their spring action design attempts to provide precision surface mating between the cooler and the case. These coolers are used under free and forced convection conditions.

Natural and forced convection heat sinks are made by many fabricators. These fabricators have tested their heat sinks extensively and usually provide data of temperature rise as a function of heat dissipated. The 400 series with transistors mounted in place is shown in Fig. 21.7. Note the provision for mounting the heat sink to the chassis. Natural convection cooling data, provided by Wakefield, are shown in Fig. 21.8. The slope of the curves in Fig. 21.8 is the thermal resistance in °C/W.

Heat sinks exclusively for forced convection applications are also available. Figure 21.9 shows the Wakefield 500 series. Forced convection thermal resistances and head loss data for the 500 series are shown in Fig. 21.10.

21.5 THE PIN FIN HEAT SINK

It may be noted that longitudinal fins are used in almost all of the heat sinks described and discussed in the previous section. Employment of thin pins or spines as a heat

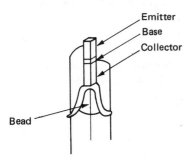

FIG. 21.5 Sketch showing one type of transistor construction.

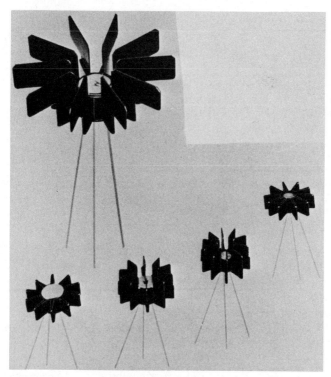

FIG. 21.6 Wakefield featherweight transistor coolers. (Courtesy of Wakefield Engineering Co., Wakefield, Mass.)

sink results in considerably better heat transfer characteristics because of the promotion of turbulence in the coolant passages between the pins and the additional surface made available.

Drexel [6] has conducted exhaustive experiments on both round and diamond-shaped pin fin coolers. In 1961 he proposed

$$\frac{hd}{k} = 1.40 \left(\frac{dG}{\mu} \right)^{0.28} \left(\frac{c_p \mu}{k} \right)^{1/3} \tag{21.5}$$

as a correlation for air flowing normal to banks of staggered round pin fins. When compared to the correlation for flow within rectangular passages, one can see the advantage of the pin fin arrangement. The increase in heat transfer is, however, accompanied by a much greater pressure loss due to the fluid turbulence.

Drexel's pin fin cooler is shown mounted in an electronic assembly in Fig. 21.11. The chassis shown in this figure contains nine transistors with a total dissipation of 338 W. Note that the cooler is tailored to the transistor and that the pins are eliminated at the points where the transistor leads come through the cooler base surface.

The thermal resistance of this pin cooler will also be

$$R_s = \frac{1}{hS\eta} \tag{21.4}$$

where S is the total surface of all the pins, η is the cylindrical spine efficiency,

$$\eta = \frac{\tanh mb}{mb} \qquad (14.5)$$

with $m = \sqrt{4h/kd}$, and h is the heat transfer coefficient obtained from Eq. (21.5). Note that the effects of the base surface of the cooler are ignored to help yield a conservative design. Note also that the efficiency given by Eq. (14.5) is tabulated in the efficiency tables [5].

21.6 ILLUSTRATIVE EXAMPLES

EXAMPLE 1:

A transistor dissipating 10 W with a base hex size of 1 in and $\frac{3}{8}$-24 stud is mounted on a circular aluminum ($k = 202.50$ W/m °C) heat sink $\frac{1}{16}$ in (0.159 cm) thick. It is necessary to use a 0.003-cm-thick washer for isolation. Thermal conductivity of the washer is 0.00433 W cm/cm² °C. The transistor junction must be held to 125°C or less, and

FIG. 21.7 Wakefield series 400 natural convection transistor coolers. (Courtesy of Wakefield Engineering Co., Wakefield, Mass.)

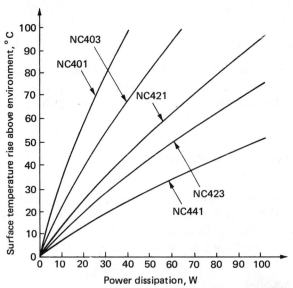

FIG. 21.8 Natural convection cooling data for Wakefield series 400 transistor coolers. (Data provided by Wakefield Engineering Co., Wakefield, Mass.)

the manufacturer has specified a value of $R_J = 0.42°C/W$. If the free convection heat transfer coefficient is 1.25 Btu/ft^2 h $°$F (7.10 W/m^2 $°$C) and the environmental temperature is 25$°$C, will a 4-in (10.16-cm)-diameter heat sink be big enough? The transistor base diameter is 1 in (2.54 cm). Take $R_{co} = 0.4°C/W$.

SOLUTION

1 $R_J = 0.42°C/W$.
2 $R_{co} = 0.4°C/W$ (given).
3 $R_w = L/kA = 0.003/0.00433(\pi/4)(2.54)^2 = 0.137°C/W$.
4 Heat sink:

$$S = 2\left(\frac{\pi}{4}\right)(d_e^2 - d_0^2) = 2\left(\frac{\pi}{4}\right)(103.23 - 6.45)$$

$$= 152.01 \text{ cm}^2 = 0.0152 \text{ m}^2$$

$$h = 7.10 \text{ W/m}^2 \text{ }°C$$

$$r_e = 5.08 \text{ cm} \quad r_0 = 1.27 \text{ cm}$$

$$\rho = \frac{r_0}{r_e} = \frac{1.27}{5.08} = 0.25$$

$$A_p = 0.159(5.08 - 1.27) = 0.606 \text{ cm}^2 = 6.058 \times 10^{-5} \text{ m}^2$$

$$(r_e - r_0)^{3/2} = (0.0508 - 0.0127)^{3/2} = 7.437 \times 10^{-3} \text{ m}^{3/2}$$

$$\phi = (r_e - r_0)^{3/2} \sqrt{\frac{2h}{kA_p}} = (7.437 \times 10^{-3}) \sqrt{\frac{2(7.10)}{(202.5)(6.058 \times 10^{-5})}}$$

$$= 0.253$$

$$\eta = 0.960 \text{ (from [5])}$$

$$R_s = \frac{1}{hS\eta} = \frac{1}{(7.10)(0.0152)(0.960)} = 9.65 \text{ °C/W}$$

5 Total resistance:

$$R_T = R_J + R_{co} + R_w + R_s$$
$$= 0.42 + 0.40 + 0.137 + 9.65 = 10.61 \text{ °C/W}$$

6 Junction temperature rise:

$$\Delta T = qR_T = 10(10.61) = 106.1°C$$

7 Junction temperature:

$$T_J = T_s + \Delta T = 25 + 106.1 = 131.1°C$$

The sink appears to be not quite big enough.

FIG. 21.9 Wakefield series 500 forced convection transistor coolers. (Courtesy of Wakefield Engineering Co., Wakefield, Mass.)

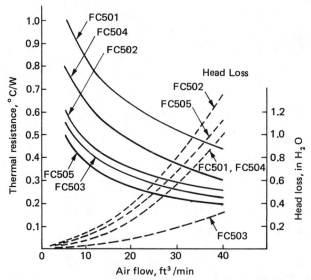

FIG. 21.10 Forced convection cooling and friction data for Wakefield series 500 transistor coolers. (Data provided by Wakefield Engineering Co., Wakefield, Mass.)

FIG. 21.11 Component board with nine transistors mounted on pin fin coolers. (Courtesy of Sperry Rand Corp.)

EXAMPLE 2:

A transistor dissipating 10 W is to be equipped with a Wakefield model 403 cooler. If the mounting resistance is 0.6°C/W, and the manufacturer specified a value of $R_J = 0.34$°C/W, find the maximum possible ambient temperature to hold 125°C on the junction.

SOLUTION

1 $R_J = 0.34$°C/W.
2 $R_m = 0.60$°C/W.
3 $R_s = 24$°C/10 W $= 2.4$°C/W, from Fig. 21.8.
4 $R_T = R_J + R_m + R_s = 3.34$°C/W.
5 $\Delta T = qR_T = 10(3.34) = 33.4$°C.
6 $T_s = T_J - \Delta T = 125 - 33.4 = 91.6$°C.

Say, 90°C.

EXAMPLE 3

In Example 2, if the surrounding temperature is 100°C, what should the dissipation be to hold 125°C on the junction?

SOLUTION

1 Assume 8 W. Then

$$R_s = \frac{20}{8} = 2.5°C/W \quad \text{from Fig. 21.8}$$

$$R_T = 0.60 + 0.34 + 2.50 = 3.44°C/W$$

$$\Delta T = qR_T = 8(3.44) = 27.52°C$$

$$T_J = T_s + \Delta T = 100 + 27.5 = 127.5°C \quad \text{too high}$$

2 Assume 7 W.

$$R_s = \frac{18}{7} = 2.57°C/W \quad \text{from Fig. 21.8}$$

$$R_T = 0.60 + 0.34 + 2.57 = 3.51°C/W$$

$$\Delta T = qR_T = 7(3.51) = 24.57°C$$

$$T_J = T_s + \Delta T = 100 + 24.6 = 124.6°C \quad \text{just about right}$$

EXAMPLE 4

A power transistor dissipating 37.5 W is mounted with an 0.003-cm washer 1 in (2.54 cm) in diameter on a pin fin cooler containing 61 aluminum pin fins $\frac{1}{8}$ in (0.317 cm) in diameter. The fins are $\frac{3}{4}$ in (1.905 cm) long and are placed in a duct carrying air. If the contact resistance is $R_{co} = 0.25$°C/W, and the junction resistance is $R_J = 0.28$°C/W, determine whether the junction will operate at 125°C or less. The mass flow of air in the duct is 1315 lb/ft² h (31,251 kg/m² h), and the average air temperature is 25°C.

SOLUTION

1 $R_J = 0.28°C/W.$
2 $R_{co} = 0.25°C/W.$
3 $R_w = 0.137°C/W$ (see Example 1).
4 $R_J + R_{co} + R_w = 0.667°C/W.$
5 Base temperature of pins:

$$T_0 = T_J - q(R_J + R_{co} + R_w)$$
$$= 125 - 37.5(0.667) = 125 - 25 = 100°C = 212°F$$

6 Film temperature in cooler:

$$T_f = \frac{100 + 25}{2} = \frac{125}{2} = 62.5°C$$

7 At film temperature:

$$\mu = 0.0661 \text{ kg/h m} \quad \text{(from Fig. 6.10)}$$
$$k\,Pr^{1/3} = 0.0234 \text{ W/m}^2 \text{ °C} \quad \text{(from Fig. 6.10)}$$

8 Heat transfer coefficient:

$$h = 1.40 \frac{k\,Pr^{1/3}}{d} \left(\frac{dG}{\mu}\right)^{0.28}$$

$$= (1.40) \left(\frac{0.0234}{0.00317}\right) \left[\frac{(0.00317)(31,251)}{0.0661}\right]^{0.28}$$

$$= 10.334(1499)^{0.28} = 80.07 \text{ W/m}^2 \text{ °C}$$

9 Surface in cooler:

$$S = 61\pi(0.00317)(0.01905) = 0.0116 \text{ m}^2$$

10 Cooler efficiency ($k = 202.5$ W/m °C):

$$m = \sqrt{\frac{4h}{kd}} = \sqrt{\frac{4(80.07)}{202.5(0.00317)}} = 22.34 \text{ m}^{-1}$$

$$mb = 0.01905(22.34) = 0.426$$

$$\eta = 0.944 \text{ [Eq. (14.5)]}$$

11 Cooler resistance:

$$R_s = \frac{1}{h\eta S} = \frac{1}{(80.07)(0.944)(0.0116)} = 1.14°C/W$$

12   Heat dissipation possible with this cooler ($77°F = 25°C$):

$$q = \frac{\Delta T}{R_s} = \frac{100 - 25}{1.14} = \frac{75}{1.14} = 65.8 \text{ W} > 37.5 \text{ W}$$

This cooler will allow the junction to operate at less than $125°C$.

21.7   COOLING BARE TUBES UNDER NATURAL CONVECTION CONDITIONS

The analysis of the heat transferred from a bare electron tube under natural convection conditions entails a trial-and-error procedure because the heat transfer coefficient is dependent on the temperature difference to the $\frac{1}{4}$ or $\frac{1}{3}$ power and the radiation involves temperatures to the fourth power. Under these conditions radiation plays a very significant role. Basically, in this type of problem, one is given a heat dissipation and must assume a value for the bulb temperature. The heat transferred by both natural convection and radiation is then calculated and added together to determine whether the total is equal to the heat dissipated.

The worksheet given as Table 21.1 can be used to excellent advantage in making these calculations. Note that four temperatures are assumed. This will usually permit at least three good points to be obtained for a graphical plot such as the one in Fig. 21.12.

Problems of this type are frequently encountered in analyzing natural convection cooling of all types of electronic equipment. The surface temperature is a function of natural convection and radiation, which in turn are governed by the temperature difference between surface and surroundings. Indeed, one may write

$$q_c = hS\theta \tag{21.6}$$

for convection,

$$q_r = \sigma SF_e F_A (T_S^4 - T_R^4) \tag{21.7}$$

for radiation, and

$$q = q_c + q_r \tag{21.8}$$

as the total heat dissipation.

The numbers in Table 21.1 and the curves in Fig. 21.12 are a solution to the problem of a vertically mounted miniature vacuum tube $2\frac{1}{2}$ in (6.35 cm) high and $\frac{3}{4}$ in (1.91 cm) in diameter dissipating 4.75 W in a still-air environment at $26.7°C$. It is presumed that a negligible amount of heat is conducted out of the tube through the electrical leads, and the tube envelope emissivity is specified as $\epsilon = 0.80$. If the tube is considered as a somewhat small body in a large enclosure, the emissivity

TABLE 21.1 Calculations for Heat Transfer from Vacuum Tube

$S = \pi dL = \pi(0.0191)(0.0635) = 0.00381$ m²
Significant dimension for natural convection: $L = 0.0635$ m
Constant in natural convection relationship: $C = 0.55$ (Table 6.9)

$q = 4.75$ W
$e = 0.80$
$p = 1.00$

Item	Dimensions	How computed	Trial I	II	III	IV
Natural convection						
t_S	°C	Assume	82.2	93.3	104.4	115.5
t_R	°C	Given	26.7	26.7	26.7	26.7
θ	°C	$\theta = t_S - t_R$	55.5	66.6	77.7	88.8
$\theta/2$	°C	Compute	27.8	33.3	38.8	44.4
t_f	°C	$t_f = \frac{1}{2}(t_S + t_R)$	54.5	60.0	65.5	71.1
k @ t_f	W/m°C	Fig. 6.10	0.0284	0.0287	0.0291	0.0294
a @ t_f	m⁻³°C⁻¹	Fig. 6.9	6.357×10^7	5.975×10^7	5.498×10^7	5.149×10^7
θ/L	°C/m	Compute	874.0	1048.8	1223.6	1398.4
$(\theta/L)^{1/4}$	(°C/m)$^{1/4}$	Compute	5.44	5.69	5.91	6.12
$(ap^2)^{1/4}$	(m⁻³°C⁻¹)$^{1/4}$	Compute	89.29	87.92	86.11	84.71
h	W/m²°C	$h = Ck(\theta/L)^{1/4}(ap^2)^{1/4}$	7.58	7.90	8.15	8.38
q_c	W	$q_c = hS\theta$	1.59	2.00	2.41	2.83
Radiation						
T_S	K	$T_S = t_s + 273$	355.2	366.3	377.4	388.5
T_R	K	$T_R = t_r + 273$	299.7	299.7	299.7	299.7
$T_S/100$	K	Compute	3.55	3.66	3.77	3.89
$T_R/100$	K	Compute	3.00	3.00	3.00	3.00
$(T_S/100)^4$	K⁴	Compute	159.18	180.03	202.87	227.81
$(T_R/100)^4$	K⁴	Compute	80.68	80.68	80.68	80.68
$(T_S/100)^4 - (T_R/100)^4$	K⁴	Compute	78.50	99.35	122.19	147.13
q_r	W	$5.67eS[(T_S/100)^4 - (T_R/100)^4]$	1.36	1.72	2.11	2.54
q	W	$q = q_c + q_r$	2.95	3.72	4.52	5.37

factor is $F_e = e = 0.80$ and the arrangement factor is $F_A = 1.00$. As a margin of safety, the surface area contributed by the top of the tube is ignored.

Figure 21.12 shows that in this application, because the value of the coefficient of heat transfer in free convection is low, convective heat transfer is not very much more significant than radiation. The radiation is, of course, governed by the arrangement factor and the tube envelope emissivity. For this reason, natural convection cooling designs that ignore radiation are likely to be unduly pessimistic.

21.8 TUBE SHIELDS

The glass envelope of an electronic tube is a hindrance rather than an aid to the flow of heat from the tube. The glass is a very poor heat conductor and is virtually opaque to thermal radiation at temperatures below 400°C. Because the glass absorbs nearly all of the heat radiated by the contained tube elements, and because it is such a poor conductor, a hot spot occurs on the tube as shown in Fig. 21.13 [7]. This hot spot usually appears opposite the plate of the tube where the dissipation is the greatest. The severe temperature gradient between the hot spot and relatively cool ends of the tube causes a thermal stress that eventually leads to the failure of the glass itself.

The hot spot also causes evolution of gas from the inner surface of the glass envelope. The rate of gas evolution is proportional to the tube bulb temperature. This is another reason for hot spot elimination; gassy tubes will not operate with any degree of predictability.

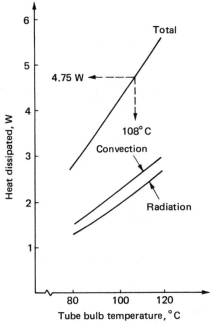

FIG. 21.12 Graphical solution for tube bulb temperature. Points for curves are obtained from calculations in Table 21.1.

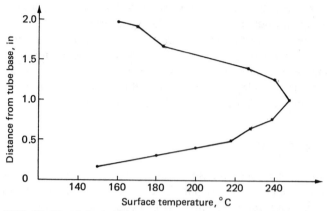

FIG. 21.13 Tube bulb temperature as a function of distance from the base of 6AQ5 vacuum tube. (From data provided by the International Electronic Research Corp., Burbank, Calif.)

Several organizations have developed tube shields that have solved the hot spot problem to a considerable degree. These tube shields have provided an efficient means of transferring heat by conduction to heat sink or chassis, reduce the severity of the temperature gradient between the hot spot and tube ends, provide a greater surface area for heat dissipation by convection, and serve as the tube mounting in severe shock and vibration environments.

Figure 21.14 shows some subminiature tube shields developed and marketed by

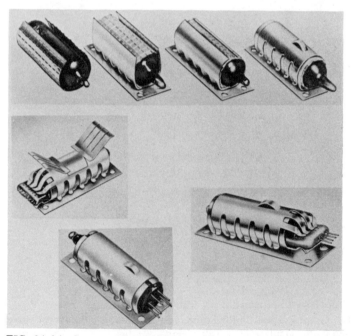

FIG. 21.14 Some subminiature tube shields. (Courtesy of International Electronic Research Corp., Burbank, Calif.)

the International Electronic Research Corporation (IERC). Note the silver liner between the tube envelope and the tube shield assembly.

Tube shields for miniature and octal tubes are shown in Figs. 21.15 and 21.16. Also shown are spring liners that adapt themselves to variations in tube envelope contour to maintain maximum contact between envelope and shield.

Typical test results [8] on a 12BY7 thermatron tube dissipating 10 W are shown in Fig. 21.17. These data were provided by IERC and show tube bulb temperature as a function of air velocity. These are forced convection tests showing that when these tubes were equipped with JAN-type tube shields, the tubes actually ran hotter than with no shield at all.

The reader should not infer that a tube shield is a cure-all for problems of high tube bulb temperature. Tube shields are helpful only when selected judiciously. Figure 21.17 shows that the lowest tube bulb temperature is obtained *with* a tube shield. These test data indicate that some shields under natural convection conditions operate at a cooler temperature than a bare tube at a 150 m/min air velocity. Figure 21.18, also based on valid test data [9], shows that the lowest bulb temperature is obtained with no shield at all. This difference in results is due to the different tube shields used, showing that the designer must exercise care in the selection and utilization of the tube shield.

FIG. 21.15 Some miniature tube shields. (Courtesy of International Electronic Research Corp., Burbank, Calif.)

FIG. 21.16 Some octal tube shields. (Courtesy of International Electronic Research Corp., Burbank, Calif.)

21.9 FORCED CONVECTION COOLING OF VACUUM TUBES

21.9.1 Vacuum Tubes in Cross Flow

In Chap. 6, a reference was made to a correlation due to Hilpert [10] for air flowing across single cylinders:

$$\text{Nu} = B(\text{Re})^n \tag{21.9}$$

where the values of B and n depend on a range of values of the Reynolds number. These values were given in Table 6.1.

Robinson et al. [11] have proposed a modification to Eq. (21.9) when the cylindrical heat sources that are encountered in electronic equipment are considered:

$$\text{Nu} = FB(\text{Re})^n \tag{21.10}$$

The factor F is an arrangement factor depending on the cylinder geometry. This factor permits utilization of the correlation for the flow of air across a single cylinder. Values of F are given in Table 21.2.

Robinson et al. [11] have further proposed values of B and n that yield a curve identical to that obtained using Table 6.1. These values of B and n are in the more limited range of $1{,}000 \leqslant \text{Re} \leqslant 100{,}000$ and are given in Table 21.3.

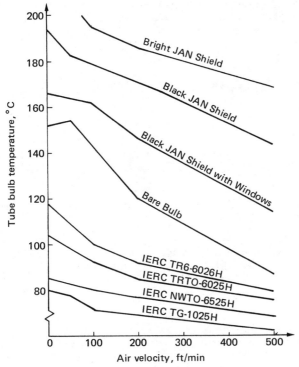

FIG. 21.17 Test results for a 12BY7 thermatron tube. (From data provided by International Electronic Research Corp., Burbank, Calif.)

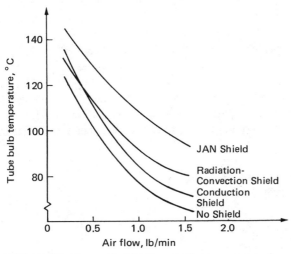

FIG. 21.18 Downwind bottom bulb temperature for seven pin miniature tubes in cross flow [9].

TABLE 21.2 Values of F to be used in Eq. (21.10)[a]

Single cylinder in free stream: $F = 1.0$
Single cylinder in duct: $F = 1 + d/w$
In-line cylinders in duct:

$$F = \left(1 + \sqrt{\frac{1}{S_T}}\right)\left\{1 + \left(\frac{1}{S_L} - \frac{0.872}{S_L^2}\right)\left(\frac{1.81}{S_T^2} - \frac{1.46}{S_T} + 0.318\right)[Re^{0.526 - (0.354/S_T)}]\right\}$$

Staggered cylinders in duct:

$$F = \left(1 + \sqrt{\frac{1}{S_T}}\right)\left\{1 + \left[\frac{1}{S_L}\left(\frac{15.50}{S_T^2} - \frac{16.80}{S_T} + 4.15\right) - \frac{1}{S_L}\left(\frac{14.15}{S_T^2} - \frac{15.33}{S_T} + 3.69\right)\right]Re^{0.13}\right\}$$

[a]Re to be evaluated at film temperature. S_L = ratio of longitudinal spacing to cylinder diameter. S_T = ratio of transverse spacing to cylinder diameter.

Welsh [9] has performed an experimental check on one of Robinson's proposals and has found excellent agreement. Using subminiature tubes (5902), he obtained a correlating equation in cross flow,

$$Nu = 0.15\,Re^{0.697} \tag{21.11}$$

for $S_T = 2.49$ and $S_L = 1.33$. In the tested region between $800 < Re < 6000$, the results coincide with Robinson's curve for $S_T = 2.333$ and $S_L = 1.250$. Welsh obtained a further check using type 805 tubes in cross flow.

Data for cross-flow cooling with air, as obtained by Welsh [9] at the Cornell Aeronautical Laboratories, are presented in Fig. 21.19. These data are for several types of tubes in various cross-flow configurations. Figure 21.19 gives the heat transfer coefficient as a function of the Reynolds number. All of the curves in Fig. 21.19 were correlated using the tube diameter as the significant dimension in the Reynolds number. For the case of curve 1, the Reynolds number is based on the major tube diameter.

The data in Fig. 21.19 are not presented in the customary manner and caution must be exercised when using them. There is no need for concern about the form of representation. Admittedly a Nusselt relationship might be more accurate, but this accuracy is only a secondary refinement, because the thermal properties of air do not vary greatly over the range of temperatures considered here. The important quantities—tube diameter and flow rate—are contained in the Reynolds number. Indeed, the Reynolds number, with its viscosity term, helps to reduce the secondary effects of the temperature level.

TABLE 21.3 Values of B and n for Use in Eq. (21.10)

Reynolds number range	B	n
1,000–6,000	0.409	0.531
6,000–30,000	0.212	0.606
30,000–100,000	0.139	0.806

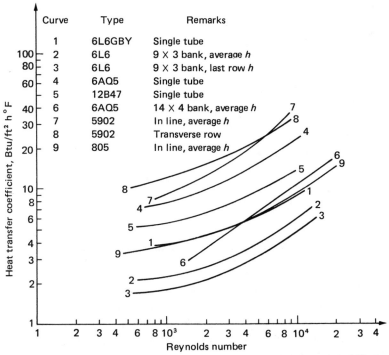

Curve	Type	Remarks
1	6L6GBY	Single tube
2	6L6	9 × 3 bank, average *h*
3	6L6	9 × 3 bank, last row *h*
4	6AQ5	Single tube
5	12B47	Single tube
6	6AQ5	14 × 4 bank, average *h*
7	5902	In line, average *h*
8	5902	Transverse row
9	805	In line, average *h*

FIG. 21.19 Data for cross flow cooling of vacuum tubes with air [9].

It is most important to note the way the Reynolds number and temperature difference are defined. All of the data in Fig. 21.19 are correlated with a Reynolds number based on the inlet film temperature, that is, the arithmetic mean of the inlet air temperature and the tube bulb or surface temperature. All data have been correlated on a temperature difference between tube bulb and inlet air. This form of data correlation will cause no trouble if care is exercised.

Robinson's data, Eq. (21.10) and Tables 21.2 and 21.3, are to be used on the basis of properties evaluated at the film temperature. Again, this will be the arithmetic mean of the inlet air and surface temperatures. Robinson's Reynolds numbers are formed using the cylinder diameter as the characteristic dimension.

All data presented here, due to Welsh and Robinson, are based on the temperature difference between surface and inlet air. When the heat transfer coefficient is obtained, the component temperature rise above ambient or environment must be calculated from

$$\theta = T_S - T_1 = \frac{q}{hS} \tag{21.12}$$

21.9.2 Vacuum Tubes in Axial (Parallel) Flow

Vacuum tubes may be cooled by air flowing parallel to the axis of the tube. An arrangement by which such parallel flow may be obtained is shown in Fig. 21.20. For

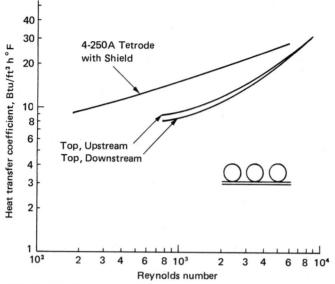

FIG. 21.20   Data for axial flow air cooling of vacuum tubes [9].

forced convection from any electron tube with air flowing upward and parallel to the axis of the tube surrounded by a shroud to confine the flow, Robinson and Jones [12] propose the correlation

$$Nu = 0.313\,Re^{0.6} \qquad\qquad (21.13)$$

where the thermal properties are evaluated at the inlet bulk air temperature. Welsh [9] has checked this equation using subminiature tubes and has obtained

$$Nu = 0.337\,Re^{0.55} \qquad\qquad (21.14)$$

Data for the 5902 tube in tube shields in parallel flow were obtained [9] at the Cornell Aeronautical Laboratories. These data, using the Reynolds number evaluated at the inlet film temperature, are plotted in Fig. 21.20. The curve for the 4-250 Beam Power Tetrode is also shown in Fig. 21.20.

21.10   MORE ILLUSTRATIVE EXAMPLES

As shown in Sec. 21.3, in forced convection applications, the effects of radiation diminish in comparison to the effects of forced convection. The examples that follow neglect radiation, which results in a somewhat more conservative design.

■   *EXAMPLE 5*

Approximate the average surface temperature of a single 6AQ5 tube placed in a duct 1 in (2.54 cm) wide by $2\frac{3}{8}$ in (6.033 cm) high. Air is the coolant, flowing at a rate of 0.00075 kg/s at 61.1°C. The tube dissipates 12.5 W.

SOLUTION

1 Tube dimensions:

Length: approximately $2\frac{1}{8}$ in (5.40 cm)
Diameter: approximately $\frac{45}{64}$ in (1.79 cm)

2 Surface area:

$$S = \pi d L + \frac{\pi}{4} d^2 = \pi(0.0179)(0.0540) + \frac{\pi}{4}(0.0179)^2$$

$$= 3.030 \times 10^{-3} + 2.505 \times 10^{-4} = 0.00328 \text{ m}^2$$

3 Free area for air flow:

$$A = (0.0603)(0.0254) - (0.0540)(0.0179) = 5.683 \times 10^{-4} \text{ m}^2$$

4 Mass flow:

$$G = \frac{W}{A} = \frac{(0.00075)(3600)}{5.683 \times 10^{-4}} = 4751.1 \text{ kg/m}^2 \text{ h}$$

5 Temperatures:

Assume: $T_S = 160.0°C$
Given: $T_1 = 61.1°C$
Compute: $T_f = \frac{1}{2}(160.0 + 61.1) = 110.6°C$

6 Reynolds number:

$$d = 0.0179 \text{ m}$$

$$\mu = 0.0774 \text{ kg/m h} \quad (\text{Fig. 6.10 at } T_f = 110.6°C)$$

$$\text{Re} = \frac{dG}{\mu} = \frac{(0.0179)(4751.1)}{0.0774} = 1096$$

7 Heat transfer coefficient:

$$h = 44.8 \text{ W/m}^2 \text{ °C} \quad (\text{Fig. 21.19, curve 4, at Re} =$$

$$1096) \, (h = 7.89 \text{ Btu/ft}^2 \text{ h °F})$$

8 Temperature rise:

$$\theta = \frac{q}{hS} = \frac{12.5}{(44.8)(0.00328)} = 85.1°C$$

9 Average surface temperature:

$$T_S = T_1 + \theta = 61.1 + 85.1 = 146.2°C$$

10 Verify assumed bulb temperature:

Calculated: $T_S = 146.2°C$
Assumed: $T_S = 160.0°C$

Recalculation is not necessary. The average bulb temperature will be 146.2°C.

■   **EXAMPLE 6**

In Example 5, how much air would be required if the average tube bulb were permitted to be 130°C?

SOLUTION

1 Tube dimensions known.
2 Surface area known.
3 Free area for flow known.
4 Temperature rise:

$$\theta = T_S - T_1 = 130 - 61.1 = 68.9°C$$

5 Film temperature:

$$T_f = \tfrac{1}{2}(130 + 61.1) = 95.6°C$$

6 Heat transfer coefficient:

$$h = \frac{q}{S\theta} = \frac{12.5}{0.00328(68.9)} = 55.3 \text{ W/m}^2 \text{ °C (9.74 Btu/ft}^2 \text{ h °F)}$$

7 Reynolds number. From curve 4, Fig. 21.20, at $h = 9.74$ Btu/ft^2 h °F,

$$\text{Re} = 2000$$

8 Air flow:

$$\text{Re} = \frac{dG}{\mu} = \frac{dW}{\mu A}$$

$$W = \frac{\mu A \text{ Re}}{d}$$

$\mu = 0.0750$ kg/h m   (Fig. 6.10 at $T_f = 95.6°C$)

$$W = \frac{0.0750(5.683 \times 10^{-4})(2000)}{0.0179} = 4.773 \text{ kg/h} = 0.00133 \text{ kg/s}$$

■   **EXAMPLE 7**

A cylindrical component 1 in (2.54 cm) in diameter and 2 in (5.08 cm) high is installed in a duct $1\tfrac{3}{8}$ in (3.493 cm) wide and $2\tfrac{1}{4}$ in (5.715 cm) wide. Air at 65°C at a flow rate of 0.056 kg/min flows through the duct. Estimate the surface temperature for a dissipation of 17.5 W.

SOLUTION

1 Cylinder dimensions:

Length: 5.08 cm
Diameter: 2.54 cm

2 Surface area:

$$S = \pi dL + \frac{\pi}{4} d^2 = \pi(5.08)(2.54) + \frac{\pi}{4}(2.54)^2 = 40.537 + 5.067$$

$$= 45.604 \text{ cm}^2 \quad \text{or} \quad 0.00456 \text{ m}^2$$

3 Free area for flow:

$$A = (5.715)(3.493) - Ld = 19.960 - 5.08(2.54) = 7.059 \text{ cm}^2$$

$$= 0.000706 \text{ m}^2$$

4 Mass flow:

$$G = \frac{W}{A} = \frac{0.056(60)}{0.000706} = 4761.6 \text{ kg/m}^2 \text{ h}$$

5 Temperatures:

Assume: $T_S = 138°C$
Given: $T_1 = 65°C$
Compute: $T_f = \frac{1}{2}(138 + 65) = 101.5°C$

6 Reynolds number:

$$d = 0.0254 \text{ m}$$

$$\mu = 0.0756 \text{ kg/h m} \quad \text{(Fig. 6.10 at } T_f = 101.5°C)$$

$$\text{Re} = \frac{dG}{\mu} = \frac{0.0254(4761.6)}{0.0756} = 1599.8 \approx 1600$$

7 Heat transfer coefficient:

$$\text{Nu} = FB(\text{Re})^n \quad \text{[Eq. (21.10)]}$$

$$B = 0.409 \text{ at Re} = 1600 \quad \text{(from Table 21.3)}$$

$$n = 0.531 \text{ at Re} = 1600 \quad \text{(from Table 21.3)}$$

$$F = 1 + \sqrt{\frac{d}{w}}$$

$$d = 2.54 \text{ cm} \quad w = 3.493 \text{ cm}$$

$$F = 1 + \sqrt{\frac{2.54}{3.493}} = 1 + \sqrt{0.727} = 1 + 0.854 = 1.853$$

Hence

$$\text{Nu} = \frac{hd}{k} = FB(\text{Re})^n = 1.853(0.409)(1600)^{0.531} = 38.1$$

$$k = 0.0317 \text{ W/m}\,^\circ\text{C} \quad (\text{Fig. 6.10 at } T_f = 101.5^\circ\text{C})$$

$$h = \text{Nu}\left(\frac{k}{d}\right) = 38.1\left(\frac{0.0317}{0.0254}\right) = 47.51 \text{ W/m}^2\,^\circ\text{C}$$

8 Temperature rise:

$$\theta = \frac{q}{hS} = \frac{17.5}{(47.51)(0.00456)} = 80.8^\circ\text{C}$$

9 Average surface temperature:

$$T_S = T_1 + \theta = 65 + 80.8 = 145.8^\circ\text{C}$$

10 Verify assumed surface temperature:

Calculated: $T_S = 145.8^\circ\text{C}$
Assumed: $T_S = 138^\circ\text{C}$

Recalculation is not necessary. The average surface temperature will be 145.8°C.

EXAMPLE 8

A chassis contains 12 vacuum tubes $\frac{3}{4}$ in (1.905 cm) in diameter and $1\frac{3}{4}$ in (4.445 cm) high mounted in an in-line bank of three transverse and four longitudinal rows. A duct carrying air at a flow rate of 0.225 kg/min covers the tubes. Tube spacings are $1\frac{1}{2}$ in (3.81 cm) in both longitudinal and transverse directions. Dimensions of the duct are width $4\frac{1}{2}$ in (11.43 cm), and height 2 in (5.08 cm). Air enters the duct at 30°C, and the 12 tubes dissipate 16 W each. Estimate the average tube surface temperature.

SOLUTION

1 Tube dimensions:

Length: 4.445 cm
Diameter: 1.905 cm

2 Surface area:

$$S = 12\left(\pi dL + \frac{\pi}{4}d^2\right) = 12\left[\pi(1.905)(4.445) + \frac{\pi}{4}(1.905)^2\right]$$

$$= 12(26.602 + 2.850) = 12(29.452) = 354.429 \text{ cm}^2 \quad \text{or} \quad 0.0353 \text{ m}^2$$

3 Free area for flow:

$$A = (5.08)(11.43) - 3Ld = 58.064 - 3(4.445)(1.905) = 58.064 - 25.403$$

$$= 32.661 \text{ cm}^2 \quad \text{or} \quad 0.00327 \text{ m}^2$$

4 Mass flow:

$$G = \frac{W}{A} = \frac{0.225(60)}{0.00327} = 4.133.3 \text{ kg/m}^2\,\text{h}$$

5 Temperatures:

Assume: $T_S = 130°C$
Given: $T_1 = 30°C$
Compute: $T_f = \frac{1}{2}(130 + 30) = 80°C$

6 Reynolds number:

$$d = 0.01905 \text{ m}$$

$$\mu = 0.0723 \text{ kg/h m} \quad \text{(Fig. 6.10 at } T_f = 80°C\text{)}$$

$$\text{Re} = \frac{dG}{\mu} = \frac{(0.01905)(4133.3)}{0.0723} = 1088$$

7 Heat transfer coefficient:

$$\text{Nu} = FB(\text{Re})^n \quad \text{Eq. (21.10)}$$

From Table 21.3,

$$B = 0.409 \text{ at Re} = 1088$$

$$n = 0.531 \text{ at Re} = 1088$$

From Table 21.2,

$$F = \left(1 + \sqrt{\frac{1}{S_T}}\right)\left[1 + \left(\frac{1}{S_L} - \frac{0.872}{S_L^2}\right)\left(\frac{1.81}{S_T^2} - \frac{1.46}{S_T} + 0.318\right)\right.$$

$$\left. \times (\text{Re}^{0.526 - 0.354/S_T})\right]$$

where $S_L = S_T = 3.81/1.905 = 2.00$.

$$1 + \sqrt{\frac{1}{S_T}} = 1 + \sqrt{\frac{1}{2}} = 1 + 0.707 = 1.707$$

$$\frac{1}{S_L} - \frac{0.872}{S_L^2} = \frac{1}{2} - \frac{0.872}{4} = 0.500 - 0.218 = 0.282$$

$$\frac{1.81}{S_T^2} - \frac{1.46}{S_T} + 0.318 = \frac{1.81}{4} - \frac{1.46}{2} + 0.318 = 0.453 - 0.730 + 0.318$$

$$= 0.041$$

$$0.526 - \frac{0.354}{S_T} = 0.526 - \frac{0.354}{2} = 0.526 - 0.177 = 0.349$$

$$(\text{Re})^{0.349} = (1088)^{0.349} = 11.48$$

$$F = 1.707[1 + (0.282)(0.041)(11.48)]$$

$$= 1.707(1 + 0.131) = 1.707(1.131) = 1.931$$

Hence

$$\text{Nu} = \frac{hd}{k} = FB(\text{Re})^n = 1.931(0.409)(1088)^{0.531} = 32.31$$

$$k = 0.0301 \text{ W/m} °C \quad (\text{Fig. 6.10 at } T_f = 175°F)$$

$$h = \text{Nu}\left(\frac{k}{d}\right) = 32.31\left(\frac{0.0301}{0.01905}\right) = 51.04 \text{ W/m}^2 \, °C$$

8   Temperature rise:

$$\theta = \frac{q}{hS} = \frac{12(16)}{(51.04)(0.0353)} = 106.6°C$$

9   Average surface temperature:

$$T_S = T_1 + \theta = 30 + 106.6 = 136.6°C$$

10   Verify assumed surface temperature:

Calculated: $T_S = 136.6°C$
Assumed: $T_S = 130°C$

Recalculation is not necessary. The average surface temperature will be 136.6°C.

21.11   NOMENCLATURE

Roman Letter Symbols

A	area, m²
b	fin or spine height, m
B	a factor defined in Eqs. (21.9) and (21.10), dimensionless
c	specific heat, J/kg °C
d	diameter, m
F	a factor defined where used, dimensionless
G	mass flow or velocity, kg/m² h
h	heat transfer coefficient, W/m² °C
k	thermal conductivity, W/m K
L	length, m
m	fin performance factor, m⁻¹
Nu	Nusselt number, dimensionless

P	perimeter, m
Pr	Prandtl number, dimensionless
q	heat flow, W
r	radius, m
R	thermal resistance, °C/W
Re	Reynolds number, dimensionless
S	surface, m^2; or spacing ratio, dimensionless
T	temperature, °C or K
w	duct width, m
W	flow rate, kg/h

Greek Letter Symbols

Δ	indicates change in variable
ϵ	emissivity, dimensionless
η	fin efficiency, dimensionless
θ	temperature difference, °C or K
μ	dynamic viscosity, kg/h m
ρ	radius ratio, dimensionless
σ	Stefan-Boltzmann constant, 5.67×10^{-8} W/m^2 K^4

Subscripts

a	indicates average condition
A	indicates arrangement factor
b	indicates base of configuration
c	indicates resistance to surrounding environment
co	indicates contact resistance
cv	indicates convection resistance
e	indicates edge condition
f	indicates film condition
J	indicates junction-to-case resistance
L	indicates longitudinal spacing ratio
m	indicates mean condition or mounting resistance
p	indicates constant-pressure condition or profile area
r	indicates radiation resistance
R	indicates receiver of radiation
s	indicates resistance to surroundings
S	indicates source of radiation
w	indicates resistance of mounting washer
0	indicates base radius
ϵ	indicates emissivity factor

21.12   REFERENCES

1   Kraus, A. D., Heat Flow Theory, *Electrical Manufacturing*, vol. 63, pp. 123–142, April 1959.
2   Kraus, A. D., The Use of Steady State Electrical Network Analysis to Solve Heat Flow Problems, ASME Paper 58-HT-14, Chicago, 1958.

3 Luft, W., Taking the Heat off Semiconductor Devices, *Electronics*, vol. 32, pp. 53–56, June 12, 1959.
4 Kraus, A. D., The Efficiency of a Transistor Cap as a Heat Dissipator, ASME Paper 58-HT-15, Chicago, 1958.
5 Kern, D. Q., and Kraus, A. D., *Extended Surface Heat Transfer*, McGraw-Hill, New York, 1972.
6 Drexel, W., Convection Cooling, *Sperry Eng. Rev.*, vol. 14, pp. 25–30, December 1961.
7 McAdams, J. C., Heat Dissipating Electron Tube Shields and Their Relation to Tube Life and Equipment Reliability, International Electronic Research Corp., Burbank, Calif., 1959.
8 International Electronic Research Corporation, Comparative Tests–Forced Air Cooling vs Heat Dissipating Tube Shields, Test Rept. No. 116, 1959.
9 Welsh, J. P., *Handbook of Methods of Cooling Air Force Ground Electronic Equipment*, RADC-TR-58-126, Rome Air Development Center–Griffiss Air Force Base, Rome, N.Y., 1959.
10 Hilpert, R., Warmeabgue von Geheizten Drähten und Rohren in Lufstrom, *Forsch. Ing.-Wes.*, vol. 4, pp. 215–224, 1933.
11 Robinson, W., Han, L. S., Essig, R. H., and Heddleson, C. F., Heat Transfer and Pressure Drop Data for Circular Cylinders in Ducts and Various Arrangements, Ohio State University Research Foundation Rept. No. 41, Columbus, Oh., 1951.
12 Robinson, W., and Jones, C., Cooling of Electronic Components by Various Methods, Ohio State University Research Foundation Rept. No. 44, Columbus, Oh., 1952.

22

■ liquid cooling of microwave equipment

22.1 INTRODUCTION

The frequent demands for high radiated power in the microwave spectrum, whether for radio broadcasts, wireless control of airborne vehicles, or ranging and detection of remote objects, has led to the development of many high-energy-density microwave devices. With the exception of high-power tube collectors, heat fluxes in microwave devices range from 5 to 30 W/cm² at maximum allowable temperatures of approximately 130 to 150°C. Gyrotrons capable of peak power levels in excess of 1000 MW and long-pulse, 400-kW, X-band radar wave guides [1], as well as water loads or "energy dumps" for such systems [2], typify the high heat flux components in conventional radiogrequency (RF) technology. Because of the inefficiencies inherent in RF wave generation, the operation of these devices is accompanied by very significant heat dissipation levels. Stripline wave guides and power diodes place similar demands on the thermal control systems of more advanced microwave equipment employing distributed RF generation and electronically steered antennae.

As suggested by Fig. 1.10, the combination of high heat fluxes and moderate temperature differences can best be accommodated by convection and/or boiling (or evaporation) heat transfer with flowing water or other favorable liquids and, indeed, liquid cooling of microwave tubes dates back to the pioneering efforts of Mouromtseff in 1942 [3]. In that year, Mouromtseff used subcooled boiling of water at a flow rate of 4 gal/min to successfully remove 11.3 kW from a single microwave power tube [3]. Klystron and cross field amplifier (CFA) tubes for high-power radar systems are also often cooled by pool boiling [4] and flow boiling of water [5], and the cooling system for the Twystron tube collector developed by Varian Associates was designed to dissipate a peak heat flux of 2000 W/cm² to 320 gal/min of high-pressure water flowing through a large number of longitudinal slots [6].

As demonstrated by London [7], the use of compact fin arrays and/or augmentation techniques (see Chap. 19) can make air cooling of RF power tubes possible. In addition, some attention has been focused on the development of heat pipe-cooled collectors. However, much of the design and development effort in this sector of electronic thermal control is devoted to liquid cooling. The fundamentals of forced convection and boiling heat transfer, on which such thermal activity is based, were examined in detail in Part II, and the application of these fundamentals to the design of thermal control systems was presented in Part III. This chapter focuses on thermal design considerations that are of special importance in microwave devices, including the selection of optimum coolants and the effect of RF absorption in flowing liquids. A detailed example of the cooling of a traveling wave tube and a comparison between two different thermomechanical, electronically steered antenna designs are presented.

22.2 SOME CRITERIA FOR COOLANT FLUID SELECTION

A coolant fluid for electronic components must, to begin with, have acceptable heat transfer properties. Selection of the fluid on this basis is discussed in Sec. 22.3. The present discussion revolves around other requirements that must be met by the fluid before it can be considered for use as a coolant.

Table 22.1 lists the criteria by which a coolant may be judged. These are discussed in the paragraphs that follow.

22.2.1 Dielectric Strength

A coolant is frequently required to contact an electronic component whose surface is maintained at some potential. When this occurs, the dielectric properties of the fluid become most important. Fluids such as water that have low dielectric strength must be dropped immediately from further consideration in direct contact or immersion cooling applications. Information on the dielectric properties of the fluid [8, 9] must be obtained and, when possible, the fluid should also be tested.[1] Manufacturers' catalogs can be consulted, or some standard test such as ASTM D-877 may be employed. In testing, one should consider the intended use of the fluid: the temperature level, the operating pressure, and the shape, spacing, and material of the components to be cooled.

22.2.2 Inertness and Compatibility

Use of a coolant that reacts chemically with any portion of the liquid loop must be discouraged. This consideration refers to all portions of the cooling system and is not confined to the components that are to be cooled. Particular attention should be given to chemical inertness and compatibility with respect to seals and gaskets. As a general rule, the inertness of coolants in the presence of metals and plastics must be considered.

---

[1] Some data may be found in Table 16.1.

TABLE 22.1 Coolant Property Criteria List

Criterion	See Section
Dielectric strength	22.2.1
Inertness and compatibility	22.2.2
Thermal decomposition and impurities	22.2.3
Effects of moisture	22.2.4
Pour and flash points	22.2.5
Flammability	22.2.6
Toxocity	22.2.7
Surface tension	22.2.8
Heat transfer properties	22.3
Specific heat	
Viscosity	
Thermal conductivity	
Density	

22.2.3   Thermal Decomposition and Impurities

A coolant must be used in accordance with the manufacturer's recommendations on service temperature and operating pressure. Many coolants decompose at higher temperatures and lose the advantageous heat transfer properties they may possess at conventional temperatures. Frequently thermal decomposition is accelerated in the presence of impurities such as carbon dioxide and oxygen [10]. Consideration may therefore have to be given to the use of a purification and deionization system or oxygen removal bed, which entail weight and power penalties to the overall system.

Most military specifications provide an upper limit on particle impurities. If such particles are not filtered from the system, they can deposit themselves on a high-heat-flux area of the component. This could cause an additional temperature rise between component and coolant and cause eventual destruction of the component. Particles may be removed from a liquid cooling system by means of a micron filter.

The problem of impurities, whether foreign matter or chemical, such as entrained catalysts or oxidizing agents, is a most important one. The effect of these impurities on system clogging and thermal decomposition of the coolant fluid must be investigated before the cooling system and coolant fluid are proposed.

22.2.4   Effects of Moisture

Moisture is a particular type of impurity responsible for problems all its own. Its effect is important because of the cost and weight penalties resulting from an attempt to design a moisture-proof system. The effects of the presence of moisture in the coolant must be investigated before the coolant is selected. Such investigation must pertain to the degradation of dielectric properties, hydrolysis, and formations of acids and salts that are likely to cause corrosion in the cooling system, as well as the danger of vapor explosions associated with bulk boiling of water droplets in the high-temperature liquid.

22.2.5   Pour and Flash Points

It is an established fact that because the viscosity of liquids decreases with temperature, the heat transfer properties of a liquid decrease with decreasing operating temperature. Although the designer should attempt to select a fluid with good viscosity characteristics, he or she should also guard against the condition where the coolant fluid will not flow.

The temperature at which a liquid will just flow is defined as the *pour point*. This usually pertains to viscous fluids. Water, for example, has the ability to flow at its melting (freezing) point. This fact is not mentioned to encourage the use of water as a coolant (it has other serious drawbacks in immersion cooling applications, such as poor dielectric characteristics), but serves to point out that the pour point has little meaning when applied to a nonviscous fluid. What is important is that operating temperatures approaching the pour point can render a cooling system inoperable. For example, a heat exchanger used to extract heat from a liquid must have a surface or wall temperature above the pour point. If not, an almost solid viscous layer will build up on the surfaces of the heat exchanger. This will result in a drastic degradation of heat exchanger effectiveness.

Coolant fluids having a high pour point must be avoided. Manufacturers' data must be consulted, and if no information can be obtained in this manner, the designer must determine the pour point (ASTM method D-97).

The *flash point* is defined as the lowest temperature at which sufficient vapor is generated to cause a flash when the vapor is brought into contact with a flame or other ignition possibility. A liquid having a low flash point must be discounted because of the associated logistic problems and handling risks.

22.2.6 Flammability

Flammability is related to the flash point, and use of a coolant fluid having this property may require a system that is far too heavy and complex to be of practical use. Flammability is a coolant property that will lead to problems during system filling and maintenance.

If the aforementioned problems can be tolerated and if tightness of the cooling system can be assured, a flammable coolant may be acceptable. For example, hydrogen, which has excellent heat transfer properties when compared with other gases, has been used for years in the cooling of central station power generators.

22.2.7 Toxicity

Toxic coolants must not be considered. Any coolant that is dangerous to personnel, either temporarily or permanently, cannot be used regardless of its heat transfer properties. A completely leakproof system is only an ideal, and even remote equipments will yield problems during filling, testing, and maintenance periods.

22.2.8 Surface Tension

Surface tension can have a profound effect on boiling heat transfer at the surface of an immersed component. Liquids having low surface tension are more likely to cause leakage problems (seals, gaskets, cracks, connectors), and use of such a liquid must be carefully considered from the overall system point of view.

22.3 FLUID SELECTION BASED ON HEAT TRANSFER PROPERTIES

The selection of a fluid on the basis of heat transfer properties is relatively simple once the heat transfer configuration is selected. Most often, because of the high rate of heat dissipation, the fluid is brought directly to the component and circulated through passages placed at an area of high heat dissipation. Hence, one is usually considering a problem of heat transfer in forced convection to a fluid in a duct or channel.

From a heat transfer viewpoint, coolants are rated by first considering the applicable heat transfer coefficient correlating equation. For heat transfer in ducts or channels, one may write a form of the McAdams equation for the heating of fluids A and B in turbulent flow [11]:

$$h_A = 0.023 \frac{k_A}{d_e} \left(\frac{\rho_A d_e V}{\mu_A} \right)^{0.8} \left(\frac{c_A \mu_A}{k_A} \right)^{0.4} \tag{22.1}$$

$$\text{and} \quad h_B = 0.023 \frac{k_B}{d_e} \left(\frac{\rho_B d_e V}{\mu_B} \right)^{0.8} \left(\frac{c_B \mu_B}{k_B} \right)^{0.4} \tag{22.2}$$

Suppose that the two fluids, A and B, are running through two passages of identical equivalent diameter and at identical velocities; then one may divide Eq. (22.1) by Eq. (22.2) to obtain

$$\frac{h_A}{h_B} = \frac{k_A}{k_B} \left(\frac{\rho_A \mu_B}{\rho_B \mu_A} \right)^{0.8} \left(\frac{c_A \mu_A k_B}{c_B \mu_B k_A} \right)^{0.4}$$

$$\frac{h_A}{h_B} = \frac{\rho_A^{0.8} k_A^{0.6} c_A^{0.4} / \mu_A^{0.4}}{\rho_B^{0.8} k_B^{0.6} c_B^{0.4} / \mu_B^{0.4}} \tag{22.3}$$

The Mouromtseff number is a dimensional grouping of fluid heat transfer properties defined as

$$\text{Mo} = \frac{\rho^a k^b c^c}{\mu^d} \tag{22.4}$$

and for the case of fluid flowing in a duct or channel under turbulent conditions where the fluid is being heated, the values of the exponents are determined by comparing Eqs. (22.3) and (22.4).

Note that the items contained in the Mouromtseff number are definite functions of temperature. There is therefore a way of comparing fluids on the basis of temperature (from the values of $\rho, k, c,$ and μ) and on the basis of configuration (the values of the exponents $a, b, c,$ and d). Hence, it can be said, for any fluid, that

$$h = f \left(\frac{\rho^a k^b c^c}{\mu^d} \right) = f(T) \tag{22.5}$$

as long as it is realized that the comparison is being made under conditions of identical physical configurations at identical velocities.

Figure 22.1 presents a plot of the Mouromtseff number for three hypothetical[2] cooling fluids, water, and air. Observe how it can definitely be established which fluid has the better heat transfer properties. Indeed, Fig. 22.1 shows that, as operating temperatures vary, the choice of the fluid may vary.

Because the heat transfer properties affect many system requirements such as fluid flow rate, fluid "fill" weight, weight and volume of pipes or ducts, power requirements, and size, weight, and volume of heat exchanger surfaces, other indices may be proposed for fluid comparison. One such index is the ducting weight index [12]:

$$\left[\left(1 + \frac{p}{f} \right)^2 - 1 \right] \left(\frac{\mu^e}{\rho^f c^g} \right) \tag{22.6}$$

---

[2] Hypothetical in the sense that these fluids do not exist.

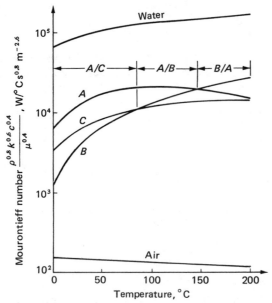

FIG. 22.1   Mouromtseff number for three hypothetical liquid coolants, air, and water. The fractions designate choice of fluid: numerator, first choice; denominator, second choice.

22.3.1   Heat Transfer Properties of Liquid Coolants

Many liquid coolants satisfy the requirements given in Sec. 22.2, and have acceptable heat transfer properties [13, 14] . For the reader's reference, these properties for three such fluids are given in Figs. 22.2 to 22.4. Reference data for the construction of these figures are given in Table 22.2.

22.4   MICROWAVE ABSORPTION IN WATER LOADS

Many high-power radar systems, in both military and commercial applications, include one or several water loads in which unwanted RF power can be dissipated. In such devices, which take the form of an impedance-matched wave-guide termination section, the microwave power is beamed through an RF window, typically an actual missile or aircraft radome, directly into a flowing stream of water. The complex interaction between the RF field and the dipole water molecules results in the absorption of the microwave energy and the generation of heat within the water.

In the absence of deliberate beam shaping, a highly nonuniform power distribution is encountered along the water load surfaces. In a perhaps typical example shown schematically in Fig. 22.5, the maximum power flux was found to occur at 35 cm from the tip and to equal nearly 700 W/cm^2, even when time averaged over many pulses. The peak power flux, during a single pulse, approached 151 kW/cm^2 at the same location [2] .

Various system constraints, including break-down voltage, voltage-standing wave

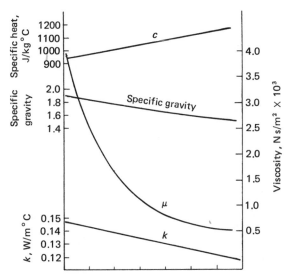

FIG. 22.2 Thermal properties, coolant fluid FC-75. (Extracted from data provided by Minnesota Mining and Manufacturing Co.)

ratio (VSWR), and permissible mechanical stress levels in the window (or radome) material, combine to dictate both a maximum allowable surface temperature and permissible temperature gradients in the window and usually require that no boiling occurs in the water channel. As a result, a precise knowledge of the temperature profile in the fluid and especially along the channel wall is crucial to the successful design of such devices.

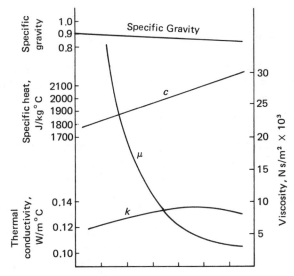

FIG. 22.3 Thermal properties, coolant fluid Coolanol 45. (Extracted from data provided by Monsanto Chemical Co.)

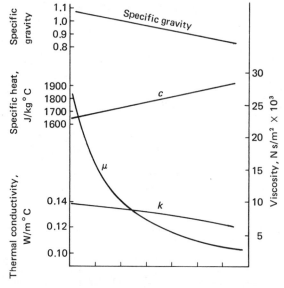

FIG. 22.4 Thermal properties, coolant fluid DC-331. (Extracted from data provided by Dow Corning Co.)

Because of the nature of the RF absorption function and its complex dependence on the velocity and temperature fields in the water channel, a general thermal solution of this problem is not presently available. The subsequent formulation and solution of the temperature field for uniform wall irradiation of a fully developed, laminar rectangular flow is meant to establish a workable approach to water load thermal analysis while presenting numerical results of relevance in a sector of the appropriate parametric field.

22.4.1   Idealized Geometry and Absorption Function

To obtain the required temperature profiles, it is necessary first to describe the manner in which microwave energy is absorbed in water and then to introduce that result into the appropriate energy equation. The solution of the energy equation with this internal heat generation term yields the desired temperature profiles.

The water load geometry considered is shown in Fig. 22.6 and consists of two parallel, thermally adiabatic surfaces separated by distance $2y_o$, enclosing a flowing stream of water, and irradiated uniformly from one side by microwave intensity I_1. Surface 1 represents the RF window, that is, the interface between the wave guide and the flowing stream, and is assumed to be totally transparent to RF radiation. As would

TABLE 22.2   Some Representative Liquid Coolants

Fluid	Manufacturer	Reference	See Figure
FC-75	Minnesota Mining and Manufacturing Co.	Catalog Bulletin	22.2
Coolanol 45	Monsanto Chemical Co.	Bulletin AV-3	22.3
DC-331	Dow Corning Co.	Bulletin 05-031	22.4

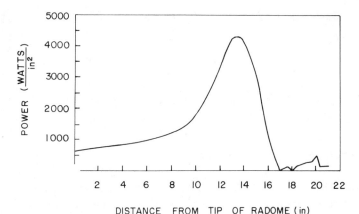

DISTANCE FROM TIP OF RADOME (in)

FIG. 22.5 Power distribution along water load surfaces.

be the case for any practical water load design, more than 90% of the incident power is assumed to be absorbed in the $2y_o$ distance separating the window and the external shroud (surface 2) so that $I_2 < 0.1 I_1$. The microwave power reaching the shroud will be partially absorbed and partially reflected back into the flowing stream. However, to facilitate the solution of the energy relations, this small thermal increment will be neglected henceforth. The physical properties (k, ρ, c, μ) of the water flowing between the parallel surfaces are assumed constant, and the present discussion is limited to fully developed laminar flow.

Under a wide variety of conditions, the volumetric heat dissipation function associated with the liquid/RF interaction decreases exponentially from the transparent wall of the liquid channel and, with the nomenclature of Fig. 22.6, is expressible as

$$W_l = 0.23\alpha' I_1 e^{-0.23 l \alpha'_{m,l}} \tag{22.7}$$

The attenuation factor, α', is itself a strong function of temperature and frequency [15] but over a modest temperature range and at a fixed electromagnetic frequency can be represented exponentially as $\alpha' = b e^{-CT}$.

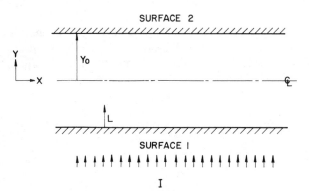

FIG. 22.6 Coordinate system for irradiated channel.

Introducing this relation into Eq. (22.7), the internal heat generation rate is given by

$$W_l = (0.23be^{-CT})(I_1 e^{-0.23lbe^{-CT_{m,l}}})$$

(22.8)

where it is to be noted that W_l is dependent on both the local temperature T and the average fluid temperature across the path length l, that is, $T_{m,l}$.

22.4.2 Thermofluid Formulation

By appropriate simplification of the Navier-Stokes (momentum conservation) equations for fully developed, constant-property laminar flow between stationary parallel surfaces, it is possible to obtain the well-known laminar velocity profile [16]

$$V = \frac{3}{2} V_m \left[1 - \left(\frac{y}{y_o} \right)^2 \right]$$

(22.9)

The general differential energy equation, which relates the rate of temperature increase of a fluid element moving in a stream to the heat conducted across its boundaries and internally generated, for two-dimensional, steady-state, fully developed flow in a rectangular channel can be expressed as

$$k \left(\frac{\partial^2 T}{\partial x^2} + \frac{\partial^2 T}{\partial y^2} \right) - \frac{3}{2} V_m \left[1 - \left(\frac{y}{y_o} \right)^2 \right] \rho c_p \frac{\partial T}{\partial x} = -W$$

(22.10)

The expression for W, as given by Eq. (22.8), with l replaced by $y_o + y$ (see Fig. 22.6), can now be inserted to complete the formulation of the energy equation:

$$k \left(\frac{\partial^2 T}{\partial x^2} + \frac{\partial^2 T}{\partial y^2} \right) - \frac{3}{2} V_m \left[1 - \left(\frac{y}{y_o} \right)^2 \right] \rho c_p \frac{\partial T}{\partial x}$$
$$= -(0.23 I_1 be^{-CT})(e^{-0.23(y_o+y)be^{-CT_{m,y}}})$$

(22.11)

With the exception of the radiationally opaque and near transparent limits, discussed in the next section, Eq. (22.11) is not easily amenable to solution, and numerical techniques must be used. To maximize the utility of the solution, Eq. (22.11) can be nondimensionalized by defining the following parameters:

$$\bar{V} = \frac{\rho c_p y_o V_m}{k} \qquad \bar{X} = \frac{X}{y_o}$$

$$\bar{T} = \frac{kT}{b I_1 y_o^2} \qquad \bar{Y} = \frac{y}{y_o}$$

$$\bar{C} = \frac{I_1 b y_o^2 C}{k} \qquad \bar{b} = y_o b$$

(22.12)

Introducing these parameters into Eq. (22.11), the energy equation takes the form

$$\frac{\partial^2 \bar{T}}{\partial \bar{X}^2} + \frac{\partial^2 \bar{T}}{\partial \bar{Y}^2} - \frac{3}{2} \bar{V}(1 - \bar{Y}^2) \frac{\partial \bar{T}}{\partial \bar{X}} = -(0.23 e^{-\bar{C}\bar{T}})(e^{-0.23\bar{b}(1+\bar{y})} e^{-\bar{C}\bar{T}m,y}) \quad (22.13)$$

An interactive computer program for solving elliptic boundary-value problems [17] was used interactively to generate the temperature fields defined by the solution of Eq. (22.13).

22.4.3 Temperature Profiles

NUMERICAL SOLUTION

The nondimensional temperature $\bar{T}$ in Eq. (22.13) can be seen to depend on the non-dimensional distance $\bar{Y}$, attenuation factor $\bar{b}$, and thermal coefficient of attenuation $\bar{C}$. Consequently, the temperature profiles across the channel, that is, $\bar{T}(\bar{Y})$, were calculated for predetermined values of $\bar{b}$ and $\bar{C}$, spanning the parametric range of interest [18]. Figure 22.7 highlights the influence of $\bar{C}$ on the temperature profile and reveals that this parameter has only a marginal influence on $\bar{T}$ in the channel. Alternatively, as can be seen in Fig. 22.8, both the temperature profile and the surface temperature are sensitive to the value of the dimensionless attenuation factor $\bar{b}$.

OPAQUE LIMIT

The form of the heat generation function, Eq. (22.8), suggests that two simple limits can be established for the temperature profile in the channel. At one limit, corresond-

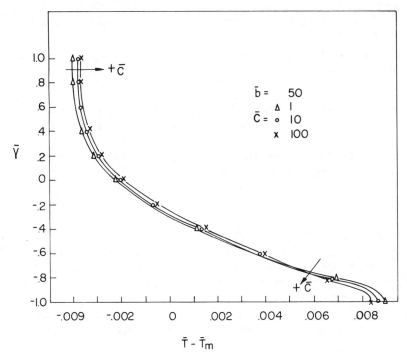

FIG. 22.7 Temperature profiles for irradiated channel, $\bar{C}$ varying.

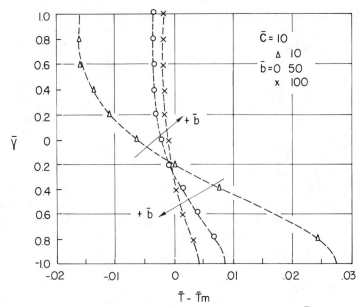

FIG. 22.8   Temperature profiles for irradiated channel, $\bar{b}$ varying.

ing to very high values of b (or $\bar{b}$), the heat generation rate is strongly asymmetric and W falls rapidly with increasing distance from the RF beam entry surface. An upper limit on the surface temperature can thus be obtained by assuming the fluid to be radiationally opaque and all heat generation to occur in a vanishingly thin fluid layer or essentially at the channel wall.

Solution of the energy equation, Eq. (22.10), with $W = 0$ and $k(\partial T/\partial y)_{-y_o} = I_1$, yields the temperature profile

$$T - T_m = \frac{I_1 y_o}{k}\left[\frac{y}{y_o} + \frac{3}{4}\left(\frac{y}{y_o}\right)^2 - \frac{1}{8}\left(\frac{y}{y_o}\right)^4 - 0.145\right] \qquad (22.14)$$

Solving Eq. (22.14) for the wall temperature and reexpressing this relation in terms of the nondimensional parameters, the opaque limit on $\bar{T}_1 - \bar{T}_m$ is found to be

$$(\bar{T}_1 - \bar{T}_m)_{\text{opaque}} = 0.74\bar{b} \qquad (22.15)$$

NEAR-TRANSPARENT LIMIT

For low values of b (or $\bar{b}$), the fluid is nearly transparent and only mildly attenuates the incident radiation. Consequently, if the second channel wall is also RF transparent, the radiation traversing the channel will result in nearly uniform heat generation. If, on the other hand, the channel walls are internally reflective, the radiation arriving at surface 2 will be reflected back toward surface 1 and the entering radiation will eventually (following many multiple reflections) be absorbed nearly uniformly in the fluid.

Returning to Eq. (22.10) and setting W equal to a constant, $W = I_1/2y_o$, the channel temperature profile is found to be

$$T - T_m = \frac{I_1 y_o}{k} \left[\frac{1}{8} \left(\frac{y}{y_o} \right)^2 - \frac{1}{16} \left(\frac{y}{y_o} \right)^4 - 0.0196 \right] \tag{22.16}$$

In addition to serving as a convenient limiting expression, Eq. (22.16) was used to check the accuracy of the numerical solution. One hundred iterations on a relatively coarse nodal matrix, of 11 nodes across and 20 nodes along the channel wall, yielded less than 8% discrepancy between the numerical solution and analytical values calculated at the nodal points. Since the number of iterations is proportional to the number of nodes, this nodal matrix was employed in all the numerical calculations.

Solving Eq. (22.16) for $y = y_o$ and reexpressing in terms of the dimensionless parameters, the lower near-transparent limit on the wall temperature is found to be $(\bar{T}_1 - \bar{T}_m)_{\text{transparent}} = 0.0428/\bar{b}$.

The two limiting equations for $(\bar{T}_1 - \bar{T}_m)$ are plotted in Fig. 22.9 and are seen to bound properly the numerically calculated values of the dimensionless wall-to-mean fluid temperature difference. However, the factor of approximately 35 separating the lower and upper limits on $(\bar{T}_1 - \bar{T}_m)$ suggests that for all but extreme values of $\bar{b}$, that is, $\bar{b} \ll 1$ or $\bar{b} \gg 100$, the numerical results must be used to determine the desired temperature.

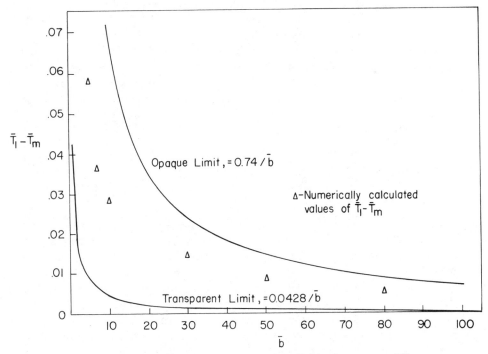

FIG. 22.9 Bounding and numerical values for surface-fluid temperature difference.

GENERAL CONSIDERATIONS

Because of the particular choice of nondimensionalizing parameter, increasing attenuation, that is, higher $\bar{b}$ values, results in smaller values of $\bar{T}_1 - \bar{T}_m$. However, a review of the pertinent equations shows that, as anticipated, increasing b leads to a progressively larger temperature difference, $T_1 - T_m$, as the solutions approach the radiationally opaque limit.

Emphasis throughout this discussion has been placed on the temperature profile relative to the mean liquid temperature in the channel. The reader is reminded that, for an axially invariant heating function, the axial gradient equals

$$\frac{dT_m}{dx} = \frac{I_1}{2\rho c_p y_o V_m} \tag{22.17}$$

Consequently, once the inlet liquid temperature is known, Eq. (22.17) suffices to calculate T_m at every location in the channel.

22.5 AN EXAMPLE OF THE DESIGN OF A TWT LIQUID COOLING SYSTEM

When cooling microwave components, one frequently encounters the problem of determining the liquid (it has been established that a liquid will be used) flow rate and the inlet temperature necessary to keep a component with a prescribed heat duty at or below a given maximum temperature. Alternatively, the problem may concern the design of a heat dissipator when a fluid flow rate, a fluid inlet temperature, a required heat dissipation, and a "not to exceed" component temperature is specified.

Figure 22.10 shows a possible configuration of the collector or anode of a traveling wave tube (TWT). Because the heat dissipation is so high, it has been determined that a four-pass liquid loop should be provided. This permits a high mass flow or velocity through the collector of the TWT with an associated high heat transfer coefficient.

Specifications are as follows:

Heat dissipation: 1500 W
Coolant: Coolanol 45
Inlet temperature: 76.7°C (170°F)
Required component temperature: 150°C (302°F)
Configuration: as in Fig. 22.10

The allowable surface temperature of 150°C (302°F) has been determined on the basis of tube degassing considerations and to provide the requisite component reliability.

A straightforward solution is easily obtained through the use of the worksheet shown in Table 22.3. In the worksheet, four flow rates are assumed and, through a logical procedure, four component temperatures are calculated. As in Chap. 10, where a similar worksheet was presented for the design of a heat exchanger, numbers have been placed at pertinent locations on the worksheet to assist the reader. These numbers refer to the dialogue that now follows:

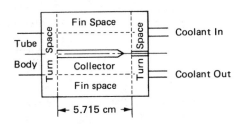

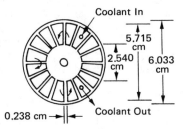

4 Fins — 0.238 cm wide — 1.588 cm high per pass
Fins are copper
4 Passes

FIG. 22.10 Finned passage for traveling wave tube collector. This configuration forms the basis for the illustrative example whose solution is detailed in Table 22.3.

(1) Per configuration; see Fig. 22.10.

(2) Fin height is $b = 1.588$ cm and fin width is $\delta = 0.238$ cm; see Fig. 22.10.

(3) $A_T = \frac{\pi}{4} (d_o^2 - d_i^2) - nb\delta = \frac{\pi}{4} [(5.715)^2 - (2.54)^2] - 16(1.588)(0.238)$

$$= 20.585 - 6.047 = 14.538 \text{ cm}^2$$

With the outer periphery assumed to be nondissipating,

$$P_T = \pi d_i - n\delta + 2nb = \pi(2.54) - 16(0.238) + 2(16)(1.588)$$
$$= 7.980 - 3.808 + 50.816 = 54.988 \text{ cm}$$

The area for each of the four passes will be

$$A = \frac{A_T}{4} = \frac{1}{4} (14.538) = 3.635 \text{ cm}^2$$

(4) The equivalent diameter d_e is

$$d_e = \frac{4A_T}{P_T} = \frac{4(14.538)}{54.988} = 1.058 \text{ cm}$$

(5) Per configuration; see Fig. 22.10.

TABLE 22.3 Worksheet for TWT Component Temperature

Item	Units	I	II	III	IV
d_o (1)	cm	5.715	5.715	5.715	5.715
d_i (1)	cm	2.540	2.540	2.540	2.540
n (1)		16	16	16	16
b (2)	cm	1.588	1.588	1.588	1.588
δ (2)	cm	0.238	0.238	0.238	0.238
A_T (3)	cm²	14.538	14.538	14.538	14.538
P_T (3)	cm	54.988	54.988	54.988	54.988
A (3)	cm²	3.635	3.635	3.635	3.635
d_e (4)	cm	1.058	1.058	1.058	1.058
L (5)	cm	5.715	5.715	5.715	5.715
S_f (6)	cm²	290.413	290.413	290.413	290.413
S_b (6)	cm²	23.841	23.841	23.841	23.841
$(d_e/L)^{1/3}$ (7)		0.570	0.570	0.570	0.570
k (8)	W/m °C	381	381	381	381
w (9)	kg/min	10	12.5	15	17.5
W (10)	kg/h	600	750	900	1050
c (11)	J/kg °C	2081	2081	2081	2081
ΔT (12)	°C	4.3	3.5	2.9	2.5
T_1 (12)	°C	76.7	76.7	76.7	76.7
T_2 (12)	°C	81.0	80.2	79.6	79.2
T_a (12)	°C	78.9	78.5	78.2	78.0
Check, c (11)	J/kg °C	2081	2081	2081	2081
T_S (12)	°C	150	150	150	150
μ (13)	kg/m s	4.382×10^{-3}	4.402×10^{-3}	4.402×10^{-3}	4.402×10^{-3}
$k\,Pr^{1/3}$ (13)	W/m °C	0.549	0.549	0.549	0.549
μ_w (14)	kg/m s	1.654×10^{-3}	1.654×10^{-3}	1.654×10^{-3}	1.654×10^{-3}
$(\mu/\mu_w)^{0.14}$ (14)		1.146	1.147	1.147	1.147
G (15)	kg/h m²	1.6506×10^6	2.0633×10^6	2.4759×10^6	2.8886×10^6
Re (15)		1107	1377	1653	1928
h (16)	W/m² °C	652.19	701.41	746.01	784.68
m (17)	m⁻¹	37.93	39.33	40.56	41.60
mb (17)		0.602	0.625	0.644	0.661
η_f (18)		0.894	0.887	0.881	0.876
$\eta_f S_f$	cm²	259.630	257.597	255.854	254.402
S_b	cm²	23.841	23.841	23.841	23.841
S_T (19)	m²	2.835×10^{-2}	2.814×10^{-2}	2.797×10^{-2}	2.782×10^{-2}
N_{tu} (20)		0.053	0.046	0.040	0.036
F (21)		1.055	1.047	1.041	1.037
FT_2	°C	85.435	83.936	82.857	82.101
$FT_2 - T_1$	°C	8.735	7.236	6.157	5.401
$F - 1$		0.055	0.047	0.041	0.037
T_S	°C	159.5	155.3	150.5	147.5

(6) $S_f = 2nbL = 2(16)(1.588)(5.715) = 290.413 \text{ cm}^2$

$$S_b = (\pi d_i - n\delta)L = [\pi(2.54) - 16(0.238)](5.715)$$

$$= (7.980 - 3.808)(5.715)$$

$$= 23.841 \text{ cm}^2$$

(7) $(d_e/L)^{1/3}$ will be required in the calculation of h:

$$\left(\frac{d_e}{L}\right)^{1/3} = \left(\frac{1.058}{5.715}\right)^{1/3} = (0.185)^{1/3} = 0.570$$

(8) Copper; $k = 381$ W/m K
(9) Four flow rates are assumed.
(10) $W = 60$ w.
(11) Thermal property; see Fig. 22.3.

(12) $T_1 = 76.5°C$ $(170°F)$ in all cases

$T_S = 150°C$ $(302°F)$ in all cases

$$\Delta T = 3600 \frac{q}{Wc}$$

$$T_2 = T_1 + \Delta T$$

$$T_a = \tfrac{1}{2}(T_1 + T_2)$$

(13) Thermal properties: see Fig. 22.3.
(14) μ_w is evaluated at $T_s = 150°C$. $(\mu/\mu_w)^{0.14}$ will be needed in the calculation of h.
(15) $G = 10^4$ W/A, because A is in cm^2. Re $= d_e G/360{,}000\mu$, because d_e is in cm and μ is in kg/m s.
(16) $h = 1.86(k/d_e)[(\text{Re})(\text{Pr})(d_e/L)]^{1/3}(\mu/\mu_w)^{0.14}$. Because d_e is in cm.

$$h = 186 \frac{k}{d_e}\left[(\text{Re})(\text{Pr})\left(\frac{d_e}{L}\right)\right]^{1/3}\left(\frac{\mu}{\mu_w}\right)^{0.14}$$

(17) $m = (2h/k\delta)^{1/2}$. With δ in cm,

$$m = \left(\frac{200h}{k\delta}\right)^{1/2} = \left[\frac{200h}{(381)(0.238)}\right]^{1/2} = 1.485\sqrt{h}$$

$$mb = 1.588 \times 10^{-2} \text{ m}$$

because $b = 1.588$ cm.
(18) Efficiencies are from Eq. (14.5); $\eta_f = \tanh mb/mb$.
(19) $S_T = (S_b + \eta_f S_f)/10{,}000$, because S_b and S_f are in cm.
(20) $N_{tu} = 3600 hS/Wc$, because W is in kg/h.
(21) For a rising temperature receiver and a small component that is approximately isothermal,

$$q = hS_T(\text{LMTD}) = Wc(T_2 - T_1)$$

with

$$\text{LMTD} = \frac{T_2 - T_1}{\ln(T_S - T_1)/(T_S - T_2)}$$

so that

$$q = \frac{hS_T(T_2 - T_1)}{\ln(T_S - T_1)/(T_S - T_2)} = Wc(T_2 - T_1)$$

Then

$$\ln\frac{T_S - T_1}{T_S - T_2} = \frac{hS_T}{Wc} = N_{tu}$$

and $\quad \dfrac{T_S - T_1}{T_S - T_2} = e^{N_{tu}} = F$

Thus

$$T_S - T_1 = F(T_S - T_2)$$

$$T_S - T_1 = FT_S - FT_2$$

$$(F - 1)T_S = FT_2 - T_1$$

and $\qquad T_S = \dfrac{FT_2 - T_1}{F - 1}$

The four surface temperatures calculated are plotted against the flow rate in Fig. 22.11. From this plot, it is seen that the surface temperature of 150°C is attained when the flow is 920 kg (2030 lb) of Coolanol per hour, which is the figure used for the heat exchanger design in Chap. 10.

22.6 A NOTE ON CONSERVATISM

The surface temperature of 150°C with a flow rate of 920 kg/h of Coolanol 45 is conservative. The temperature will not be this high because there is more surface in the configuration than was used in the calculations in Table 22.3.

The outer periphery of the configuration is physically bonded to the fins. It possesses a surface area of $(\pi d_o - n\delta)L = [\pi(5.715) - 16(0.238)](5.715) = 80.846$

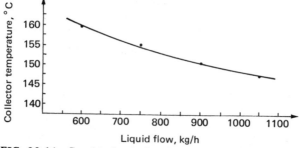

FIG. 22.11 Graphical solution for TWT collector temperature from data obtained in Table 22.3.

cm^2. This surface is not as effective as the fins, but its use in the calculations, with appropriate modification for its efficiency, will bring the surface temperature down.

To show that the previous result of 150°C at 920 kg/h of Coolanol is conservative, consider that the equivalent diameter will be smaller because the outer surface is now heat transfer area. With a previous $P_T = 54.988$ cm, the new P_T will be

$$P_T = 54.988 + \pi d_o - n\delta = 54.988 + \pi(5.715) - 16(0.238)$$

$$= 69.134 \text{ cm}$$

Then with A_T the same as before,

$$d_e = \frac{4A_T}{P_T} + \frac{4(14.538)}{69.134} = 0.841 \text{ cm}$$

At 920 kg/h,

$$G = 10^4 \left(\frac{920}{3.635}\right) = 2.531 \times 10^6 \text{ kg/m}^2 \text{ h}$$

and $\quad$ $Re = \dfrac{d_e G}{360,000\mu} = \dfrac{0.841(2.531 \times 10^6)}{360,000(4.402 \times 10^{-3})}$

or $\quad$ $Re = 1343$

The value of h is obtained from

$$h = \frac{186k(Pr)^{1/3}}{d_e} (Re)^{1/3} \left(\frac{d_e}{L}\right)^{1/3} \left(\frac{\mu}{\mu_w}\right)^{0.14}$$

or $\quad$ $h = \dfrac{186}{0.831} (0.549)(1343)^{1/3} \left(\dfrac{0.841}{5.715}\right)^{1/3} (1.147)$

The result is

$$h = 811.22 \text{ W/m}^2 \,{}^\circ\text{C}$$

One will now be able to see the power of the input-admittance-cascade algorithm approach to a finned configuration described in Sec. 14.7. Observe in Fig. 22.12 that there are two admittances in parallel: one for the base surface ($q_b/\theta_b|_b = Y_b$),

$$Y_b = hS_b \text{ W/}^\circ\text{C}$$

and one for the finned array of 16 fins ($q_b/\theta_b|_f = Y_f$),

$$Y_f = 16Y_1 \text{ W/}^\circ\text{C}$$

where $Y_1 = q_b/\theta_b|_1$ is for a single entity of a single fin plus two fins in cluster. The total input admittance for the configuration is

$$Y_T = Y_b + Y_f = hS_b + 16Y_1$$

or $$Y_T = \frac{811.22(23.841)}{10,000} + 16Y_1 = 1.934 + 16Y_1 \qquad (22.18)$$

It now remains to calculate Y_1.

The finned part of the configuration contains, as has been tacitly assumed in the previous paragraph, 16 repeating sections of fins as shown in Fig. 22.12. Each repeating section contains two fins in cluster, which represent the periphery with

$$b = \frac{\pi d_{o,av}}{32} = \frac{\pi(6.033 + 5.715)}{2(32)} = 0.577 \text{ cm}$$

and $\delta = 6.033 - 5.715 = 0.318$ cm

These fins are at the tip of each of 16 fins, with

$$b = 1.588 \text{ cm}$$

and $\delta = 0.238$ cm

All fins in each repeating section are 5.715 cm long.

For the two fins in cluster with $h = 811.22$ W/m^2 °C,

$$Y_o = (hk\delta)^{1/2} L = \left[2(811.22)(381)\left(\frac{0.318}{100}\right) \right]^{1/2} \left(\frac{5.715}{100}\right)$$

$$= 1.7917 \text{ W/°C}$$

and $$mb = \left(\frac{h}{k\delta}\right)^{1/2} b = \left[\frac{2(811.22)}{381(0.318/100)}\right]^{1/2} \left(\frac{0.577}{100}\right) = 0.1492$$

Only γ_{21} and γ_{22}, which are displayed in Eq. (14.29), are needed.

$$\gamma_{22} = \cosh mb = \cosh (0.1492) = 1.0112$$

and $$\gamma_{21} = -\frac{\sinh mb}{Z_o} = -Y_o \sinh mb$$

$$= -(1.7917) \sinh (0.1492) = -0.2683 \text{ W/°C}$$

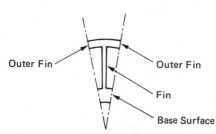

FIG. 22.12   Repeating section of finned passage in TWT collector.

Then, by Eq. (14.31),[3]

$$\left.\frac{q_b}{\theta_b}\right|_p = \frac{\gamma_{21}}{-\gamma_{22}} = \frac{-0.2683}{-1.0112} = 0.2654 \text{ W/}^\circ\text{C}$$

and twice this value is to be applied in Eq. (14.31) to the fin that is attached to the base of the configuration (the exterior of the TWT collector):[4]

$$\left.\frac{q_b}{\theta_b}\right|_c = 0.5308 \text{ W/}^\circ\text{C}$$

For the single fin, as dictated by the elements of $[\Gamma]$ in Eq. (14.29),

$$\gamma_{11} = \gamma_{22} = \cosh mb$$

$$\gamma_{12} = -Z_o \sinh mb = \frac{-\sinh mb}{Y_o}$$

and $\quad \gamma_{21} = \dfrac{-\sinh mb}{Z_o} = -Y_o \sinh mb$

With

$$Y_o = (2hk\delta)^{1/2}L = \left[2(811.22)(381)\left(\frac{0.238}{100}\right)\right]^{1/2}\left(\frac{5.715}{100}\right)$$

$$= 2.1965 \text{ W/}^\circ\text{C}$$

and $\quad mb = \left(\dfrac{2h}{k\delta}\right)^{1/2}b = \left[\dfrac{2(811.22)}{381(0.238/100)}\right]^{1/2}\left(\dfrac{1.588}{100}\right) = 0.6717$

it is easy to compute the γ's:

$$\gamma_{11} = \gamma_{22} = 1.2342$$

$$\gamma_{12} = -0.3300 \text{ }^\circ\text{C/W}$$

$$\gamma_{21} = -1.5857 \text{ W/}^\circ\text{C}$$

Now, by Eq. (14.31) with $q_b/\theta_b|_c = 1.0587$,

$$\left.\frac{q_b}{\theta_b}\right|_1 = Y_1 = \frac{\gamma_{21} - \gamma_{11}(q_b/\theta_b|_c)}{-\gamma_{22} + \gamma_{12}(q_b/\theta_b|_c)}$$

$$= \frac{-1.5857 - 1.2342(0.5308)}{-1.2342 - 0.3300(0.5308)} = \frac{-2.2408}{-1.4094} = 1.5899 \text{ W/}^\circ\text{C}$$

[3] With subscript p for periphery.
[4] With subscript c for cluster.

The entire input admittance of the finned array now comes from Eq. (22.18):

$$Y_T = 1.934 + 16(1.5899)$$

or $$Y_T = 1.934 + 25.439 = 27.372 \text{ W/}°\text{C}$$

This means that for 1500 W, a θ_b of

$$\theta_b = \frac{q_b}{Y_T} = \frac{1500}{27.372} = 54.8°\text{C}$$

is required and, hence, $T_S = \theta_b + T_a$ will be the surface temperature. With 920 kg/h flowing at an inlet temperature of 76.7°C,

$$\Delta T = T_2 - T_1 = \frac{q}{Wc} = 3600 \frac{1500}{920(2801)} = 2.1°\text{C}$$

so that

$$T_2 = 2.1 + 76.7 = 78.8°\text{C}$$

and $$T_a = \tfrac{1}{2}(76.7 + 78.8) = 77.7°\text{C}$$

This makes T_S

$$T_S = \theta_b + T_a = 54.8 + 77.7 = 132.5°\text{C}$$

a marked decrease over the 150°C previously cited.

22.7 ADDITIONAL REMARKS PERTAINING TO THE EXAMPLE

1 Observe in Table 22.3 that a flow of 1050 kg/h resulted in a Reynolds number of 1928. Additional calculations much above 1050 kg/h would be meaningless unless it is recognized that above Re = 2300, the flow goes into transition from laminar to turbulent flow. In this case, the equation governing h [see item (16) in the dialogue] would not apply.

2 A pressure loss is associated with the excellent heat transfer in the TWT collector. This pressure drop would require estimation before a pump could be specified. Friction and turn losses are involved.

3 The reader can well imagine the futility of attempting to cool the collector with air.

4 There is a temperature drop across the collector metal. In this case, with a 150°C surface temperature,

$$\Delta T = T_i - 150 = \frac{q \ln r_S/r_i}{2\pi k L}$$

or with a 0.250-cm-diameter hole,

$$T_i = 150 + \frac{1500 \ln (1.27/0.125)}{2\pi(381)(5.715/100)}$$

or $\quad T_i = 150 + 25.4 = 175.4°C$

5 If, perchance, the shop had bored the inner hole off-center, a complex calculation requiring a conformal map (see Sec. 4.10) would be required.

22.8 THERMAL CONTROL OF ELECTRONICALLY STEERED ANTENNAS

The advent of ferrite phase shifters, capable of introducing phase delays in traveling microwaves, has made possible major modifications in the design of RF antennas. When properly activated, an array of ferrite-carrying, phase-shifter RF tubes can be made to radiate a microwave beam to any desired location, and beam steering, through all points in the forward hemisphere, can be accomplished by coordination of the electrical signals sent to the ferrite elements. The electronic steering capability thus gained reduces and/or eliminates the need for mechanical steering of the RF antenna but, in parallel, introduces a relatively stringent requirement for the thermal control of the phase shifter elements.

The following discussion of electronically steered antenna concepts and their associated thermal control strategies is meant to highlight the thermal features of these microwave systems and identify some viable approaches to the desired degree of thermal control. Although the description of the system and the cooling concept are necessarily brief, it is important to note that the specific thermal control strategies described evolved from a detailed design procedure and reflect a balance among the various system constraints. The physical variations between the two concepts and the hierarchy of physical and thermal constraints appropriate to each of the two concepts channeled the design process toward the particular thermal control strategy adopted.

22.8.1 Hybrid Antennas

ARRAY DESCRIPTION
The hybrid antenna consists of a mechanically steered, elliptical, flat plate array of phase-shifting microwave tubes as shown in Fig. 22.13. The array contains 372 radiating/receiving elements and, with its associated microwave equipment, is supported on a pedestal structure that provides two axes of motion. Rotary joints at the elevation and azimuth axes provide RF and coolant continuity. The array is composed of six aluminum extrusions running parallel to the major axis of the array. As shown in Fig. 22.14, each extrusion is bored to accept a double row of antenna elements; and a coolant passage is provided in the center. Appropriate inlet and exit headers are provided at opposite ends of the array to distribute and collect the coolant flowing through the six extrusions.

The antenna elements are arranged in clusters of four and fed from a Duroid stripline power divider board attached to the rear surface of the array. Each four-

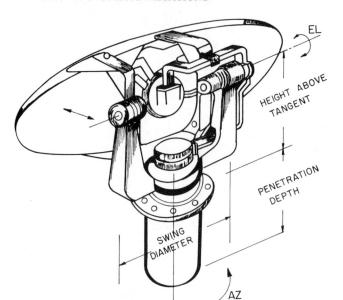

FIG. 22.13   Mechanically scanned elliptical array.

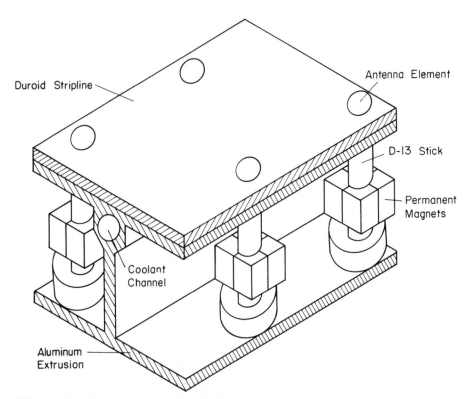

FIG. 22.14   Element carrying extrusion.

element stripline cluster accepts 120 W average power from a corporate wave guide and "magic tee" divider network. The individual elements consist of a ferrite section sandwiched between two sections of D-13 dielectric material. These are adhesively bonded and subsequently enclosed in a 6- to 8-mil electroplated casing of copper. Four L-shaped permanent magnets are bonded to each element in the region of the ferrite material.

The elliptical array is designed to provide in excess of 75 db_W in effective radiated power (ERP = gain $\times$ 10 $\log_{10}$ Radiated power/1 W). A total of 12 kW of RF power are to be generated in the antenna transmitter aboard the aircraft. Previous calculations suggest that 0.6 kW will be dissipated in the RF rotary joints of the waveguide feed network and approximately 1.4 kW in the array. In the absence of environmental cooling loads, this results in a 2.0-kW thermal load for the array cooling system. Additional dissipation in the transmitter and associated equipment can be expected to total approximately 12 kW, but these components are not within the array thermal control system. For purposes of this study, environmental loads have been neglected.

THERMAL CONSIDERATIONS

The primary thermal dissipation in the array occurs within the Duroid stripline power divider board. Approximately 0.3 db of the RF power incident at the stripline is dissipated, resulting in the generation of 9 W of heat in the stripline of a four-element cluster. An additional 1 W per element is dissipated in the ferrite section of each element, and some 3 W per double cluster (eight elements) are dissipated in the magic tee power divider. The total RF dissipation in the 372-element array is, then, approximately 1.4 kW or an average of 3.6 W per element.

Although only 1 W per element or 28% of the dissipation occurs in the ferrites, successful electrical operation of the ferrites demands that they be maintained below 93°C and that the variation in ferrite temperature across the array in this temperature range not exceed 11°C. Duroid stripline is, on the other hand, relatively insensitive to temperature and can perform successfully at temperatures approaching 250°C. The ferrites are therefore the thermally critical components in the elliptical array and play a major role in establishing the requirements of the thermal control system.

THERMAL CONTROL SYSTEM

The thermal constraints imposed by the temperature sensitivity of the ferrites and the need to reduce the array profile/volume and, hence, the swept area of the radome, combine to dictate the use of a liquid cooling system for the thermal control of the gimbaled elliptical array. Although forced convection air cooling of airborne systems is generally desirable, it could not be achieved in this event without substantially increasing the array volume above the design goal. Air cooling the elliptical array would necessitate handling some 7.5 m^3/min (at atmospheric pressure) and would lead to larger extrusions, large peripheral plenums, and large rotary joints. Alternatively, the use of liquid cooling results in small coolant passages, which can be contained within the array geometry. The associated rotary joints are small, and a flow of less than 5.75 l/min provides the necessary cooling.

A thermal network analysis of the aluminum extrusion shown in Fig. 22.14, based on a 6-mil copper plating on the outer surface of the D-13 stick and using the conservative assumption that all the element dissipation (1 W) occurs uniformly in the ferrite, yields a thermal resistance from the ferrite to the coolant channel wall of approximately 36 °C/W. A similar analysis, assuming uniform power dissipation in the stripline board (9 W/cluster), establishes the thermal resistance from the Duroid to the coolant channel wall at 2.65 °C/W. The ferrite can therefore be expected to operate at approximately 36°C above the channel wall temperature and the Duroid stripline at approximately 24°C above the channel wall. Any variation in the channel wall temperature across the array will be reflected directly in the Duroid and ferrite temperatures. In view of the thermal requirement of the ferrites, therefore, the coolant channel wall must not exceed 57°C and must not vary more than 11°C across the array.

The decision to use liquid cooling for the array can be implemented by the use of either a forced convection system or a flow boiling system. In forced convection, high rates of transfer at the channel wall and a relatively small temperature rise in the coolant across the array can be achieved by high-velocity flow in the channel. In flow boiling, on the other hand, the channel wall can be expected to exceed the boiling point of the fluid by approximately 5 to 10°C, and the fluid temperature will remain very close to its boiling point for a relatively small flow rate. Preliminary calculations indicate that the required thermal control in the elliptical array could be realized with either an ethylene glycol/water forced convection system, or a FC-88 flow boiling system. Because of the higher density of the FC-88, its greater tendency to leak through seals and rotary joints, and the more complex heat exchange equipment required for the flow boiling system, the ethylene glycol/water system was selected for this task.

DESIGN TEMPERATURES AND FLOW RATE

In forced convection, the local coolant channel wall temperature is determined by the coolant temperature as well as the heat transfer coefficient h and heat flux q'' at the channel wall, according to

$$T_{\text{channel wall}} = T_{\text{coolant}} + \frac{q''}{h}$$

For a given channel diameter and choice of fluid, h in fully developed turbulent flow is directly proportional to the fluid flow rate to the 0.8 power and for the ethylene glycol/water in a nominal 0.65-cm (0.250-in)-diameter channel is given approximately by

$$h = 45.3(W)^{0.8}$$

where h is in W/m² °C and W is in kg/h.

The coolant flow rate is in this case established by the need to maintain less than a 11°C rise in the coolant flowing through a single extrusion. The two central extrusions carry the largest number of elements (~100) and, hence, dissipate the greatest amount of heat per extrusion (~0.36 kW). Substituting in the governing equation below and solving for W,

$$W = \frac{q_{\text{dic}}}{c_p \, \Delta T} = \frac{360}{(3.14 \times 10^3)(11)} = 0.0104 \text{ kg/s} = 37.5 \text{ kg/h}$$

yields a flow requirement of 37.5 kg/h or 0.833 l/min for each of the two central extrusions. In view of the modest flow requirement and the desire to minimize temperature variation across the array, a flow rate of 0.833 l/min is to be specified for each extrusion channel. This can be accomplished with appropriate pressure control measures and results in a total flow requirement for the array of 4.93 l/min of ethylene glycol/water.

Substituting W into the equation for the heat transfer coefficient now yields an h of 823 W/m² °C. Returning to the channel wall temperature expression with this value of h and the channel surface heat flux equal to 1.05×10^4 W/m² leads to

$$T_{\text{channel wall}} - T_{\text{coolant}} = 13°C$$

To satisfy the earlier requirement of a maximum channel wall temperature of 57°C, the coolant must not exceed 44°C when exiting the central array extrusion and must therefore enter the array at no more than 33°C.

Upon exiting the array, the total flow of 4.93 l/min at an average temperature of 39.5°C is directed through the microwave rotary joints to remove the 0.6 kW dissipated in these joints. This load results in an additional 3°C temperature rise in the coolant. The ethylene glycol/water thus experiences a total temperature rise of 9.5°C and exits the array cooling system at 42.5°C.

The heated ethylene glycol/water is now returned to the aircraft through rotary joints and is passed through an air-liquid heat exchanger, which uses 21°C cabin air to cool the 4.93 l/min of ethylene glycol/water to 33°C. A standard military specification heat exchanger can accomplish this task.

22.8.2 Active Distributed Arrays

ARRAY DESCRIPTION

The active, distributed, circular array differs substantially from the elliptical array discussed in Sec. 22.6.2. The array contains 1518 elements in 759 two-element, ceramic modules, one of which is shown in Fig. 22.15, and is electronically steered. Approximately 90 mW of transmit RF power impinge on each module, but the prime RF power generation for the array is provided by the two amplifier diodes associated

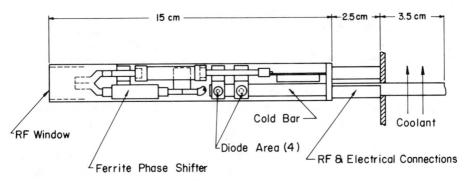

FIG. 22.15 Active element for RF antenna [19].

with each element. Two approximately 1-m-diameter arrays are mounted flush on either side of the aircraft fuselage.

The ceramic modules are approximately 15 cm long, and each contains a central 0.65-cm-diameter heat pipe in intimate contact with the four high-dissipation amplifier diodes. The cylindrical heat pipe extends 6.5 cm beyond the end of the module. It passes through a 3-cm region of electrical/RF connections and plugs into a liquid-cooled manifold.

The distributed array is designed to provide 66 db W in ERP, some 10 db W lower than the elliptical array. Because of the inherent low efficiency of the amplifier diodes, this results in the generation of 20 W per module or nearly 15 kW of heat in the array. It must be noted that the distributed array dissipation is thus an order of magnitude higher than the dissipation within the elliptical array. However, in the distributed antenna design, dissipations external to the array are small, since only 70 W of RF power are generated in the aircraft. Alternatively, in the elliptical antenna design, approximately 12 kW of transmitter dissipation aboard the aircraft must be added to the 2 kW dissipated within the array. Consequently, the thermal loads for the entire elliptical antenna system and the entire distributed antenna system are comparable. For purposes of this study, the environmental loads and any variation in these loads for the two antenna designs have been neglected.

THERMAL CONSIDERATIONS

The primary and only significant dissipation in the distributed array occurs at the amplifier diodes. As shown in Fig. 22.15, there are two such diodes for each element. The first dissipates approximately 1 W and the second 9 W for a total of 10 W per antenna element or 20 W per module. To ensure proper RF functioning of the elements, the diode junction must be kept below 200°C. However, adjacent to the diodes, on the low-power side, are several ferrite phase shifters that have negligible dissipation but that must be kept below 93°C with no more than an 11°C variation across the array. Previous operating experience with distributed array modules of this type indicates that a 75°C internal heat sink temperature could provide a safe operating environment for all the components. The scope of the distributed array thermal control system is then determined jointly by the diodes, which establish the load requirement, and the ferrites, which establish the operating temperature of the internal heat sink.

THERMAL CONTROL SYSTEM

The high heat dissipation at the amplifier diodes and the need to transport the dissipated heat approximately 12 cm to a coolant manifold, while maintaining a 75°C heat sink temperature, strongly suggests the use of an integral heat pipe in the two-element module.

In keeping with the physical constraints of the module and the desirability of a plug-in capability at the coolant manifold, a 0.65-cm stainless steel, cylindrical heat pipe, 12 cm long, was selected. It is anticipated that the diodes will be brazed to the heat pipe and the ferrites bonded to the heat pipe at the appropriate locations. The exposed end of the heat pipe will be inserted into a well provided in an aluminum manifold. The 0.32-cm-wall thickness well will be lined with a 1-mm polyethylene layer to eliminate thermally critical air gaps and ensure firm positioning of the module. The thermal resistance from the external heat pipe surface to the coolant, through the polyethylene and well wall, can be shown to approximate 0.6 °C/W.

Evaluation of published heat pipe data indicate that a 1.5-mm-wall thickness stainless steel, water-filled heat pipe with a feltmetal wick could transport the 20 W dissipated in the diodes to the coolant manifold in up to a 2g environment. At this power dissipation level, a temperature difference of 25°C can be expected between the heat pipe surface in the module and the heat pipe surface at the coolant manifold well. Taking proper account of the heat pipe and well thermal resistance, the need to maintain the module end of the heat pipe at 75°C can now be translated into maintaining the coolant side of the manifold well at less than 38°C.

The use of air cooling to remove the heat from the manifold end of the heat pipe is once again an initially attractive approach. However, the low density and specific heat of air, combined with a level of array dissipation that is an order of magnitude above standard antenna arrays, essentially rules out the use of air cooling. To maintain less than a 11°C variation across the array would necessitate the use of nearly 70 m³/min (at atmospheric pressure) and would result in a prohibitive pressure drop through the array. It is therefore necessary to circulate a liquid coolant through the manifold.

As discussed previously, liquid cooling in antenna arrays can, in general, be achieved by either forced convection at high velocity or flow boiling at low velocity. Unlike the case of the elliptical array, where either approach is viable, the distributed array cannot be cooled efficiently by forced convection. The use of forced convection heat transfer to remove the high heat flux at the well surface ($q'' \sim 2.6 \times 10^4$ W/m²) over the surface area of 270 wells (~7.7 cm²/well) would necessitate the use of more than 150 l/min of ethylene glycol/water. Finning the well surface could decrease this flow rate, but only at a substantial increase in the cost of fabricating and assembling the array.

Alternately, the use of 17 l/min of 3M's FC-88 flowing through the manifold and providing flow boiling heat transfer at the unfinned well surface could provide the necessary thermal control. At the specified heat flux, the well surface temperature will be at the desired temperature of approximately 38°C or 6°C above the 32°C saturation temperature (at 1 atm) of the FC-88. The 17 l/min of FC-88 entering the array will experience no significant temperature rise while flowing through the array. Approximately 33% of the mass flow will, however, undergo a change in phase from liquid to vapor and thus absorb 3.3 kW of dissipated heat per gallon per minute of flow.

The use of an FC-88 flow boiling system requires the use of special care in sealing the liquid chambers, but satisfactory sealing is quite feasible in the absence of rotary joints. The weight of FC-88 in the array at approximately 57 kg is nearly 1.7 times the weight of ethylene glycol/water in the same array, but the weight of the entire cooling system, including a pump, heat exchanger, and plumbing, is certainly comparable to, if not less than, the total weight of any forced convection system. Consequently, the desired thermal control of the distributed array can be achieved simply and efficiently with an FC-88 flow boiling system.

22.9   NOMENCLATURE

Roman Letter Symbols

A	area, m^2
b	a constant defined where used; or fin height, m
c	specific heat, J/kg °C
C	a constant, K^{-1}
d	diameter, m
f	friction factor, dimensionless
F	a factor defined where used, dimensionless
h	heat transfer coefficient, W/m^2 °C
I	microwave intensity, W/m^3
k	thermal conductivity, W/m °C
l	length of path, m
L	length, m
LMTD	logarithmic mean temperature difference, °C
m	fin performance factor, m^{-1}
n	number of fins, dimensionless
N_{tu}	number of transfer units, dimensionless
p	pressure, kg/m^2
P	perimeter, m
Pr	Prandtl number, dimensionless
r	radius, m
Re	Reynolds number, dimensionless
S	surface, m^2
T	temperature, K or °C
V	velocity, m/s
w	flow rate, kg/min
W	volumetric heat dissipation, W/m^3; or flow rate, kg/h
x	length coordinate, m
y	length coordinate in space between two adiabatic surfaces, m
Y	fin input admittance, W/°C

Greek Letter Symbols

α	attenuation factor defined in Eq. (22.7), dimensionless
γ	element of thermal transmission matrix, dimensions vary
δ	fin width, m
η	fin efficiency, dimensionless
μ	dynamic viscosity, kg/h m
ρ	density, kg/m^3

Subscripts

a	indicates an average condition
A	indicates a fluid
b	indicates base or prime surface
B	indicates a fluid
c	indicates a cluster of fins

e	indicates an equivalent condition
f	indicates a film condition
i	indicates an inner radius or diameter
l	indicates path length
m	indicates a mean condition
o	indicates a half-spacing, an outer radius or diameter, or a characteristic value
p	indicates a constant-pressure peripheral condition
S	indicates a surface condition or property
T	indicates a total condition
w	indicates a wall property

Superscripts

$''$	indicates a heat flux
$-$	indicates normalized values [see Eq. (22.12)]

22.10 REFERENCES

1 Allen, J. L., High Power Microwave Devices, Report of NSF Workshop, "Research Directions of Heat Transfer in Electronic Equipment," Atlanta, Georgia, October 1977.

2 Markowitz, A., RF Water Load Thermal Analysis, Raytheon Company Missile Systems Div. Rept. BR-5318, Bedford, Mass., June 1969.

3 Mouromtseff, I. E., Water and Forced Air Cooling of Vacuum Tubes, *Proc. IRE*, vol. 30, pp. 190–205, 1942.

4 Paradis, L. R., Simplified Transmitter Cooling System, 8th Int. Circuit Packaging Symp. (WESCON), San Francisco, August 1967.

5 Gucker, G. B., Long Term Frequency Stability for a Reflex Klystron Without the Use of External Cavities, *Bell System Tech. J.*, vol. 41, pp. 945–958, May 1962.

6 London, A. L., *Coolant Path Hydraulic Design for High Power TWYSTRON Tubes*, Varian Associates, Palo Alto, Calif., 1971.

7 London, A. L., Air Coolers for High Power Vacuum Tubes, *Trans. IRE*, vol. ED-1, pp. 9–26, April 1954.

8 Devins, J. C., and Sharbaugh, A. H., The Fundamental Nature of Electrical Breakdown, *Electro-Technology*, vol. 67, pp. 104–122, February 1961.

9 Devins, J. C., and Sharbaugh, A. H., Electrical Breakdown in Solids and Liquids, *Electro-Technology*, vol. 68, pp. 97–116, October 1961.

10 Grill, B., Liquid Coolants for Microwave Tubes, *Electrical Products* (published by Industrial Media, Tunbridge Wells, U.K.), April 1962.

11 Jakob, M., *Heat Transfer*, Wiley, New York, 1949.

12 Knights, A. F., Choice of Fluids for Cooling Electronic Equipment, *Electro-Technology*, vol. 71, pp. 57–63, June 1963.

13 Martin, L. J., Mell, C. W., and Milek, J. T., Advanced Heat Transfer Fluids, Hughes Aircraft Co., Ground Systems Group, WADC Technical Rept. TR-16-186, July 1961.

14 Barsness, D. A., Extreme Temperature Range Organic Coolants, WADC Technical Rept. TR-61-795, Parts I and II, Wright Air Development Center, Dayton, Oh., September 1961.

15 Von Hippel, A., The Dielectric Relaxation Spectra of Water, Ice and Aqueous Solutions and Their Interpretation, Technical Rept. II, Laboratory for Insulation Research, Massachusetts Institute of Technology, Cambridge, Mass., 1967.

16 Rohsenow, W. M., and Choi, H., *Heat, Mass and Momentum Transfer*, Prentice-Hall, Englewood Cliffs, N. J., 1961.

17 Tillman, C. C., Jr., EPS: An Interactive System for Solving Elliptic Boundary Value Problems, Massachusetts Institute of Technology Rept. MAC-TR-62, Cambridge, Mass., 1969.

18 Bar-Cohen, A., Exponentially Decaying Internal Heat Generation in Fully Developed Laminar Rectangular Channel Flow, *Int. J. Heat Mass Transfer*, vol. 23, pp. 181–183, 1981.

19 Markowitz, A., Thermal Control of SHF Antennas–Conceptual Design, Raytheon Company, MSD, Internal Memo BSH-53, Bedford, Mass., 1971.

23

■ microelectronics and printed circuit boards

23.1 INTRODUCTION

It is contended that the success of the Apollo Project, which placed a man on the moon in the late 1960s, was due to many major technological breakthroughs, not least of which was the reduction of the size of electronic components and the ability to dispose of dissipated heat at a temperature consistent with overall system reliability. Indeed, throughout the history of electronic device manufacture and application, there has been a consistent desire to make electronic components smaller; and the change of electronic components from vacuum tubes to miniature and then subminiature tubes and thence to semiconductor devices has satisfied this desire to a great extent.

The trend toward size reduction with the increased utilization of transistors and then microelectronics was first a boon and then a nightmare for the thermal analyst. It was very pleasant in the beginning to find that a vacuum tube airborne electronic system dissipating 3000 W was to be redesigned using transistors *without filament power* so that the dissipation dropped to about 2100 W. This was helpful in terms of the overall system even though reliability requirements for low junction temperatures and the smaller size of the transistor added to the thermal control problem at the component level. This was the start of the nightmare, which has reached a point that is well summarized by Chu [1], who pointed out that, at the periphery of the sun, where the temperature is about 6000°C, the heat flux is about 10^7 W/m². In a microcircuit chip that may have a dissipation of 60 W/in², the heat flux converts to 93,000 W/m² (about 10^5 W/m²), and this dissipation must be accomplished at junction temperatures of 125 to 150°C. The need for precise thermal analysis and thermal system design (only two orders of magnitude from the solar dissipation at a stringent operating temperature) is patently obvious.

An integrated circuit may be defined as a group of inseparably connected circuit elements fabricated in place on and within a substrate. The silicon integrated circuit (SIC), invented by Kilby in 1958 [2], is a complete circuit of silicon-based components on a silicon wafer or substrate. As shown in Table 23.1 [3], there has been a steady increase in the evolution of the SIC. Reliability, speed of operation, and production yields have improved steadily; while cost, power consumption, and size have been reduced drastically.

It is apparent that SICs, which originally consisted of a single circuit on a silicon wafer with dimensions of about 3 × 5 × 0.1 mm, can now be produced in large quantities. Furthermore, more complex circuitry for greater functional complexity can, as the technology continues to improve, be placed on a single chip. As pointed

TABLE 23.1   Evolution of Integrated Circuits [3]

Monolithic integrated circuit	1958
Planar SIC	1959
Commercial monolithic IC (RTL)	1961
Diode-transistor logic	1962
Transistor-transistor logic	1962
Emitter-coupled logic	1962
Metal oxide semiconductor IC (MOS)	1962
Complementary MOS	1963
First linear ICs	1964
MOS memory chips	1968
Charge-coupled devices (CCD)	1969
MOS calculator chips	1970
Microprocessor	1971
Integrated injector logic (I^2L)	1972
Very-large-scale integration (VLSI)	1975

out by Millman [4], a chip of this size might contain (1978) some 30,000 components (transistors, diodes, resistors, or capacitors), which averages out to about 2000 per square millimeter. The term *microelectronics* is used to describe high-density chips. Some additional nomenclature in the industry, along with the dates the listed component counts were achieved, can now be listed [4]:

1960: small-scale integration (SSI), fewer than 100 components per chip

1966: medium-scale integration (MSI), more than 100 but fewer than 1000 components per chip

1969: large-scale integration (LSI), more than 1000 but fewer than 10,000 components per chip

1975: very large-scale integration (VLSI), more than 10,000 components per chip

As pointed out by Seely and Chu [5], there are three levels of thermal resistance in the path from the component junction to the environment: the component-, package-, and system-level resistances. These are all of significance to the thermal analyst.

The component-level resistance is the internal resistance of the microelectronic entity that exists between the junctions and the outside surface of the case. The next sections of this chapter deal with component-level resistance and, in these, an assessment of the heat flow path from junctions to case and some typical examples are provided.

The package-level resistance consists of the resistance to the flow of heat from the surface of the case to some reference point in the entire system. The system reference point may be the temperature of the air surrounding the component, the edge of a printed circuit board (PCB) to which several of the components are mounted, or, indeed, the wall of a cold plate heat exchanger that accommodates several such printed circuit boards. The printed circuit board and its capability (or lack of capability) to transfer heat is also a subject for detailed treatment elsewhere in this chapter.

System-level resistance refers to the resistance to the flow of heat from the package-level reference point to the ultimate sink or environment. A typical example of this is the resistance from the dissipating surface of a cold plate with due cognizance of the air temperature rise within the cold plate.

23.2 REASONS FOR THERMAL CONTROL

There are four reasons why precise thermal control of semiconductor devices in general and microcircuits in particular is necessary.

23.2.1 Reliability

It is an established fact that the reliability of an electronic component is a strong inverse function of junction or component temperature. This fact is markedly apparent from an examination of failure rate data such as presented in Chap. 2. Thus, there is a place for thermal analysis and subsequent preliminary assessment of reliability as part of the iterative design procedure of an electronic circuit, device, or system.

23.2.2 Manufacturing Guidance

A knowledge of temperature levels can dictate the method of fabrication as well as the selection of materials to be used in fabrication. For example, suppose that it is desired to solder a semiconductor device (a microdiscrete package as discussed in Sec. 23.3) to the substrate of a thick-film hybrid package. The temperature at which the solder melts is well established. A knowledge of the joint temperature with reasonable accuracy can influence the choice of bonding technique and often results in considerable reduction in the manufacturing cost of the device.

23.2.3 Bias Stabilization

Bias stabilization refers to the operation of a semiconductor device at a stable and predictable point. For example, consider Fig. 23.1, which shows a transistor connected in a common-emitter configuration. Suppose that a prescribed value of collector-to-emitter voltage (v_{CE}) is required to provide a logic operation.

In the loop composed of the bias voltage V_{CC}, the voltage drop across R_L, and the voltage drop v_{CE}, Kirchhoff's voltage law must apply, so that

$$v_{CE} = V_{CC} - i_C R_L \tag{23.1}$$

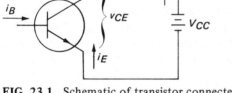

FIG. 23.1 Schematic of transistor connected in common-emitter configuration.

Observe that if the collector current i_C is allowed to increase beyond a certain value given by

$$i_C = \frac{V_{CC} - v_{CE}}{R_L} \tag{23.2}$$

where v_{CE} is the prescribed value of collector-to-emitter voltage required by the logic operation, v_{CE}, by Eq. (23.1) must decrease because V_{CC} and R_L are fixed. Thus, it may be noted that, if i_C increases because of high junction temperature, the desired value of v_{CE} can no longer be obtained.

Circuit designers can, of course, compensate for this through the use of ingeneous biasing circuits. This, however, does not alter the fact that operation at high junction temperatures can cause functional failures in the electronic circuitry.

23.2.4 Thermal Runaway

At thermal equilibrium, the power dissipated by the transistor junction, primarily in the common-emitter configuration,

$$P = i_C v_{CE} \tag{23.3}$$

is just equal to the heat transmitted from the junction to the surrounding environment. It has been observed that high junction temperature causes an increase in the collector current i_C. This causes an increase in the power dissipation, which, in turn, because of the fixed thermal resistance between junction and environment, causes an increase in junction temperature. Again, the collector current increases; so does the power dissipation, and so does the junction temperature.

It is seen that a regenerative heating cycle may occur. This phenomenon is called *thermal runaway* and may lead to a catastrophic failure of the device.

Although this *can* happen, experience has shown that before it does, communications will be received from the reliability people complaining about low reliability. Indeed, the manufacturing department will advise that solder joints are melting and the circuit designers will be observed to be wondering why the circuit is not functioning in accordance with the design specification.

23.3 HEAT TRANSFER FROM JUNCTION TO CASE

The determination of junction temperature in a microcircuit that involves the evaluation of the temperature drop between junction and case can be a relatively simple calculation, or it can become quite involved and require a comprehensive computer-aided analysis. The complexity of the calculations are a function of the number of modes of heat transfer considered and the number of components in the microelectronic entity.

Simplified analyses can, however, be executed, and this is easily demonstrated.

Consider Fig. 23.2*a*, which depicts[1] a 600 × 800 mil (1.5240 × 2.0320 cm) alumina substrate with four 50 × 50 mil (0.1270 × 0.1270 cm) semiconductor chips eutectically bonded to the substrate. The chips, labeled *A*, *B*, *C*, and *D*, are symmetrically placed and dissipate unequally:

Chip *A*   3.24 W

Chip *B*   2.81 W

Chip *C*   2.52 W

Chip *D*   3.03 W

A side view showing all thicknesses is shown in Fig. 23.2*b*. The Kovar steel carrier is attached to an aluminum heat sink held at 55°C. Observe that the chip spacing will preclude thermal interplay between chips. Because chip *A* is the heaviest dissipator, its temperature will now be considered.

Materials with their thermal conductivities are shown in Table 23.2. The heat sources on the semiconductor chips have a diameter of 35 mils (0.0889 cm). Although there is an air gap between the alumina substrate and the Kovar lid of 140 mils (0.3556 cm), the analysis to be presented here does not take air gap conduction and convection and radiation into effect. Thus the solution strategy that will yield a conservative results is as follows:

---

[1] Although the configuration described is not atypical of integrated circuit packages, the reader should realize that it is, in this case, meant primarily to serve as a vehicle for displaying analytical techniques.

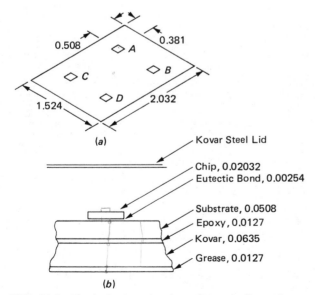

FIG. 23.2 Typical microcircuit package configuration.

TABLE 23.2   Physical and Thermal Data for Fig. 23.2

	Thickness		Thermal conductivity (100°C)	
Material	mils	cm	Btu/ft h°F	W/cm°C
Silicon chip	8	0.02032	65.7	1.137
Eutectic bond	1	0.00254	171	2.958
Alumina substrate	20	0.05080	17.0	0.294
Conductive epoxy	5	0.01270	1.04	0.0180
Kovar carrier	25	0.06350	8.2	0.1419
Silicon grease	5	0.01270	1.21	0.0209

1 Neglect the conduction/convection/radiation mode through the air gap between the alumina substrate and the Kovar steel lid.
2 Use the area of the alumina substrate as the basis for the calculations, ignoring the fact that the area of the Kovar steel carrier is somewhat larger than that of the substrate and that the heat will spread somewhat in the Kovar as it is conducted toward the aluminum heat sink.
3 Account for lateral conduction of chip and thick-film dissipation in the alumina substrate and Kovar carrier. This "thermal spreading" effect is analogous to the so-called constriction effect discussed in Chap. 4. The term *constriction effect* is used in the subsequent discussion. The constriction effect just under the chip dissipator and the thick film resistor is evaluated using Eqs. (4.42) and (4.45).

The electrothermal analog circuit with all of the constriction effects indicated is shown in Fig. 23.3.

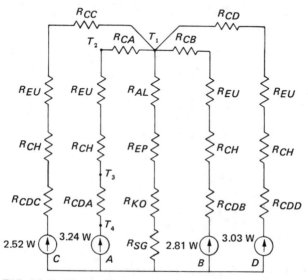

FIG. 23.3   Electrothermal analog circuit for the package of Fig. 23.2.

The evaluation of the thermal resistances may now proceed in accordance with Eq. (4.16a):

$$R = \frac{L}{kA} \qquad (4.16a)$$

1 Chip resistance R_{CH} for a single chip is[2]

$$R_{CH} = \frac{0.02032}{1.137(0.1270)(0.1270)} = 1.108 \, °C/W$$

2 Eutectic bond resistance R_{EU} for the entire substrate is

$$R_{EU} = \frac{0.00254}{2.958(1.524)(2.032)} = 0.0003 \, °C/W$$

3 Alumina R_{AL} for the entire substrate is

$$R_{AL} = \frac{0.0508}{0.294(1.524)(2.032)} = 0.0558 \, °C/W$$

4 Conductive epoxy R_{EP} for the entire substrate is

$$R_{EP} = \frac{0.0127}{0.0180(1.524)(2.032)} = 0.2278 \, °C/W$$

5 Kovar steel carrier R_{KO} for the area under the entire substrate is

$$R_{KO} = \frac{0.0635}{0.1419(1.524)(2.032)} = 0.1445 \, °C/W$$

6 Silicon grease R_{SG} for the area under the entire substrate is

$$R_{SG} = \frac{0.0127}{0.0209(1.524)(2.032)} = 0.1962 \, °C/W$$

The temperature T at the surface of the alumina substrate can now be established by using items 3 through 6:

$$T_1 - 55 = (R_{AL} + R_{EP} + R_{KO} + R_{SG})q$$

where q is the total dissipation of chips A, B, C, and D. Thus, with $q = 11.60$ W,

$$T_1 - 55 = (0.0558 + 0.2278 + 0.1445 + 0.1962)(11.60)$$

---

[2] The thermal conductivity of silicon varies greatly with temperature. This explains the discrepancy between values in Table 23.3 at 100°C and in Table 4.1 at 27°C.

or $T_1 = 55 + 0.6243(11.60) = 62.2°C$

As shown in Fig. 23.3, T_1 is the temperature at the surface of the substrate. This temperature is the driving force for the conduction of heat from all four chips to a heat sink at 55°C through the resistances provided by the configuration if the heat were distributed uniformly over the surface of the substrate.

It is to be noted that the inclusion of air layer conduction/convection and radiation from substrate to lid will result in a somewhat reduced value of T_1.

Attention is now focused on semiconductor chip A, which is the heaviest dissipator among the four chips. The actual temperature under chip A is T_2, which is the temperature required to overcome the constriction effect for heat flow from chip A into the substrate. This constriction effect is indicated by R_{CA} in Fig. 23.3. With

$$\Delta T_C = T_2 - T_1$$

then

$$R_{CA} = \frac{\Delta T_C}{q} = \frac{\Delta T_C}{3.24} \quad (°C)$$

where T_C is calculated from Eq. (4.45) (see Fig. 4.17).

$$\Delta T_C = \Delta T_{C1} + \Delta T_{C2} + \Delta T_{C3} \tag{4.45}$$

with $\Delta T_{C1} = \dfrac{q}{2\pi^2 k} \dfrac{b}{ac} \displaystyle\sum_{m=1}^{\infty} \dfrac{\sin(m\pi a/b)}{m^2}$

$\Delta T_{C2} = \dfrac{q}{2\pi^2 k} \dfrac{c}{db} \displaystyle\sum_{m=1}^{\infty} \dfrac{\sin(m\pi d/b)}{m^2}$

and $\Delta T_{C3} = \dfrac{q}{2\pi^2 k} \dfrac{2}{ad} \displaystyle\sum_{m=1}^{\infty} \displaystyle\sum_{n=1}^{\infty} \dfrac{\sin(n\pi d/c)\sin(m\pi a/b)}{mn[(m\pi/b)^2 + (n\pi/c)^2]^{1/2}}$

With $a = d = 0.0635$ cm, $b = 0.3810$ cm, $c = 0.5080$ cm, $q = 3.24$ W, and $k = 0.294$ W/cm °C:

$$\Delta T_{C1} = \frac{3.24}{2\pi^2(0.294)} \frac{0.3810}{(0.0635)(0.5080)} (0.8645)$$

or $\Delta T_{C1} = 5.70°C$ with the last term representing a summation over m from 1 to 50;

$$\Delta T_{C2} = \frac{3.24}{2\pi^2(0.294)} \frac{0.5080}{(0.0635)(0.3810)} (0.7600)$$

or $\Delta T_{C2} = 8.91°C$ with the last term representing a summation over m from 1 to 50;

$$\Delta T_{C3} = \frac{3.24}{2\pi^2(0.294)} \frac{2}{(0.0635)(0.0635)} (0.0933)$$

or $\Delta T_{C3} = 25.84°C$ with the last term representing a summation over $m = 1$ to 12 for $n = 1$ to 35. Thus

$$\Delta T_C = 5.70 + 8.91 + 25.84 = 40.5°C$$

which makes

$$T_2 = T_1 + \Delta T_C = 62.2 + 40.5 = 102.7°C$$

and if there is any interest,

$$R_{CA} = \frac{40.5}{3.24} = 12.50 \,°C/W$$

In accordance with the solution strategy, the next step is to determine the surface temperature of the silicon chip. This is the temperature T_3 in Fig. 23.3, and its value is calculated from

$$T_3 = T_2 + (R_{EU} + R_{CH})3.24$$
$$= 102.7 + (0.0003 + 1.1080)(3.24) = 102.7 + 3.60$$

or $\quad T_3 = 106.3°C$

Then, by calculating the constriction effect for the component on the chip, the sought-for maximum chip temperature is obtained. For a circular dissipator on an infinite conducting chip (Fig. 4.14), Eq. (4.42) applies:

$$\Delta T_C = \frac{q}{2\sqrt{\pi}ak} \qquad\qquad (4.42)$$

Here $a = 17.5$ mil (0.04445 cm) and $k = 1.137$ W/cm $°C$ for silicon and

$$\Delta T_C = \frac{3.24}{2\sqrt{\pi}(0.04445)(1.137)} = 18.08°C$$

and $\quad T_4 = T_3 + \Delta T_C = 106.3 + 18.1 = 124.4°C$

This is the calculated maximum temperature in the configuration. Observe that from this maximum temperature to the heat sink, the temperature difference is $69.4°C$.

23.4 THE ADJUSTMENT FOR CONVECTIVE AND RADIATIVE EFFECTS

23.4.1 The Substrate Level

If heat transfer from substrate to lid through the air gap is to be accounted for by simple analytic formulations, it is necessary to make two basic assumptions:

1 The lid is at a constant temperature and, indeed, in the ensuing computations, it
   will be taken at 55°C.
2 The temperature source is the average temperature of the substrate: T_1 in Fig. 23.3.

Heat will flow across the air gap by conduction and/or convection in parallel
with radiation. As shown in Sec. 6.5, the Grashof number based on the gap dimension
dictates the actual mechanism and for horizontal air gaps, Eq. (6.81),

$$Nu_s = 0.195(Gr_s)^{1/4} \qquad\qquad (6.81)$$

may be employed *if* the Grashof number

$$Gr_s = \frac{g\beta\rho^2 \, \Delta T s^3}{\mu^2}$$

lies between 10^4 and 4×10^5. This will obviously not be the case for small gaps.

When s is small, such that for horizontal gaps $Gr_s < 1,700$, the heat flow is primarily by conduction (see Sec. 6.5) and the Nusselt number is equal to unity,

$$Nu = \frac{hs}{k} = 1.00$$

which shows that

$$h = \frac{k}{s}$$

Thus, to establish the conductance across the gap:

1 Below $Gr_s = 1,700$, use purely conduction mode employing the relationships

$$q = hS \, \Delta T$$

   or $$q = \frac{k}{s} \, S \, \Delta T$$

2 Between $1,700 < Gr < 10,000$, use Fig. 23.4 to establish U:

$$U = \frac{k}{s} \, Nu$$

   and $$q = US \, \Delta T$$

3 Above $Gr = 10,000$, use Eq. (6.81) to establish h and then employ

$$q = hS \, \Delta T$$

The radiation across the air gap may be calculated from Eq. (7.14):

$$q = \sigma F_A F_e S(T_1^4 - T_R^4)$$

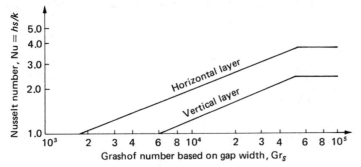

FIG. 23.4 Heat transfer through enclosed plane air layers [17, 18].

where $T_R = 273 + 55 = 328°C$. Here one may assume that the Kovar steel lid is in full view of the substrate, so that $F_A = 1.0$ and for parallel planes (see Table 7.1),

$$F_\epsilon = \frac{1}{(1/\epsilon_1) + (1/\epsilon_2) - 1}$$

For $\epsilon_1 = \epsilon_2 = 0.1$,

$$F_\epsilon = \frac{1}{(1/0.1) + (1/0.1) - 1} = \frac{1}{19}$$

and with $\sigma = 5.67 \times 10^{-8}$ W/m² K⁴,

$$q = 5.67(1.0) \left(\frac{1}{19}\right) S \left[\left(\frac{T_s}{100}\right)^4 - (3.28)^4\right]$$

or with $S = 600 \times 800$ mil $= 0.48$ in² $(3.097 \times 10^{-4}$ m²$)$,

$$q = 9.241 \times 10^{-5} \left[\left(\frac{T_1}{100}\right)^4 - 115.74\right] \tag{23.4}$$

It will now be shown that the heat flow by radiation and air layer conduction-convection has a negligible effect on the substrate surface temperature. Using the calculated value of $T_1 = 62.2°C$ (Sec. 23.3), the radiation heat flow can be computed using Eq. (23.4):

$$q = 9.241 \times 10^{-5} \left[\left(\frac{62.2 + 273}{100}\right)^4 - 115.74\right]$$

$$= 9.241 \times 10^{-5} [(3.352)^4 - 115.74]$$

$$= 9.241 \times 10^{-5} (10.506) = 0.000971 \text{ W}$$

Observe that if the inside of the Kovar steel lid were treated so that its emissivity is 0.8, then

$$F_\epsilon = \frac{1}{(1/0.8) + (1/0.1) - 1} = \frac{1}{10.25}$$

and the heat flow by radiation would be

$$q = \frac{19}{10.25}(0.000971) = 0.0018 \text{ W}$$

In either case the heat flow by radiation is negligible but might not be if the substrate temperature were higher, the lid temperature were lower, and the surfaces were treated to enhance radiative transport (higher F_ϵ).

The heat flow across the air gap requires calculation of the Grashof number. Recall that the Rayleigh number is the product of the Grashof and Prandtl numbers,

$$\text{Ra} = \text{Gr} \, \text{Pr}$$

and that Eq. (6.78) points out that

$$\text{Ra} = aL^3 \, \Delta T$$

at atmospheric pressure. This allows for a simple calculation of the Grashof number:

$$\text{Gr} = \frac{aL^3 \, \Delta T}{\text{Pr}}$$

At an average temperature of $\frac{1}{2}(55 + 62.2) = 58.6°\text{C}$, Figs. 6.9 and 6.10 give

$$a = 5.75 \times 10^7 \text{ m}^{-3}\,°\text{C}^{-1}$$
$$k = 0.0278 \text{ W/m}\,°\text{C}$$

and $\text{Pr} = 0.707$

Thus

$$\text{Gr}_s = \frac{5.75 \times 10^7 (0.003556)^3 (62.2 - 55)}{0.707} = 26.3$$

and hence Nu = 1.00. Thus conduction predominates and

$$q = \frac{kS}{s} \Delta T = \left(\frac{0.0278}{0.003556}\right)(3.097 \times 10^{-4})(7.2) = 0.0174 \text{ W}$$

which is also negligible.

One may draw the conclusion that at substrate temperatures compatible with typical junction temperature goals for reasonable reliability, air layer conduction-convection and radiation effects within the package are negligible. Reasonable attention means that excessive thicknesses of conductive epoxy between the substrate and the metal carrier and grease between the carrier and the heat sink should be minimized. Observe in Sec. 23.3 that the total thermal resistance between substrate surface and heat sink is composed of $R_{AL} = 0.0558$, $R_{EP} = 0.2278$, $R_{KO} = 0.1445$, and $R_{SG} = 0.1962$, all in $°C/W$. The total thermal resistance is $0.6243\,°C/W$, and of this total, the epoxy and grease contributes $0.2278 + 0.1962 = 0.4240$ or $0.4240/0.6243 = 0.679$, which is about 68% of the total. This means that 68% of the total temperature drop or $0.68(7.2) = 4.9°C$ occurs across the conductive epoxy and grease layers. It is obvious that when a heavy dissipator is under consideration, the thicknesses of these layers should be minimized.

23.4.2 The Chip Level

The heat flow by radiation and air layer conduction-convection from the surface of the silicon semiconductor chip to the Kovar steel lid is also negligible, as will now be shown. The same procedure that was used in Sec. 23.4.1 may be employed. Here, the chip surface area is 50×50 mils (1.613×10^{-6} m^2) and the gap thickness is $140 - 9 = 131$ mils (0.3327 cm). The chip surface temperature is T_3 in Fig. 23.3, and in Sec. 23.3 it was found that $T_3 = 106.3°C$.

For radiation with $F_e = 1/19$, $F_A = 1.0$, and using $T_3 = 106.3 + 273 = 379.3$ K, the heat flow by radiation from Eq. (7.14) is

$$q = 5.67(1.0) \left(\frac{1}{19}\right) (1.613 \times 10^{-6})[(3.793)^4 - 115.74]$$

or $q = 0.000044$ W

which is indeed negligible, but negligible because of the 50×50 mil surface area.

For air layer conduction-convection at an average temperature of $T_a = \frac{1}{2}(106.2 + 55) = 80.60°C$,

$$a = 4.30 \times 10^7 \text{ m}^{-3}\,°C^{-1}$$

$$k = 0.0294 \text{ W/m}\,°C$$

and   Pr $= 0.705$

Now

$$\mathrm{Gr}_s = \frac{(4.30 \times 10^7)(0.003327)^3(106.2 - 55)}{0.705} = 115.0$$

which again shows that conduction predominates. Thus

$$q = \frac{kS}{s}\,\Delta T = \left(\frac{0.0294}{0.003327}\right)(1.613 \times 10^{-6})(51.2) = 0.00073 \text{ W}$$

which is again negligible, partly because of the markedly reduced surface area.

Before drawing an adverse conclusion regarding the importance or lack thereof of including air layer and radiative effects in microelectronic thermal analysis, consider the following facts:

1 Surface area is required for the transfer of heat. Chip sizes vary, and what may seem negligible in one case may not be negligible in another.
2 Temperature levels and temperature differences play a significant role. For two surfaces of the same size, the previous calculations show that a comparison of $106.2°C$ and $62.2°C$ would yield, at a $55°C$ lid temperature,

$$\frac{(3.793)^4 - 115.74}{(3.352)^4 - 115.74} = \frac{91.02}{10.50} = 8.67$$

This means that the surface at $106.2°C$ would tend to transfer 8.67 times the amount of heat, if all other things were equal.

Indeed, in the air layer calculations with the Grashof number a function of $\Delta T^{1/4}$,

$$\left(\frac{106.2 - 55}{62.2 - 55}\right)^{1/4} = 1.63$$

which means that the air layer heat flow is 1.63 times greater at a surface temperature of $106.2°C$, all other things being equal.

3 If radiation and air layer conductances are considered, then the electrothermal analog circuit would look like Fig. 23.5, where provision is made for a temperature drop from lid to heat sink.

23.4.3 High Packaging Density

The complexity of the circuit shown would increase with a heavier population of heat dissipating devices and, indeed, the validity of the constriction resistance concept would be questionable as the effects of one dissipating component begin to interfere, due to proximity, with the effects of another. In such cases a computer-aided thermal analysis is called for and one might as well let the computer handle the air layer effects. One can be assured that the computer can handle, in a well-conceived program, all effects, whether negligible or not.

Thus, although the calculations in this and the foregoing subsection have shown that air layer effects are negligible for the example chosen, inclusion of these effects is strongly recommended.

23.5 A NOTE ON THE SO–CALLED SPREADING ANGLE

The heat flow by conduction from the surface of the substrate to the heat sink under the metal carrier is a three dimensional phenomenon. In Secs. 23.3 and 23.4 use was made of the average dissipation in a one-dimensional calculation and adjustments were made through the use of constriction effect.

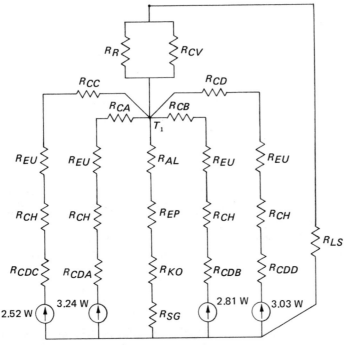

FIG. 23.5 Modification of electrothermal analog circuit of Fig. 23.3 to account for air gap heat transfer and radiation from substrate to lid of package.

The literature abounds with references to a spreading angle to account for the multidimensional heat flow. This allows the use of an average cross-sectional area, as Fig. 23.6 clearly shows. Observe that in a conductive medium of thickness δ, the average cross-sectional area can be calculated using the angle ϕ, which is the spreading angle:

$$A_a = A_1 + 2\left(\frac{\delta}{2}\cot\phi\right)$$

or $\quad A_a = A_1 + \delta\cot\phi$ $\hfill (23.5)$

Note, however, that A_a is the arithmetic mean of A_1 and A_2:

$$A_a = \tfrac{1}{2}(A_1 + A_2)$$ $\hfill (23.6)$

and $\quad A_2 = A_1 + 2\delta\cot\phi$ $\hfill (23.7)$

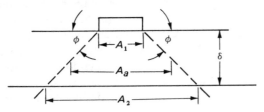

FIG. 23.6 Illustration of the spreading angle.

Solve Eq. (23.7) for $\cot \phi$,

$$\cot \phi = \frac{A_2 - A_1}{2\delta}$$

and put this result into Eq. (23.5),

$$A_a = A_1 + \frac{A_2 - A_1}{2}$$

or $A_a = \frac{1}{2}(A_1 + A_2)$

which is Eq. (23.6). This shows that if the arithmetic average of A_1 and A_2 (note that A_2 depends on ϕ) is to be used, the value of the spreading angle has no bearing on the calculation.

The absurdity of this and, indeed, the use of an arithmetic average for the conduction area, is easily demonstrated. Take as limits $A_1 \to 0$ and $A_2 \to \infty$. Then

$$A_a \approx \frac{0 + \infty}{2} \to \infty$$

which suggests a flow of heat at no temperature difference (infinite cross-sectional area), which is clearly in violation of the principles of thermodynamics. This corresponds to a spreading angle of zero degrees.

One can only conclude that because A_2 depends on the spreading angle ϕ, the calculation of A_a by these equations is vulnerable to assumption and may lead to erroneous results.

23.6 HEAT FLOW IN PRINTED CIRCUIT BOARDS

In printed circuit boards where face cooling is negligible and where primary cooling is provided at the board edges, it is necessary to evaluate in detail the thermal penalty associated with conduction of the dissipated heat to the board edges.

A complex heat flow configuration such as a multilayered printed circuit board may be treated as a resistive network (electrothermal analog). When the heat flow is multidimensional, approximate solutions can be obtained by estimating the heat transmission, through the use of the appropriate resistances, in each direction independently. The error inherent in this technique is inversely proportional to the size of the section under consideration (see Sec. 23.3). That is, the procedure is sound when applied to an array of small subvolumes, but the precision decreases as the subvolumes become progressively larger. For preliminary analyses, the effective conductances or resistances should be evaluated to determine the dominant heat flow path(s). In this manner, the accuracy of expedient approximations can be evaluated. Unfortunately, there are no all-inclusive guidelines, and the dominant heat flow path(s) in each case must be assessed recursively.

Recall that the conductive resistance is

$$R = \frac{L}{kA} \tag{4.16a}$$

and its reciprocal is the conductance,

$$K = \frac{kA}{L} = \frac{kW\delta}{L} \tag{23.8}$$

where δ is the thickness of the heat flow path. If n_c is the number of cutouts (holes for interconnection) and the diameter of each cutout is d_c, than a plane conduction factor may be defined as

$$\eta_p = \frac{W - n_c d_c}{W} \tag{23.9}$$

where W is the width of the plane. Use of this plane conduction factor allows the modification of Eq. (23.8) to account for the cutouts:

$$K_p = \frac{\eta_p kA}{L} \tag{23.10}$$

If the cut-out pattern varies in the direction of primary heat flow, an overall conductance may be computed by breaking the plane into a series of sections (in the direction of heat flow) and using

$$\frac{1}{K_p} = \frac{1}{\sum_{i=1}^{n} L/\eta_p kA} \tag{23.11}$$

where n is the number of sections. This, however, does not include the effects of cut-out pattern on transverse heat flow. It may be anticipated that progressively higher in-plane thermal resistances will be encountered as cut-out alignment varies from an ordered to a staggered distribution.

Smith [6] has studied printed circuit board gradients as a function of uniformly distributed power dissipation, plane conduction efficiency, number of 1-oz copper planes n_p, and board width. His results are shown in Figs. 23.7 and 23.8. The term *ounces of copper* refers to copper plane thickness, each ounce representing a 1.4-mil (0.003556-cm) thickness of copper. These numbers derive from the fact that 1 ft^2 of copper sheet 1.4 mils thick weighs 1 oz.

EXAMPLE
A 4 $\times$ 6 in (10.16 $\times$ 15.24 cm) circuit board dissipates 8 W of uniformly distributed power. The board is grounded (attached to an enclosure) along each of the 4-in edges to surfaces at identical temperatures. The board contains three 2-oz and four 1-oz

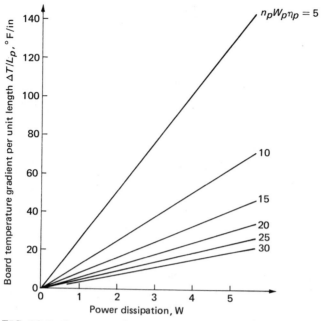

FIG. 23.7 Summary of printed circuit board thermal gradient studies, low power [6].

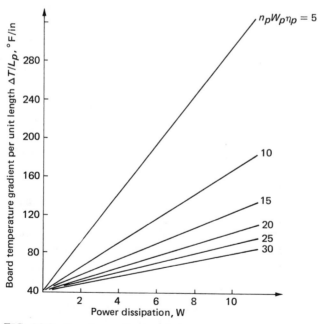

FIG. 23.8 Summary of printed circuit board thermal gradient studies, high power [6].

copper ground and circuit planes whose plated-through hole patterns are reasonably uniform in the direction of heat flow. The conduction factor calculated from Eq. (23.9) is 0.45. Use Fig. 23.8 to determine the center-to-edge temperature rise for the board.

SOLUTION
Here $W = 10.16$ cm, $\eta_p = 0.45$, and $n_p = 3(2) + 4(1) = 10$. Thus

$$\eta_p W n_p = 0.45(10.16)(10) = 45.72$$

and at 8 W, Fig. 23.8 shows that $\Delta T/L = 17.5\,°F/in$ or $12.4\,°C/cm$. With $L = 7.62$ cm, the temperature rise is

$$\Delta T = 12.4(7.62) = 94.5°C$$

It is to be noted that this result, like others based on Figs. 23.7 and 23.8, ignores the contribution of convection and radiation to the dissipation from printed circuit boards.

23.7 THE APPARENT CIRCUIT BOARD THERMAL CONDUCTANCE

In analyzing heat conduction in printed circuit boards, there is a tendency on the part of some analysts to ignore the thermal contribution of the epoxy board relative to the high-thermal-conductivity copper claddings. It should be recognized that heat transfer is governed by the thermal conductance, which, in this configuration, depends on the cross-sectional area for heat flow as well as the material thermal conductivity and length of heat flow path. To a first approximation, for parallel paths, the apparent conductance of a printed circuit board may be calculated by summing the conductance of the epoxy and copper:

$$K = \frac{kA}{L}\bigg|_{copper} + \frac{kA}{L}\bigg|_{epoxy} \tag{23.12}$$

From this relation, it may be seen that for a given conduction path length, it is the ratio of kA's, epoxy to copper, that must be examined to determine the appropriateness of neglecting the contribution of the epoxy. Clearly, as the thickness of the copper layer (or its total cross-sectional area) decreases, thermal conduction in the epoxy becomes more significant.

For example, consider the side view of a typical printed circuit board, shown in side view in Fig. 23.9, which contains dissipating components on one face. The dimensions are shown in Fig. 23.9a and an electrothermal analog circuit for the flow of heat by conduction is shown in Fig. 23.9b. It is presumed that heat transfer by conduction is the only mode; no heat flows off the board faces by convection or radiation, and the area of interest is 1 cm in length and width.

Then for the 1-oz copper layer (0.00356 cm) with no cutouts,

$$R_1 = R_4 = \frac{L}{kA} = \frac{1.00}{3.80(1.00)(0.00356)} = 7.4004\,°C/W$$

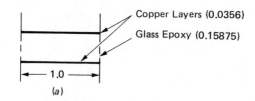

(a)

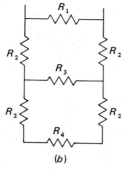

(b)

FIG. 23.9 Thermal network (electrothermal analog circuit) for side view of printed circuit board between two nodes with single-side dissipation.

For the two epoxy resistances,

$$R_2 = \frac{L}{kA} = \frac{0.07938}{(0.0026)(1.0)(1.0)} = 30.5288 \,°\text{C/W}$$

and $R_3 = \frac{L}{kA} = \frac{1.0}{(0.0026)(0.15875)(1.0)} = 2422.7741 \,°\text{C/W}$

The total thermal resistance from point A to point B is a simple exercise in combining resistances.

$$R_a = 2R_2 + R_4 + 2(30.5288) + 7.4004 = 65.4581 \,°\text{C/W}$$

$$R_b = \frac{R_3 R_a}{R_3 + R_a} = \frac{(2422.7741)(68.4581)}{2422.7741 + 68.4581} = 66.5769 \,°\text{C/W}$$

$$R_c = 2R_2 + R_b = 2(30.5288) + 66.5769 = 127.6346 \,°\text{C/W}$$

and finally,

$$R_T = \frac{R_1 R_C}{R_1 + R_C} = \frac{(7.4004)(127.6346)}{7.4004 + 127.6346} = 6.9948 \,°\text{C/W}$$

For this configuration the board conductance defined by $K = 1/R_T$ is

$$K = \frac{1}{6.9948} = 0.1430 \,\text{W/}°\text{C}$$

whereas the conductance for the copper alone is $K = 1/R_1$:

$$K = \frac{1}{7.4004} = 0.1351 \text{ W/}°\text{C}$$

Consequently, the percent improvement by considering the entire board is

$$100 \ \frac{0.1430 - 0.1351}{0.1351} = 5.79\%$$

and there is a slight improvement in the conductance when the entire circuit board is considered as an entity.

Suppose, however, that the copper layers on the board faces contain cutouts in them so that $\eta_p = 0.60$. Then $R_1 = R_4 = 7.4004/0.60 = 12.334 °\text{C/W}$, and the calculations now yield

$$R_a = 2R_2 + R_4 = 2(30.5288) + 12.3340 = 73.3917 °\text{C/W}$$

$$R_b = \frac{R_3 R_a}{R_3 + R_a} = \frac{(2422.7741)(73.3917)}{2422.7741 + 73.4581} = 71.2338 °\text{C/W}$$

$$R_c = 2R_2 + R_b = 2(30.5288) + 71.2338 = 132.2915 °\text{C/W}$$

and $\quad R_T = \frac{R_1 R_C}{R_1 + R_C} = \frac{(12.3340)(132.2915)}{12.3340 + 132.2915} = 11.2821 °\text{C/W}$

For this case, the total conductance is

$$K = \frac{1}{R_T} = 0.0886 \text{ W/}°\text{C}$$

which may be compared to the conductance of the copper alone:

$$K = \frac{1}{R_1} = \frac{1}{12.3340} = 0.0811 \text{ W/}°\text{C}$$

and the percent improvement realized by including the contribution of the epoxy is 9.32, as shown by

$$100 \left(\frac{0.0886 - 0.0811}{0.0811} \right) = 9.32\%$$

Observe that any favorable effect of the epoxy laminate on the overall conduction process becomes more pronounced as the copper layers become more populated with hole cutouts.

23.8 THE RESISTANCE PER SQUARE METHOD

Dickerson [7] has provided an interesting method for the evaluation of the maximum circuit board temperature under four common mounting conditions. The method is based on the thermal resistance per square ($^\circ$C/W) after exhaustive computer analyses of many, many circuit board configurations. Note in Eq. (23.8) that if $L = W, K = k\delta$, and thermal resistance for conduction heat flow becomes simply

$$R_{\text{sq}} = \frac{1}{k\delta} \; ^\circ\text{C/W} \tag{23.13}$$

where δ is the thickness of the board.

Because of multilayer and heat distribution effects, the calculation of this resistance per square for a printed circuit board becomes quite complex. For this reason, Dickerson used numerical calculations to determine this resistance for several board configurations. His results are shown in Fig. 23.10. To use this figure, the analyst must determine which of the five configurations shown is most similar to the board to be analyzed. Next, the appropriate circuit board constant, ψ, is selected from the logarithmic scale along the top of the figure. With this parameter, the thermal resistance per square can be determined from

$$R_{\text{sq}} = R_{\text{sq}b} - \psi(R_{\text{sq}b} - R_{\text{sq}c}) \; ^\circ\text{C/W} \tag{23.14}$$

where $R_{\text{sq}b}$ and $R_{\text{sq}c}$ are the resistances per square of the naked board and copper layers, respectively (both $^\circ$C/W).

Once the value of R_{sq} has been computed, the analyst may determine the temperature rise between the center of the board and the board edges, ΔT_m. Dickerson provides four cases, which are summarized in the display in Fig. 23.11. For case (a), which considers two board edges as nondissipating (to the enclosure, of course), no heat transfer from the faces of the board and uniform power dissipation over one (not both) board face is

$$\Delta T_{\text{max}} = T_m - T_e = \frac{R_{\text{sq}}qL}{qW} \; ^\circ\text{C} \tag{23.15}$$

with q the dissipation over the entire face of the board in watts and L and W as shown in Fig. 23.11, both in inches.

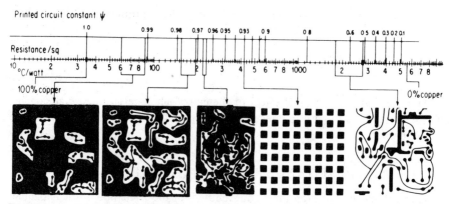

FIG. 23.10 Printed circuit board copper coverage and thermal resistance data [7].

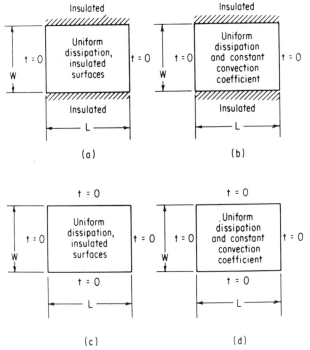

FIG. 23.11 Boundary conditions for printed circuit board analysis [7].

For case (b), which is identical to case (a) except that there is a constant heat transfer coefficient (h in Btu/ft^2 h °F) at both faces, allowing for heat dissipation to the surroundings,

$$\Delta T_{max} = T_m - T_e = 137 \frac{q}{hLW} \left\{ 1 - \frac{1}{\cosh\left[0.0426L(hR_{sq})^{1/2}\right]} \right\} °C \qquad (23.16)$$

Cases (c) and (d) consider all four board edges dissipating to the enclosure. For case (c) without convection from either face,

$$\Delta T_{max} = T_m - T_e = \frac{R_{sq}qW}{8L} \left[1 - \frac{1}{\cosh\left(1.57L/W\right)} \right] °C \qquad (23.17)$$

and for case (d), which includes heat transfer from both faces,

$$\Delta T_m = T_m - T_e = 137 \frac{q}{hLW} (1 - A + B) °C \qquad (23.18)$$

where

$$A = 0.785 \left(\frac{1}{\cosh\left\{ \left[0.00725hR_{sq} + (\pi^2/L^2)\right]^{1/2} (W/2) \right\}} \right)$$

and

$$B = \frac{1}{\cosh\{[0.00725hR_{sq} + (\pi^2/W^2)]^{1/2}(L/2)\}}$$

23.9 THE PRINTED CIRCUIT BOARD WITH NONUNIFORM HEAT INPUT

In Chap. 14 it was shown that the heat loss from a longitudinal fin of rectangular profile varies from the edge to the base of the fin. The most important variation results from the array of temperature excesses $\theta = t - t_s$ and heat fluxes that must coexist between surroundings and fin.

Consider a printed circuit component comprising a plastic or glass epoxy board or card and its integral holder, as shown in Fig. 23.12. Over different sections of the card height, copper electric resistances are deposited. Consider further than the card may be divided into uniform or nonuniform increments of height and that the heats liberated in the increments are constant with time and may or may not be equal. Since the incremental heights and heat inputs do not vary as any related function of the fin height, they are thus discrete. If a convection coefficient is present, it is conceivable that heat may flow to the base of the card, its edge, or both. Some of the plastic materials selected for copper circuit deposition, for reasons of cost, light weight, or unique molding requirements, have low softening temperatures. In such cases it is desirable to determine whether the heat generated by the circuit at some height can raise the card above its softening temperature.

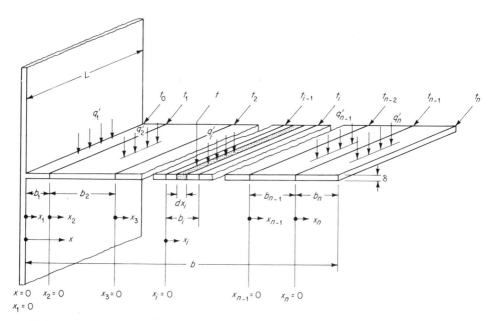

FIG. 23.12 Possible arrangement of printed circuit board with integral holders (card guides), one of which is at the left. The terminology and coordinate system is shown.

23.9.1 Longitudinal Fin of Rectangular Profile
with Discrete, Nonuniform Heat Input

The longitudinal fin of rectangular profile with discrete, nonuniform heat input has been analyzed by Smith [8] for the case with no external convective or radiative coefficient. In the configuration shown in Fig. 23.12, q_i' represents the heat input per unit area over each side of the fin for the ith area under consideration. Since the different incremental areas receive different heat inputs, the overall loading of the fin is nonuniform. The differential equation for the temperature profile in any of the incremental areas may be obtained from an energy balance on the differential element dx_i. Assuming steady, unidirectional flow, the difference between the heat entering and that leaving the element by conduction is

$$-dq = kA \frac{d^2 t}{dx_i^2} dx_i$$

where dq is negative because the net heat flow to the base has been assumed to decrease in the direction of positive x. This may be equated to the heat gained by the fin, so that

$$-kA \frac{d^2 t}{dx_i^2} dx_i = 2q_i'L \, dx_i$$

Observe that there is no convection heat flow from the board in this case. With the cross-sectional area taken as $A = \delta_0 L$, the differential equation for the temperature profile of the ith incremental area may be written as

$$\frac{d^2 t}{dx_i^2} + \frac{2q_i'}{k\delta_0} = 0 \tag{23.19}$$

Integrating twice for the general solution gives

$$\frac{dt}{dx_i} + \frac{2q_i'}{k\delta_0} x_i + C_1 = 0$$

and $$t + \frac{q_i' x_i^2}{k\delta_0} + C_1 x_i + C_2 = 0 \tag{23.20}$$

Equation (23.20) is the general solution, and the arbitrary constants C_1 and C_2 may be evaluated from the boundary conditions:

At $x_i = 0$ $t = t_{i-1}$ (23.21)

At $x_i = b_i$ $t = t_i$ (23.22)

Applying Eq. (23.21) in Eq. (23.20) gives

$$C_2 = -t_{i-1}$$

Applying Eq. (23.22) in Eq. (23.20) with this value of C_2 gives

$$t_i - t_{i-1} + \frac{q_i' b_i^2}{k\delta_0} + C_1 b_i = 0$$

$$C_1 = -\frac{(t_i - t_{i-1})}{b_i} - \frac{q_i' b_i}{k\delta_0}$$

The particular solution to Eq. (23.19) is

$$t = t_{i-1} + \frac{x_i}{b_i}(t_i - t_{i-1}) + \frac{q_i' x_i}{k\delta_0}(b_i - x_i) \tag{23.23}$$

It is seen that when $x_i = 0$, $t = t_{i-1}$ and when $x_i = b_i$, $t = t_i$, as they should. It is also seen that Eq. (23.23) gives the temperature at any point in the incremental area governed by the coordinate x_i as long as the end-point temperatures t_{i-1} and t_i are known.

The end-point temperatures may be found by considering the heat flow in each incremental area. At the origin of the ith incremental area, the heat flow in the card is the sum of the heat flows in all the incremental areas remote from the point under consideration:

$$kA \left.\frac{dt}{dx_i}\right|_{x_i=0} = 2L \sum_{i=i}^{i=n} q_i' b_i$$

With $A = \delta_0 L$, this may be rearranged to give

$$\left.\frac{dt}{dx_i}\right|_{x_i=0} = \frac{2}{k\delta_0} \sum_{i=i}^{i=n} q_i' b_i \tag{23.24}$$

By taking the derivative of Eq. (23.23), evaluating at $x_i = 0$, and then using the result in Eq. (23.24), one obtains

$$\frac{t_i - t_{i-1}}{b_i} + \frac{q_i' b_i}{k\delta_0} = \frac{2}{k\delta_0} \sum_{i=i}^{n} q_i' b_i$$

or $$t_i = t_{i-1} - \frac{q_i' b_i^2}{k\delta_0} + \frac{2b_i}{k\delta_0} \sum_{i=i}^{n} q_i' b_i \tag{23.25}$$

Equation (23.25) permits evaluation of one incremental-area end-point temperature if the other end-point temperature is known. To compute the temperature profile, Eq. (23.23) is used after the end-point temperatures have been obtained from Eq. (23.25).

When a convection or radiation coefficient is present, it may be treated quite readily by a convergence procedure. First obtain the temperature profiles for the case with no convective or radiative coefficient. Then estimate corrected temperatures and incremental heat inputs for the effects of convection or radiation and repeat the calculation until successive estimates and calculations converge.

EXAMPLE

A glass epoxy printed circuit board similar in configuration to that shown in Fig. 23.12 is to be installed in a closed box and subjected to a number (5) of simultaneous discrete heat inputs that are due to component dissipation. The allowable maximum surface temperature of the epoxy material has been selected to be 125°C. The board has dimensions 7.5 × 25 cm with heat flowing along the 25 cm direction and the board midline taken at 12.5 cm. It is 0.305 cm thick and has an effective thermal conductivity of 0.0385 W/cm°C. The chassis (mounting at the fin base) is to be maintained at 71°C, and it is desired to find out if the maximum temperature of 125°C will be exceeded and, if so, at what height. The five discrete heat inputs along with their dimensions of applicability are displayed in Table 23.3.

SOLUTION

To find the required temperature distribution, it is necessary to:

1. Establish the incremental-area end-point temperatures. This is done by using Eq. (23.25) in the step-by-step procedure outlined in Table 23.4.
2. Pinpoint the temperature of 125°C by utilizing Eq. (23.23).

Table 23.4 shows that the limiting temperature of 125°C is reached in the third incremental area, and it is now required to find the precise location of the 125°C isotherm. From Eq. (23.23),

$$t = t_{i-1} + \frac{x_i}{b_i}(t_i - t_{i-1}) + \frac{q_i' x_i}{k\delta_0}(b_i - x_i) \tag{23.23}$$

a quadratic in x_i is easily obtained:

$$\frac{q_i' x_i^2}{k\delta_0} - \frac{q_i' b_i x_i}{k\delta_0} - \frac{x_i}{b_i}(t_i - t_{i-1}) + (t - t_{i-1}) = 0$$

TABLE 23.3 Two-Sided Heat Inputs and Pertinent Dimensions for Example

i	b_i (cm)	q' (W/cm^2); two sides
1	2.50	0.0080
2	3.75	0.0060
3	2.25	0.0048
4	2.00	0.0040
5	2.00	0.0030

TABLE 23.4 Computation of Incremental-Area End-Point Temperatures

$k\delta_0 = (0.0385)(0.305) = 0.01174 \text{ W/}^\circ\text{C}$

i	t_{i-1}	b_i	$q_i' b_i$	$q_i' b_i^2$	$q_i' b_i^2/k\delta$	$\Sigma\, q_i' b_i$	$2b_i$	$(1/k\delta)\,\Sigma\, q_i' b_i$	$(2b_i/k\delta)\,\Sigma\, q_i' b_i$	t_i
1	71.0	2.50	0.0200	0.0500	4.26	0.0673	5.00	5.73	28.67	95.40
2	95.40	3.75	0.0225	0.0844	7.19	0.0473	7.50	4.03	30.21	118.42
3	118.42	2.25	0.0108	0.0243	2.07	0.0248	4.50	2.11	9.50	125.86
4	125.86	2.00	0.0080	0.0160	1.36	0.0140	4.00	1.19	4.77	129.26
5	129.26	2.00	0.0060	0.0120	1.02	0.0060	4.00	0.51	2.04	130.29

$$t_i = t_{i-1} - \frac{1}{k\delta_0}\, q_i' b_i^2 + \frac{2b_i}{k\delta_0} \sum_{i=1}^{n} q_i' b_i$$

Insertion of the proper numbers from Tables 23.3 and 23.4 gives

$$0.4088x_i^2 - 0.9197x_i - 3.3067x_i + 6.58 = 0$$

or $$x_i^2 - 10.3393x_i + 16.0970 = 0$$

The result is $x_i = 1.91$ cm, which puts the line of 125°C a distance of 2.50 + 3.75 + 1.91 = 8.16 cm from the base of the board.

23.10 CONDUCTION HEAT FLOW FROM CIRCUIT BOARD TO ENCLOSURE

Printed circuit boards may be physically connected to the mounting enclosure by hard-mounting them with screws to ribs or brackets that may be considered to be integral with the enclosure. The thermal resistance of this connection and its associated temperature drop must be taken into account in predicting component temperatures. Considerable aid in determining these thermal resistances may be found in Chap. 9.

Another hard-mounting method is to screw the board onto hexagonal, rectangular, or cylindrical prisms known as standoffs. This is probably the least effective method, because the total contact area is quite small unless many standoffs are used. The use of many standoffs defeats the idea of high maintainability to some extent, but there is a slight advantage in that a standoff can possibly be employed at a heavy dissipating portion of the board. Here, too, the analyst is confronted with contact conductance (two contacts) and intrastandoff temperature drop problems.

Plug-in boards are mounted with card guides or wedges to allow for a heat conduction path and physical alignment with the electrical connector. Many types of card guides are available, and the vendor will specify the thermal resistance in °C/W per unit length. Values of thermal resistance vary from 30°C/W per centimeter to 15°C/W per centimeter. Wedges, or wedge clamps as they are often called, are somewhat more effective with typical values of 2.5 to 5.8 °C/W per centimeter, but these require a physical operation such as physically turning a screw to expand the wedge at each installation.

23.11 CONDUCTION COOLING FOR AN LSI PACKAGE

A key challenge in the design of large computers is maintaining temperature levels within specified limits. Computer technology developments give rise to large increases in heat generation per unit volume of the machine. At the same time, circuit designers require closer control of circuit temperatures and temperature gradients.

The objective of a heat-transfer design is to allow thermal energy to flow from the computer heat source to the energy heat sink (usually outside air) within the constraints of given temperature levels. The objective of the design is always the same. The thermal design will vary in the means of implementation.

All so-called small IBM computers are conventionally air-cooled. Despite the fact that many of the newer small computers contain relatively high-powered components, air cooling is feasible because the number of components and the total heat load are modest. By contrast, large machines, such as the IBM 3081, contain large numbers of high heat flux chips, resulting not only in a high local heat density but also in a relatively high total heat load and in a high heat load per unit floor area. The aim of thermal design of the 3081 is to use air-cooling where feasible and water-cooling when it is necessary [9, 10]. The choice between air and water-cooling depends on:

1 The heat density within the computer at the chip and package level and the specified chip junction and ambient temperature constraints.
2 The number of components contained in the computer and the magnitude of the resulting total heat load.
3 The heat load per unit floor area for the entire processor/storage complex.

The IBM 3081 uses air-cooling for the main memory, channels, and certain power supplies, and water-cooling to remove heat from the logic packages and certain power supplies. The balance of this section considers the water-cooled thermal conduction module (TCM) and the unique manner in which heat is carried away from the chip.

At the heart of the 3081 processor complex is a hermetically sealed, water-cooled package (Fig. 23.13) called the thermal conduction module (TCM). Each TCM contains a 90 mm² multi-layer ceramic substrate, designed to hold up to 118 semiconductor chips [11]. Each logic chip contains up to 704 circuits, for a maximum capacity of 45,000 circuits per module. As many as nine modules may be mounted on a 700 mm wide × 600 mm high performance printed circuit board [12]. As a result of the high packaging density achieved, peak heat fluxes of 20 W/cm² and 4 W/cm² were encountered at the chip and module levels, respectively. By contrast, earlier IBM systems such as the System/370 Model 168 and the IBM 3033 typically had peak heat

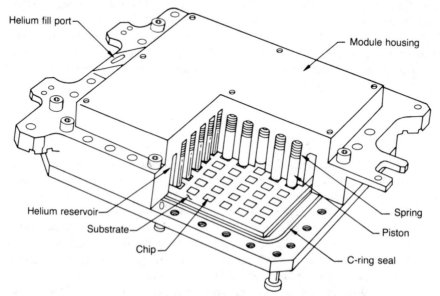

FIG. 23.13 Thermal conduction module (TCM).

fluxes of 1.5-2.5 W/cm² at the chip level and 0.3-0.6 W/cm² at the module level. Because of the high heat fluxes of the 3081, it was evident that air cooling could not provide an optimal solution to the logic cooling problem.

It was hoped that the water-carrying cold-plate surface could be brought as close to the chip heat source as possible. At the same time, it was necessary to allow for variations in chip heights and locations resulting from the manufacturing tolerances. Additionally, allowances had to be made for nonuniform thermal expansion or contraction to assure a good thermal path. To accomplish this, the helium-filled thermal conduction module, originally patented as the gas encapsulated module [13], was conceived. This cooling assembly has an individual piston contacting each chip providing a thermal path to the housing.

Although an individual TCM contains multiple chips and thermal paths to the cold plate, for purposes of describing the paths it is convenient to consider a single chip cell with its corresponding thermal analog circuit shown in Fig. 23.14. Each of the thermal resistances shown may be mathematically derived [14], but the intent here is to describe them with typical values.

The heat dissipating devices are on the side of the chip opposite the piston, and therefore, heat flow to the piston will encounter a conduction-constriction resistance R_c (0.4 °C/W) crossing the chip. Although the piston contacts the chip, most of the heat actually flows from the chip to the piston by means of thermal conduction across the gas gap surrounding the contact point. To minimize the magnitude of this thermal resistance, R_{c-p} (3.0 °C/W), the module is evacuated and then backfilled with helium gas (which has a thermal conductivity about six times that of air). The heat then spreads from around the contact region to the main body of the piston across a conduction-spreading resistance R_t (1.0 °C/W). Heat flows along the piston and across the surrounding helium-filled gas gap by thermal resistance R_{p-h} (2.2 °C/W) and is calculated by treating the piston and the surrounding housing structure as thermally coupled fins. The heat then flows across the module housing thermal conduction resistance R_h (1.6 °C/W) to the interface with the cold plate.

These several thermal resistance components may then be added together to form the total internal thermal resistance R_{int}:

$$R_{int} = R_c + R_{c-p} + R_t + R_{p-h} + R_h = 8.0 °C/W$$

Using this calculated internal resistance and the system water supply temperature of 24°C, maximum allowable chip power may be calculated as a function of module power and the module external thermal resistance from the housing to the water, R_{ext}. Thus:

$$T_J = \Delta T_{J-C} + P_c R_{int} + P_m R_{ext} + T_w$$

giving $P_c \leqslant 7.16 - (P_m R_{ext}/8.0)$ for $\Delta T_{J-C} = 3.0°C$, maximum $T_J = 85°C$ and $T_w = 24°C$. The results of this relation are shown in Fig. 23.15 for a range of R_{ext} (0.02-0.04 °C/W).

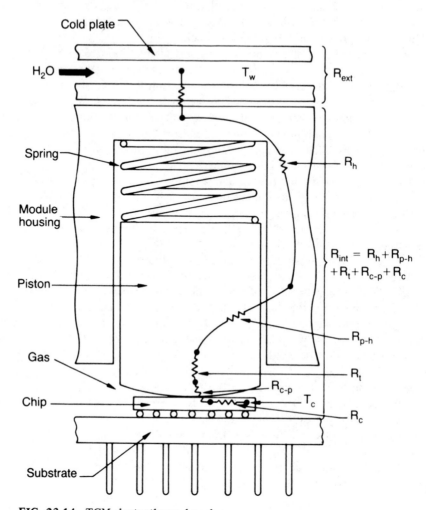

FIG. 23.14   TCM electrothermal analog.

23.12   VERY HIGH SPEED INTEGRATED CIRCUITS (VHSIC)

Future trends in electronic hardware for aircraft, missile, and satellite applications will require significant increases in capability as follows:

- Increased performance (greater accuracy, increased complexity, vastly increased functions)
- Increased operational lifetime (fault tolerance, improved reliability)
- Improved productibility (modular design and fabrication, testability)
- Greatly decreased size, weight, and total power

However, commensurate with these advances, the new hardware that will result from the implementation of VHSIC microelectronics will exhibit volumetric power densities that will severely challenge electrical and mechanical engineers to achieve successful thermal management. The objectives of this section are to summarize the potential VHSIC hardware thermal management problems and recommend several innovative approaches or technologies that may offer solutions to the thermal problems.

23.12.1   VHSIC Thermal Management Requirements

As pointed out by Block et al. [15], chip powers will vary from one watt to up to five watts. The chip size will be 0.2 to 0.25 in (5.08 to 6.35 mm) on a side yielding a maximum chip area of 0.0625 in^2 (40.3 mm^2). This results in a chip power density of 16 W/in^2 to 80 W/in^2 (2.5 to 12.4 W/cm^2). The chips will be packaged in a ceramic chip carrier (C^3) similar to the package configuration shown in Fig. 23.16. The C^3 will be approximately 1.0 in (25.4 mm) on a side and constructed of alumina. On a C^3 area basis, the typical VHSIC-C^3 will possess power densities up to 6 W/in^2 or up to 1 W/cm^2. In the absence of physical details at this stage of C^3/VHSIC circuit development, a thermal resistance between the chip junction(s) and the C^3 case (ΔT_{JC}) has been estimated at 15 °C/W. This value should be achievable and the discussion in this and subsequent sections of this chapter are based on this value.

In a typical functional packaging approach for an electronic board, up to six of

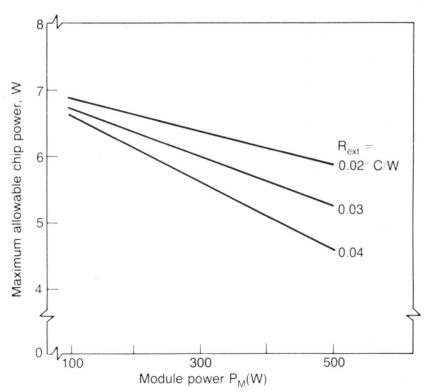

FIG. 23.15 Maximum chip power as a function of module power to hold $T_J = 85°C$ with $T_w = 24°C$.

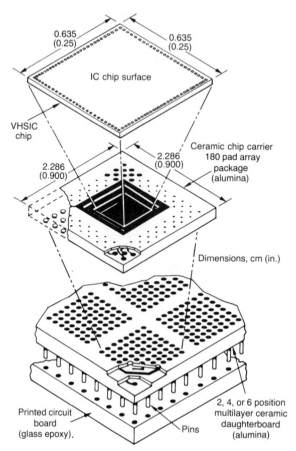

FIG. 23.16 Exploded view of VHSIC packaging concept.

these C^3s will be mounted to an alumina interconnect board (daughterboard). Thus, a 2 in × 3 in (5.08 cm × 7.62 cm) or larger VHSIC module may exhibit a power dissipation of from 6 to 30 W, posing a significant challenge to today's thermal management techniques. Because of thermal expansion mismatch between Al_2O_3 (alumina) and a G-10 fiberglass epoxy board, the VHSIC-C^3 module cannot be mounted directly to a printed circuit board (PCB), but must be attached to an intermediate ceramic interconnect that is connected to the PCB by pins soldered through plated-through holes as shown in Fig. 23.16.

Future hardware that implements VHSIC technology must meet the military specifications governing long term reliability and thermal environments. The requirements defined in MIL-STD-38510 were assumed in the study [15]. Thus, a chip junction temperature limit of 125°C was defined as the limit for long-term component reliability. An environmental temperature of 71°C, based on MIL-E-5400 and MIL-STD-810, was selected for the package heat sink environment.

In summary, the future implementation of VHSIC chips may well involve packaging chips in C^3 (Fig. 23.16) having power densities of 40 to 80 W/in² (6.2 to

12.4 W/cm^2) on a chip basis, 1 to 6 W/in^2 (0.155 to 0.93 W/cm^2) on a C^3 basis, and total package power densities of between 500 and 675 W/ft^3. Thus, even though VHSIC offers great advances in improving the performance and reducing the size, weight, and power of electronic hardware in future missions, the power densities involved result in heat dissipation rates that are beyond the present state of the art in thermal management of similar hardware.

23.12.2 Candidate Thermal Control Concepts for VHSIC Hardware

An evaluation of thermal management approaches for VHSIC components requires a review of the available thermal control techniques and other heat removal concepts that are not in general use at this time. The proposed use of ceramic chip carriers (C^3) mated to ceramic interconnects and then to printed circuit boards (PCB) narrows the choice and establishes a common reference for evaluating the efficacy of various thermal control concepts.

The primary features of each of these concepts can be defined by a vector-like quantity defined as follows:

F_1 —Heat Transfer Medium
F_2 —Thermal Mode
F_3 —Heat Removal Surface
F_4 —Ultimate Heat Sink

The interrelations between these quantities and the thermal management approaches considered for VHSIC are given in Fig. 23.17. Complete consideration of all available thermal control techniques would require analysis along the four axes defined.

The cited study [15] has addressed the use of coolants and conductive heat carriers as the heat transfer medium. Natural and forced convection (of each of the relevant coolants), direct heat conduction, and change-of-phase processes were the thermal modes examined. Heat exchange with the coolant or heat carrier was con-

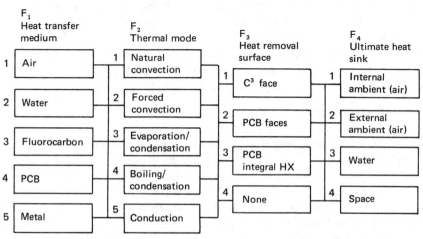

FIG. 23.17 Matrix of possible VHSIC thermal control approaches.

sidered to occur on the face of the C^3 package, on the surfaces of the motherboard ceramic or PCB, or within an integral heat exchanger bonded to the PCB. Finally, depending on the application of interest, heat removal from the total VHSIC electronics package can be achieved by once-through flow of the coolant or by heat rejection through a secondary heat exchanger to air, water, or an extraplanetary space radiator.

A rigorous application of this form of analysis would require investigation of 400 "potential concepts"; however, multiple concepts can be unified where possible—for example, direct air cooling of both C^3 and PCB surfaces. Further, the concepts can be filtered to remove inherently contradictory combinations such as the use of ambient air in a boiling/condensing mode or the use of metal carrier plates in a forced convection mode; the remaining concepts can then be subjected to a preliminary analysis to establish their suitability for more detailed consideration.

The following combinations of the features presented in Fig. 23.17 serve as an illustration:

1 Direct air heat transfer medium (F_1, 1)
   Natural convection mode of heat transfer (F_2, 1)
   Directly on C^3 face (also could use PCB face) (F_3, 1)
   Internal ambient air as heat sink (F_4, 1)
2 Metal plate heat transfer medium (F_1, 5)
   Conduction mode of heat transfer (F_2, 5)
   Directly to a space radiator (F_4, 4)

An immersion cooling system is given by the following combinations of features:

- Fluorocarbon heat transfer medium (F_1, 3)
- Boiling/condensing mode of heat transfer (F_2, 4)
- C^3 face heat removal surface (F_3, 1)
- External ambient heat sink (F_4, 2)

In the interest of brevity, selections within the concept matrix of Fig. 23.17 were made to yield only a limited number of VHSIC thermal control strategies. It was assumed that for all but the once-through air flow scheme, the ultimate heat sink was an 80°C heat rejection surface.

In succeeding subsections, several candidate concepts are described and analyzed in some detail. In a later section, these candidate concepts are compared and evaluated in terms of their ability to meet VHSIC thermal requirements.

DIRECT FORCED AIR COOLING

Simplicity and ease of maintenance makes reliance on direct convective air heat exchange a most attractive approach. The high heat fluxes and low junction-to-ambient temperature differences necessary for VHSIC packaging appear to eliminate the natural convection option and restrict attention to an approach using forced air convection. This configuration is shown in Fig. 23.18. The heat flow is from the VHSIC chip into the C^3 base and then either down through the alumina interconnect to the PCB or up through the C^3 sides to the top and cover. The forced flow of cooling air removes heat from the C^3 package, the alumina interconnect, and the PCB. A preliminary analysis using direct forced air cooling with a film heat transfer coefficient

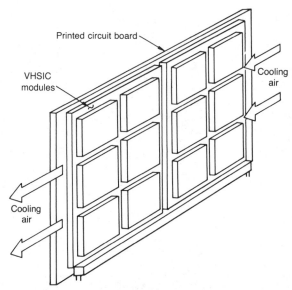

Printed circuit board

VHSIC
modules

Cooling
air

Cooling
air

FIG. 23.18 Air cooling on PCB and C^3 face.

(h) of 7.9 W/ft^2 $^\circ$C (85 W/m^2 $^\circ$C) yields a case-to-internal coolant thermal resistance of 11 $^\circ$C/W. Combining this with the assumed θ_{JC} of 15 $^\circ$C/W, a total junction-to-internal air resistance of 26 $^\circ$C/W is obtained.

In an open cooling system with 71°C air inlet, an average internal ambient air temperature of 76°C is reasonable. A closed cooling system consisting of a compact crossflow heat exchanger to cool the internal air would provide an average surface temperature of 80°C with 71°C external air. For such a cooling system, the average internal air temperature is likely to be at least 85°C. The compact heat exchanger and blowers needed to operate such a closed cycle system are estimated to add 3.75 lb (1.7 kg) and require fan power of 130 W at the system level.

IMMERSION COOLING

Dramatic improvement in the case-to-internal coolant thermal resistance can be obtained by immersion of the VHSIC carrying PCBs in a dielectric fluid with a boiling point at the desired temperature. Heat transfer by natural convection is obtained for low heat fluxes when the dielectric fluid bathes the C^3 and PCBs as shown in Fig. 23.19. For higher heat fluxes, heat transfer will be obtained by boiling of the coolant. It then transfers the heat by natural convection and/or condensation to the package heat rejection surfaces. It is to be noted that a gravitational field is required for the immersion convection/boiling process to occur.

A high convective heat transfer coefficient of 35 Btu/ft^2 h $^\circ$F (200 W/m^2 $^\circ$C) can be obtained by using a dielectric fluorocarbon as the coolant. This will reduce the case-to-coolant resistance of the VHSIC baseline C^3 package to 5.5 $^\circ$C/W for low C^3 heat dissipations (less than 2.5 W). Adding the assumed junction-to-case resistance of 15 $^\circ$C/W to this value gives an overall junction-to-internal coolant resistance of 20.5 $^\circ$C/W. This thermal resistance is applicable when low powers are being used and heat transfer is by natural convection.

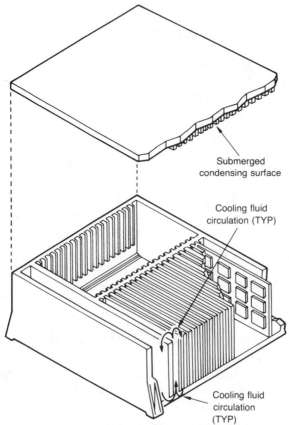

FIG. 23.19   Immersion cooling on PCB and C^3 face.

At higher C^3 power levels greater than 2.5 W, boiling is likely to occur and the case-to-coolant resistance decreases to 4 °C/W for dissipations up to 5 W and 3 °C/W up to 10 W. Taking into account the assumed value of ΔT_{JC} of 15 °C/W for a sealed C^3, boiling does not significantly alter the junction-to-internal coolant resistance. Thus, the junction-to-coolant resistance approaches a value of 18 °C/W at the highest heat dissipation rates. The presence of vapor bubbles in the coolant does substantially increase the heat transfer coefficient between the coolant and the package cooling surfaces. While in the boiling regime, condensation and bubble pumping of the dielectric fluid could be expected to maintain the internal coolant at an average temperature of approximately 85°C for an 80°C cooling surface temperature without the use of additional thermal design measures. Operation in the purely convective mode would require finning of the surface or circulation of the coolant through a heat exchanger to achieve the same internal temperature.

Thus, immersion cooling of the proposed baseline VHSIC packaging configuration could provide an average internal coolant temperature similar to that encountered in direct (closed system) forced air cooling. However, the higher heat transfer coefficients attainable in the immersion cooling mode would reduce the junction-to-coolant thermal resistance. To obtain these higher coefficients, the VHSIC unit must be filled with fluorocarbon fluid that produces an estimated weight penalty of 20 lb (9.1 kg) for the baseline configuration. At the higher power levels, it may be possible to reduce this weight by spacing the PCBs closer together. Conversely, the weight would increase at very low heat fluxes where heat transfer is by natural convection; this weight increase would be 1 lb (0.45 kg) for finning the heat rejection surface or 5 lb (2.3 kg) for a circulation pump and 1.5 lb (0.68 kg) for a liquid-to-air heat exchanger.

Redesign of the C^3 to allow boiling directly off the VHSIC chip surfaces would eliminate the impediment to heat flow posed by the junction-to-case resistance and could result in junction temperatures as low as from 10°C to 20°C above the coolant boiling point. This technique would require chip protection in the form of a conformal coating and/or a porous or screen-like C^3 lid and a fully compatible process/materials system to ensure chip integrity.

AIR COOLED INTEGRAL HEAT EXCHANGER

Direct air cooling of the VHSIC electronic components offers many thermal advantages but system constraints and specific military applications often eliminate this choice, thus forcing the use of air cooled integral heat exchangers. An integral heat exchanger mounted directly to the PCB and using forced air as the coolant is shown in Fig. 23.20.

The air-cooled integral heat exchanger requires that heat dissipated in the C^3 package be conducted through the ceramic interconnect and PCB to the heat exchanger. The baseline configuration yields a thermal resistance from the junction to the surface of the integral heat exchanger of 20°C/W. Using a finned cold plate as the integral compact heat exchanger can provide an average heat exchanger surface

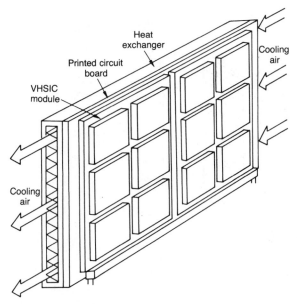

FIG. 23.20 Air cooled PCB integral cold plate heat exchanger.

temperature of 87°C for an inlet air of 71°C. The average surface temperature increases to 96°C when recirculated air through a secondary compact heat exchanger is used.

Each integral heat exchanger in the VHSIC baseline configuration weighs approximately 0.2 lb (0.09 kg), yielding a total weight increase of 9.5 lb (4.3 kg) for a once-through cooling concept. In a closed cycle mode, an additional weight penalty of 2.75 lb (1.25 kg) for the external air/air heat exchanger and 1 lb (0.45 kg) for the blower must be added.

LIQUID COOLED INTEGRAL HEAT EXCHANGER

The forced circulation of water through an integral heat exchanger bonded to the printed circuit board is shown in Fig. 23.21. This heat transfer system can be expected to provide lower junction temperatures than the air cooled integral heat exchanger technique. The junction-to-heat exchanger surface thermal resistance for this type of heat exchanger will be the same as for the air cooled integral heat exchanger and can be estimated at 20°C/W. However, the high heat transfer coefficients obtained in the liquid cooled integral PCB heat exchanger and on the waterside of the water/air external heat exchanger make a 90°C average surface temperature possible for a 71°C external air temperature. This cooling technique requires the addition of an external heat exchanger, pump, and 48 integral heat exchangers to implement this thermal control strategy and would result in an additional 21 lb (9.5 kg) to the VHSIC package.

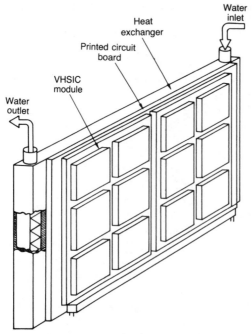

FIG. 23.21   Water cooled integral cold plate heat exchanger.

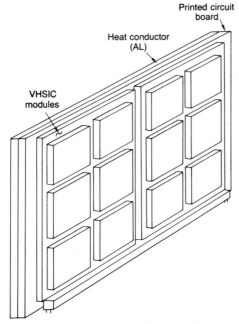

FIG. 23.22 Conduction cooled PCB.

CONDUCTION COOLED PRINTED CIRCUIT BOARD

A possible configuration of the Metal/Conduction/PCB Integral Heat Exchanger/ External Ambient concept "vector" is depicted in Fig. 23.22. In this thermal packaging configuration, heat must be conducted from the PCB surface through a thermally conductive epoxy layer into a metallic heat-sink plate, then through the PCB mounting to the chassis, and then by chassis conduction to the cooled external heat exchanger.

The junction-to-metal surface thermal resistance can again be estimated at 20 °C/W. Card retainer wedges on the two side support edges of the PCB results in a contact thermal resistance of 0.5 °C/W. This must be added to a "spreading" resistance (card-cold wall) of 0.5 °C/W to establish the card retainer wedge-to-external heat exchanger surface resistance.

The thermal resistance not yet considered is conduction through the PCB metal backing plate. This resistance is a function of metal heat sink conduction properties and the heat sink plate size. Since aluminum is the usual choice of material for such a plate, it is sufficient to consider only plate thickness as a variable. Assuming a heat dissipation of 25.2 W for each PCB, the surface temperature will be approximately 30°C above the wedge temperature for a 0.04 in (1 mm) thick plate and 6°C peak temperature (4°C average) above the wedge temperature for a 0.2 in (5 mm) thick plate.

For an assumed 80°C package skin temperature, a 0.2 in (5 mm) thick aluminum carrier plate will provide a maximum heat sink plate temperature of 115°C and an average of 113°C for the C³ heat dissipation associated with the baseline design.

Recent design studies appear to suggest that similar, or somewhat better, thermal results could be obtained with the use of flat heat pipes instead of the metallic backing plate [16]. While this approach could lead to a considerable reduction in the added weight, it is unclear at this time whether heat pipe technology can meet the VHSIC reliability requirements.

23.12.3 Comparison of Thermal Management Approaches

The thermal management approaches to cooling representative electronic packages containing VHSIC modules and chips are compared in Fig. 23.23. To put these results in proper perspective, the following considerations relative to these results should be noted:

- The analysis was based on VHSIC chips with equal power dissipations. In reality, an electronic package will contain chips with a variety of power dissipations.
- A ΔT_{JC} of $15\,°C/W$ was assumed for the C^3 package.

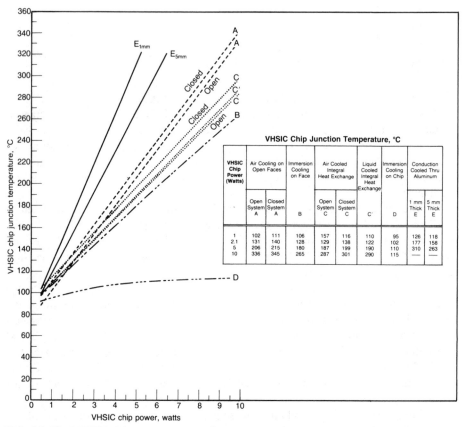

VHSIC Chip Junction Temperature, °C

VHSIC Chip Power (Watts)	Air Cooling on Open Faces		Immersion Cooling on Face	Air Cooled Integral Heat Exchange		Liquid Cooled Integral Heat Exchange	Immersion Cooling on Chip	Conduction Cooled Thru Aluminum	
	Open System A	Closed System A	B	Open System C	Closed System C	C'	D	1 mm Thick E	5 mm Thick E
1	102	111	106	157	116	110	95	126	118
2.1	131	140	128	129	138	122	102	177	158
5	206	215	180	187	199	190	110	310	263
10	336	345	265	287	301	290	115	—	—

FIG. 23.23 VHSIC chip junction temperature as a function of chip power for several thermal management approaches using junction-to-case resistance of $15\,°C/W$.

- The analysis involved only PCBs having two 2 × 3 in VHSIC modules, each containing six C^3 s.
- The thermal management approaches (i.e., flow rates, heat exchanger efficiencies) were designed only for heat transfer sufficiency without significant design optimization.
- The data and results for the 2.1 W/chip power dissipation is considered the system package baseline. Also, a maximum junction temperature of 125°C remains a requirement for long-term, high reliability electronics performance.

Figure 23.23 presents the analysis results for VHSIC junction temperatures with a package heat sink temperature of 71°C. The data compare five basic thermal management approaches in controlling the heat transfer from VHSIC chips that dissipate from 1 to 10 W each. The following conclusions are evident:

- At 1 W chip dissipation, all thermal management approaches are competitive and the chip junction temperature can be maintained at or below 125°C in an electronics package dissipating a total of 1200 W. Thermal management techniques employed today in aerospace hardware can be applied with success.
- At an average chip power dissipation of 2.1 W either immersion cooling on C^3 face or chip, or liquid cooling with a PCB integral heat exchanger can achieve the required VHSIC chip junction temperature (125°C). Direct package air cooling and air cooled integral heat exchangers can achieve approximately 130°C junction temperatures.
- Above the 2.1 W per chip level, the only design approach that can properly maintain acceptable VHSIC chip temperatures is immersion cooling with boiling directly off the chip itself.
- The foregoing is based on 15°C/W junction to case thermal resistance. The reader will find it easy to adjust the curves in Fig. 23.23 for values departing from 15°C/W.

23.13 CLOSURE

One can observe that a great deal of effort has been and continues to be expended in the thermal management arena. Innovative thinking is definitely required if the building of hardware is to keep pace with the ever burgeoning microelectronic technology. It is hoped that the survey contained in this chapter shows where we have been, where we presently are, and what issues we must address if there are to be minimal thermal and reliability problems in the future.

23.14 NOMENCLATURE

Roman Letter Symbols

a	natural convection property, $m^{-3} K^{-1}$; or a length, m
A	indicates a microelectronic chip; or a cross-sectional area, m^2
b	a length, m
B	indicates a microelectronic chip
c	a length, m; or specific heat, J/kg °C
C	indicates a microelectronic chip

d	a length, m; or diameter, m
D	indicates a microelectronic chip
F	a factor defined where used
g	acceleration of gravity, m/s^2
Gr	Grashof number, dimensionless
i	current, A
k	thermal conductivity, $W/m\,°C$
K	thermal conductance, $W/°C$
L	length of heat flow path, m
n	number of cutouts, dimensionless
Nu	Nusselt number, dimensionless
p	pressure ratio, dimensionless
P	power, W
Pr	Prandtl number, dimensionless
q	heat flow, W
R	thermal resistance, $°C/W$; or electrical resistance, Ω
Ra	Rayleigh number, dimensionless
s	gap spacing, m
S	surface area, m^2
t	temperature, $°C$
T	temperature, $°C$ or K
v	voltage (variable), V
V	voltage (steady), V
W	width of heat flow path, m
x	length coordinate, m

Greek Letter Symbols

β	volumetric coefficient of expansion, K^{-1}
δ	thickness of heat flow path, m
Δ	indicates change in variable
η	plane conduction factor, dimensionless
μ	dynamic viscosity $kg/m\,s$
ρ	density, kg/m^3
σ	Stefan-Boltzmann constant, $W/m^2\,K^4$
ϕ	spreading angle, degrees or radians
ψ	circuit board constant defined by Eq. (23.14), dimensionless

Subscripts

a	indicates average value
A	indicates arrangement factor
AL	indicates alumina substrate
c	indicates constriction effect or cutout or collector
ca	indicates constriction effect at alumina surface
cc	indicates bias quantity
ce	indicates collector to emitter quantity
CH	indicates chip resistance

EU   indicates eutectic bond
DP   indicates epoxy
JC   indicates junction to case
KO   indicates Kovar
L   indicates load
m   indicates maximum value
max   indicates maximum value
p   indicates constant-pressure condition or plane conduction factor
SG   indicates silicon grease
S   indicates silicon
sq   indicates per square resistance
T   indicates total quantity
ϵ indicates emissivity factor

23.15 REFERENCES

1 Chu, R. C., Direct Liquid Cooling, Report of Research Workshop, "Directions of Heat Transfer in Electronic Equipment," NSF Grant ENG-7701297 (1977).

2 Kilby, J. S., Invention of the Integrated Circuit, *IEEE Trans. Electron Devices*, vol. ED-23, pp. 648–654, July 1976.

3 Glaser, A. B., and Subak-Sharpe, G. E., *Integrated Circuit Engineering*, Addison-Wesley, Reading, Mass., 1977.

4 Millman, J., *Microelectronics*, McGraw-Hill, New York, 1979.

5 Seely, J. H., and Chu, R. C., *Heat Transfer in Microelectronic Equipment*, Marcel Dekker, New York, 1972.

6 R. K. Smith, Personal communication, 1975.

7 Dickerson, P., Convenient Thermal Analysis Techniques for Printed Circuit Board Assemblies, Proc. Natl. Electronics Package Conf. (NEPCON), Long Beach, Calif., January 1967.

8 R. K. Smith, Personal communication, 1966.

9 Antonetti, V. W., Cooling High End IBM Computers, *IBM Technical Report*, TR 00.3067, October, 1980.

10 Antonetti, V. W., Simons, R. E., and Arent, G. R., Cooling a Hot Computer, *Electromechanical Design*, pp. 34–37, September 1973.

11 Clark, B. T., Designing the Thermal Conduction Module for the IBM 3081 Processor, Proc. 31st Electronic Components Conf., Atlanta, Georgia, May 11, 1981.

12 Bonner, R. F., Asselta, J. A., and Haining, F. W., High Performance Printed Circuit Board for the IBM 3081 Processor, Proc. 31st Electronic Components Conf., Atlanta, Georgia, May 11, 1981.

13 Chu, R. C., Gupta, O. R., Hwang, U. P., and Simons, R. E., Gas Encapsulated Cooling Module, U.S. Patent 3,993,123, November 23, 1976.

14 Chu, R. C., Hwang, U. P., and Simons, R. E., Conduction Cooling for an LSI Package: A One-dimensional Approach, *IBM Journal of Research*, January, 1982.

15 Block, R. F., Silverstein, M., Arnold, J. T., and Bar-Cohen, A., Thermal Management of VHSIC Based Hardware, Proc. 1st Annual Conf. International Electronics Packaging Society, pp. 314–328, Cleveland, Ohio, November, 1981.

16 Feldmanis, C. J., Cooling Techniques and Thermal Analyses of Circuit Board Mounted Electronic Equipment, in *Heat Transfer in Electronic Equipment*, edited by M. D. Kelleher and M. M. Yovanovich, HTD vol. 20, American Society of Mechanical Engineers, New York, 1981.

17 Mull, W., and Rieher, H., Der Wärmeschutz von Luftschichten, *Gesundh.-Ing. Beihefte*, vol. 28, Berlin, Germany, 1930.

18 DeGraaf, J. G. A., and von der Held, E. F. M., The Relation between the Heat Transfer and the Convection Phenomena in Enclosed Plane Air Layers, *Appl. Sci. Res.*, Sec. A, vol. 3, pp. 393–410, 1953.

■ author index

■ subject index